Trichothecenes
and
Other
Mycotoxins

Trichothecenes and Other Mycotoxins

Proceedings of the International Mycotoxin Symposium
held in Sydney, Australia, August 1984
organized by the Mycotoxicology Committee of the International
Society for Plant Pathology

Edited by

John Lacey
Plant Pathology Department
Rothamsted Experimental Station

JOHN WILEY & SONS
CHICHESTER · NEW YORK · BRISBANE · TORONTO · SINGAPORE

Library of Congress Cataloging in Publication Data:

International Mycotoxin Symposium (1984 : Sydney, N.S.W.)
Trichothecenes and other mycotoxins.

(Progress in pesticide biochemistry and toxicology; (v. 4)
Includes index.
1. Mycotoxins—Congresses. 2. Trichothecenes—Congresses. I. Lacey, John. II. International Society for Plant Pathology. Mycotoxicology Committee. III. Title. IV. Series.
QP801.P38P76 vol. 4 632′.95 s 85–16907
[QP631.M9] [589.2]
ISBN 0 471 90751 0

British Library Cataloguing in Publication Data:

International Mycotoxin Symposium (*1984* : *Sydney*)
Trichothecenes and other mycotoxins : proceedings of the International Mycotoxin Symposium held in Sydney, Australia, August 1984.—(Progress in mycotoxins research)
1. Mycotoxins
I. Title II. International Society for Plant Pathology. *Mycotoxicology Committee*. III. Lacey, John IV. Series
574.2′326 QP632.M9
ISBN 0 471 90751 0

Typeset by Photo·Graphics, Honiton, Devon.
Printed and bound in Great Britain

Organizing Committee

I.S.P.P. MYCOTOXICOLOGY COMMITTEE J. Lacey (U.K., Chairman)
W. F. O. Marasas (South Africa)
C. J. Mirocha (U.S.A.)

WORLD VETERINARY ASSOCIATION L. G. Morehouse (U.S.A.)
G. H. Nelson (U.S.A.)

LOCAL ORGANIZATION J. Pitt (Australia)

I.S.P.P. MYCOTOXICOLOGY COMMITTEE C. J. Mirocha (U.S.A., Chairman)
G. A. Bean (U.S.A.)
H. K. Frank (Germany)
M. Jemmali (France)
H. Kurata (Japan)
J. Lacey (U.K.)
W. F. O. Marasas (South Africa)

Acknowledgements

The Organizing Committee are most grateful to all who contributed to the success of the Symposium through their papers, posters and discussion and through their assistance in the provision of services, social events and comfortable accommodation in Sydney. Generous financial assistance towards the cost of the meeting was received from Anax Pty. Ltd., Ansett Airlines of Australia, Australian Chicken Meat Research Committee, Australian Meat Research Committee, Australian Pig Industry Research Committee, Australian Poultry Research Advisory Committee, Australian Wool Corporation, Chemical Resources Pty. Ltd., Kemin Industries Pty. Ltd., Waters Associates—a Division of Millipore Ltd., and is gratefully acknowledged.

We thank P. Andersen, P. K. C. Austwick, G. Ayerst, C. Booth, A. E. Buckle, G. R. Cayley, R. O. Clements, T. A. Connors, J. E. L. Correy, F. Elling, B. Flannigan, J. Gilbert, C. D. Green, J. F. Grove, P. Harries, I. Harvey, M. Hetmanski, R. A. Hill, B. Jarvis, M. C. Johnson, K. H. Kiessling, P. Krogh, H. J. Kurtz, A. Lyne, N. Magan, P. Mantle, J. K. Miller, M. O. Moss, A. C. Pier, W. S. Pierpoint, J. Robb, R. Schoental, J. E. Smith and D. R. Threlfall for refereeing papers and for giving thoughtful, constructive and helpful criticism.

Figures 1 to 15 in Chapter 1 are published with the permission of the University of Pennsylvania Press.

Contents

VETERINARY AND MEDICAL IMPLICATIONS OF MYCOTOXINS

List of Contributors

R. AMANO	*Department of Microbiology, National Institute of Hygienic Sciences, Kamiyoga, Setagaya, Tokyo 158, Japan.*
R. G. ARORA	*Department of Pathology, Faculty of Veterinary Medicine, Swedish University of Agricultural Sciences, Box 7028, S-750 07 Uppsala, Sweden.*
R. L. ASQUITH	*Animal Science Department, Horse Research Center, University of Florida, 2825 NW 100th Street, Ocala, Florida 3 2675, U.S.A.*
L. A. BABIUK	*Department of Veterinary Microbiology, Western College of Veterinary Medicine, University of Saskatchewan, Saskatoon, Saskatchewan S7N OW0, Canada.*
A. BAETZ	*U.S. Department of Agriculture, Agricultural Research Service, National Animal Disease Center, P.O. Box 70, Ames, Iowa 50010, U.S.A.*
P. A. BASSON	*State Veterinarian, P.O. Box 81, Grootfontein, South West Africa.*
G. F. BATH	*Regional Veterinary Laboratory, Private Bag X528, Middleburg 5900, South Africa.*
G. A. BEAN	*Department of Botany, University of Maryland, College Park, Maryland 20742, U.S.A.*
JOAN W. BENNETT	*Department of Biology, Tulane University, New Orleans, Louisiana 70118, U.S.A.*
T. N. BHAVANISHANKER	*Groundnut Improvement Program, International Crops Research Institute for the Semi-Arid Tropics (ICRISAT), Patancheru P.O., Andhra Pradesh 502324, India.*

Angelika R. Bieger — *Institute for Biochemistry, Joh. Gutenberg University, Joh.-Joachim-Becher-Weg 30, D-6500 Mainz, W. Germany.*

B. J. Blaney — *Queensland Department of Primary Industries, Animal Research Institute, Yeerongpilly, Queensland, Australia.*

P. D. Blankenship — *U.S. Department of Agriculture, Agricultural Research Service, National Peanut Research Laboratory, 600, Forrester Drive, Dawson, Georgia 31742, U.S.A.*

L. H. Brennecke — *U.S. Army Medical Research Institute of Infectious Diseases, Fort Detrick, Frederick, Maryland 21701, U.S.A.*

G. M. Buening — *Veterinary Diagnostic Laboratory, Iowa State University, Ames, Iowa 50011, U.S.A.*

D. L. Bunner — *U.S. Army Medical Research Institute of Infectious Diseases, Fort Detrick, Frederick, Maryland 21701, U.S.A.*

L. W. Burgess — *Department of Plant Pathology and Agricultural Entomology, University of Sydney, Sydney 2006, N.S.W., Australia.*

L. P. Bush — *Department of Agronomy, University of Kentucky, Lexington, Kentucky 40546, U.S.A.*

Luzviminda P. Celino — *National Crop Protection Center, U.P. at Los Banos College, Laguna, Philippines.*

W. B. Chapman — *Leatherhead Food R.A., Randalls Road, Leatherhead, Surrey KT22 7RY, U.K.*

J. A. W. Coetzer — *Section of Pathology, Veterinary Research Institute, P.O. Box 12567, Onderstepoort 0110, South Africa.*

R. J. Cole — *U.S. Department of Agriculture, Agricultural Research Service, National Peanut Research Laboratory, 600, Forrester Drive, Dawson, Georgia 31742, U.S.A.*

Sara J. Cooper — *Leatherhead Food R.A., Randalls Road, Leatherhead, Surrey KT22 7RY, U.K.*

T. M. COSGRIFF — *U.S. Army Medical Research Institute of Infectious Diseases, Fort Detrick, Frederick, Maryland 21701, U.S.A.*

R. G. CUERO — *Department of Bioscience and Biotechnology, Applied Microbiology, University of Strathclyde, Glasgow G1 1XW, U.K.*

C. C. J. CULVENOR — *C.S.I.R.O. Division of Animal Health, Private Bag No. 1, P.O., Parkville, Victoria 3052, Australia.*

R. R. DAHLGREN — *University of Wyoming, Department of Microbiology and Veterinary Medicine, Laramie, Wyoming 82071, U.S.A.*

G. N. DANIELS — *Veterinary Diagnostic Laboratory, College of Veterinary Medicine, Ames, Iowa 50011, U.S.A.*

W. V. DASHEK — *Department of Biology, West Virginia University, Morgantown, West Virginia 26506, U.S.A. Present address: Atlanta University, 223 Chestnut Street S.W., Atlanta, Georgia 31304, U.S.A.*

MARGARET E. DI MENNA — *Ruakura Agricultural Research Centre, Ministry of Agriculture and Fisheries, Hamilton, New Zealand.*

R. E. DINTERMAN — *U.S. Army Medical Research Institute of Infectious Diseases, Fort Detrick, Frederick, Maryland 21701, U.S.A.*

J. W. DORNER — *U.S. Department of Agriculture, Agricultural Research Service, National Peanut Research Laboratory, 600, Forrester Drive, Dawson, Georgia 31742, U.S.A.*

K. P. DOSE — *Institute for Biochemistry, Joh. Gutenberg University, Joh.-Joachim-Becher-Weg 30, D-6500 Mainz, W. Germany*

J. A. EDGAR — *C.S.I.R.O. Division of Animal Health, Private Bag No. 1, P.O., Parkville, Victoria 3052, Australia.*

B. FLANNIGAN — *Department of Brewing and Biological Sciences, Heriot-Watt University, Edinburgh EH1 1HX, U.K.*

J. L. Frahn — *C.S.I.R.O. Division of Animal Health, Private Bag No. 1, P.O., Parkville, Victoria 3052, Australia.*

J. M. Frank — *Department of Microbiology, University of Surrey, Guildford, Surrey GU2 5XH, U.K.*

Susan C. E. Friend — *Department of Veterinary Paraclinical Science, University of Melbourne, Werribee, Victoria 3030, Australia.*

C. P. Gorst-Allman — *National Chemical Research Laboratory, Council for Scientific and Industrial Research, P.O. Box 395, Pretoria 0001, Republic of South Africa.*

P. B. Hamilton — *Department of Poultry Science, North Carolina State University, Raleigh, North Carolina 27695, U.S.A.*

Y.-Y. Hao — *Institute of Botany, Academia Sinica, Taipeh, Taiwan, Republic of China.*

I. C. Harvey — *Plant Health Station, Ministry of Agriculture and Fisheries, Lincoln, New Zealand.*
Present address: *Plant Protection Centre, Ministry of Agriculture and Fisheries, P.O. Box 24, Lincoln, New Zealand.*

C. R. Hassler — *Battelle Memorial Institute, Fort Detrick, Frederick, Maryland 21701, U.S.A.*

R. A. Hill — *Plant Pathology Department, University of Georgia, Coastal Plain Experiment Station, Tifton, Georgia 31793, U.S.A.*
Present address: *Ruakura Agricultural Research Centre, Ministry of Agriculture and Fisheries, Hamilton, New Zealand.*

Ailsa D. Hocking — *C.S.I.R.O. Division of Food Research, P.O. Box 52, North Ryde, N.S.W. 2113, Australia.*

B. S. Hook — *Veterinary Diagnostic Laboratory, Iowa State University, Ames, Iowa 50011, U.S.A.*

D. P. H. Hsieh — *Department of Environmental Toxicology, University of California, Davis, California 95616, U.S.A.*

M. ICHINOE — *Department of Microbiology, National Institute of Hygienic Sciences, Kamiyoga, Setagaya, Tokyo 158, Japan.*

K. ISHII — *Department of Toxicology and Microbial Chemistry, Faculty of Pharmaceutical Sciences, Science University of Tokyo, Shinjuku-ku, Tokyo 172, Japan.*

M. V. JAGO — *C.S.I.R.O. Division of Animal Health, Private Bag No. 1, P.O., Parkville, Victoria 3052, Australia.*

B. JARVIS — *Leatherhead Food R.A., Randalls Road, Leatherhead, Surrey KT22 7RY, U.K.*

B. B. JARVIS — *Department of Chemistry, University of Maryland, College Park, Maryland 20742, U.S.A.*

M. C. JOHNSON — *United States Department of Agriculture, Agriculture Research Service, Department of Plant Pathology, University of Kentucky, Lexington, Kentucky 40546, U.S.A.*
Present address: *Department of Biological Sciences, Georgetown College, Georgetown, Kentucky 40324, U.S.A.*

A. J. JONES — *Department of Chemistry, The Faculties, Australian National University, Canberra, A.C.T. 2600, Australia.*

T. S. KELLERMAN — *Section of Toxicology, Veterinary Research Institute, P.O. Box 12567, Onderstepoort 0110, South Africa.*

H. KURATA — *Department of Microbiology, National Institute of Hygienic Sciences, Kamiyoga, Setagaya, Tokyo 158, Japan.*

W. F. KWOLEK — *United States Department of Agriculture, Agricultural Research Service, Northern Regional Research Center, 1815, N. University Street, Peoria, Illinois 61604, U.S.A.*

J. LACEY — *Rothamsted Experimental Station, Harpenden, Herts. AL5 2JQ, U.K.*

G. C. M. Latch — *Plant Diseases Division, D.S.I.R., Palmerston North, New Zealand.*

Y.-H. W. Lee — *Institute of Biochemistry, National Yang-Ming Medical College, Taipei, Taiwan, Republic of China.*

Y.-W. Lee — *Department of Plant Pathology, University of Minnesota, St. Paul, Minnesota 55108, U.S.A.*

E. B. Lillehoj — *United States Department of Agriculture, Agricultural Research Service, Southern Regional Research Center, 1100, Robert E. Lee Boulevard, New Orleans, Louisiana 70179, U.S.A.*

N. Lisker — *Department of Stored Products, Institute for Technology and Storage of Agricultural Products, The Volcani Institute, P.O.B. 6, Bet Dagan 50250, Israel.*

G. C. Llewellyn — *Department of Biology, Virginia Commonwealth University, Richmond, Virginia 23284, U.S.A.*

W. E. Lloyd — *Veterinary Diagnostic Laboratory, College of Veterinary Medicine, Iowa State University, Ames, Iowa 50011, U.S.A. Present address: VET-A-MIX, Inc., Box A, Shenandoah, Iowa 51601, U.S.A.*

L. G. Lomax — *Veterinary Diagnostic and Investigational Laboratory, P.O. Box 1389, Tifton, Georgia 31794, U.S.A.*

Lilia S. P. Madamba — *Institute of Chemistry, UP at Los Banos, College, Laguna, Philippines.*

N. Magan — *Rothamsted Experimental Station, Harpenden, Herts. AL5 2JQ, U.K.*

W. F. O. Marasas — *National Research Institute for Nutritional Diseases, South African Medical Research Council, P.O. Box 70, Tygerberg 7505, South Africa.*

D. McDonald — *Groundnut Improvement Program, International Crops Research Institute for the Semi-Arid Tropics (ICRISAT), Patancheru P.O., Andhra Pradesh 502324, India.*

M. E. McLoughlin — *Iowa State University, College of Veterinary Medicine, Ames, Iowa 50011, U.S.A.*

W. W. McMillian — *U.S. Department of Agriculture, Southern Grains Insect Research Laboratory, Coastal Plain Experiment Station, Tifton, Georgia 31793, U.S.A.*

V. K. Mehan — *Groundnut Improvement Program, International Crops Research Institute for the Semi-Arid Tropics (ICRISAT), Patancheru P.O., Andhra Pradesh 502324, India.*

J. A. Minné — *Section of Toxicology, Veterinary Research Institute, P.O. Box 12567, Onderstepoort 0110, South Africa.*

C. J. Mirocha — *Department of Plant Pathology, University of Minnesota, St Paul, Minnesota 55108, U.S.A.*

L. G. Morehouse — *Veterinary Medical Diagnostic Laboratory, University of Missouri-Columbia, Columbia, Missouri 65211, U.S.A.*

J. D. Morgan — *Department of Pharmacology and Toxicology, Virginia Commonwealth University, Richmond, Virginia 23284, U.S.A.*

P. H. Mortimer — *Ruakura Agricultural Research Centre, Ministry of Agriculture and Fisheries, Hamilton, New Zealand.*

J. G. Morton — *Department of Brewing and Biological Sciences, Heriot-Watt University, Edinburgh EH1 1HX, U.K.*

M. O. Moss — *Department of Microbiology, University of Surrey, Guildford, Surrey GU2 5XH, U.K.*

R. J. Naylor — *Department of Brewing and Biological Sciences, Heriot-Watt University, Edinburgh EH1 1HX, U.K.*

P. E. Nelson — *Fusarium Research Center, Department of Plant Pathology, Pennsylvania State University, University Park, Pennsylvania 16802, U.S.A.*

H. A. Neufeld — *U.S. Army Medical Research Institute of Infectious Diseases, Fort Detrick, Frederick, Maryland 21701, U.S.A.*

MARY NORVAL	*Department of Bacteriology, University of Edinburgh, Edinburgh EH9 3JG, U.K.*
G. D. OSWEILER	*Veterinary Diagnostic Laboratory, Iowa State University, Ames, Iowa 50011, U.S.A.*
JUDITH G. PACE	*U.S. Army Medical Research Institute of Infectious Diseases, Fort Detrick, Frederick, Maryland 21701, U.S.A.*
MURIEL S. PALMGREN	*Tulane School of Public Health and Tropical Medicine, Environmental Health Sciences, 1430, Tulane Avenue, New Orleans, Louisiana 70112, U.S.A.*
G. W. PARKER	*U.S. Army Medical Research Institute of Infectious Diseases, Fort Detrick, Frederick, Maryland 21701, U.S.A.*
N. PASTER	*Department of Stored Products, Institute of Technology and Storage of Agricultural Products, A.R.O., The Volcani Center, P.O.B. 6, Bet Dagan 50250, Israel.*
G. PAVANASASIVAM	*Department of Chemistry, University of Maryland, College Park, Maryland 20742, U.S.A.*
R. J. PAWLOSKY	*Department of Plant Pathology, University of Minnesota, St Paul, Minnesota 55108, U.S.A.*
A. L. PAYNE	*C.S.I.R.O. Division of Animal Health, Animal Health Research Laboratory, Private Bag No. 1, P.O. Parkville, Victoria 3052, Australia.*
A. C. PIER	*Department of Microbiology and Veterinary Medicine, University of Wyoming, Laramie, Wyoming 82071, U.S.A.*
J. I. PITT	*C.S.I.R.O. Division of Food Research, P.O. Box 52, North Ryde, N.S.W. 2113, Australia.*
A. S. RAO	*Department of Botany, Nagarjuna University, Guntur, A.P., India 522 510.*
C. S. REDDY	*Department of Veterinary Biomedical Sciences, College of Veterinary Medicine, University of Missouri-Columbia, Columbia, Missouri 62511, U.S.A.*

J. L. RICHARD	*U.S. Department of Agriculture, Agriculture Research Service, National Animal Disease Center, P.O. Box 70, Ames, Iowa 50010, U.S.A.*
JEAN ROBB	*Microbiology, East of Scotland College of Agriculture, Edinburgh EH9 3JG, U.K.*
CECILIA ROUX	*Plant Protection Research Institute, Private Bag X134, Pretoria 0001, South Africa.*
T. H. SANDERS	*U.S. Department of Agriculture, Agriculture Research Service, National Peanut Research Laboratory, 600, Forrester Drive, Dawson, Georgia 31742, U.S.A.*
H. B. SCHIEFER	*Toxicology Research Centre, University of Saskatchewan, Saskatoon, Saskatchewan S7N OW0, Canada.*
E. W. SHAW	*Institute of Botany, Academic Sinica, Nankang, Taipei, Republic of China.*
M.-S. SHIAO	*Institute of Botany, Academia Sinica, Taipei, Taiwan, Republic of China.*
M. R. SIEGEL	*Department of Plant Pathology, University of Kentucky, Lexington, Kentucky 40546, U.S.A.*
B. L. SMITH	*Mycotoxic Diseases Group, Ruakura Animal Research Station, Private Bag, Hamilton, New Zealand.*
H. M. STAHR	*Veterinary Diagnostic Laboratory, College of Veterinary Medicine, Iowa State University, Ames, Iowa 50011, U.S.A.*
P. S. STEYN	*National Chemical Research Laboratory, Council for Scientific and Industrial Research, P.O. Box 395, Pretoria 0001, Republic of South Africa.*
T. A. TOUSSON	*Fusarium Research Center, Department of Plant Pathology, Pennsylvania State University, University Park, Pennsylvania 16802, U.S.A.*
JERMING TSENG	*Institute of Botany, Academia Sinica, Nankang, Taipei, Republic of China.*
T. C. TSENG	*Institute of Botany, Academia Sinica, Nankang, Taipei, Republic of China.*

S. UCHIYAMA — *Department of Microbiology, National Institute of Hygienic Sciences, Kamiyoga, Setagaya, Tokyo 158, Japan.*

Y. UENO — *Department of Toxicology and Microbial Chemistry, Faculty of Pharmaceutical Sciences, Science University of Tokyo, Shinjuku-ku, Tokyo 172, Japan.*

G. C. A. VAN DER WESTHUIZEN — *Plant Protection Research Institute, Private Bag X134, Pretoria 0001, South Africa.*

M. VIJAYALAKSHMI — *Department of Botany, Nagarjuna University, Guntur, A.P., India 522 510.*

R. VLEGGAAR — *National Chemical Research Laboratory, Council for Scientific and Industrial Research, P.O. Box 395, Pretoria 0001, Republic of South Africa.*

S. J. WALKER — *Biology Department, West Virginia University, Morgantown, West Virginia 26506, U.S.A.*

R. W. WANNEMACHER — *U.S. Army Medical Research Institute of Infectious Diseases, Fort Detrick, Frederick, Maryland 21701, U.S.A.*

R.-D. WEI — *Institute of Biochemistry, National Yang-Ming Medical College, Taipei, Taiwan, Republic of China.*

Y.-H. WEI — *Institute of Biochemistry, National Yang-Ming Medical College, Taipei, Taiwan, Republic of China.*

P. L. WESSELS — *National Chemical Research Laboratory, Council for Scientific and Industrial Research, P.O. Box 395, Pretoria 0001, Republic of South Africa.*

N. W. WIDSTROM — *U.S. Department of Agriculture, Agriculture Research Service, Southern Grains Insect Research Laboratory, Coastal Plains Experiment Station, Tifton, Georgia 3179.*

A. P. WILLIAMS — *Leatherhead Food R.A., Randalls Road, Leatherhead, Surrey KT22 7RY, U.K.*

D. M. WILSON — *Plant Pathology Department, University of Georgia, Coastal Plains Experiment Station, Tifton, Georgia 31793, U.S.A.*

GILLIAN M. WOOD	*Leatherhead Food R.A., Randalls Road, Leatherhead, Surrey KT22 7RY, U.K.*
G. F. YUAN	*Institute of Botany, Academia Sinica, Nankang, Taipei, Republic of China.*
ZHU TONG-XIA	*Department of Plant Pathology, University of Minnesota, St Paul, Minnesota 55108, U.S.A.*
M. S. ZUBER	*Department of Agronomy, University of Missouri-Columbia, Columbia, Missouri 62511, U.S.A.*

Preface

Mycotoxicology is the study of the toxic metabolites of fungi, of their occurrence and production in food and feeds, and of their effects on man and animals. This study was initiated in the U.S.S.R., U.S.A. and South Africa in the 1920s and 1930s when fungal growth on straw and grain was implicated in such diseases as stachybotryotoxicosis in horses, barley scab poisoning and mouldy corn toxicosis in pigs and in alimentary toxic aleukia in man. This led to a vigorous and systematic study of mycotoxins in the U.S.S.R. but elsewhere their study was sporadic until about 1960. By then, the toxicity of *Penicillium citreonigrum* (*P. citreoviride*) in toxic mouldy rice had been demonstrated in Japan and, within a period of 2 years, facial eczema of sheep in New Zealand was shown to be caused by toxins from *Pithomyces chartarum* and the deaths of 100,000 turkey poults was attributed to aflatoxins from *Aspergillus flavus*. As a consequence there was a worldwide surge in research into all aspects of mycotoxins, which still continues.

Increasing numbers of toxins are being isolated from fungi, and their structures and toxicities determined. Some have been found in natural substrates and the conditions for their formation have been defined. Yet there is still a reluctance to establish their true significance to human and animal health or to consider them as causes of disease, except as the last resort when other possible causes have been eliminated. This is a view that needs to be changed.

Until recently, aflatoxins have dominated the mycotoxin literature and most attention has been given to the acute toxicity of mycotoxins. Now the widespread occurrence of deoxynivalenol in American and Canadian grain crops in recent years, and the controversy surrounding the alleged use of *Fusarium* toxins in warfare, has led to increased research into the trichothecenes. It is also evident that acute disease caused by mycotoxins is not the major concern that it was once thought to be. The covert threat of subacute doses of mycotoxins in decreasing productivity and affecting reproductive performance is being increasingly recognized. Mycotoxins may increase susceptibility to other infections, decrease growth rates, cause bone malformations and have teratogenic and carcinogenic effects. These effects are more insidious and even more difficult to correlate with the natural occurrence of mycotoxins. As a consequence, a multidisciplinary approach is needed with extremely careful surveys to establish the occurrence and economic significance of mycotoxins. Careful work is also necessary to define the conditions which limit toxin production, so that occasions when small amounts of toxin may be present can be recognized more readily.

In 1983 the 4th International Congress of Plant Pathology, the 5th International *Fusarium* Workshop and the 22nd World Veterinary Congress were all held in Australia. Their juxtaposition provided a good opportunity for drawing together the many disciplines involved in mycotoxin research. The World Veterinary Association readily agreed to collaborate with the Mycotoxicology Committee of the International Society for Plant Pathology and Dr Lester Burgess extended an invitation to meet jointly with the *Fusarium* Workshop in Sidney. About 120 people participated in the Symposium, which lasted 4 days.

In common with current research trends, the programme had a strong emphasis on trichothecenes and other *Fusarium* toxins. However, a meeting in Australasia could not ignore the important contribution New Zealand and Australia have made to our appreciation of the effects of toxin-producing fungi in pastures. The third element in the programme consisted of papers on mycotoxins produced by *Aspergillus* and *Penicillium* species, which have long dominated mycotoxin research. Within this programme the importance of subacute doses of mycotoxins was referred to repeatedly. This book draws together most of these papers to give an account of recent progress in research into trichothecenes and other mycotoxins, which together made a fascinating Symposium.

JOHN LACEY

TAXONOMY AND OCCURRENCE OF TOXIGENIC FUNGI

Trichothecenes and Other Mycotoxins
Edited by J. Lacey

CHAPTER 1

Taxonomy of toxigenic fusaria[1]

Walter F. O. Marasas, Paul E. Nelson and T. A. Tousson

INTRODUCTION

The genus *Fusarium* contains important mycotoxin-producing species. Interest in these is growing world-wide as a growing number of naturally occurring mycotoxins are identified which pose a threat to human and animal life. These toxigenic fusaria have been implicated in such human diseases as alimentary toxic aleukia, Urov (= Kashin-Beck) disease, akakabi-byo (= scabby grain intoxication) and oesophageal cancer and in animal diseases such as haemorrhagic, oestrogenic, emetic and feed-refusal syndromes, fescue foot, degnala disease, mouldy sweet potato toxicosis, bean hulls poisoning, and equine leukoencephalomalacia. Recent allegations that trichothecene toxins produced by *Fusarium* species are components of the chemical warfare agent 'Yellow Rain' (Mirocha *et al.*, 1982; Rosen and Rosen, 1982), have increased the awareness of, and interest in, these highly toxic compounds.

Taxonomy of the genus *Fusarium* is complex and difficult to apply, partly because different taxonomic systems have been used in different countries. The taxonomy of the genus is further complicated by the extreme variability of *Fusarium* species in culture and the fact that they mutate and degenerate rapidly, particularly under conditions of repeated subculturing on common laboratory media. This situation has led to great confusion in the extensive literature on *Fusarium* mycotoxicology because the same fungus is known under a variety of different names, because different fungi are lumped together under names such as *F. tricinctum* (Corda) Sacc. emend. Snyd. & Hans. and *F. roseum* Lk. emend. Snyd. & Hans., because several *Fusarium*

1 The material in this chapter is reproduced from *Fusarium Species: An Illustrated Manual for Identification* by P. E. Nelson, T. A. Toussoun and W. F. O. Marasas (1983) and from *Toxigenic Fusarium Species: Identity and Mycotoxicology* by W. F. O. Marasas, P. E. Nelson and T. A. Toussoun (1984), by permission of the Pennsylvania State University Press, University Park, Pennsylvania, U.S.A.

toxins have been named for mis-identified producing species, because elaborate chemical and pathological studies have been reported in the literature and attributed to incorrectly named species, and because many toxigenic *Fusarium* strains have become degenerate and lost their toxigenic ability due to maltreatment in laboratories that do not specialize in the maintenance of *Fusarium* cultures. Consequently it has become impossible to relate toxicological studies in different laboratories to each other because of the existing confusion with regard to the taxonomy and nomenclature of toxigenic Fusaria.

At the Conference on Mycotoxins in Human and Animal Health held at the University of Maryland during 4–8 October 1976, a Panel on Trichothecene Toxins (Smalley *et al.*, 1977) unanimously recommended that

> urgent attention should be given to the development of a uniform, internationally acceptable taxonomic system for trichothecene-producing Fusaria. This should be done in part by the establishment of a central collection of toxicologically important strains of *Fusarium* reported in the literature as trichothecene producers and preserved at central locations.

The above recommendations resulted in an HEW/PHS contract, No. 223–78–2018, with the authors of this paper and culminated in the preparation of the following two books: P. E. Nelson, T. A. Tousson and W. F. O. Marasas (1983), Fusarium *Species: an Illustrated Manual for Identification*, and *Toxigenic* Fusarium *Species: Identification and Mycotoxicology*, by W. F. O. Marasas, P. E. Nelson and T. A. Toussoun (1984).

This chapter presents a brief summary of the main conclusions of our study of the taxonomy of some of the most important toxigenic *Fusarium* species.

MATERIALS AND METHODS

A worldwide collection of published toxigenic strains of *Fusarium*, known as the International Toxic *Fusarium* Reference Collection (ITFRC), has been established over a period of 20 years by requesting cultures from recognized culture collections and also directly from authors of papers dealing with *Fusarium* mycotoxicology. All these cultures have been single-spored, identified according to the system proposed by Nelson *et al.* (1983), and lyophilized. These lyophilized cultures are being maintained in the culture collections of the *Fusarium* Research Center at The Pennsylvania State University and of the National Research Institute for Nutritional Diseases of the South African Medical Research Council.

RESULTS

Toxigenic *Fusarium* strains in the international toxic *Fusarium* reference collection (ITFRC)

Approximately 200 *Fusarium* strains reported to be toxigenic in the literature were collected. These were identified as belonging to 20 species in 10 Sections (Table 1). Approximately 50 per cent of these strains were incorrectly named in the literature according to our current taxonomic concepts. Many of these incorrectly named strains belong to important toxigenic species of *Fusarium* such as *F. sporotrichioides* Sherb., *F. poae* (Peck.) Wollenw., *F. equiseti* (Corda) Sacc., *F. graminearum* Schwabe, and *F. moniliforme* Sheldon. This chaotic situation is further complicated by the fact that many of the strains in

Table 1 Toxigenic *Fusarium* strains in the International Toxic *Fusarium* Reference Collection (ITFRC)

Section	Species	Number of toxigenic strains	Number of mis-identified strains
Eupionnotes	*F. merismoides*	1	0
Arachnites	*F. nivale*	1	0
	F. larvarum	1	0
Sporotrichiella	*F. sporotrichioides*	43	28
	F. chlamydosporum	3	3
	F. poae	8	2
	F. tricinctum	8	0
Roseum	*F. avenaceum*	5	1
Arthrosporiella	*F. semitectum*	8	8
Gibbosum	*F. equiseti*	15	10
	F. acuminatum	4	2
Discolor	*F. sambucinum*	6	5
	F. culmorum	11	1
	F. graminearum	31	11
Liseola	*F. moniliforme*	32	12
	F. proliferatum	1	1
	F. subglutinans	4	0
	F. anthophilum	1	0
Elegans	*F. oxysporum*	9	3
Martiella	*F. solani*	7	5
Total		199	92 (= 46%)

the ITFRC were received in a badly degenerated condition due to unsuitable handling and maintenance in culture. Although every attempt was made to improve the condition of these cultures by single-sporing, culturing on suitable media and lyophilization, many of these strains had probably lost their toxigenic abilities prior to deposition in the ITFRC.

The 20 toxigenic species of *Fusarium* (Table 1) are each discussed by Marasas *et al.* (1984) with respect to incidence and distribution, association with human and/or animal diseases, toxicity to experimental animals, mycotoxins produced, toxigenic strains represented in the ITFRC, and the names used in the literature for the strains included in each taxon.

Table 2 *Fusarium* species reported in the literature to produce trichothecenes

Section		Species
Eupionnotes		*F. episphaeria*
Spicarioides		*F. decemcellulare*
Arachnites		*F. nivale*
Sporotrichiella		*F. sporotrichioides* *F. poae* *F. tricinctum*
Roseum		*F. avenaceum*
Arthrosporiella		*F. anguioides* *F. concolor* *F. diversisporum* *F. incarnatum* *F. semitectum*
Gibbosum		*F. equiseti* *F. scirpi* *F. acuminatum*
Discolor		*F. heterosporum* *F. sambucinum* *F. sulphureum* *F. culmorum* *F. graminearum*
Lateritium		*F. lateritium*
Liseola		*F. moniliforme* *F. moniliforme* var. *subglutinans*
Elegans		*F. oxysporum*
Martiella		*F. solani*
Totals	12	25

Table 3 *Fusarium* species authenticated to produce trichothecenes by means of cultures represented in the International Toxic *Fusarium* Reference Collection (ITFRC)

Section		Species
Arachnites		*F. nivale*
Sporotrichiella		*F. sporotrichioides* *F. chlamydosporum* *F. poae*
Arthrosporiella		*F. semitectum*
Gibbosum		*F. equiseti* *F. acuminatum*
Discolor		*F. sambucinum* *F. culmorum* *F. graminearum*
Elegans		*F. oxysporum*
Totals	6	11

An examination of the *Fusarium* strains in the ITFRC revealed *inter alia* that the 25 species in 12 Sections reported in the literature to produce trichothecenes (Table 2), represent only 11 species in 6 Sections (Table 3) according to our current concepts of these taxa.

It is hoped that this comparative taxonomic study of published toxigenic *Fusarium* strains will contribute to the clarification of the existing confusion in the mycotoxicological literature with regard to the taxonomy and nomenclature of toxigenic fusaria.

Important toxigenic *Fusarium* species

Five of the 20 recognized toxigenic species of *Fusarium* (Table 1) will be discussed briefly, both as examples of the mycotoxicological importance of these fungi and of the taxonomic and nomenclatural confusion that surrounds them. Characters used to separate these species are illustrated in Figures 1–15.

F. sporotrichioides

Serious confusion has been introduced into the mycotoxicological literature by the use of the name *F. tricinctum* (Corda) Sacc. emend. Snyd. & Hans. in the sense of Snyder and Hansen (1945) which encompasses at least four

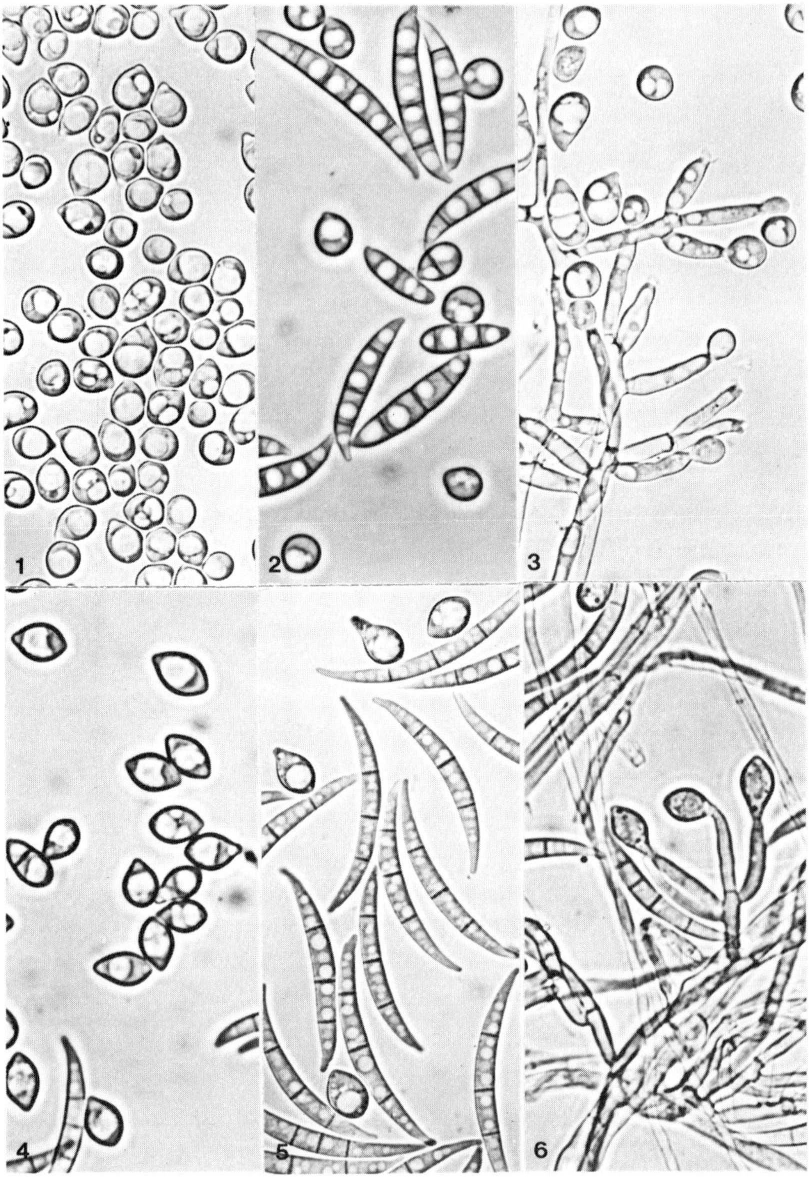

Figures 1–6 Illustrations of the characteristics used to separate *Fusarium poae* (Peck) Wollenw. and *F. tricinctum* (Corda) Sacc. from other species in the section Sporotrichiella. **1–3**: *F. poae*. Microconidia are globose or oval in shape, and 0–1 septate (1). Macroconidia are sickle-shaped, and 3–5 septate (2) and generally are rare. Conidiophores bearing microconidia are unbranched and branched monophialides, and are short and fat (3). **4–6**: *F. tricinctum*. Microconidia are lemon- to pear-shaped (4) or spindle-shaped, and are 0–1 septate. Macroconidia are sickle-shaped and are 3–5 septate (5). Conidiophores bearing microconidia are unbranched and branched monophialides (6). (All ×1000.)

distinct species (Nelson *et al.*, 1983). We have identified many toxigenic strains referred to as *F. tricinctum* in the literature, as *F. sporotrichioides*. Toxigenic strains of *F. sporotrichioides* have also been mis-identified as *F. episphaeria* (Tode) Snyd. & Hans., *F. moniliforme*, *F. nivale* (Fr.) Ces., *F. oxysporum* Schlecht., *F. poae*, and *F. solani* (Mart.) Appel & Wollenw.

Association with human and animal diseases Alimentary toxic aleukia; haemorrhagic syndrome; fescue foot; bean hulls poisoning; akakabi-byo (= scabby grain intoxication).

Mycotoxins produced Butenolide; 13 trichothecenes including the naturally occurring diacetoxyscirpenol, T-2 toxin, deoxynivalenol and nivalenol; zearalenone.

Toxigenic strains in the ITFRC Forty-three, including the follows:
F. nivale NRRL 3249
F. nivale Fn-2B
F. episphaeria Fn-M
F. oxysporum "niveum" Melon-1
F. solani M-1-1
F. moniliforme NRRL 3197
F. poae NRRL 3287
F. tricinctum Strain T-2 (= NRRL 3299) and 18 other strains referred to as *F. tricinctum*.

F. poae

Toxigenic strains of *F. poae* have also frequently been referred to as *F. tricinctum* in the sense of Snyder and Hansen (1945).

Association with human and animal diseases Alimentary toxic aleukia; haemorrhagic syndrome; Urov (= Kashin-Beck) disease.

Mycotoxins produced Butenolide; trichothecenes such as diacetoxyscirpenol and T-2 toxin.

Toxigenic strains in the ITFRC Eight, including the following:
F. tricinctum B-24
F. tricinctum M-334

F. equiseti

One of the major sources of confusion in the mycotoxicological literature on *Fusarium* is the use of the name *F. roseum* Lk. emend. Snyd. & Hans. in the

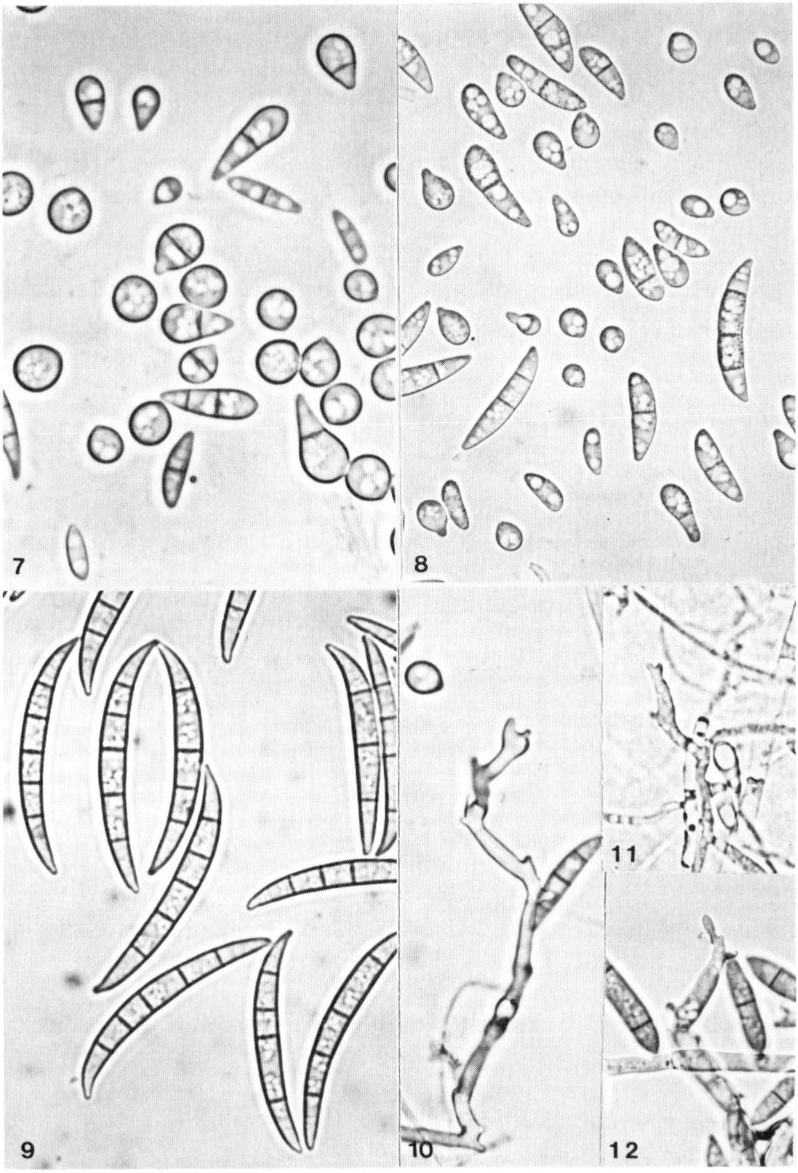

Figures 7–12 Illustrations of the characteristics used to separate *Fusarium sporotrichioides* Sherb. from other species in the section Sporotrichiella. Microconidia are oval to pear-shaped or spindle-shaped, and are 0–1 septate (7,8). Macroconidia are sickle-shaped and 3–5 septate (9). Conidiophores bearing microconidia are unbranched and branched monophialides and polyphialides (10–12). (All ×1000.)

sense of Snyder and Hansen (1945) which encompasses all of the species in the Sections Roseum, Arthrosporiella, Gibbosum, and Discolor (Nelson *et al.*, 1983). We have identified many toxigenic strains referred to as *F. roseum* or *F. roseum* 'Gibbosum' in the literature, as *F. equiseti*.

Association with human and animal diseases Degnala disease; fescue foot; tibial dyschondroplasia.

Mycotoxins produced Butenolide; eight trichothecenes including diacetoxy-scirpenol, nivalenol and T-2 toxin; zearalenone.

Toxigenic strains in the ITFRC Fifteen, including the following:
F. roseum Alaska 2-2
F. roseum 'Gibbosum' NRRL 5537
F. scirpi NRRL 3212
F. anguioides NRRL 3020
F. concolor NRRL 3214
F. diversisporum NRRL 3211

F. graminearum

Toxigenic strains of *F. graminearum* (= *Gibberella zeae*) have also frequently been referred to in the literature as *F. roseum* or *F. roseum* 'Graminearum' in the sense of Snyder and Hansen (1945). Certain toxigenic strains of *F. graminearum* have also been mis-identified as *F. culmorum* (W. G. Smith) Sacc., *F. tricinctum* and *F. sporotrichioides*.

Association with human and animal diseases Oestrogenic, emetic and feed refusal syndromes; akakabi-byo (= scabby grain intoxication).

Mycotoxins produced Eight trichothecenes including deoxynivalenol and nivalenol; zearalenone.

Toxigenic strains in the ITFRC Thirty-one, including the following:
F. roseum No. 117 (= ATCC 28114)
F. culmorum F-99 (= ATCC 34910)
F. roseum 'Culmorum' 70-K-11
F. tricinctum A-B-2
F. sporotrichioides NRRL 5908

F. moniliforme

Toxigenic *Fusarium* strains have frequently been referred to as *F. moniliforme* (= *Gibberella fujikuroi*) without differentiation of this species from

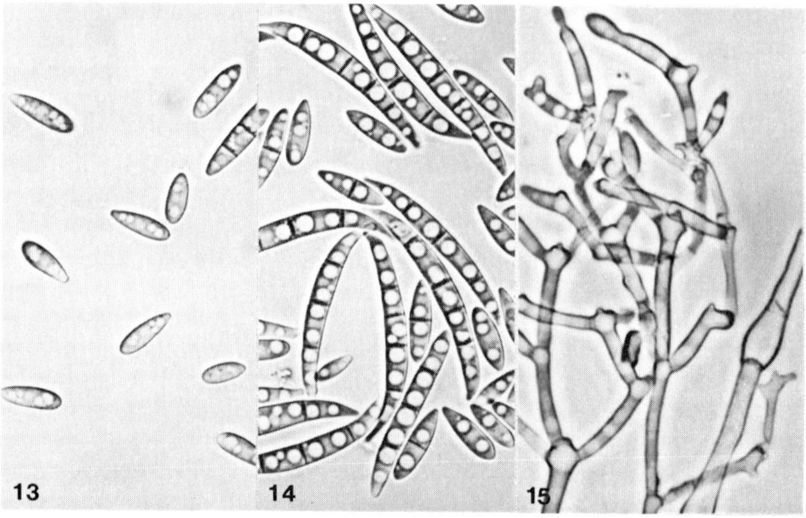

Figures 13–15 Illustrations of the characteristics used to separate *Fusarium chlamydosporum* Wollenw. from other species in the section Sporotrichiella. Microconidia are spindle-shaped but not globose, and 0–1 septate (13,14). Macroconidia are sickle-shaped and are 3–5 septate (14), and generally are rare. Conidiophores bearing microconidia are unbranched and branched monophialides and polyphialides (15). (All ×1000.)

other species in the Section Liseola such as *F. subglutinans* (Wollenw. & Reink.) Nelson, Toussoun & Marasas and *F. proliferatum* (Matsushima) Nirenberg. Toxigenic strains of *F. moniliforme* have also been referred to as *F. verticillioides* (Sacc.) Nirenberg.

Association with human and animal diseases Human oesophageal cancer; equine leukoencephalomalacia; abnormal bone development; mouldy sweet potato toxicosis.

Mycotoxins produced Fusaric acid; fusarins; fusariocins; gibberellins; moniliformin; zearalenone. Numerous authors have reported negative results for trichothecene production by *F. moniliforme*. Isolated reports that this fungus produces diacetoxyscirpenol and T-2 toxin have not been confirmed.

The mycotoxin responsible for equine leukoencephalomalacia is unknown.

Toxigenic strains in the ITFRC Thirty-two, including the following:

F. verticillioides MRC 826

F. solani (= *F. javanicum*) mouldy sweet potato toxicosis

CONCLUSIONS AND RECOMMENDATIONS

It is clear from our study of 200 published toxigenic strains of *Fusarium* that a serious state of confusion exists in the mycotoxicological literature on *Fusarium*. We have tried to salvage as much as possible from this almost hopeless situation by maintaining a unique collection of cultures, which took 20 years to assemble, under optimal conditions. We have also provided a critical review of the literature on *Fusarium* mycotoxicology and detailed references to the literature on the 200 strains in the collection in the form of a book. The major problems that we have identified and recommendations to solve them are as follows:

1. *Fusarium* cultures are not maintained under optimal conditions in most laboratories where mycotoxicological research is being conducted and also not in some major culture collections. We recommend that every *Fusarium* strain used in mycotoxicological investigations should be deposited as soon as possible in a recognized culture collection and preferably also in the ITFRC at the *Fusarium* Research Center at The Pennsylvania State University, University Park, Pennsylvania 16802, U.S.A.
2. *Fusarium* cultures used in mycotoxicological studies are not identified with sufficient care. The existing confusion arises partly from the use of different taxonomic systems in different countries, and partly from identifications made by non-specialists. We recommend that all *Fusarium* strains used in mycotoxicological investigations should be submitted to a specialist *Fusarium* taxonomist for identification. Cultures submitted to the *Fusarium* Research Center at The Pennsylvania State University will be identified according to Nelson *et al.* (1983). It is hoped that this system will be accepted internationally as a uniform system for the identification of toxigenic Fusaria.
3. *Fusarium* cultures are not numbered with sufficient care. The mycotoxicological literature is replete with statements such an 'An isolate of *Fusarium roseum* was used in these studies'. During the past 20 years we encountered endless problems in attempting to determine which isolates were used in publications dealing with unnumbered toxigenic *Fusarium* strains—particularly in cases where the same unnumbered isolate is referred to under different names in different publications. We recommend that every *Fusarium* strain used in mycotoxicological investigations should be referred to by a specific, traceable number in each publication and preferably the accession number(s) of the culture collection(s) where the strain has been deposited should also be supplied.

The application of these recommendations can only result in a reduction in the amount of confusion rampant in the mycotoxicological literature on *Fusarium*. We sincerely hope that it will be possible to paint a brighter picture of the status of taxonomy and nomenclature of toxigenic fusaria at the next Mycotoxin Symposium than has been possible in this paper.

REFERENCES

Marasas, W. F. O., Nelson, P. E., and Tousson, T. A. (1984). *Toxigenic* Fusarium *Species: Identification and Mycotoxicology*. The Pennsylvania State University Press, University Park.

Mirocha, C.J., Watson, S., and Hayes, W. (1982). 'Occurrence of trichothecenes in samples from Southeast Asia associated with "Yellow Rain" '. *Proc. V Internat. IUPAC Symp. on Mycotoxins and Phycotoxins*. Vienna, 1–3 September 1982, pp. 130–133.

Nelson, P. E., Tousson, T. A., and Marasas, W. F. O. (1983). Fusarium *Species. An Illustrated Manual for Identification*. The Pennsylvania State University Press, University Park.

Rosen, R. T., and Rosen, J.D. (1982). 'Presence of four *Fusarium* mycotoxins and synthetic material in "yellow rain". Evidence for the use of chemical weapons in Laos', *Biomed. Mass Spectrom.*, **9**, 443–450.

Smalley, E. B., Joffe, A., Palyusik, M., Kurata, H., and Marasas, W. F. O. (1977). 'Panel on trichothecene toxins', in *Mycotoxins in Human and Animal Health* (eds. J. V. Rodricks, C. W. Hesseltine, and M. A. Mehlman), pp. 337–340. Pathotox Publishers, Park Forest South, Illinois.

Snyder, W. C., and Hansen, H. N. (1945). The species concept in *Fusarium* with reference to Discolor and other sections. *Amer. J. Bot.*, **32**, 657–666.

Trichothecenes and Other Mycotoxins
Edited by J. Lacey

CHAPTER 2

Mycotoxigenic species of *Fusarium* associated with grain diseases in eastern Australia

L. W. Burgess

INTRODUCTION

Many of the common mycotoxigenic *Fusarium* species have been isolated from grain in eastern Australia. These include *Fusarium graminearum* Schwabe Group 2 (Francis and Burgess, 1977), *F. moniliforme* Sheldon, *F. subglutinans* (Wollenw. & Reinking) Nelson, Toussoun & Marasas, and *F. sporotrichioides* Sherb. Until recently, however, there had been no confirmed reports of mycotoxicoses caused by *Fusarium* mycotoxins in farm animals or poultry in eastern Australia. Indeed there have been no major surveys of grain for contamination with any of the important *Fusarium* mycotoxins.

Various reports, mostly unpublished, indicate that *Fusarium* mycotoxins have probably caused problems in the past. In 1937, for example, Pullar and Lerew described vulvovaginitis in swine, most likely caused by the ingestion of corn contaminated by mycotoxins produced by *F. graminearum* Group 2. Recent problems in pig herds fed zearalenone-contaminated maize produced on the Atherton Tableland are described by Blaney in chapter 9 of this volume.

This chapter briefly discusses the main pre-harvest diseases of wheat, maize and sorghum grain in eastern Australia caused by *Fusarium* species, and indicates where there is a potential risk of contamination by *Fusarium* mycotoxins. Australian isolates of the species mentioned have been described by Burgess and Liddell (1983).

WHEAT

Dryland wheat is the most important crop in eastern Australia and between five and six million hectares are grown annually in the wheat belt which extends from southern Queensland through New South Wales into Victoria (Figure 1). The wheat belt is located between the Great Dividing Range and the arid interior of the continent. It can be divided into two major climatic regions with the transition zone being in the Dubbo area of New South Wales (Figure 1) (Burgess *et al.*, 1981). The southern region is characterized by a Mediterranean-type climate, with a winter-dominant rainfall pattern and relatively mild winter temperatures. The climate at Horsham is typical of this

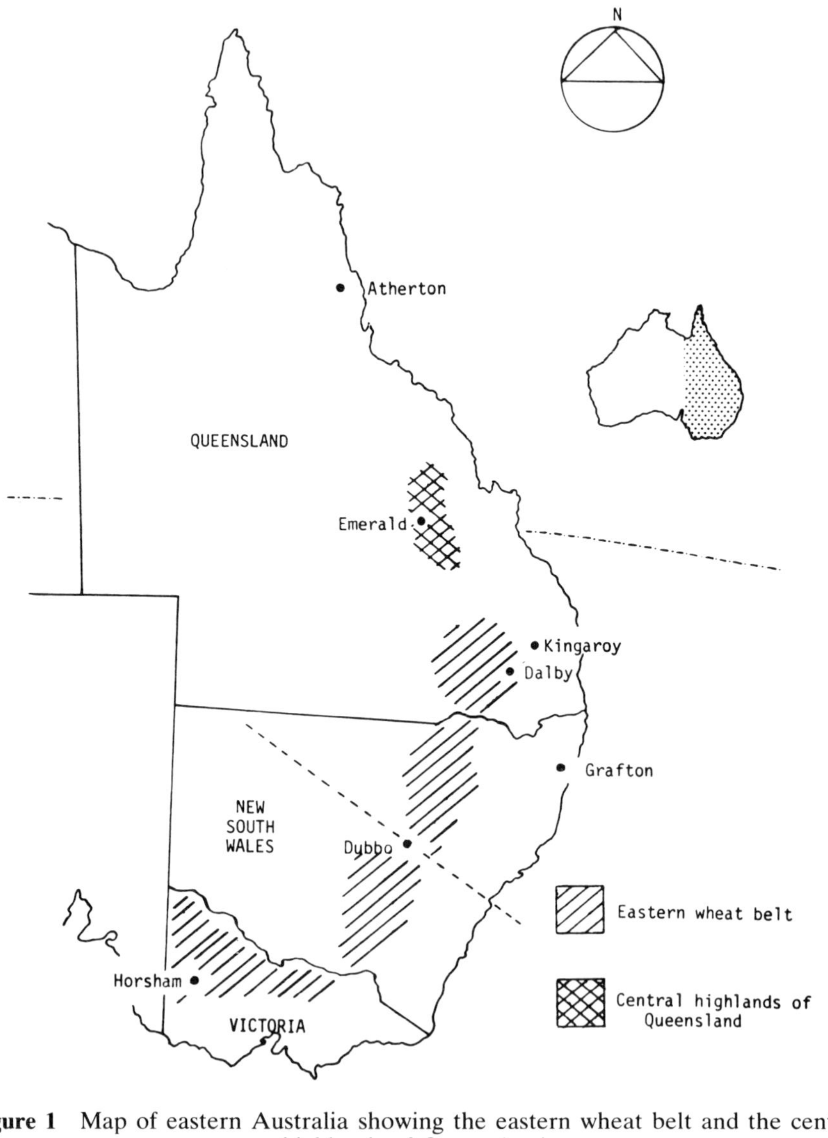

Figure 1 Map of eastern Australia showing the eastern wheat belt and the central highlands of Queensland

region (Table 1). The northern region of the wheat belt is characterized by a summer-dominant rainfall pattern and mild winter temperatures as illustrated by the climatic data for Dalby (Table 1).

All commercial cultivars are spring wheats which are grown during winter. Crops are normally planted between late April and mid-June and harvested between early October and December.

In the southern region, wheat is grown in rotation with barley and oats and subterraneum clover-grass pastures, while in the northern region it is grown in monoculture in the drier areas or in rotation with grain sorghum or sunflowers where the rainfall is higher and more reliable along the eastern edge. Head scab caused by *F. graminearum* Group 2 rarely occurs in the southern region because it is usually too dry between flowering and maturity. However, in the northern region wet conditions often prevail between flowering and the wheat harvest and head scab does occur but at very low levels. Possibly inoculum levels are low because maize, a favoured host for *F. graminearum* Group 2, is not grown in rotation with wheat. It seems very unlikely that *Fusarium* mycotoxins will be a problem in southern wheat and, unless there are significant changes in management practices, it is also unlikely that they will be a problem in northern wheat.

In contrast to the situation in the main wheat belt, head scab is a serious problem in two minor wheat areas. Head scab has long been a problem in the Kingaroy area of Queensland, where a very small acreage is grown, but it is a new problem on the north coast of New South Wales where wheat production has expanded in the past 2 years. Approximately 2000 ha was grown in 1982,

Table 1 Climatic data for key locations in eastern Australia as indicated in Figure 1

Location	Mean daily temperature (°C) max/min		Summer rainfall (mm)	Winter rainfall (mm)	Mean annual rainfall (mm)
	Jan	July	Nov.–Apr.	May–Oct.	
Atherton	28/19	21/11	1071	189	1260
Emerald	34/21	22/7	457	174	631
Dalby	32/19	19/4	430	243	673
Kingaroy	29/17	18/4	530	247	777
Grafton	30/20	20/5	623	407	1030
Dubbo	32/18	15/3	309	275	584
Horsham	31/14	14/4	170	279	449

Derived from *Climatic Averages, Australia*, Australian Government Publishing Service, Canberra (1975).

still a relatively small acreage, with the grain being mainly used locally for livestock and poultry feed. Head scab will almost certainly remain a serious problem because weather conditions usually favour disease development and *F. graminearum* Group 2 occurs widely as a pathogen of maize, a major crop in both areas. The fungus forms abundant perithecia on maize stubble providing ascospore inoculum to initiate head scab epidemics in the subsequent wheat crop in the same or adjacent fields. Wheat produced in these two areas should be monitored for scabby grain and the presence of mycotoxins produced by *F. graminearum* Group 2, particularly zearalenone and vomitoxin. Until we have detailed information to the contrary, grain from these areas must be considered at risk from contamination by *Fusarium* mycotoxins.

MAIZE

Maize is a relatively minor crop in eastern Australia. It is mainly grown under dryland conditions on the north coast of New South Wales, the south-east of Queensland and the Atherton Plateau in north Queensland (Figure 1) in areas with a relatively high summer rainfall (Table 1). Small areas are also grown under irrigation in various drier areas of the wheat belt.

A wide range of *Fusarium* spp. have been associated with cob rot and grain damage in dryland maize in the wetter areas. Various reports indicate that *F. graminearum* Group 2, *F. subglutinans* and *F. moniliforme* are the most common species associated with maize grain from eastern Australia. However several saprophytic species have also been isolated from maize grain in my laboratory and at other institutions. These include *F. equiseti* (Cda.) Sacc., *F. semitectum* Berk. & Rev. and *F. chlamydosporum* Wollenw. & Reinking (Connole *et al.*, 1981). Field observations indicate that cob rot and grain damage occur in most seasons but that severity fluctuates in relation to rainfall during the pre-harvest period.

Fusarium graminearum Group 2 is probably the most common species associated with severe grain damage. It is also the most common *Fusarium* associated with stalk rot of dryland corn and is particularly common on the Atherton Plateau (Francis and Burgess, 1975) where it also occurs in the kikuyu pastures. Thus it is not surprising that zearalenone has been detected in maize from the Atherton Plateau (Blaney, chapter 9 in this volume).

There is little information on the *Fusarium* species associated with maize grain produced under irrigation in the wheat belt. The few isolations which have been made in our laboratory suggest that *F. moniliforme* is common in some seasons.

The evidence presented above indicates that there is a considerable risk of contamination of maize grain by *Fusarium* mycotoxins in all areas.

GRAIN SORGHUM

Grain sorghum is grown mainly as a dryland summer crop in the summer-dominant rainfall region of the wheat belt and the central highlands (Emerald area) of Queensland (Figure 1). The crop is normally harvested in late summer or autumn. In some seasons harvesting is delayed by prolonged wet conditions which favour the occurrence of head blight and pre-harvest grain damage. There have, however, been no systematic surveys of the fungi associated with these problems in the various growing areas. Various minor studies indicate that a wide range of fungi are associated with head blight and damaged grain. *Fusarium moniliforme* and *Alternaria* spp., however, are particularly common on rachis tissue and damaged grain (Burgess *et al.*, 1981: Trimboli and Burgess, unpublished data). *Fusarium moniliforme* is also commonly associated with basal stalk rot, the most important disease of grain sorghum in New South Wales and Queensland, and is thus extremely common in residues which can provide abundant inoculum to infect aerial plant parts. *F. graminearum* Group 2 has not been isolated from sorghum grain so that *F. moniliforme* is the most likely *Fusarium* to produce mycotoxins in this substrate.

CONCLUSION

Contamination by *Fusarium* mycotoxins is most likely to be a problem in maize and sorghum grain and in wheat grain produced in a small area of coastal New South Wales. The main Australian wheat crop is unlikely to be affected.

REFERENCES

Burgess, L. W., Dodman, R. L., Pont, W., and Mayers, P. (1981). 'Fusarium diseases of wheat, maize and grain sorghum in eastern Australia'. In *Fusarium: Diseases, Biology* and *Taxonomy* (eds P. E. Nelson, T. A. Toussoun, and R. J. Cook), pp. 64–76. Pennsylvania State University Press, University Park.

Burgess, L. W., and Liddell, C. M. (1983). *Laboratory Manual for Fusarium Research*. Department of Plant Pathology and Agricultural Entomology, University of Sydney, Sydney 2006, NSW, Australia.

Connole, M. D., Blaney, B. J., and McEwan, T. (1981). 'Mycotoxins in animal feeds and toxic fungi in Queensland 1971–80', *Aust. Vet. J.*, **57**, 314–318.

Francis, R. G., and Burgess, L. W. (1975). 'Surveys of Fusaria and other fungi associated with stalk rot of maize in eastern Australia', *Aust. J. Agric. Res.*, **26**, 801–807.

Francis, R. G., and Burgess, L. W. (1977). 'Characteristics of two populations of *Fusarium roseum* 'Graminearum' in eastern Australia', *Trans. Brit. Mycol. Soc.*, 68, 421–427.

Pullar, E. M., and Lerew, W. M. (1937). 'Vulvovaginitis of swine'. *Aust. Vet. J.*, **13**, 28–31.

Trichothecenes and Other Mycotoxins
Edited by J. Lacey

CHAPTER 3

Trichothecene-producing *Fusarium* in barley and wheat in Japan

M. Ichinoe, S. Uchiyama, R. Amano, and H. Kurata

INTRODUCTION

It is well documented that cereal grains infected by fungi of the genus *Fusarium* have caused outbreaks of mycotoxicoses, such as alimentary toxic aleukia, Akakabi-toxicosis and scabby grain toxicosis in humans and feed refusal and emetic syndromes in farm animals. The toxins implicated in these mycotoxicoses have been identified as 12,13-epoxytrichothecenes and of these, so far T-2 toxin, diacetoxyscirpenol, nivalenol, and deoxynivalenol (vomitoxin) have been found to occur naturally in diseased maize, barley, wheat and other cereal grains.

In the past decade the relationship between the different species of *Fusarium* and their production of toxic trichothecenes has been studied extensively (Ueno *et al.*, 1973; Joffe and Yagen, 1977; Pathre and Mirocha, 1978; Marasas *et al.*, 1979; Suzuki *et al.*, 1980; Scott *et al.*, 1980). However, there is still much confusion due to the use of different taxonomic systems for fusaria in the literature.

The purpose of the study reported here was to determine the distribution of *Fusarium* species in barley and wheat grains in central and northern Japan, and to establish the ability of different *Fusarium* species to produce toxic trichothecenes.

MATERIALS AND METHODS

Samples

Samples of approximately 100 g of freshly harvested barley and wheat grains were collected from different sources, and were kept in paper bags at 10 °C until examined. A total of 60 barley samples came from 20 fields in Ishikawa Prefecture (central Japan) during the first weeks of harvest in July 1978 and

1982. Five samples of barley and four samples of wheat harvested during 1978 in Miyagi Prefecture (north-east Japan) were obtained from farmers' stocks and 39 samples of wheat grain were in Hokkaido (northern Japan) in 1978 and 1981.

Isolation of *Fusarium*

Fifty seeds from each sample were surface sterilized for 2 min in 1% aqueous sodium hypochlorite. They were then washed four times in 50 ml sterile water and plated on potato–dextrose agar (PDA) containing 100 μg chloramphenicol/ml medium to suppress bacteria. Five seeds were placed in each of ten plates and were then incubated at 25 °C. Fungi growing out from the seeds of those plates examined after 4–10 days, and which resembled *Fusarium*, were subcultured on PDA slants. Single spore isolation and identification of the isolates were carried out following Booth (1971).

Screening of *Fusarium* for trichothecene production

Fusarium isolates for mycotoxin screening were cultured in 400 ml of Czapek–Dox broth (Difco) + 0.5 per cent peptone at 25 °C for 2 weeks, following Ueno *et al.* (1973). The mycelial mat was removed by filtration and the culture filtrate (300 ml) was then treated with active charcoal, and eluted with 200 ml of methanol. The methanol elute was concentrated to dryness *in vacuo* and then redissolved in 10 ml of hot methanol and added to 10 ml of chloroform. It was then filtered through Toyo No. 2 filter paper into a round-bottomed flask. After removing insoluble materials, the solution was concentrated *in vacuo* to give a crude extract, which was used for thin-layer chromatography (TLC) and for toxicity tests using chicken embryos.

The crude extract was analysed for the presence of known trichothecene mycotoxins using TLC plate. Mycotoxin standards used in TLC were nivalenol, fusarenon-X, deoxynivalenol, 3-acetyldeoxynivalenol, T-2 toxin, HT-2 toxin, diacetoxyscirpenol and neosolaniol kindly given by Y. Ueno, Tokyo University of Science, and T. Yoshizawa, Kagawa University.

Individual crude extracts and mycotoxin standards were dissolved in chloroform–methanol (3 : 1) and then spotted on a precoated silica gel plate (20 × 20 cm, activated, 0.25 mm thick, E. Merck, Art. 5715). Solvent systems employed for TLC of trichothecenes were chloroform–methanol (97 : 3) and benzene–acetone (3 : 1). For detection of nivalenol, fusarenon-X, deoxynivalenol and 3-acetyldeoxynivalenol, the developed TLC plate was sprayed with 20% $AlCl_3$ in ethanol, and then heated at 110 °C for 10 min. The plate was examined by comparing the blue fluorescence under long-wave ultraviolet light with the mycotoxin standards. A spray application of 20% H_2SO_4 solution was used for detection of T-2 toxin, HT-2 toxin, diacetoxyscirpenol and neosolaniol, which appear as blue-grey fluorescent spots of T-2 toxin, HT-2 toxin and neosolaniol and purple/brown spots of diacetoxyscirpenol when heating at 110 °C for 10 min.

Quantitative estimation of trichothecenes in extracts

The culture extracts of TLC-positive samples were examined by two different procedures. Samples suspected of containing nivalenol were dissolved in 15 ml of water and passed through a resin (Amberlite XAD-2) column following Kamimura *et al.* (1978). The column was washed with 100 ml of water, then eluted with 100 ml of methanol. The eluant was then concentrated and the residue was used for gas–liquid chromatography (GLC) after trimethyl silylization. Samples thought to contain trichothecenes, other than nivalenol, were dissolved in 30 ml methanol–1% NaCl solution (55 : 45), then washed with 30 ml of *n*-hexane. The aqueous methanol layer was extracted twice with 25 ml of chloroform. The chloroform extract was examined by GLC after trimethyl silylization. The extracts were converted to trimethyl silyl ethers with 0.4 ml of a mixture of 1-(trimethylsilyl)-imidazole, trimethylchlorosilane and chloroform (1 : 0.2 : 9). The reaction solution was analysed with a GLC equipped with a hydrogen flame-ionization detector. Two kinds of column were used: (i) 3.5% silicone SE-30 on Gas Chrom Q (2 m); and (ii) 2.5% silicone OV-17 on Gas Chrom Q (1.5 m); the operating conditions were as follows: column temperature, 220–230 °C; flow rate of carrier gas, 40–60 ml/min.

Chicken embryo bioassay for toxicity

Chicken egg yolk sac inoculation was used to assay the toxicity of crude extracts from *Fusarium* cultures following McLaughlin *et al.* (1963) and Lafont and Lafont (1979). Fertile White Leghorn chicken eggs were obtained commercially and incubated in a temperature-controlled incubator at 37 °C and 80% relative humidity. Eggs were candled regularly for viability. Only eggs with a viable embryo after 4 days of incubation were used in these experiments.

To determine toxicity, approximately 30 mg of each extract was dissolved in 1 ml of methanol and then 0.05 ml of the solution was injected into the yolk sacs of five eggs. The developing embryos were observed for toxic effects for 48 h after injection. Control groups of eggs were similarly injected with solvent alone. If the mortality of chicken embryos was > 80% in this assay, further eggs were tested with a 10-fold dilution of the extract.

Prior to the present study, a preliminary toxicity test was undertaken to determine the effects of different doses of purified trichothecenes on developing chicken embryos after yolk sac injection. The embryotoxicity dose range was found to be between 0.1 and 0.5 μg/egg for T-2 toxin, HT-2 toxin and diacetoxyscirpenol; 2.5–5 μg/egg for nivalenol and fusarenon-X. The chicken embryos were less sensitive to deoxynivalenol and 3-acetyldeoxynivalenol at a concentration of 100 μg/egg or more.

RESULTS AND DISCUSSION

Mycological evaluation

Examination of the mycoflora of 65 barley and 43 wheat grain samples harvested from three different geographic areas in Japan showed *Alternaria*, *Epicoccum* and *Fusarium* to be the most common genera that could be isolated, followed by *Drechslera*, *Curvularia* and *Cladosporium*. Nearly all samples of grain were contaminated by *Fusarium* but the degree of contamination varied between the different sampling areas in Japan.

As shown in Table 1, *F. graminearum* was the dominant species in barley from Ishikawa, especially in samples of the 1982 crop. *F. equiseti* and *F. acuminatum* were also isolated frequently in both crop years. By contrast, the most common *Fusarium* species from Miyagi samples was *F. semitectum*, followed by *F. graminearum* and *F equiseti*. *F. graminearum* and *F. poae* were also common species in Hokkaido. *F. nivale* was only isolated from Hokkaido samples. *F. avenaceum* was not often isolated from grain from any area.

As a result of this survey it was concluded that *F. graminearum*, the causative fungus of head blight of barley and wheat, is the predominant species found on Japanese cereal grains. However, *F. sporotrichioides* and *F. tricinctum*, both well-known toxigenic species of *Fusarium*, were uncommon.

Trichothecene production by *Fusarium* species

A total of 105 *Fusarium* isolates were examined for production of trichothecene mycotoxins in liquid medium using TLC and GLC analysis (Table 2). Species differ widely in their ability to produce trichothecenes. For instance nivalenol and fusarenon-X were produced by 30% of *F. graminearum* isolates but only 10% of *F. graminearum* produced deoxynivalenol and its monoacetate. Diacetoxyscirpenol was produced by 40% of *F. equiseti*, 20% of *F. semitectum* and by all *F. acuminatum* and *F. sporotrichioides* isolates. The last two species were also the only species to produce T-2 toxin and HT-2 toxin, and they also produced neosolaniol, another 8-keto type trichothecene. A single isolate of *F. equiseti* produced diacetoxyscirpenol, nivalenol and fusarenon-X. No isolate of *F. avenaceum*, *F. poae* or *F. tricinctum* tested by chemical analysis produced any trichothecene.

Table 1 Distribution of *Fusarium* species on barley and wheat grain in central and northern Japan

	Ishikawa				Miyagi		Hokkaido			
	Barley 1978		Barley 1982		Barley and wheat 1978		Wheat 1978		Wheat 1981	
Species	Sample (24)	Grain (%)	Sample (36)	Grain (%)	Sample (9)	Grain (%)	Sample (21)	Grain (%)	Sample (18)	Grain (%)
F. graminearum	16	8.1	36	20.9	9	6.2	7	2.0	9	5.4
F. equiseti	10	2.8	13	1.2	2	3.6	1	tr	1	tr
F. semitectum	4	0.5	1	tr	4	7.1	—		—	
F. acuminatum	8	1.8	9	0.6	2	0.4	—		—	
F. avenaceum	1	tr	6	0.3	4	2.2	4	0.5	5	1.0
F. sporotrichioides	3	0.8	1	tr	—		2	0.2	3	0.3
F. poae	2	0.2	—		3	0.9	15	4.5	6	2.8
F. nivale	—		—		—		—		9	1.6
F. tricinctum	1	tr	3	0.2	—		1	tr	—	
F. oxysporum	1	tr			1	0.2	—		—	
F. moniliforme	1	tr	—		—		—		—	
(Total no. of isolates)		(178)		(427)		(98)		(78)		(103)

Sample: number of samples yielding species
Grain: percentage of seeds yielding species
tr: trace, i.e. less than 0.1%.

Table 2 Trichothecene production by *Fusarium* species isolated from barley and wheat grains

Species	No. of strains examined	No. of positive strains	Trichothecenes produced[a]
F. graminearum	40	12	NIV, F-X
		4	DON, 3AcDON
F. equiseti	10	4	DAS, NS
		1	DAS, NIV, F-X
F. semitectum	10	2	DAS
F. avenaceum	10	0	
F. acuminatum	10	10	T-2, HT-2, DAS, NS
F. sporotrichioides	10	10	T-2, HT-2, DAS, NS
F. poae	10	0	
F. tricinctum	5	0	

[a] NIV, nivalenol; F-X, fusarenon-X; DON, deoxynivalenol; 3AcDON, 3-acetyldeoxynivalenol; DAS, diacetoxyscirpenol; NS, neosolaniol; T-2, T-2 toxin; HT-2, HT-2 toxin.

The GLC quantification and chicken embryo test results with representative isolates of *Fusarium* are shown in Tables 3 and 4. Isolates of *F. graminearum* were divided into two groups, one producing nivalenol and fusarenon-X and other deoxynivalenol and 3-acetyldeoxynivalenol. In the first group, all 12 isolates produced fusarenon-X (0.30–5.33 μg/ml) but only five also produced nivalenol (0.03–0.80 μg/ml). In the second group of *F. graminearum* isolates production of deoxynivalenol was closely correlated to 3-acetyldeoxynivalenol. The toxicity of the two groups of *F. graminearum* culture extracts to chicken embryo was low to moderate, and was greatest when there was more than 1.0 μg/ml fusarenon-X.

All five isolates of *F. equiseti* produced from 0.60 to 23.7 μg/ml of diacetoxyscirpenol and two also produced neosolaniol (data not presented). One of the five isolates also produced small amounts of nivalenol and fusarenon-X, 0.30 μg/ml and 0.23 μg/ml, respectively. All isolates were highly toxic to chicken embryos including one producing little detectable trichothecene. Two of four *F. semitectum* isolates produced diacetoxyscirpenol but all isolates were highly toxic to chicken embryos.

Trichothecene production by *F. acuminatum* and *F. sporotrichioides* is shown in Table 4. Only one isolate of both species failed to produce few toxins and they only failed to produce HT-2 toxin. T-2 toxin was produced in

Table 3 Presence of trichothecenes and chicken embryo toxicity in extracts of *Fusarium* isolates from barley and wheat grains (I)

Species	Strain no.	Trichothecenes (μg/ml)[a] NIV	F-X	DON	3AcDON	DAS	Mortality of chicken embryo ×1	×1/10
F. graminearum	F-1088	—	2.47	—	—	—	4/5	0/5
	F-1093	—	1.52	—	—	—	4/5	0/5
	F-1097	—	0.30	—	—	—	0/5	
	F-1101	0.47	1.17	—	—	—	4/5	0/5
	F-1103	0.10	0.63	—	—	—	1/5	
	F-1104	0.30	5.33	—	—	—	5/5	0/5
	F-1112	0.80	0.77	—	—	—	2/5	0/5
	F-1114	0.03	0.50	—	—	—	0/5	
	F-1118	—	—	0.93	1.07	—	0/5	
	F-1185	—	—	2.70	6.33	—	2/5	0/5
	F-1406	—	—	0.08	0.15	—	0/5	
	F-1411	—	—	0.04	0.05	—	0/5	
F. equiseti	F-1016	—	—	—	—	23.7	5/5	5/5
	F-1023	—	—	—	—	3.10	5/5	3/5
	F-1084	—	—	—	—	4.33	5/5	3/5
	F-1121	0.30	0.20	—	—	3.67	5/5	3/5
	F-1128	—	—	—	—	0.60	4/5	0/5
F. semitectum	F-996	—	—	—	—	—	4/5	0/5
	F-1031	—	—	—	—	—	5/5	2/5
	F-1133	—	—	—	—	2.13	5/5	3/5
	F-1138	—	—	—	—	0.80	5/5	3/5
F. avenaceum	F-1035	—	—	—	—	—	4/5	0/5
	F-1040	—	—	—	—	—	1/5	

[a] Trichothecenes: NIV, nivalenol; F-X, fusarenon-X; DON, deoxynivalenol; 3AcDON, 3-acetyldeoxynivalenol; DAS, diacetoxyscirpenol.

the range of 1.0 to 40.3 μg/ml, HT-2 toxin from trace to 3.60 μg/ml, diacetoxyscirpenol from 0.23 to 8.67 μg/ml, and neosolaniol from 0.13 to 12.7 μg/ml. Even diluted to one-tenth of the original concentration, extracts were highly toxic to chicken embryos. Although none of the isolates of *F. poae* tested produced trichothecenes, five were moderately to highly toxic to chicken embryos.

Cultural characteristics of toxic *Fusarium*

The eleven species of *Fusarium* used in this study were identified according to Booth (1971). Those that produced trichothecenes, i.e. *F. graminearum*, *F. acuminatum*, and *F. sporotrichioides*, were characterized further using single-spore isolates grown on potato-sucrose agar (PSA) and carnation leaf agar (CLA) (Toussoun and Nelson, 1968).

Table 4 Presence of trichothecenes and chicken embryo toxicity in extracts of *Fusarium* isolates from barley and wheat grains (II)

Species	Strain no.	Trichothecenes[a]				Mortality of chicken embryo	
		T-2	HT-2	DAS	NS	×1	×1/10
F. acuminatum	F-986	28.0	1.37	8.67	12.0	5/5	5/5
	F-990	14.7	1.50	6.00	7.33	5/5	5/5
	F-994	14.0	tr	2.33	3.67	5/5	5/5
	F-1025	2.0	—	0.23	1.67	5/5	4/5
	F-1026	17.0	0.90	6.67	10.3	5/5	5/5
F. sporotrichioides	F-998	1.0	0.13	0.67	0.13	5/5	4/5
	F-1006	8.3	—	3.67	5.33	5/5	4/5
	F-1049	40.3	tr	6.33	12.7	5/5	5/5
	F-1063	21.3	3.60	3.67	6.00	5/5	5/5
	F-1174	20.0	2.16	5.33	6.00	5/5	5/5
F. poae	F-1004	—	—	—	—	4/5	0/5
	F-1005	—	—	—	—	4/5	4/5
	F-1029	—	—	—	—	4/5	0/5
	F-1052	—	—	—	—	4/5	2/5
	F-1055	—	—	—	—	5/5	5/5
F. tricinctum	F-1010	—	—	—	—	0/5	
	F-1051	—	—	—	—	0/5	
	F-1452	—	—	—	—	0/5	

[a] μg/ml in culture filtrate: T-2, T-2 toxin; HT-2, HT-2 toxin; DAS, diacetoxyscirpenol; NS, neosolaniol.

Japanese isolates of *F. graminearum* could be divided into two groups producing, respectively, nivalenol and deoxynivalenol. However the two groups could not be separated on colony appearance and growth, or conidial dimensions, when cultured on PDA and CLA (Ichinoe *et al.*, 1984). Yoshizawa *et al.* (1979) were the first to recognize two groups producing different trichothecenes among isolates of *F. graminearum* from barley from southern Japan. Recently Suzuki *et al.* (1980) also isolated nivalenol- and deoxynivalenol-producing strains of *F. graminearum* from wheat and barley in central Japan.

Isolates of *F. acuminatum* on PSA formed colonies with dense, floccose aerial mycelium which was white to buff-coloured, becoming honey to flesh coloured in the centre. The colony reverse was rosy buff becoming red in the centre. Isolates of *F. acuminatum* produced abundant macroconidia, which long, slender, curved, 3–7-septate and 43–63 × 3.5–4.0 μm. With few exceptions, the dimensions and septation of the conidia were within the range described by Booth (1971) for *F. acuminatum*. This species is known to produce the trichothecenes T-2 toxin and neosolaniol (Suzuki *et al.*, 1980; Burmeister *et al.*, 1981).

All isolates of *F. sporotrichioides* were recognized by the presence of polyphialidic sporogenous cells. Cultural characteristics of most isolates on PSA plates agreed well with Booth's description of *F. sporotrichioides* but one strain, F-1049, formed subglobose microconidia and was considered to be intermediate between *F. poae* and *F. sporotrichioides*. Spore morphology of this strain is similar to that of NRRL 3287, known as *F. poae* (Ellison and Kotosonis, 1973). A critical examination of strain NRRL 3287 received from Professor Y. Ueno, showed that it had polyphialidic conidiophores and so was considered a variant of *F. sporotrichioides* rather than *F. poae*. However, when Joffe and Palti (1975) studied the taxonomic position of T-2 toxin-producing *Fusarium* isolates previously identified as *F. tricinctum*, they concluded that NRRL 3287 belonged to *F. poae*.

Trichothecene production of *F. nivale*

Nivalenol and fusarenon-X are important toxic trichothecenes in Japan. Nivalenol was first isolated from *F. nivale* isolate Fn-2B by Tatsuno *et al.* (1968). Later fusarenon-X was discovered by Ueno *et al.* (1969) in the same culture of *F. nivale*.

In northern Japan, *F. nivale* is frequently found from winter wheat as a causal fungus of snow mould (Tanaka *et al.*, 1983). As shown in Table 1, *F. nivale* was only found in our mycological survey in wheat grains collected from Hokkaido in 1981. However, our isolates of *F. nivale* have a tendency to produce abundant mycelium without forming spores, and failed in single spore isolation. To supplement our results with isolates from this survey, we examined the ability of *F. nivale* isolates to produce trichothecenes.

F. nivale isolates ARC-744, -745, -746, -785, -786, -787, and -788 were obtained from H. Kato, National Agriculture Research Center, Tsukuba, Ibaraki in 1982. *F. nivale* Fn-2B and two strains of *F. graminearum*, F-1104 and F-1114, were also included as positive controls in tests of trichothecene production in culture in liquid or rice media using methods described by Suzuki *et al.* (1980).

The results of this experiment are summarized in Table 5. No trichothecenes were detected by TLC analysis of extracts of any culture of the ARC strains or of one other strain, F-828, isolated from imported wheat grain. The isolates of known trichothecene-producing strains, Fn-2B, F-1104 and F-1114, all produced nivalenol and fusarenon-X under the same conditions on both media.

Because *F. nivale* Fn-2B differed from other *F. nivale* isolates in its ability to produce trichothecenes, cultures were evaluated taxonomically. The taxonomic position of the strain is still in doubt because colonies on PSA grew rapidly, produced reddish pigment and produced few macroconidia, but

Table 5 Comparison of trichothecene production by *Fusarium nivale* and *F. graminearum* on liquid and solid media

		Trichothecenes[a]			
		Liquid (µg/ml)		Rice (µg/g)[b]	
Species and strain no.	Source	NIV	F-X	NIV	F-X
F. nivale					
ARC-744[c]	Rice, leaf spot	—	—	—	—
ARC-745	Wheat, snow blight	—	—	—	—
ARC-746	Wheat, snow blight	—	—	—	
ARC-785	Wheat, snow blight	—	—	—	—
ARC-786	Wheat, snow blight	—	—	—	—
ARC-787	Wheat, snow blight	—	—	—	—
ARC-788	Wheat, snow blight	—	—	—	—
F-828	Wheat, imported	—	—	—	—
F. nivale?					
Fn-2B	Barley grain	—	0.28	6.50	0.80
F. graminearum					
F-1101	Barley, scab	0.47	1.17	0.98	0.42
F-1114	Barley, scab	0.03	0.50	0.39	0.20

[a] NIV, nivalenol; F-X, fusarenon-X.
[b] Determined by Suzuki's method (1980).
[c] Obtained from National Agriculture Research Center.

many variously shaped microconidia, while on CLA both macro- and microconidia were abundant. Microconidia were formed on simple phialides on lateral branches of the aerial mycelium. They were ovoid or spindle-shaped and measured: 0-septate, 6.0–12.0 × 4.0–5.5 μm; 1-septate, 9.0–16.0 × 3.5–5.5 μm; 2-septate, 12.5–21.0 × 4.0–5.5 μm. The cultural characters of strain Fn-2B are quite different from those of typical *F. nivale* and more like *F. tricinctum* than *F. nivale*. However, because of its ability to produce the toxic metabolites, nivalenol and fusarenon-X, it may probably be considered an abnormal variant of *F. graminearum*.

ACKNOWLEDGEMENTS

These experiments were partly supported by Grant in Aid from the Ministry of Health and Welfare, Japan. We thank Professor Yoshio Ueno, Tokyo University of Science, and Dr Takumi Yoshizawa, Kagawa University, for standard samples of trichothecenes, and Dr Hajime Kato, National Agriculture Research Center, Tsukuba, for strains of *F. nivale*.

REFERENCES

Booth, C. (1971). *The genus* Fusarium. Commonwealth Mycological Institute, Kew.

Burmeister, H. R., Ellis, J. J., and Vesonder, R. F. (1981). 'Survey for fusaria that produce an antibiotic that causes conidia of *Penicillium digitatum* to swell', *Mycopathologia*, **74**, 29–33.

Ellison, R. A., and Kotosonis, F. N. (1973). 'T-2 toxin as an emetic factor in moldy corn', *Appl. Microbiol.*, **26**, 540–543.

Ichinoe, M., Hagiwara, H., and Kurata, H. (1984). 'Distribution of trichothecene-producing fungi in barley and wheat fields in Japan'. In *Toxigenic Fungi—Their Toxins and Health Hazard* (eds H. Kurata and Y. Ueno), pp. 190–198. Kodansha Ltd, Tokyo, and Elsevier, Amsterdam.

Joffe, A. Z., and Palti, J. (1975). Taxonomic study of Fusaria of the Sporotrichiella section used in recent toxicological work. *Appl. Microbiol.*, **29**, 575–579.

Joffe, A. .Z., and Yagen, B. (1977). 'Comparative study of the yield of T-2 toxin produced by *Fusarium poae*, *F. sporotrichioides* and *F. sporotrichioides* var. *tricinctum* strains from different sources'. *Mycopathologia*, **60**, 93–98.

Kamimura, H., Nishijima, M., Saito, K., Takahashi, S., Ibe, A., Ochiai, S., and Naoi, Y. (1978). 'Mycotoxins in foods, VIII. Analytical procedure of trichothecene mycotoxins in cereals', *J. Food Hyg. Soc. Jpn.*, **19**, 443–448.

Lafont, J., and Lafont, P. (1979). 'Sensibilité de l'embryon de poulet à des mycotoxines', *Bull. Acad. Vet. de France*, **52**, 119–124.

Marasas, W. F. O., Leistner, L., Hofmann, G., and Eckardt, C. (1979). 'Occurrence of toxigenic strains of *Fusarium* in maize and barley in Germany', *Eur. J. Appl. Microbiol. Biotechnol.*, **7**, 289–305.

McLaughlin, J. Jr., Marliac, J. P., Verrett, M. J., Mutchler, M. K., and Fitzhugh, O. G. (1963). 'The injection of chemicals into yolk sac of fertile eggs prior to incubation as a toxicity test', *Toxicol. Appl. Pharmacol.*, **5**, 760–771.

Pathre, S. V., and Mirocha, C. J. (1978). 'Analysis of deoxynivalenol from cultures of *Fusarium* species', *Appl. Environ. Microbiol.*, **35**, 992–994.

Scott, P. M., Harwig, J., and Blanchfield, B. J. (1980). 'Screening *Fusarium* strains isolated from over-wintered Canadian grains for trichothecenes' *Mycopathologia,* **72**, 175–180.

Suzuki, T., Kurisu, M., Hoshino, Y., Ichinoe, M., Nose, N., Tokumaru, Y., and Watanabe, A. (1980). 'Production of trichothecene mycotoxins of *Fusarium* species in wheat and barley harvested in Saitama Prefecture', *J. Food Hyg. Soc. Jpn.,* **21**, 43–49.

Tanaka, F., Saito, I., Miyajima, K., Tsuchiya, Y., and Tsuboki, K. (1983). 'Occurrence of thiophanate-methyl tolerant isolates of *Fusarium nivale* (= *Gerlachia nivalis*), a causal fungus of snow mold of winter wheat, in Japan', *Ann. Phytopath. Soc. Japan*, **49**, 565–566.

Tatsuno, T., Saito, M., Enomoto, M., and Tsunoda, H. (1968). 'Nivalenol, a toxic principle of *Fusarium nivale*', *Chem. Pharm. Bull.*, **16**, 2519–2520.

Toussoun, T. A., and Nelson, P. E. (1968). *A Pictorial Guide to the Identification of* Fusarium *Species According to the Taxonomy system of Snyder and Hansen.* Pennsylvania State University Press, University Park and London.

Ueno, Y., Ueno, I., Tatsuno, T., Ohkubo, K., and Tsunoda, H. (1969). 'Fusarenon-X, a toxic principle of *Fusarium nivale* culture filtrate', *Experientia*, **25**, 1062.

Ueno, Y., Sato, N., Ishii, K., Sakai, K., Tsunoda, H., and Enomoto, M. (1973). 'Biological and chemical detection of trichothecene mycotoxins of *Fusarium* species', *Appl. Microbiol.*, **25**, 699–704.

Yoshizawa, T., Matsuura, Y., Tsuchiya, Y., Morooka, N., Kitani, K., Ichinoe, M., and Kurata, H. (1979). 'On the toxigenic fusaria invading barley and wheat in southern Japan', *J. Food Hyg. Soc. Jpn.*, **20**, 21–26.

Trichothecenes and Other Mycotoxins
Edited by J. Lacey

CHAPTER 4

Toxigenic fungi associated with sunflower seeds during development

M. Vijayalakshmi and A. S. Rao

INTRODUCTION

Sunflower (*Helianthus annuus* L.) is one of the world's principal oil seed crops, and its importance in India has increased with the introduction of such improved cultivars as 'Sunrise' from Canada and E.C.68413, E.C.68414, E.C.68415 and E.C.69874 from the U.S.S.R. The seeds may be invaded by a number of fungi (Noble and Richardson, 1968; Neergaard, 1977; Zad, 1978), including plant pathogens and field and storage saprophytes. Maggio *et al.* (1976) reported qualitative differences between samples of freshly harvested and stored seeds. In the only previous study of mycoflora on developing sunflower seeds, Vijay Kumar and Dwivedi (1981) found a gradual increase in seed infection during development with changing dominant species. Our study was started to observe the infection of seeds successively during development on the plants, post-harvest drying and storage. The toxigenic potential of the fungi isolated from the seeds was also determined.

MATERIALS AND METHODS

Infection during seed development

The cultivars E.C.68413, E.C.68414 and E.C.68415 were planted in small field plots during 1979 and 1980 rabi (January to April) and kharif (August to November). The rabi season is warm and dry, and crops are grown under irrigation, while the kharif crops are grown under rain-fed conditions. Seven samples of 100 florets/seeds from the periphery of flowerheads were taken weekly between flowering and harvest. The florets/seeds were surface-sterilized with mercuric chloride prior to plating on Czapek–Dox agar and the plates were incubated at room temperature (26 ± 2 °C) for 1 week. Colonies were then classified, counted, isolated and identified.

Infection during drying and in storage

Seed lots were dried after harvest in one of two ways: by spreading them in pans and placing under direct sun (sun-dried) or in open shade (shade-dried). Both the percentage of seeds infected and moisture content of the seeds were determined daily till they attained safe moisture level ($< 6\%$). Moisture content was determined according to the specifications of the USDA Marketing Service (1959). Freshly harvested and sun-dried seeds were stored in gunny bags and also at four relative humidity levels viz., 66%, 75%, 88% and 95%, maintained by using saturated salt solutions as specified by Lange (1967). Samples were withdrawn fortnightly for 3 months and seed infection, seed moisture level and percentage of seed germination were recorded simultaneously.

Toxicity of fungal isolates

About 40 different fungi isolated from sunflower seeds were cultured on Czapek–Dox broth for 15 days. The culture filtrates were then collected aseptically and tested for phytotoxicity, by assaying their effect on seed germination (Ludwig, 1957), elongation of roots (Luke and Wheeler, 1955) and shoots of rice seedlings, and on seed germination, root elongation and wilting of sunflower seedlings.

Toxicity to *Bacillus megaterium* and *B. subtilis*, known to be sensitive to aflatoxins (Burmeister and Hesseltine, 1966) was also tested by the seeded plate method. Inhibition zones were measured around filter paper discs saturated with concentrates of culture filtrates extracted with hot chloroform from cultures raised in Roux bottles on yeast extract–sucrose broth (YES broth).

Identification of toxic metabolites

For the identification of aflatoxins and other mycotoxins, concentrated chloroform extracts from 1 litre of culture filtrate of each fungus grown on YES broth for 15 days were used. TLC analysis employing toluene–ethyl acetate–90% formic acid (6 : 3 : 1) as the developing solvent system (Scott *et al.*, 1970) was carried out. The plates were observed under ultraviolet illumination and the colours of fluorescent spots were noted. Both the fluorescing and non-fluorescing areas were eluted separately in ethanol and their toxicity was tested by observing the wilting of sunflower cut shoots.

For the identification of tenuazonic acid (TeA), the culture filtrates were acidified and extracted in ethyl acetate according to the procedure of Meronuck *et al.* (1972).

On the basis of R_f values and colours of fluorescence of the spots selected toxins were co-chromatographed with authentic samples of known mycotoxins. For aflatoxins, a confirmatory test was carried out with trifluoroacetic acid (Jones, 1977) and for TeA, colour reaction with ethanolic ferric chloride (Rosett *et al.*, 1957) and UV spectral characteristics were determined employing water or ethanol as carrier solvents (Stickings, 1959).

Production of toxins in sunflower seeds

Samples of sunflower seeds were analysed for aflatoxins and TeA, immediately after harvest, after drying to safe moisture level and after storage for 12 weeks. Toxin production was also estimated in autoclaved seeds inoculated with the toxin-producing isolates. Samples were inoculated with spore suspensions and incubated in Erlenmeyer flasks for 15 days. The methods of extraction and analysis of aflatoxins and TeA were those of Karmelic *et al.* (1973) and Meronuck *et al.* (1972) respectively.

Bioassay of mycotoxins

The confirmation of mycotoxic activity was initiated on 1 day-old chicks which were fed for 10 days on corn meal containing 50 ml of culture filtrate of the test fungus per 100 g of the feed, cultured on YES broth for 15 days. The symptoms and mortality were recorded daily up to 10 days.

RESULTS

Infection during seed development

At the time of flower head opening, 16–35% of samples were colonized by one or more fungi, increasing to 100% in about 4 weeks. The fungal count recorded on each variety in each season ranged between 11 and 24. *Alternaria alternata* was usually the predominant fungus, colonizing 30.7% of the seeds of the three varieties in the four seasons (Table 1). In addition, *Curvularia lunata*, *Fusarium moniliforme*, *F. sporotrichioides*, *F. oxysporum*, *Penicillium citrinum*, *Cladosporium herbarum*, *Corynespora cassiicola*, *Aspergillus flavus*, *A. luchuensis*, *A. niger* and *Rhizopus stolonifer* also appeared frequently. The 1979 kharif season was particularly wet and, for once, *Fusarium* spp. and *P. citrinum* were isolated from more seeds than *A. alternata* and also a greater variety of other species was isolated from the samples than in any other season.

Infection during drying and in storage

The seeds exposed to sun-drying attained safe moisture relatively faster (3–4 days) than those exposed to shade-drying (5–6 days). The proportion of

Table 1 Fungi recorded on developing seeds (after surface sterilization) from flowerhead opening up to harvest, in four seasons

Fungus	No. of samples yielding species (out of 84 isolations)	Average percentage of seeds invaded
Alternaria alternata	81	30.7
Aspergillus candidus	3	0.1
A. flavus	30	2.4
A. fumigatus	2	0.1
A. glaucus	1	<0.1
A. luchuensis	23	2.1
A. nidulans	7	0.3
A. niger	17	1.7
A. tamarii	5	0.4
A. terreus	4	0.2
Chaetomium arcuatum	1	0.1
Chaetomium sp.	1	<0.1
Cladosporium herbarum	33	3.5
Cunninghamella sp.	1	0.1
Corynespora cassiicola	23	2.8
Curvularia andropogonis	1	0.1
C. lunata	51	9.3
C. maculans	11	0.8
C. pallescens	1	<0.1
C. tuberculata	2	0.1
Diplodia sp.	2	0.2
Drechslera hawaiiensis	14	0.9
D. rostrata	6	0.3
Fusarium dimerum	3	0.1
F. heterosporum	2	0.1
F. lateritium	7	0.5
F. moniliforme	63	9.2
F. oxysporum	38	3.0
F. sambucinum	2	0.2
F. semitectum	14	0.8
F. solani	2	0.1
F. sporotrichioides	27	2.8
Macrophomina phaseolina	6	0.4
Nigrospora sphaerica	6	0.7
Penicillium citrinum	56	9.1
Phoma herbarum	10	0.7
Rhizoctonia solani	3	0.2
Rhizopus stolonifer	14	1.4
Trichothecium roseum	7	0.7
Non-sporulating forms	17	1.1

fungal colonies developing from the seeds declined only slightly during drying from 110–125 to 90–100 colonies per 100 seeds. *A. alternata* remained predominant in both kharif and rabi seasons, while *F. moniliforme* and *F. oxysporum* declined slightly in numbers barely large enough to be significant (Table 2).

The proportion of seeds yielding fungi declined from an initial 100% to 29–44% when stored in gunny bags and at lower humidity levels (66% and 75%). However at high humidity levels, (88% and 95%) seed infection remained at 100% throughout and the fungal count per 100 seeds gradually increased as the storage time extended (Table 3).

Some field fungi, including *A. alternata*, *C. lunata*, *F. moniliforme* and *F. oxysporum*, declined in storage. Storage fungi, especially species of *Aspergillus* (*A. flavus*, *A. glaucus*, *A. luchuensis*, *A. niger* and *A. sydowi*) increased in incidence at 88% and 95% RH levels, but the relative abundance of different species varied in different seasons. Incidence of *Aspergillus* spp. decreased during storage at lower humidity levels and in gunny bags. Only occasionally was a slight increase in their incidence noticed on seeds stored in gunny bags.

Seed moisture content gradually increased to 8.6%, from 5.2%, when stored in gunny bags at lower humidity levels. At 88% and 95% RH levels increased to about 12% and 19% respectively.

The percentage of seed germination recorded ranged between 80 and 90 at 66% and 75% RH levels and also in gunny bags, while it declined to less than 15% at 88% and 95% RH levels. Seeds were visibly mouldy at these levels.

Toxicity of fungal isolates

Phytotoxicity and antibacterial activity of the fungal isolates which exhibited moderate to high levels of toxicity are presented in Table 4. *A. glaucus* was most toxic to the test plants, followed by *A. alternata*, *P. citrinum*, *A. niger*, *A. flavus* and *F. moniliforme*. It is evident from the results that the culture filtrates had greater effect in reducing root elongation of rice seedlings than that of shoot elongation. *Macrophomina phaseolina*, *P. citrinum*, *A. candidus*, *F. moniliforme*, *A. alternata* and *A. flavus* produced metabolites that were highly inhibitory to the test bacteria.

Identification of toxic metabolites

TLC analysis of fungal culture filtrates and their toxicity to cut sunflower shoots are given in Table 5. Aflatoxin-B_1 was produced by *A. flavus*, citrinin by *P. citrinum* and *A. terreus*, and TeA by *A. alternata*. The confirmatory test for aflatoxin-B_1 with trifluoroacetic acid gave positive results, and for TeA the confirmatory test was carried out by determining the UV absorption spectrum

Table 2 Changes in frequency of occurrence of the dominant fungi during drying of sunflower cultivar E.C.68414 in the four seasons (percentage of seeds yielding fungus)

	1979 Rabi			1980 Rabi			1979 Kharif			1980 Kharif		
Fungus	H*	Sun	Shade	H	Sun	Shade	H	Sun	Shade	H	Sun	Shade
Alternaria alternata	90	64	92	64	62	65	26	44	57	51	52	48
Fusarium moniliforme	9	6	6	9	6	7	29	26	28	20	18	10
Curvularia lunata	0	0	0	0	0	0	0	0	0	10	8	9
Penicillium citrinum	5	0	0	5	0	0	3	0	0	0	0	0
Corynespora cassiicola	10	0	0	14	0	0	0	0	0	0	0	0
Fusarium oxysporum	5	4	6	5	0	0	18	15	13	12	7	8
Aspergillus flavus	0	4	0	0	16	18	3	0	0	0	0	4

H* = At harvest; Sun = After drying in direct sunlight; Shade = After drying in open shade.

Table 3 Fungal infection of seeds of sunflower cultivar E.C.68414 during storage (average of four seasons) (percentage of seeds yielding fungus)

		After 12 weeks of storage				
Fungus	Initial (after drying)	Gunny bags	66% RH	75% RH	88% RH	95% RH
Alternaria alternata	53	37	30	32	26	15
Aspergillus candidus	0	1	0	0	4	1
A. flavus	8	8	8	6	25	37
A. glaucus	0	0	1	<1	8	19
A. luchuensis	0	<1	1	1	7	<1
A. nidulans	0	<1	0	0	2	2
A. niger	3	4	3	7	20	30
A. sydowi	0	0	0	<1	1	5
A. tamarii	0	1	<1	<1	3	3
A. terreus	0	0	0	0	1	1
Chaetomium arcuatum	0	0	<1	1	1	0
Cladosporium herbarum	3	0	1	0	<1	0
Corynespora cassiicola	3	<1	<1	<1	0	0
Curvularia lunata	8	4	7	4	4	2
C. maculans	0	<1	0	0	1	1

in water or ethanol as carrier solvent systems and the results (peaks at 239, 279 nm in water and at 217, 277 nm in ethanol) coincide with those reported by Stickings (1959), who characterized this toxin. In addition, ethanolic solution of TeA produces a brilliant orange-red colour when treated with ethanolic ferric chloride (Rosett *et al.*, 1957).

The toxins produced by *Fusarium moniliforme* and *F. sporotrichioides* did not correspond to T-2 toxin or zearalenone and their identity could not be established. Similarly, it was not possible to confirm the identity of the toxic metabolites produced by the other fungi.

Production of toxins in sunflower seeds

Aflatoxins were not produced in seeds up to harvest, after drying, during storage in gunny bags or when stored at 66% and 75% RH levels. Respectively 20–23 μg/kg and 25–30 μg/kg aflatoxin-B_1 were found after 12 weeks storage at 88% and 95% RH. Inoculated seeds contained 23 μg/kg of this toxin after only 15 days incubation at room temperature. TeA was found only in inoculated seeds.

Table 4 Phytotoxicity and microbial toxicity of culture filtrates of selected fungal isolates

Fungus	Phytotoxicity						Microbial toxicity[b]	
	Germination (%)		Shoot elongation[a]	Root elongation[a]		Wilting	*B.subtilis*	*B.megaterium*
	S	R	R	S	R	S		
Alternaria alternata	8	13	65	nil	nil	+	2.95	2.07
Aspergillus candidus	43	50	100	18	25	+	2.96	2.96
A. flavus	12	18	53	nil	nil	+	2.95	1.57
A. fumigatus	48	54	92	10	18	+	1.57	1.86
A. glaucus	nil	2	68	nil	nil	+	2.35	1.57
A. luchuensis	84	90	92	73	71	–	0.31	nil
A. niger	9	17	61	nil	nil	+	nil	nil
A. tamarii	59	74	62	nil	nil	+	0.75	0.75
A. terreus	42	39	84	24	30	+	2.07	1.86
Curvularia lunata	54	61	100	41	50	–	0.93	0.31
Fusarium moniliforme	11	13	46	nil	nil	+	2.95	2.08
F. sporotrichioides	24	32	68	15	21	+	1.81	1.13
Macrophomina phaseolina	51	77	100	19	30	+	3.92	3.61
Penicillium citrinum	7	9	69	16	17	+	3.61	2.35

S—Sunflower; R—Rice.
[a] Expressed as percentage over control (distilled water).
[b] Area of inhibition zone (cm^2) on seeded plates.

Table 5 Fluorescence and R_f values of toxic areas and the mycotoxins of the selected fungi

Fungus	Toxic area no.	Nature of fluorescence	R_f values	Mycotoxins
Alternaria alternata	1	Blue	0.23	
	2	Dark brown elongated spot[a]	0.25–0.36	Tenuazonic acid
Aspergillus candidus	1	Green	0.09	
	2	Green	0.17	
A. flavus	1	Blue	0.26	Aflatoxin B_1
	2	Blue	0.32	
	3	Blue	0.40	
A. fumigatus	1	Yellow	0.05	
	2	Blue	0.26	
	3	Blue	0.45	
A. glaucus	1	Yellow	0.19	
	2	Cream	0.55	
A. luchuensis	1	Green	0.05	
	2	Cream	0.64	
A. niger	1	Light green	0.04	
A. tamarii	1	Dark brown elongated spot[a]	0.25–0.36	
A. terreus	1	Yellow	0.08–0.57	Citrinin
Curvularia lunata	1	Non-fluorescent	0.29	
	2	Non-fluorescent	0.66	
Fusarium moniliforme	1	Light green	0.03	
	2	Non-fluorescent	0.40	
	3	Non-fluorescent	0.57	
F. sporotrichioides	1	Light green	0.03	
	2	Non-fluorescent	0.47	
Macrophomina phaseolina	1	Brown streak[a]	0.04–0.25	
	2	Yellow	0.64	
Penicillium citrinum	1	Non-fluorescent	0.05	
	2	Yellow	0.08–0.57	Citrinin
	3	Blue	0.93	

[a] Visible colour.

Bioassay of mycotoxins

Chicks fed on corn meal with culture filtrates of four of the toxin-producing fungi developed tremorgenic symptoms and mortality rate increased during the 10-day period of observation. With *A. flavus* mortality was 86% greater than with untreated control chicks; with *A. alternata* it was 57% greater; and with *P. citrinum* 40% greater. *Fusarium moniliforme* caused no toxic symptoms but the mortality rate increased over controls by 14%.

DISCUSSION

The sunflower produces a large inflorescence which becomes contaminated by deposition of airborne spores. About 50 species of fungi were isolated from unsterilized developing seeds between the disc floret stage and harvest, of which 40 occurred on surface-sterilized seeds, indicating possible penetration of seed tissues. Most of the species recorded were field fungi (Christensen and Kaufman, 1965). *Alternaria alternata* was predominant and appeared particularly well adapted to development on this plant. Seasonal differences were noticed in the prevalence of several fungi, which might be due to variations in the weather conditions as suggested by Kolk (1970). *A. alternata* was more abundant in the dry season than species of *Aspergillus*. In the rainy season *Fusarium* spp., particularly *F. moniliforme*, *F. oxysporum* and *F. sporotrichioides* with *Penicillium citrinum*, were predominant. Vijay Kumar and Dwivedi (1981) found *Alternaria*, Curvularia, Drechslera and *Rhizopus* consistently during the four stages of sunflower seed development and these fungi have also predominated in studies of market samples (Chohan and Kaur, 1975; Anilkumar and Urs, 1976; Abraham *et al.*, 1976; Singh *et al.*, 1977; Zad, 1978; Kanwar *et al.*, 1979; Jovicevic, 1980; Chandra *et al.*, 1981; Wadhwani, 1982). Clearly these fungi had invaded the seeds prior to harvest. The present study additionally indicates that fungi can penetrate the interior of the seed during development.

During drying, the percentage of seeds yielding fungi slightly declined. However, seed infection by *A. alternata* increased in 1979 kharif, which might be due to infection or contamination during harvest as suggested by Neergaard (1977). The percentage of seeds infected gradually declined further during storage at 66% and 75% RH levels and also in gunny bags, and only a slight increase in seed moisture was recorded under these conditions. However at 88% and 95% RH levels, changes in seed moisture content were very significant. It increased up to 12% at 88% RH and up to 19% at 95% RH. At 75, 80 and 85% RH and 5–30 °C, Christensen (1969, 1972) found that invasion of stored sunflower seeds by storage fungi increased with increasing moisture content, temperature and storage time. It is possible that initial low percentage of infection by aspergilli might have increased with storage time at higher RH levels, either due to concomitant contamination, as suggested by Neergaard (1977), or activation of dormant infection.

Most of the fungi isolated from the developing seeds were phytotoxic to varying degrees as well as inhibitory to test bacteria. Of these, 14 isolates were moderately to highly toxic. Analysis and toxicity tests of culture filtrates indicated the possibility of several toxic metabolites being produced by different fungi; among them aflatoxin-B_1, citrinin and TeA were identified. The phytotoxic nature of aflatoxins and citrinin has been reported earlier by several workers (Wright, 1951; Reiss, 1971). The phytotoxic properties of *Fusarium* spp. were also reported previously by Marasas *et al.* (1971) and Cole *et al.* (1973). However, the phytotoxic nature of TeA has not been reported before.

Culture filtrates containing aflatoxin-B_1, citrinin and TeA were also toxic to young chicks. Toxicity of aflatoxins to chicks is well known, but there were no reports for citrinin and TeA.

Aflatoxin-B_1 was found in sunflower seeds stored at high RH (88% and 95%). Natural occurrence of aflatoxins in sunflower was previously reported from three countries—Hungary, Germany and Italy (Anonymous, 1977). Llewellyn and Eadie (1974) found no aflatoxin in sunflower seeds from central Virginia. However, inoculation with *A. flavus* resulted in toxin production. Nagarajan *et al.* (1974) also found aflatoxin in sunflower seeds inoculated with toxin-producing strains of *A. flavus* and *A. parasiticus*, but only in low concentrations, which they attributed to the hard, glossy nature of the seed coat. However, production of TeA in inoculated seeds had not been reported before. The production of aflatoxin-B_1 in uninoculated samples stored at high RH emphasizes the danger of toxin formation if seeds become exposed to high moisture in storage.

REFERENCES

Abraham, K. J., Menon, M. R., and Nair, M. C. (1976). 'Microflora associated with sunflower seeds', *Agric. Res. J. Kerala*, **14**, 97–98.

Anilkumar, T. B., and Urs, S. D. (1976). 'Seed mycoflora of sunflower and its control', *Mysore J. Agric. Sci.*, **10**, 424–427.

Anonymous (1977). *Global Perspectives on Mycotoxins*. Joint FAO/WHO/UNEP Conference on mycotoxins. Nairobi–Kenya, 19–27 September 1977.

Burmeister, H. R., and Hesseltine, C. W. (1966). 'Survey of the sensitivity of micro-organisms to aflatoxins', *Appl. Microbiol.*, **14**, 403–404.

Chandra, S., Narang, M., and Srivastava, R. K. (1981). Studies on seed mycoflora of oil seeds in India. Part I. Qualitative and quantitative estimations, *Int. Biodet. Bull.*, **17**, 71–75.

Chohan, J. S., and Kaur, J. (1975). 'Seed mycoflora of sunflower and control of seed-borne pathogens', *Ind. J. Mycol. Pl. Pathol.*, **5**, 210–211.

Christensen, C. M. (1969). 'Factors affecting invasion of sunflower seeds by storage fungi', *Phytopathology*, **59**, 1699–1702.

Christensen, C. M. (1972). 'Moisture content of sunflower seeds in relation to invasion by storage fungi', *Pl. Dis. Reptr.*, **56**, 173–175.

Christensen, C. M., and Kaufmann, H. H. (1965). 'Deterioration of stored grains by fungi', *Ann. Rev. Phytopath.*, **3**, 69–84.

Cole, R. J., Kirksey, J. W., Cutler, H. G., Doupnik, B. L., and Peckhan, J. L. (1973). 'Toxin from *Fusarium moniliforme*: effects on plants and animals', *Science*, **179**, 1324.

Jones, B. D. (1977). 'Aflatoxin and related compounds—Extraction and analysis'. In: *Mycotoxic Fungi, Mycotoxins, Mycotoxicoses* (eds T. D. Wyllie and L. G. Morehouse), Vol. I, pp. 201–237. Marcel Dekker, New York.

Jovicevic, B. (1980). ('A contribution to the knowledge of harmful microflora in wheat, corn and sunflower seed and seedlings'), *Zastita bilja*, **31**(2), br. 152, 101–119.

Kanwar, Z. S., Khanna, P. K., and Thareza, M. L. (1979). 'Seed mycoflora of sunflower and their control', *Pesticides*, **13**, 41–44.

Karmelic, J., Israel, M., Benado, S., and Leon, C. (1973). 'Analysis of aflatoxins in sunflower seeds and sunflower meal', *J. Assoc. Off. Anal. Chem.*, **56**, 219–222.

Kolk, H. (1970). 'Seedling diseases of cereals in Sweden', *Proc. Int. Seed Test. Assoc.*, **35**, 51–67.

Lange, N. A. (1967) *Handbook of Chemistry*. McGraw-Hill, New York.

Llewellyn, G. C., and Eadie, T. (1974). 'Evaluating the potential for occurrence of aflatoxins in black sunflower seeds', *J. Assoc. Off. Anal. Chem.*, **57**, 858–860.

Ludwig, R. A. (1957). 'Toxin production by *Helminthosporium sativum* and its significance in disease development', *Can. J. Bot.*, **35**, 291–303.

Luke, H. H., and Wheeler, H. E. (1955). 'Toxin production by *Helminthosporium victoriae*', *Phytopathology,* **4**, 453–458.

Maggio, D. D., Canella, M., and Sodini, G. (1976). 'Studi micologici E biochimici su varieta italiane di girasole': 1, *Inf. tore fitopatol.*, **10**, 7–12.

Marasas, W. F. O., Smalley, E. B., Bamburg, J. R., and Strong, F. M. (1971). 'Phytotoxicity of T-2 toxin produced by *Fusarium tricinctum*', *Phytopathology,* **61**, 1488.

Meronuck, R. A., Steele, J. A., Mirocha, C. J., and Christensen, C. M. (1972). 'Tenuazonic acid, a toxin produced by *Alternaria alternata*', *Appl. Microbiol.*, **23**, 613–617.

Nagarajan, V., Bhat, R. V., and Tulpule, P. G. (1974). 'Aflatoxin production in sunflower (*Helianthus annuus*) seed varieties', *Curr. Sci.*, **43**, 603–604.

Neergaard, P. (1977). *Seed Pathology* (Vols. I and II), Macmillan, London.

Noble, M., and Richardson, M. J. (1968). *An Annotated List of Seed-borne Diseases*. CMI, England.

Reiss, J. (1971). ('Inhibition of germination of *Lepidium sativum* by aflatoxin B_1 and rubratoxin B'), *Biochem. Physiol. Pflanzen*. **162**, 363.

Rosett, T., Sankhala, R. H., Stickings, C. E., Taylor, M. E. U., and Thomas, R. (1957). 'Studies in the biochemistry of micro-organisms. 103. Metabolites of *Alternaria tenuis* Auct. Culture filtrate products', *Biochem. J.*, **67**, 390–400.

Scott, P. M., Lawrence, J. W., and Van Walbeek, W. (1970). 'Detection of mycotoxins by thin-layer chromatography: application to screening of fungal extracts', *Appl. Microbiol.*, **20**, 839–842.

Singh, D., Mathur, S. B., and Neergaard, P. (1977). 'Histopathology of sunflower seeds infected by *Alternaria tenuis*', *Seed Sci*. Technol., **5**, 579–586.

Stickings, C. E. (1959). 'Studies in the biochemistry of micro-organisms. 106. Metabolites of *Alternaria tenuis* Auct: The structure of tenuazonic acid', *Biochem. J.*, **72**, 332–340.

USDA Marketing Service (1959). Service and Regulatory Announcements No. 147. Methods for determining moisture content as specified in the U.S. Official Grain Standards, 3 pp.
Vijay Kumar, and Dwivedi, R. S. (1981). 'Mycoflora associated with floral parts of sunflower', *Ind. Phytopath.*, **34**, 314–317.
Wadhwani, K. (1982). 'Thermophilous and thermotolerant fungi associated with sunflower seeds', *Ind. Phytopath.*, **35**, 162–163.
Wright, J. M. (1951). 'Phytotoxic effects of some antibiotics', *Ann. Bot.*, **15**, 493.
Zad, J. (1978). 'A note on the mycoflora of sunflower seeds in Iran', *Seed Sci. Technol.*, **6**, 953–956.

Trichothecenes and Other Mycotoxins
Edited by J. Lacey

CHAPTER 5

Microbial and toxic constituents of grain dust and their health implications

Muriel S. Palmgren

INTRODUCTION

Grain shipping is of great economic importance in southern Louisiana and since the recent slump in oil shipment, grain has been the major item exported from these ports. In 1982, 34.6 million short tons of grain and grain products were exported from terminal grain elevators located along the Mississippi River near New Orleans. Some of these elevators are more than 20 years old, while others are less than 5 years old. Their dust control systems vary widely in efficiency with some, even in older elevators, being very efficient. Within each elevator, some areas are dustier than others, e.g. where grain is unloaded from barges or trains, cleaned and weighed, and loaded onto ships. No work area was totally without dust, although many had relatively low levels.

Grain dust is heterogeneous, consisting of broken and abraded particles of grain and also a variety of materials that contaminate grain, e.g. insects, insect parts, mites, fungi, and bacteria, their waste products and metabolites, plant parts, soil, and pesticides. This dust may become suspended in the air as grain is harvested, transported, moved from one container to another, or cleaned. Dust that settles in work areas may be resuspended through the activities of workers. Therefore, people working on farms and in storage and transport facilities may often inhale grain dust.

Inhalation of grain dust was recognized as a cause of illness many years ago when Ramazzini (1713) described the general ill health, respiratory disorders, and shortened lifespan of grain handlers and millers. Since then, studies have sought to determine the composition of grain dust, to describe the effects that grain dust has on lungs and the general health of those exposed, and to determine the causes of the symptoms.

CHARACTERIZATION OF GRAIN DUST

Micro-organisms are consistently present in grain dust, including fungi and bacteria of various species. The predominant species of fungi may vary dependent on the history of the grain. For example, dust collected during harvest contained different fungi from that collected after the grain had been stored. Dust collected behind harvesters contained large numbers of both phytopathic and saprophytic fungi, as well as soil, bacteria, and fragments of straw, grain, and fungal hyphae (Lacey, 1980). Numbers of fungal spores ranged from 3.5 to 200 million/m^3 of air and commonly isolated genera are listed in Table 1. After storage, genera common before harvest were seldom isolated, while other species often became abundant in the grain and were frequently in large numbers in airborne dust of storage facilities and grain elevators (Whidden *et al.*, 1980; Lacey, 1980).

Up to 958 million viable fungal spores/m^3 of air have been found in a country grain elevator in Canada without mechanical exhaust ventilation (Farant *et al.*, 1973). Later Farant and Moore (1978) reported total and respirable ($<$10 μm in diameter) dust concentrations that were very high (Table 2), which could explain the high concentration of spores in the first report. Lower concentrations of fungal spores and dust were found in terminal grain elevators in Duluth, MN—Superior, WI with mechanical ventilation (Table 2) (Whidden *et al.*, 1980). In a similar study of terminal elevators located along the Mississippi River near New Orleans (Palmgren and DeLucca, in review), the maximum dust concentration found (1.22 g/m^3) may be unrepresentative (Table 2). This was collected in the baghouse of a dust control system that a worker entered to solve a problem. The atmosphere was extremely dusty and the man was covered with dust when he emerged. Consequently, this was not a routine dust exposure, nor did it represent that person's exposure, since adequate respiratory protection was worn. The highest concentration of fungal spores was likewise in this sample (Table 2). The next highest concentrations were 371 mg of dust/m^3 of air and 5.5 million spores/m^3 of air.

Fungi are not the only micro-organisms present in grain dust. Grain dust from Polish elevators contained few fungi, and bacteria predominated (Dutkiewicz, 1978). *Aspergillus glaucus*, *Aspergillus fumigatus*, *Geotrichum candidum*, and 29 other fungal species were present in concentrations of less than 10,000/m^3 of air while a bacterium, *Enterobacter agglomerans* (= *Erwinia herbicola*), contributed 50,000 to 100,000 of the 129,000 viable micro-organisms/m^3 of air. In grain cleaning rooms of mills, 109,000 out of 189,000 viable micro-organisms/m^3 of air were *E. agglomerans* while dust that had settled on floors or equipment contained 303 million/g. *E. agglomerans* comprised 75% of all the Gram-negative bacteria and nearly 50% of all micro-organisms in the dust.

Table 1 Percentage frequency of fungi commonly isolated from grain dust

	During harvest		During storage					
	Lacey[a]		Farant[b]		Lacey[a]		Whidden[c]	
	V[d]	NV[e]	V	NV	V	NV	V	NV
Penicillium	93	36[f]	69	84–100[f]	100	—[g]	79	—
Aspergillus	<10		43		94	—	94	—
Cladosporium	100	100	46	25–95	82	100	81	68
Mucorales	55	<10	20	50–97	87	—	34	30
Yeasts	32	33	8	—	98	—	80	—
Verticillium	100	67	—	—	—	—	—	—
Alternaria	90	99	29	6–80	52	100	20	7
Botrytis	66	67	3	89–100	3	—	—	—
Ustilago	—	55	—	—	—	98	—	86

[a] Lacey, 1980.
[b] Farant *et al.*, 1973. In this study viable counts represent percentage frequency of the most frequently isolated species of the genera listed; non-viable counts were not tabulated for the group of elevators. The range of frequencies of the individual elevators are listed.
[c] Whidden *et al.*, 1980.
[d] V = Viable counts; fungi grown and then counted.
[e] NV = Non-viable counts; direct microscopic counts.
[f] Value represents combined count of *Penicillium* and *Aspergillus* spores because most are indistinguishable microscopically.
[g] Not reported.

Table 2 Concentration of fungal spores and dust in the air of grain of grain elevators

	Duluth–Superior[a] (Range)	New Orleans[b] (Range)	Canada[c] (Range)
Fungal spores (No./m^3)			
Viable	ND[d]–9.4×10^4	1.0×10^{10}[e]	ND–9.85×10^8
Non-viable	ND –3.6×10^7	—	—
Resp. dust (mg/m^3)	ND–8.26	ND–1223[e]	ND–76.3
Total dust (mg/m^3)	0.14–39.1	—	0.18–781

[a] Whidden *et al.*, 1980.
[b] Palmgren and DeLucca, 1983.
[c] Farant *et al.*, 1973; Farant and Moore, 1978.
[d] ND = None detected.
[e] Next highest values were 5.5 million spores/m^3 and 371 mg of dust/m^3 of air.

Bacteria were also recovered in the Duluth–Superior and New Orleans studies, but less consistently. Up to 260,000 bacteria and yeasts were isolated from the dust in Duluth–Superior grain elevators, but some samples contained no bacteria (Whidden *et al.*, 1980). In the New Orleans grain elevators there were few bacteria in respirable dust (DeLucca *et al.*, 1984). However, in dust that had settled on floors and equipment, bacteria were numerous (1.9 to 53.4 million/g of dust); of the Gram-negative bacteria, 61% were *E. agglomerans* (DeLucca *et al.*, 1984). These findings corroborate the findings of Dutkiewicz (1978), who speculated that endotoxins produced by the high concentrations of Gram-negative bacteria might represent a hazard to workers. DeLucca *et al.* (1984) measured both endotoxin levels and numbers of Gram-negative bacteria in grain dust. Endotoxin was detected in samples of airborne dust with 35% yielding 3.75 endotoxin units (0.75 mg) per m^3 of air and one sample, 37.5 endotoxin units (7.5 mg) per m^3 of air. Endotoxin was also detected in all 20 settled dust samples ranging from 112.5 to 927.5 endotoxin units per g of dust (≡ 22.5 to 187.5 mg per g).

Toxic chemicals that contaminate grain may also be found in grain dust. Residues of insecticide that had been applied before harvest or during transportation and storage of the grain have been detected in dust from Australian grain elevators (Murray, 1979). Since 1963, all Australian wheat had been treated before storage with insecticides, such as malathion, dichlorvos, fenitrothione or bioresmethrin. Since most insecticide remains on the surface and little penetrates into the endosperm, high concentrations of insecticide accumulated in the dust produced by abrasion of the grain's surface. For example, grain treated with malathion at 8 mg/kg yielded dust containing 150–250 mg/kg. Similarly, 2–4 weeks after treating with 6–12 mg/kg of dichlorvos, 2 mg/kg remained on the grain while dust from this grain contained up to 400 mg/kg and persisted for several months (Murray, 1979). Residues of insecticides in dust from New Orleans grain elevators were lower than those reported in Australia. A pilot study of malathion and diazinon residues in grain dust reported that all samples contained 0.17 to 32 mg of malathion/kg dust, but no detectable diazinon (Palmgren and Lee, 1984). Differences in insecticides used and the amounts applied may account for the lower level found in the New Orleans study. Further surveys of insecticides in dust from U.S. elevators would determine whether there is a problem.

Grain may become contaminated naturally with other toxic chemicals. For instance, mycotoxins produced by fungi that grow on grain before harvest or during storage could end up in dust produced by abrasion of the grain. Dust was produced experimentally from aflatoxin-contaminated corn during grinding in the laboratory and as it was moved by augers out of a bin into a holding wagon and back again (Burg *et al.*, 1981). Both the airborne dust and the bulk corn samples were assayed for aflatoxin content. Airborne concentrations of the four aflatoxins during grinding ranged from 4.45 to 38.8 mg/m^3 of air and during transfer to and from the bin, from trace quantities to 107 mg/m^3 of air with an average aflatoxin level of 3.886 mg/g dust and 0.138 mg/g dust for the respective operations. Particles deposited in the first stage of the Andersen sampler appeared to contain more aflatoxins than all of the rest combined. The average aflatoxin concentration in dust from grinding was higher than in the corn from which it came, while that produced during transfer of the grain was only about half of the concentration in the grain (Burg *et al.*, 1981). It was recommended that precautions should be taken if there was 20 mg of aflatoxin/g of grain. Grain does not usually contain that much aflotoxin when shipped for export because it is of high quality and is monitored for aflatoxin. Consequently, only two of 49 settled dust samples from the New Orleans grain terminals yielded detectable aflatoxin, and neither exceeded 10 mg/g of dust (Palmgren *et al.*, 1983b; Palmgren and Fleischhacker, submitted). However, most samples did contain other mycotoxins (Table 3). Only four (8%) contained no detectable amount of the three mycotoxins—aflatoxin, secalonic acid D and zearalenone—while 14 of 41 samples (34%) contained both secalonic acid D and zearalenone. Unconfirmed results suggest that very low levels of ochratoxin A may also be present in a few samples and moderate levels of deoxynivalenol in most samples.

CLINICAL OBSERVATIONS OF THE HEALTH OF GRAIN HANDLERS

Febrile responses have been reported in humans at doses of 2 mg of inhaled endotoxin/kg of body weight (Wolff, 1973) with 10 mg/kg body weight

Table 3 Mycotoxins in settled grain dust samples[a]

	Positive samples		Frequency	
	Mean (ng/g)	Range (ng/g)	No. positive/ total	Percentage
Aflatoxin-B_1	15	10–20	2/49	4
Secalonic acid D	280	30–200	28/41	68
Zearalenone	154	trace–500	33/48	69

[a] Combined results from Palmgren *et al.*, 1983b; Palmgren and Fleischhacker, 1984.

producing pulmonary dysfunction (Muittari *et al.*, 1980). Worker exposure as estimated from the respirable dust samples collected in New Orleans area grain elevators, was below those doses, ranging from 0 to 0.1 mg/kg body weight (DeLucca *et al.*, 1983). However, workers without respiratory protection might be exposed to enough endotoxin to elicit febrile or pulmonary reactions if concentrations of dust or endotoxins in dust were unusually high.

The effects that mycotoxins produce in people who inhale small doses is undetermined. However, there have been a few reports of medical problems in people after exposure to agricultural dust containing large numbers of fungal spores and perhaps large quantities of mycotoxins. For example, one group of workers unloading mouldy grain from railcars for 4 hours subsequently developed malaise, headache, stinging bitterness in the mouth, pain in the eyes, chills, fever ranging up to 39.5 °C, cough, quickened pulse, and dry, whistling rales in the lungs (Samsonov, 1960). Symptoms lasted 4–7 days. Spores of *Mucor*, *Aspergillus*, and *Pencillium* species were isolated from sputum of workers and the air of the workplace, but no mycelium was isolated from their sputum.

In an attempt to duplicate the syndrome in animals, guinea pigs were exposed to an aerosol of spores from *Dendrodochium toxicum*, *Stachybotrys alternans (S. atra*) and *Aspergillus niger* (Samsonov and Samsonov, 1960). Histological examination of the lungs revealed local serious inflammation, necroses, and ulcerations, while bronchopneumonic foci occurred in the pulmonary parenchyma. Atelectasis was observed with compensatory emphysema and there was no evidence of hyphae from a mycotic infection. The authors suggested that a toxic agent in the spores was responsible for the observations in both the guinea pigs and the workers (Samsonov and Samsonov, 1960). No attempt was made to isolate and administer the toxic agent to reproduce the syndrome. An immunological reaction has also been suggested as the causative factor of these symptoms (Austwick, personal communication, 1983). Thus, the aetiology appears to be in doubt.

Emanuel *et al.* (1975) reported illness in farmers cleaning out mouldy silage from silos in Wisconsin and termed it pulmonary mycotoxicosis. The farmers suffered burning eyes, nose, throat, and chest immediately after exposure to massive concentrations of spores, followed by chills, fever, and a dry, irritating cough during the night. Lung biopsy cultured on blood agar yielded five different fungi including *Fusarium* and *Penicillium* species. Histological examination of the biopsy tissue revealed a multifocal, acute process with most bronchioles filled with an exudate of neutrophils and many histiocytes. Two children of one farmer were playing near the silo as he cleaned it, and also became ill with fever and cough, but for only a short time. The symptoms thus appeared to be dose-dependent.

Inhalation of Brazilian peanut meal contaminated with aflatoxin has been linked with two deaths caused by pulmonary adenomatosis (Dvorackova, 1976). In one analysis for aflatoxin, lung tissue yielded a blue fluorescent spot at the same R_f as aflatoxin-B_1 that also turned yellow after spraying with 50% sulphuric acid, as does aflatoxin. Aflatoxin was also implicated in the deaths of two newborns from Reye's syndrome (encephalopathy with fatty degeneration of the viscera) after both mothers had handled contaminated fodders throughout their pregnancies (Dvorackova *et al.*, 1977). Reye's syndrome was also diagnosed as the cause of death of two children, aged 8 and 12 months, whose mothers worked at a poultry farm where there was an outbreak of aflatoxicosis (Dvorackova *et al.*, 1977). Aflatoxin-B_1 was detected in the livers of both newborns and also in the liver of the one older child for which there was a specimen.

Other evidence that grain dust is harmful comes from epidemiological studies of grain workers from various parts of the world. In Canada, Warren and Manfreda (1980) questioned grain handlers about respiratory symptoms and tested pulmonary function. The same 51 grain workers were examined twice, 6 years apart. However, control workers were studied only the second time. The prevalence of chronic respiratory symptoms in grain workers changed little in the 6 years, but was greater in grain workers than controls. Dyspnoea (shortness of breath) occurred in 51% of the grain workers and grain fever (a delayed, nocturnal fever) in 27%. These results were also evident in another longitudinal study where both Canadian grain workers and civic workers were studied 2½ years apart (Chan-Yeung *et al.*, 1981). Each time the prevalence of respiratory symptoms was greater in the grain workers than in the civic workers. Similar findings were reported in one-time studies of grain workers in the United States and Egypt (Rankin *et al.*, 1979; Moselhi *et al.*, 1979).

Grain workers and non-grain workers also different significantly in their pulmonary functions. The prevalence of airway obstruction was markedly different as measured by the ratio of forced expiratory volume at 1.0 second and forced vital capacity (FEV_1/FVC) (Warren and Manfreda, 1980). When grain workers were tested twice; seven grain workers (14%) had abnormal FEV_1/FVC values both times, seven (14%) had abnormal values only the second time, while 37 were normal both times. Control workers were studied only once. In the other group of Canadian grain workers examined twice (Chan-Yeung *et al.*, 1981), the expected decrease in pulmonary function due to increasing age was more rapid in grain workers than in civic workers each time. Airway obstruction was also more prevalent in grain workers in the United States than in city worker controls (Rankin *et al.*, 1979).

Smoking seems to be a major risk factor in producing both respiratory symptoms and lung dysfunction in grain workers. However, non-smoking workers also exhibit adverse changes due to grain exposure (Dosman *et al.*, 1981). For instance, the prevalence of chronic bronchitis was greater in non-smoking grain workers (23%) than in non-smoking control workers (3%) and similar differences have also been found in other studies in Canada and the United States (Chan-Yeung *et al.*, 1981; Rankin *et al.*, 1979). Respiratory symptoms were more prevalent in both groups of grain workers than in either of the smoking or non-smoking controls (Warren and Manfreda, 1980). The increased risk of chronic bronchitis was calculated to be greater from grain handling (a factor of 4.4) than from smoking (a factor of 2.9) (Rankin *et al.*, 1979). In contrast, lung function was affected by grain handling at the same or smaller magnitude than by smoking in one study (Rankin *et al.*, 1979), but airway obstruction in the Egyptian study was believed to be more harmful than smoking, and even to potentiate the effects of smoking (Moselhi *et al.*, 1979).

EXPERIMENTAL STUDIES ON GRAIN DUST AND ITS CONSTITUENTS

The effects of grain dust have been studied both in animals and human volunteers. Mice exposed to grain dust for 8 hours a day, 5 days a week for 16 weeks or continuously (three 8-hour periods) for 2–28 days developed clusters of macrophages in the alveolar sacs and alveoli, but no fibrosis (Armanious *et al.*, 1982). Total dust levels were 5.4 g/m^3 after 30 minutes and decreased to 1.2 g/m^3 after 8 hours. The respirable portion of the dust (<10 μm in diameter) was only 4% of the total dust and part of that cannot be inhaled by mice. The quantity of dust actually inhaled was difficult to assess, especially when a massive proportion of particles were of a size that would clog the nasal passages of mice.

Crude extracts of grain dust were inhaled by human volunteers in doses of up to 5 ml of a 1/10 dilution of the extract (Warren *et al.*, 1974). Either immediate, or both immediate and late, respiratory symptoms developed in seven of the eight subjects with skin hypersensitivity to grain dust, but not in the controls. However, systemic symptoms, such as malaise, headache, and fever, were not correlated with positive skin reactions to grain dust. It seemed that although immunological reactions were responsible for the respiratory symptoms, they were not responsible for the systemic symptoms. Different results were obtained with 11 grain workers in the United States. Respiratory response to durum wheat dust extract was unrelated to skin hypersensitivity and there were no systemic reactions (doPico *et al.*, 1981, 1982). Sodium chromoglycate inhibited the bronchial response to the dust extract and histamine or another mediator was postulated to occur in response to direct irritation. Another mechanism suggested was non-specific bronchial reactivity, but the constituents of the dust that might be responsible have not been determined.

Our studies of the constituents of grain dust (Palmgren *et al.*, 1983a, b; Palmgren and Fleischhacker, submitted) suggest mycotoxins are commonly present, but at low levels. Also being studied are the effects of secalonic acid D (SAD) on respiratory infection in mice. Mice are injected with SAD after being exposed to an aerosol of influenza virus, both at less than the LD_{20} dose. Preliminary data suggest that the time to death was unaltered, but mortality following joint treatment was significantly greater than the sum from the separate treatments. Thus, mycotoxins in grain dust could exacerbate the effects of infectious agents.

ACKNOWLEDGEMENTS

E. B. Smalley, R. C. Caldwell, J. K. Phillp, S. Mai, A. Tsiasto, and W. E. Burkholder participated with me in the Wisconsin study (National Institute of Occupational Safety and Health Contract No. 210–77–0150). Characterization of grain dust was performed with the help of A. J. DeLucca II, L. S. Lee, A. Ciegler, K. G. Ehrlich and T. C. Lee, while toxicity studies are conducted with the aid of D. S. Fleischhacker, T. G. Akers, and P. J. Gerone under U.S. Department of Agriculture Cooperative Agreement No. 58–7B30–0–216.

REFERENCES

Aramanious, M., Cotton, D., Wright, J., Dosman, J., and Carr, I. (1982). 'Grain dust and alveolar macrophages: An experimental study of the effects of grain dust on the mouse lung', *J. Pathol.*, **136**, 265–272.

Burg, W. R., Shotwell, O. L., and Saltzman, B. E. (1981). 'Measurements of airborne aflatoxins during the handling of contaminated corn', *Amer. Ind. Hyg. Assoc. J.*, **42**, 1–11.

Chan-Yeung, M., Schulzer, M., MacLean, L., Dorken, E., Tan, F., Lam, S., Enarson, D., and Grzybowski, S. (1981). 'A follow-up study of the grain elevator workers in the port of Vancouver', *Arch. Environ. Health*, **36**, 75–81.

DeLucca, A. J. II, Godshall, M. A., and Palmgren, M. S. (1984). 'Gram-negative bacterial endotoxins in grain elevator dusts', *Amer. Ind. Hyg. Assoc. J.* **45**, 336–339.

doPico, G. A., Jacobs, S., Flaherty, D., and Rankin, J. (1982). 'Pulmonary reaction to durum wheat, a constituent of grain dust', *Chest*, **81**, 55–61.

doPico, G. A., Reddan, W. G., Tsiatis, A., Flaherty, D., Jacobs, S., and Rankin, J. (1981). 'Pulmonary reaction to grain dust and its constituents', *Chest* **80** (Supplement), 57S–60S.

Dosman, J., Cotton, D., Graham, B., Li, R., Froh, F., and Barnet, G. (1981). 'Chronic bronchitis and decreased forced expiratory flow rates in lifetime nonsmoking grain workers', *Chest* (Supplement), **80**, 33S.

Dutkiewicz, J. (1978). 'Exposure to dust-borne bacteria in agriculture. I. Environmental studies', *Arch. Environ. Health* **33**, 250–257.

Dvorackova, I. (1976). 'Aflatoxin inhalation and alveolar cell carcinoma', *Brit. Med. J.*, **3**, 691.

Dvorackova, I., Kusak, V., Vesely, D., Vesela, J., and Nesnidal, P. (1977). 'Aflatoxin and encephalopathy with fatty degeneration of viscera (Reye)', *Ann. Nutr. Alim.*, **31**, 977–990.

Emanuel, D. A., Wenzel, F. J., and Lawton, B. R. (1975). 'Pulmonary mycotoxicosis', *Chest*, **67**, 293–297.

Farant, J.-P., and Moore, C. F. (1978). 'Dust exposures in the Canadian grain industry', *Amer. Ind. Hyg. Assoc. J.* **39**, 177–194.

Farant, J.-P., Moore, C. F., and Windish, J. P. (1973). *Airborne Fungal Spores in Canadian Grain Elevators*. Health and Welfare Canada Report No. HHAW73–2.

Lacey, J. (1980). 'The microflora of grain dusts', in *Occupational Pulmonary Disease: Focus on Grain Dust and Health* (eds J. A. Dosman and D. J. Cotton), pp. 417–440. Academic Press, New York.

Moselhi, M., El-Gazzar, R., Abdel-Kader, H. M., El-Sadik, M., and El-Dakhakhny, M. (1979). 'Clinical and biological investigations among grain handlers, *J. Egypt. Pub. Health Assoc.*, **54**, 396–414.

Muittari, A., Rylander, R., and Salkinoja-Salonen, M. (1980). 'Endotoxin and bath water fever', *Lancet*, **ii**, 89.

Murray, W. J. (1979). 'Methods of overcoming merchandising problems associated with excessive pesticide residues in grain dust', in *Proceedings of the International Symposium on Grain Dust* (eds B. S. Miller and Y. Pomeranz), pp. 309–320. Division of Continuing Education, Kansas University, Manhattan, KS.

Palmgren, M. S., and DeLucca, A. J., II (1984). 'Survey of fungi in grain dust from grain elevators in the New Orleans area', *Amer. Ind. Hyg. Assoc. J.* (in review).

Palmgren, M. S., and Fleischhacker, D. S. (1984). 'Survey of mycotoxins in grain dust from New Orleans area grain elevators', *Amer. Ind. Hyg. Assoc. J.* (in review).

Palmgren, M. S., and Lee, T. C. (1984). 'Malathion and diazinon levels in grain dust from New Orleans area grain elevators', *Amer. Ind. Hyg. Assoc. J.* **45**, 168–171.

Palmgren, M. S., DeLucca, A.J. II, and Ciegler, A. (1983a). 'Fungi and bacteria in grain dust from New Orleans area grain elevators', in *Aerosols in the Mining and Industrial Environment* (eds V. A. Marple and Y. H. Li), pp. 549–557. Ann Arbor Press, Ann Arbor, MI.

Palmgren, M. S., DeLucca, A. J. II, Lee, L. S., and Ciegler, A. (1983b). 'Preliminary study of mycoflora and mycotoxins in grain dust from New Orleans area grain elevators', *Amer. Ind. Hyg. Assoc. J.*, **44**, 485–488.

Ramazzini, B. (1713). *DeMorbus Artificum Diatriba* (transl. W. C. Wright), Chicago, IL (1940).

Rankin, J., doPico, G. A., Reddan, W. G., and Tsiatis, A. (1979). 'Respiratory disease in grain handlers', in *Proceedings of the International Symposium on Grain Dust* (eds B. Miller and Y. Pomeranz), p. 91. Division of Continuing Education, Kansas State University, Manhattan, KS.

Samsonov, P. F. (1960). 'Respiratory mycotoxicoses (pneumomycotoxicoses)', in *Mycotoxicoses of Man and Agricultural Animals* (ed. V. I. Bilay), pp. 131–140. U.S. Joint Publication Research Service, Office of Technical Services, U.S. Department of Commerce, Washington, D.C.

Samsonov, P. F. and Samsonov, A. P. (1960) 'The respiratory mycotoxicoses (pneumomycotoxicoses) experimentally', in *Mycotoxicoses in Man and Agricultural Animals* (ed. V. I. Bilay), pp. 140–151. U.S. Publication Research Service, Office of Technical Services, U.S. Department of Commerce, Washington, D.C.

Warren, C. P. W. and Manfreda, J. (1980). 'Respiratory abnormalities in grain workers 1971–1977: A preliminary analysis', in *Occupational Pulmonary Disease: Focus on Grain Dust and Health* (eds J. A. Dosman and D. J. Cotton), pp. 417–440. Academic Press, New York.

Warren, P., Cherniack, R. M., and Tse, K. S. (1974). 'Hypersensitivity reactions to grain dust', *J. Allergy Clin. Immunol.*, **53**, 139–149.

Whidden, M. P., Smalley, E. B., Caldwell, R. W., Phillip, J. K., Mai, S. H., and Burkholder, W. E. (1980). *Microbial Flora and Fauna of Respirable Grain Dust from Grain Elevators*. NIOSH Contract 210–77–0150 Final Report, NIOSH, ALOSH, Morganstown, W.V.

Wolff, S. M. (1973). 'Biological effects of bacterial endotoxins in man', in *Bacterial Lipopolysaccharides* (eds E. H. Kass and S. M. Wolff), pp. 251–256. University of Chicago Press, Chicago, IL.

NATURAL OCCURRENCE OF TOXINS

Trichothecenes and Other Mycotoxins
Edited by J. Lacey

CHAPTER 6

Natural occurrence of *Fusarium* mycotoxins in grains and feeds in Taiwan

T. C. Tseng, G. F. Yuan, Jerming Tseng, E. W. Shaw, and C. J. Mirocha

INTRODUCTION

Species of *Fusarium* are widespread in nature, occurring as saprophytes in many plant materials and as pathogens of various field crops. The major mycotoxins produced by these fungi are known as zearalenone and the trichothecenes.

Mirocha *et al.* (1976) have reported the natural occurrence of zearalenone in cereal grains and feeds involved in mycotoxicoses. Usually its oestrogenic properties were evident, causing infertility in cattle and swine. Other surveys have shown its presence in maize, barley, grain sorghum and sesame meal (Eriksen, 1968; Hesseltine, 1974; Stoloff, 1976; Shotwell, 1977; Bennett and Shotwell, 1979; Shotwell *et al.*, 1980; Thiel *et al.*, 1982).

Trichothecenes have also been found in maize, associated with death of cows in a Wisconsin (U.S.A.) dairy herd (Hsu *et al.*, 1972); also Morooka *et al.*, (1972) isolated two toxic substances identified as deoxynivalenol and nivalenol from barley and wheat infected with *Fusarium* in the Kagawaken region of Japan. Eppley (1974) found that 54% of 173 maize samples from various grain elevators in the midwest United States contained a skin-irritating factor, presumably T-2 toxin. Deoxynivalenol has been identified as the emetic principle in field maize infected with *Fusarium* (Vesonder *et al.*, 1976). Maize and barley have frequently been found to contain deoxynivalenol, sometimes in large amounts (Yoshizawa and Morooka, 1977; Vesonder *et al.*, 1978; Ishii *et al.*, 1975; Jemmal *et al.*, 1978) and a toxic metabolite has been isolated from mouldy sorghum infected with *F. incarnatum* (Rukmini and Bhat, 1978).

Recently, Ueno (1983), in reviewing the occurrence of trichothecenes in various parts of the world, noted that surveys of their natural occurrence in agricultural commodities and feeds are still few. Most analyses have come from isolated outbreaks on farms or from selected samples. The only general surveys for deoxynivalenol were of wheat in Canada in 1980 and 1981 (Trenholm *et al.*, 1983) and of maize in the midwest United States (Vesonder and Ciegler, 1979) in 1972. The occurrence of toxin-producing *Fusarium* species, their mycotoxins, and associated toxicoses in Taiwan is little known. Since the hazards to animal health posed by *Fusarium* mycotoxins have only recently become known, this report deals with the occurrence in Taiwan of some *Fusarium* mycotoxins (zearalenone, T-2 toxin, diacetoxyscirpenol and deoxynivalenol) in imported maize, domestic polished rice and in feeds.

MATERIALS AND METHODS

Sample collection and preparation

Samples of imported maize (total = 311) from Thailand, South Africa and the United States were supplied by Agricultural Product, Utilization Department, Union Industrial Research Laboratory after collection by Kaushung and Taichung Bureau of Commodity Inspection and Quarantine from 1978 to 1981. Polished rice samples (total = 428) intended for human consumption and mixed feeds (total = 160) for poultry and livestock were also purchased from various districts of Taiwan (Figure 1).

The samples (3 kg) were collected and immediately stored at 4 °C before analysis. All samples were ground in a Krups 75 mill (Ribert Krups 5650 Solingen 19, Germany), passed through a 2 mm screen, and mixed thoroughly. The representative samples (100 g) were taken and separately assayed for selected *Fusarium* mycotoxins, including zearalenone, T-2 toxin, diacetoxyscirpenol and deoxynivalenol.

Chemical analyses

Solvents for sample extraction and clean-up were purchased from Wako Chemical Industries Ltd, Japan (first class reagent grade) and ALPS Chemical Co., Ltd, Taiwan (ultra pure). Those for HPLC analysis were from E. Merck, Germany (GR grade). Standards of zearalenone, T-2 toxin, diacetoxyscirpenol and deoxynivalenol were prepared at the University of Minnesota by C. J. Mirocha.

Fusarium mycotoxins from maize kernels, polished rice and feeds were analysed following the methods of Mirocha *et al.* (1976, 1977) with the modifications described below:

Zearalenone analysis

Extraction and clean-up Extracts were prepared from 25 g of ground maize, polished rice or feed.

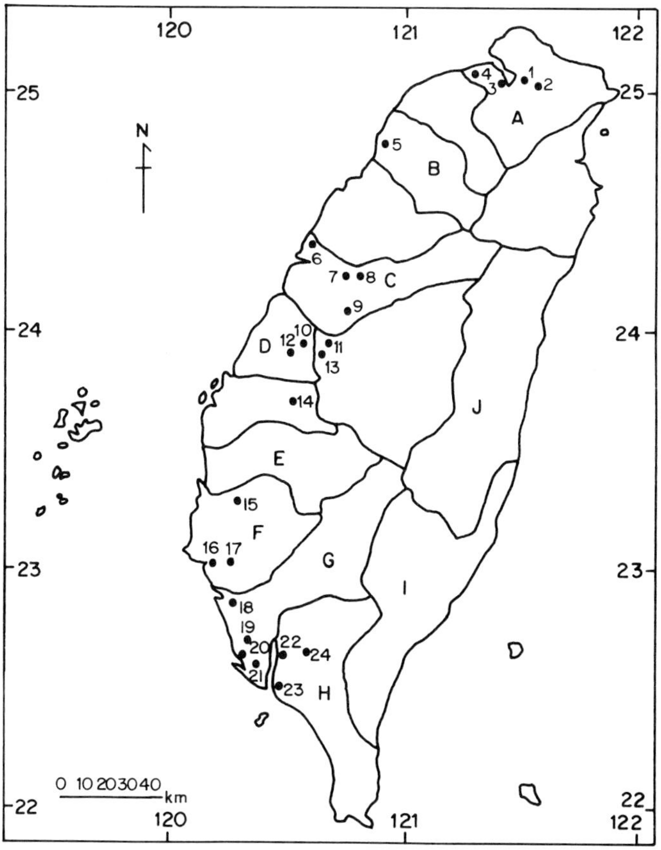

Figure 1 Sampling locations of polished rice and feeds in various districts of Taiwan. *Rice*: A.—Taipei; B.—Hsinchu; C.—Taichung; D.—Chunghua; E.—Chiayi; F.—Tainan; G.—Kaohsiung; H.—Pington; I.—Taitung; J.—Hualien. *Feed*: 1.—Taipei; 2.—Nankang; 3.—Sachchung; 4.—Linkou; 5.—Hsinchu; 6.—Tachia; 7.—Shinkang; 8.—Tungshin; 9.—Taiping; 10.—Yuanlin; 11.—Tsaotung; 12.—Yungching; 13.—Nantou; 14.—Touliu; 15.—Hsinying; 16.—Tainan; 17.—Yungkang; 18.—Luchu; 19.—Tsoying; 20.—Kaohsiung; 22.—Fengshan; 22.—Pingtung; 23.—Hsinyuan; 24.—Neipu

Thin-layer chromatography Precoated E. Merck silica gel TLC plates (20 × 20 cm; 0.25 mm thickness, without fluorescence indicator) previously scribed into 18 lines were spotted with 30 μl, 50 μl and 100 μl of test extracts and 5 μg of standard zearalenone. The plates were then developed with chloroform–ethanol (97 : 3, v : v). Zearalenone fluoresced blue-green under a UV lamp. Identification was confirmed as previously described (Mirocha *et al.*, 1977).

Spectrodensitometer For quantitative analysis of maize and polished rice the zearalenone positive samples were re-spotted on TLC plates, developed as above, air-dried and scanned using a SD-300 Spectrodensitometer (Schoeffel Instrument Corporation, New Jersey, U.S.A.). This was equipped with a scanning thin-layer plate stage with a search unit (Mode OPM 30 quartz prism monochromator), light source (200–700 nm, set at 313 nm) and interference wedge monochromators (400–700 nm set at 443 nm). A density computer (SDC 300) received signals from the SD-300 scanning units, converting these into absorbance. Zearalenone was quantified by peak height measurement against standards, giving a limit of detection of 10 ng.

High-pressure liquid chromatography Quantitative analysis of feeds was performed with a Waters Associates Model ALC/GPC-204 instrument equipped with M-6000A pump, U6K universal injector, U-Bondapack C18 (4 mm i.d. × 30 cm) and Model 440 absorbance detector (236 nm filter) set at sensitivity of 0.01 absorption unit full scale (AUFS).

The standard was injected into the instrument at 2 ml/min using a CH_3CN : H_2O (60 : 60, v : v) solvent system. Zearalenone in feed samples was quantified by comparing peak heights against a calibration curve prepared using serial dilutions of a standard dissolved in methanol. The limit of detection is 1 ng.

Recovery Quadruplicate 25 g samples of uncontaminated ground maize, polished rice and feed were individually spiked with 40 μg of zearalenone in 1 ml methanol in sealed containers and analysed. Average recoveries were 53% for maize, 71% for polished rice, 79% for pig feed, 75.5% for chicken feed and 75% for feedstuff.

T-2 toxin and diacetoxyscirpenol analyses

Extraction and clean up Extracts were prepared from 25 g of ground maize, polished rice or feed.

Thin-layer chromatography The concentrated extracts were spotted onto TLC plates together with authentic standards and developed with ethyl acetate–toluene (3 : 1, v : v). T-2 toxin and diacetoxyscirpenol fluoresced under a 365 nm UV lamp. Identification was confirmed as described previously (Mirocha *et al.*, 1976).

Gas–liquid chromatography A Hitachi Model 163 gas chromatograph equipped with FID was used for quantitative analysis. Nitrogen was used as carrier gas with a flow rate adjusted to 30 ml/min. The column (15.3 × 0.3 cm) consisted of 10% SE-30 adsorbed on chromosorb w (100/120 mesh). Temperatures of both the injector and detector were set at 250 °C, with a temperature programme of 200 °C to 280 °C at 7.5 °C/min. The toxins were analysed as their trimethylsilyl ester derivatives by using Tri-Sil-TBT as silylation reagents (Pierce Chemical Co., Box 117, Rockford, Illinois 61105, U.S.A.). Twenty microlitres of concentrated extracts were placed in 2.5 ml vials, and 50 μl of a standard solution of T-2 toxin (0.5 μg/μl) or diacetoxyscirpenol in acetone in another vial. After evaporating to dryness under nitrogen, 20 μl of Tri-Sil-TBT were added to each vial and left for 20 min. at room temperature. The peak enhancement technique was used to confirm the identity of T-2 or diacetoxyscirpenol. Sensitivity of GLC method for identification of T-2 toxin and diacetoxyscirpenol was less than 0.04 μg (40 ng).

Recovery Twenty-five grams of uncontaminated maize or polished rice were spiked with 50 μg of T-2 toxin or diacetoxyscirpenol. The average recoveries from these tests with T-2 toxin in maize, polished rice and feed were, respectively, 64%, 70% and 71%, and with diacetoxyscirpenol 55%, 61% and 60%.

Deoxynivalenol analysis

Extraction and clean-up Ground maize, polished rice and feed (50 g) were extracted with methanol containing 1% NaCl (55 : 45, v : v) and 100 ml petroleum ether (60–70 °C) in a blender and filtered through Whatman No.1 filter paper. The residue was rinsed with a further 10 ml of the methanol mixture and extracts combined. The filtrate was allowed to separate and after further partitioning with ether, the methanol–NaCl was extracted with chloroform three times. The combined $CHCl_3$ extract was evaporated to dryness over a steam bath, redissolved in acetone and transferred to an 8 ml vial and again evaporated to dryness. The residue was reconstituted with 2 ml methanol–H_O (2 : 3, v : v) pushed through a prepared Sep-pak column into a 2 ml vial, and after drying in N_2 was redissolved in acetone for TLC and GLC analyses.

Thin-layer chromatography Concentrated extracts were applied to precoated TLC plates and developed in chloroform:acetone (90 : 10, v : v). Plates were air-dried, sprayed with *p*-anisaldehyde solution (7 ml methanol, 1 ml glacial acetic acid, 0.5 ml concentrated sulphuric acid, 0.15 ml *p*-anisaldehyde). Plates were dried with a hot-air blower and deoxynivalenol showed up immediately as a yellow spot.

For quantitative analysis of deoxynivalenol, gas–liquid chromatography was used as for T-2 toxin and diacetoxyscirpenol. Average recoveries of toxin, from three tests in maize, polished rice and feed, were 76%, 78% and 75%, respectively.

RESULTS

Altogether 428 rice samples intended for human consumption were collected from various districts of Taiwan (Figure 1) from 1978 to 1981 and analysed for zearalenone, T-2 toxin, diacetoxyscirpenol and deoxynivalenol. None of these *Fusarium* mycotoxins was found in any of these samples. The incidence and levels of zearalenone in imported maize from Thailand, South Africa and the United States are shown in Table 1. Of 82 imported samples, eight contained zearalenone. Nearly 10% of 73 samples of maize from the United States were positive, containing 49–303 μg zearalenone per kg. Zearalenone was detected in only one of the eight samples from South Africa but was not found in the single sample from Thailand.

Table 1 Incidence and levels of zearalenone in imported maize for human consumption

Commodity	Source	Samples analysed	Concentration (μg/kg)	No. of samples
Maize	U.S.A.	73	ND	65
			49	1
			106–170	4
			244	1
			303	1
	Thailand	1	ND	1
	South Africa	8	ND	7
			303	1

ND: Not detected.

Table 2 Incidence and levels of T-2 toxin in imported maize for human consumption

Commodity	Source	Samples analysed	Concentration (μg/kg)	No. of samples
Maize	U.S.A.	118	ND	109
			78–98	2
			134	1
			213–217	3
			317	1
			415	1
			650	1
	South Africa	6	ND	0

ND: Not detected.

Maize samples from South Africa and the United States were also examined for T-2 toxin and diacetoxyscirpenol (Table 2). Of the 124 samples tested, only nine were definitely contaminated with T-2 toxin, all from the United States, with levels of 78–650 μg T-2 toxin per kg. No diacetoxyscirpenol was found in any sample.

Deoxynivalenol was found in imported maize from both South Africa and the United States. Seven of the eight positive samples came from the United States with levels ranging between 95 and 312 μg deoxynivalenol per kg (Table 3). The one positive sample from South Africa was contaminated with 140 μg deoxynivalenol per kg.

Table 3 Incidence and levels of deoxynivalenol in imported maize for human consumption

Commodity	Source	Samples analysed	Concentration (μg/kg)	No. of samples
Maize	U.S.A.	100	ND	93
			95	1
			124–185	3
			219–264	2
			312	1
	South Africa	5	ND	4
			140	1

ND: Not detected.

No T-2 toxin, diacetoxyscirpenol or deoxynivalenol was found in any of 160 feed samples, including swine feeds, chicken feeds and various other feedstuffs. However, some were contaminated with zearalenone (Table 4). Four of 58 swine feeds (6.9%) were contamined with 162–1203 μg zearalenone per kg, the highest level is a feed from Tachia county.

The occurrence of zearalenone in samples of chicken feeds is illustrated in Table 5. Eight of the 84 samples were contaminated with zearalenone at levels of 203–1974 μg/kg. The greatest incidence was in starter feed, although the maximum levels found in starter and broiler feeds were the same. Other feedstuffs which included duck feed, maize meal, wheat meal, soybean meal and miscellaneous feed stuff, were also examined. However, two of the 16 feedstuffs contained zearalenone, at levels of 126 and 152 μg/kg (Table 6).

DISCUSSION

Although the survey of polished rice in 1978 and 1981 suggested that this staple was not an important source of zearalenone, T-2 toxin, deoxynivalenol

Table 4 Occurrence of zearalenone in domestic swine feeds prepared from maize imported from the United States and South Africa

Kind of mixed feed	Location[a]	No. of samples Examined	No. of samples Contaminated	Level (μg/kg)
Growing pig	3, 6, 7, 10 11, 24	6	0	
Developing pig	1, 3, 6, 9, 10, 18	7	1	1203
Starter pig	1, 4, 6, 7, 10, 11, 12, 14, 20	16	0	
Sow	5, 6, 7, 20	4	2	243–300
Broiler pig	3, 10, 15	4	0	
Miscellaneous	1, 3, 5, 7, 9, 10 11, 13, 14	21	1	162

[a] Figures indicate the location in Figure 1.

Table 5 Occurrence of zearalenone in domestic chicken feeds prepared from maize imported from the United States and South Africa

Kind of mixed feed	Location[a]	No. of samples Examined	No. of samples Contaminated	Level (μg/kg)
Growing chicken	1, 2, 5, 8, 10 11, 13, 19, 20 21, 22, 23, 24	18	0	
Developing chicken	1, 2, 5, 7, 10 11, 20, 21, 22 23	11	0	
Starter chicken	1, 2, 3, 5, 10 11, 13, 15, 16 19, 20, 22, 24	26	4	354–1974
Broiler chicken	1, 6, 9, 10, 13 15, 17, 20, 21, 22	19	2	405–1238
Laying hen	9, 10, 17	4	0	
Miscellaneous	1, 6, 10	6	1	203

[a] Figures indicate the location in Figure 1.

Table 6 Occurrence of zearalenone in domestic feedstuffs prepared from maize imported from the United States and South Africa

Kind of feedstuff	Location[a]	No. of samples Examined	Contaminated	Level ($\mu g/kg$)
Duck	1, 7	2	0	
Maize meal	13	2	0	
Wheat meal	8	1	0	
Soybean meal	8, 13	2	0	
Fish meal	8	2	0	
Miscellaneous	1, 6, 14, 15	9	2	126–152

[a] Figures indicate the location in Figure 1.

and diacetoxyscirpenol in Taiwan, 10% of maize samples imported from the United States and sampled at the same time contained the first three of these *Fusarium* toxins. Also, one maize sample from South Africa contained zearalenone and another deoxynivalenol. No zearalenone was found in the only sample from Thailand. However, too few maize samples came from Thailand and South Africa to draw any conclusions, and further study is required. Animal feeds contained zearalenone but no other *Fusarium* mycotoxins. The relationship of these mycotoxins to mycotoxicoses of farm animals is well established and there is an increased awareness of the hazards they pose to animal and human health in many countries. Thus, imported maize and domestic feeds need continued inspection for contamination with *Fusarium* and other mycotoxins.

Studies of the incidence of *Fusarium* mycotoxins, specially trichothecenes, in agricultural commodities in the past have been handicapped for lack of suitable analytical methods. We have modified slightly the original methods (Mirocha *et al.*, 1976, 1977) used for the extraction, clean-up and quantitative analysis of *Fusarium* mycotoxins and have used high-pressure liquid chromatography to analyse zearalenone. The spectrodensitometric method developed to quantify zearalenone could be useful in future studies.

ACKNOWLEDGEMENTS

We are indebted to Drs Y. J. Lin and Y. S. Shieh for their valuable suggestions and technical assistance. This project was fully supported by National Science Council, Republic of China, grant No. NSC–69B–0409003 (10), under the US-ROC cooperative science programme with Professor C. J. Mirocha.

REFERENCES

Bennett, G. A., and Shotwell, O. L., (1979). Zearalenone in cereal grains. *J. Am. Oil Chem. Soc.*, **56**, 812–879.

Eppley, R. M., (1974). Sensitivity of brine shrimp (*Artemia salina*) to trichothecenes. *J. Assoc. Off. Anal. Chem.*, **57**, 618–620.

Eriksen, E. (1968). Oestrogene faktoreri; muggemt Korn. Vulvovaginitis hos svin. *Nord. Vet. Med.*, **20**, 396–401.

Hesseltine, C. W. (1974). Natural occurrence of mycotoxins in cereals. *Mycopath. Appl.*, **53**, 141–153.

Hsu, I. C., Smalley, E. B., Strong, F. M., and Ribelin, W. E. (1972). Identification of T-2 toxin in mouldy corn associated with a lethal toxicosis in dairy cattle. *Appl. Microbiol.*, **24**, 684–690.

Ishii, K., Ando, Y., and Ueno, Y., (1975). Toxicological approaches to the metabolites of Fusaria. IX. Isolation of vomiting factor from mouldy corn infected with *Fusarium* species. *Chem. Pharm. Bull.*, **23**., 2162–2164.

Jemmali, M., Ueno, Y., Ishii, K., Frayssinet, C., and Etienne, M., (1978). Natural occurrence of trichothecenes (nivalenol, deoxynivalenol, T_2) and zearalenone in corn. *Experientia*, **34**, 1333–1334.

Mirocha, C. J., Pathre, S. V., and Christensen, C. M., (1977). Chemistry of *Fusarium* and *Stachybotrys* mycotoxins. In *Mycotoxic Fungi, Mycotoxins, Mycotoxicoses* (eds T. D. Wyllie and L. G. Morehouse), pp. 365–420. Marcel Dekker Inc., New York and Basel.

Mirocha, C. J., Pathre, S. V., Schauerhamer, B., and Christensen, C. M. (1976). Natural occurrence of *Fusarium* toxins in feedstuff. *Appl. Env. Microbiol.*, **32**, 533.

Morooka, N., Uratsuji, N., Yoshizawa, T., and Yamamoto, H. (1972). Studies on the toxin substances in barley infected with *Fusarium* spp. *Hyg. Soc. Japan*, **13**, 368–375.

Rukmini, C., and Bhat, R. V. (1978). Occurrence of T-2 toxin in *Fusarium*-infested sorghum from India. *J. Agr. Food Chem.*, **26**, 647–649.

Shotwell, O. L. (1977). Assay methods for zearalenone and its natural occurrence. In *Mycotoxins in Human and Animal Health* (eds J. V. Rodricks, C. W. Hesseltine, and M. A. Mehlmen), pp. 403–413. Pathotox Publishers Inc., Park Forest.

Shotwell, O. L., Bennett, G. A., Goulden, M. L., Plattner, R. D., and Hesseltine, C. W. (1980). Survey for zearalenone, aflatoxin and ochratoxin in U.S. grain sorghum from 1975 and 1976 crops. *J. Assoc. Off. Anal. Chem.*, **63**, 922–926.

Stoloff, L. (1976). Occurrence of mycotoxins in food and feed. In *Mycotoxins and other Fungal Related Problems* (ed. J. V. Rodricks). Adv. Chem. Ser. 149, Am. Chem. Soc., 23–50.

Thiel, P. G., Meyer, C. J., and Marasas, W. F. O. (1982). Natural occurrence of moniliformin together with deoxynivalenol and zearalenone in Transkeian corn. *J. Agr. Food Chem.*, **30**, 308–312.

Trenholm, H. L., Cochrane, W. P., Cohen, H., Elliot, J. I., Farnworth, E. R., Freind, D. W., Hamilton, Robert, M. G., Standish, J. F., and Thompson, B. K. (1983). Survey of vomitoxin contamination of 1980 Ontario white winter wheat crop: results of survey and feeding trials. *J. Assoc. Off. Anal. Chem.*, **66**, 92–101.

Ueno, Y. (ed.) (1983). *Trichothecenes*. Kodansha Ltd, Tokyo.

Vesonder, R. F., and Ciegler, A. (1979). Natural occurrence of vomitoxin in Austrian and Canadian corn. *Eur. J. Appl. Microbiol. Biotechnol.*, **8**, 273–340.

Vesonder, R. F., Ciegler, A., Rogers, R. F., Burbridge, K. A., Bothast, R. J., and Jensen, A. H. (1978). Survey of 1977 crop year preharvest corn for vomitoxin. *Appl. Env. Microbiol.*, **36**, 885–888.

Vesonder, R. F., Ciegler, A., Jensen, A. H., Rohwedder, W. K., and Weisleder, D. (1976). Co-identity of the refusal and emetic principle from *Fusarium*-infected corn. *Appl. Env. Microbiol.*, **31**, 280–285.

Yoshizawa, T., and Morooka, N. (1977). Trichothecenes from mould-infested cereals in Japan. In *Mycotoxins in Human and Animal Health* (eds J. V. Rodricks, C. W. Hesseltine, and M. A. Mehlman), pp. 309–321. Pathotox Publishers Inc., Park Forest.

Trichothecenes and Other Mycotoxins
Edited by J. Lacey

CHAPTER 7

Zearalenone production in groundnut*

V. K. Mehan, T. N. Bhavanishanker and D. McDonald

INTRODUCTION

Fusarium spp. are important components of the seed mycoflora of healthy groundnut pods (Jackson, 1965; Joffe, 1969; McDonald, 1969), and are also implicated in pod rots (Mercer, 1977; Mehan *et al.*, 1981). Some species of this genus are well-known producers of mycotoxins such as zearalenone, diacetoxyscirpenol and neosolaniol (Mirocha *et al.*, 1976; Ueno, 1977). Zearalenone, an oestrogenic mycotoxin, has mainly been found in cereal grains and their products (Mirocha and Christensen, 1974). Zearalenone has been reported to occur naturally in sesame (Mirocha *et al.*, 1976) and in groundnuts (Mehan and McDonald, 1982). This chapter reports the natural occurrence of zearalenone and the effects of cultural conditions on its production in groundnut seeds by a toxigenic strain of *F. oxysporum*.

MATERIALS AND METHODS

Partially rotted pods were collected from the 1981 and 1982 rainy season and the 1981/82 and 1982/83 post-rainy season groundnut crops at ICRISAT Center farm at Patancheru near Hyderabad, India. The pods were hand-shelled and seeds with pink to purple-stained testae typical of *Fusarium* infection were selected for testing. A total of 204 samples from four seasons were tested for the presence of zearalenone.

Zearalenone extraction

Seed samples of 25 g each were finely ground in a Waring blender at full speed for 5 min. Zearalenone was extracted with dichloromethane and purified using acetonitrile : petroleum ether partition as described by Mirocha *et al.* (1967).

* Submitted as ICRISAT Conference Paper No. 163 by the International Crops Research Institute for the Semi-Arid Tropics (ICRISAT), Patancheru P.O. 502 324, A.P., India.

Detection, confirmation and quantitation zearalenone

Purified extracts were spotted on activated silica gel G (250 μm thickness) coated thin-layer plates. A known quantity of a standard solution of zearalenone in chloroform was also applied to the plates for comparison. The plates were developed with toluene : ethyl acetate : chloroform : 90% formic acid (7 : 5 : 5 : 2) in an unlined, unequilibrated tank (Gimeno, 1980). Zearalenone spots in sample extracts were identified by their correspondence with the standard toxin spots in R_f value and fluorescence under UV light: bluish-green at 365 nm and greenish at 254 nm.

Zearalenone spots of sample extracts were confirmed after additional chemical tests. The toxin spots turned yellowish-brown when the developed thin-layer chromatoplates were sprayed with freshly prepared 50% H_2SO_4 in methanol and then heated in a drying oven at 100 °C for 10 min (Mirocha *et al.*, 1967). The fluorescence of the toxin spots was enhanced when the thin-layer plates were sprayed with freshly prepared 15% aluminium chloride in methanol (Eppley *et al.*, 1974). Spots eluted in methanol gave a positive reaction with an aqueous solution of ferric chloride. When sample extracts eluted in absolute ethanol were scanned for absorption maxima between 220 and 330 nm in a Beckman spectrophotometer, the toxin exhibited characteristic peaks at 236, 274 and 314 nm, corresponding to that of standard zearalenone. The amount of zearalenone was calculated by determining absorbance at 236 nm after eluting the zearalenone spots from the plates with absolute ethanol. The amount of zearalenone was computed from a standard curve prepared by determining the absorbance at 236 nm of different concentrations of standard zearalenone.

Testing *Fusarium* species for production of zearalenone

The *Fusarium* cultures used in this investigation were obtained from the seeds and shells of rotted groundnut pods. Thirteen *Fusarium* isolates belonging to five different species were obtained from rotted pods from the 1978/79 post-rainy season groundnut crop at the ICRISAT Center. Single-spore isolates of different *Fusarium* species were maintained on Czapek–Dox agar slants at 4 °C. The following isolates were tested for their ability to produce zearalenone: five each of *F. solani* and *F. oxysporum*, and one each of *F. moniliforme*, *F. nivale* and *F. fusarioides*. In another experiment, 54 isolates belonging to *F. solani*, *F. semitectum*, *F. oxysporum* and *F. moniliforme*, obtained from groundnut shells and seeds from the 1982 rainy season and the 1982/83 post-rainy season crops, were also screened for their capacity to produce zearalenone.

Inoculum of different isolates of *Fusarium* species was raised in shake cultures in 50 ml Czapek–Dox broth in 250 ml conical flasks for 10 days at 25 °C. One millilitre of inoculum (6.1×10^4 conidia/ml) of each isolate was used to separately inoculate 25 g samples of autoclaved polished rice (with 45% moisture content before autoclaving) in 250 ml Erlenmeyer flasks. Inoculated samples were incubated at 25 °C for 2 weeks, then at 15 °C for 8 weeks. After incubation the mouldy rice samples in the flasks were dried overnight at 50 °C. The dried material was then ground in a blender and zearalenone was extracted and analysed as described earlier.

Effects of cultural conditions on *in vitro* production of zearalenone in groundnuts

Using a toxigenic isolate of *F. oxysporum* (F6), the effects of moisture, temperature and incubation period on *in vitro* production of zearalenone in groundnut seeds were investigated. Mature groundnut seeds (cv. Robut 33–1), free of internally borne fungi, were used. Seed samples were surface-sterilized by soaking them in an aqueous solution of 1% sodium hypochlorite for 3 min and then rinsing thrice in sterile distilled water. Seed moisture content was adjusted by soaking the seeds in sterile distilled water for different periods. Seed lots (25 g each) were adjusted to three levels of moisture content, i.e., 20, 35 and 50%. The seed testae were scarified using a sterile needle. Each seed lot was then transferred to a sterile 250 ml conical flask and inoculated with 2 ml of the spore suspension (6.1×10^4 conidia/ml) of a 10-day-old culture of the toxigenic strain of *F. oxysporum*. After inoculation, seeds in the flasks were subjected to eight different incubation temperatures: (i) 18 °C for 3 weeks, (ii) 18 °C for 6 weeks, (iii) 27 °C for 3 weeks, (iv) 27 °C for 6 weeks, (v) 25 °C for 1 week then 15 °C for 2 weeks, (vi) 25 °C for 2 weeks then 15 °C for 4 weeks, (vii) a diurnal temperature cycle of 30 °C by day (6 a.m.–6 p.m.) and 18 °C by night (6 p.m.–6 a.m.) for 3 weeks and (viii) a diurnal temperature cycle of 30 °C by day and 18 °C by night (6 p.m.–6 a.m.) for 3 weeks and (viii) a diurnal temperature cycle of 30°C by day and 18°C by night for 6 weeks. Each treatment was replicated thrice.

After incubation, samples were dried at 50 °C for 24 h and ground finely in a Waring blender at full speed for 5 min. Zearalenone was extracted and analysed as described earlier.

RESULTS AND DISCUSSION

Of the 204 seed samples tested, one sample each from the 1981 rainy season and the 1981/82 post-rainy season and two samples from the 1982/83 post-rainy season groundnut crops were contaminated with zearalenone ranging from 0.72 to 1.84 mg/kg (Table 1). None of the 94 samples from the 1982 rainy season crop contained zearalenone. Of 13 strains tested in 1981, only one isolate of *F. oxysporum* (F6) produced zearalenone (at a level of 148 mg/kg) in rice culture. No isolate of any other species of *Fusarium* tested (*F.*

Table 1 Natural occurrence of zearalenone in *Fusarium*-infected groundnut seeds

Crop/year	No. of samples tested	No. of samples containing zearalenone	Level of zearalenone (mg/kg)
Rain-fed, 1981	12	1	1.5
Irrigated, 1981/82	60	1	1.8
Rain-fed, 1982	94	0	—
Irrigated, 1982/83	38	2	0.7, 1.1

solani, *F. moniliforme*, *F. fusarioides*, and *F. nivale*), nor any of 54 isolates tested in 1983, including *F. solani*, *F. semitectum*, *F. oxysporum* and *F. moniliforme*, produced zearalenone. Eugenio *et al.* (1970) failed to find zearalenone production by *F. solani*, *F. nivale*, *F. moniliforme* and *F. oxysporum*, but Mirocha *et al.* (1969, 1976) reported its production by some isolates of *F. oxysporum* and *F. moniliforme*. The small number of isolates tested in all these studies precludes any generalization as to the likelihood of occurrence of isolates able to produce zearalenone.

Zearalenone was produced in groundnut seeds at all moisture contents, incubation temperatures and incubation periods tested (Table 2). Significant differences were found in the quantities of zearalenone produced with different treatments. The production of zearalenone increased significantly with increasing seed moisture content and period of incubation. Of the three moisture contents tested, fungal growth was more vigorous at 50% seed moisture content that at either 35% or 20% which did not differ markedly. Similar effects of moisture content on fungal growth were noted by Sherwood and Peberdy (1972) while investigating zearalenone production in wheat, barley and oats by a toxigenic isolate of *F. graminearum*. Fungal growth was not affected by temperature differences. Toxin production was highest at 50% seed moisture content irrespective of temperature treatment (Table 2).

At constant temperatures, toxin production was significantly greater at 18 °C than at 27 °C. Treatments in which temperature was altered, either after one-third of the incubation period from 25 °C to 15 °C, or diurnally from 30 °C by day to 18 °C at night, gave higher levels of zearalenone than did the constant-temperature treatments. Incubation at 15 °C, after a period at 25 °C, enhanced zearalenone production but the lower temperature did not appear to be essential as considerable amounts of the toxin were produced during constant incubation at 27 °C. Similar observations have been reported by Sherwood and Peberdy (1972) and Naik *et al.* (1978) for zearalenone production in cereals by toxigenic isolates of *F. graminearum*. Schroeder and

Table 2 Effects of temperature, moisture content and incubation period on zearalenone production in groundnut seeds by *Fusarium oxysporum*

		Zearalenone production (mg/kg)			
				Incubation temperatures	
Incubation period (weeks)	Seed moisture content (%)	Constant at 27 °C	Constant at 18 °C	At 25 °C for first third of period, then at 15 °C	At 30 °C during day and 18 °C at night[b]
3 weeks	20	3.1	6.2	10.0	8.0
	35	4.9	10.1	16.1	14.1
	50	6.0	11.9	20.1	17.6
6 weeks	20	5.1	9.8	11.9	10.9
	35	7.8	13.1	20.6	18.8
	50	9.8	16.2	26.6	21.3
SE ±		0.50			
CV (%)		4.10			

[a] Mean of three replications.
[b] Day period = 6 a.m. to 6 p.m.; night period = 6 p.m. to 6 a.m.

Hein (1975) also reported that the toxigenic isolate (S-74-lc) of *F. graminearum* produced considerably more zearalenone in cracked corn or grain sorghum when incubated at a constant 25 °C than when changed to 10 °C after growth was established at 25 °C. These results suggest that varied temperature stress and relatively low temperatures are not essential for production of zearalenone, and that wide variations in optimum temperatures for zearalenone production exist among the isolates of *Fusarium* species. There is a need to assess the toxin-producing potential of many more isolates of *F. oxysporum*, *F. moniliforme*, *F. solani* and *F. roseum* from groundnuts.

REFERENCES

Eppley, R. M., Stoloff, L., Trucksess, M., and Chung, M. W. (1974). 'Survey of corn for *Fusarium* toxins', *J. Assoc. Off. Anal. Chem.* **57**, 632–635.

Eugenio, C. P., Christensen, C. M., and Mirocha, C. J. (1970). 'Factors affecting production of the mycotoxin F-2 by *Fusarium roseum*', *Phytopathology*, **60**, 1055–1057.

Gimeno, A. (1980). 'Improved method for thin-layer chromatographic analysis of mycotoxins', *J. Assoc Off. Anal. Chem.*, **63**, 182–186.

Jackson, C. R. (1965). 'A list of fungi reported in peanut pods and kernels', *Univ. Georgia College Agric. Mimeo. Series*. N.S. 234, 6 pp.

Joffe, A. Z. (1969). 'The mycoflora of fresh and stored groundnut kernels in Israel', *Mycopathol. Mycol. Appl.*, **39**, 255–264.

McDonald, D. (1969). 'The influence of the developing groundnut fruit on soil mycoflora in Nigeria', *Trans. Brit. Mycol. Soc.*, **53**, 393–406.

Mehan, V. K., and McDonald, D. (1982). 'Mycotoxin-producing fungi in groundnuts—potential for mycotoxin contamination', In *Proc. International Symposium on Mycotoxins and Phycotoxins*, pp. 98–101, 1–3 September 1982, Vienna, Austria.

Mehan, V. K., McDonald, D., and Rao, V. R. (1981). 'Pod rot disease of peanut at ICRISAT, India', *Proc. Am. Peanut Res. Educ. Soc.*, **13**, 91.

Mercer, P. C. (1977). 'A pod rot of peanut in Malawi', *Plant Dis. Rep.*, **61**, 51–55.

Mirocha, C. J., and Christensen, C. M. (1974). 'Oestrogenic mycotoxins synthesized by *Fusarium*', in *Mycotoxins* (ed. I. F. H. Purchase), pp. 129–148. Elsevier, Amsterdam.

Mirocha, C. J., Christensen, C. M., and Nelson, G. H. (1967). 'Estrogenic metabolite produced by *Fusarium graminearum* in stored corn', *Appl. Microbiol.*, **15**, 497–503.

Mirocha, C. J., Christensen, C. M., and Nelson, G. H. (1969). 'Biosynthesis of the fungal estrogen F-2 and a naturally occurring derivative by *Fusarium moniliforme*', *Appl. Microbiol.* **17**, 482–483.

Mirocha, C. J., Pathre, S. V., Schauerhamer, B., and Christensen, C. M. (1976). 'Natural occurrence of *Fusarium* toxins in feed stuffs', *Appl. Environ. Microbiol.*, **32**, 553–556.

Naik, D. M., Busch, L. V., and Barron, G. L. (1978). 'Influence of temperature on the strain of *Fusarium graminearum* Schwabe in zearalenone production', *Can. J. Plant Sci.*, **58**, 1095–1097.

Schroeder, H. W., and Hein, H. (1975). 'A note on zearalenone in grain sorghum', *Cereal Chem.*, **52**, 751–752.

Sherwood, R. F., and Peberdy, J. F. (1972). 'Factors affecting the production of zearalenone by *Fusarium graminearum* in grain', *J. Stored Prod. Res.*, **8**, 71–75.

Ueno, Y. (1977). 'Trichothecenes: overview address', in *Mycotoxins in Human and Animal Health* (eds J. Rodricks, C. W. Hesseltine and M. A. Mehlman), pp. 189–207. Pathotox Publishers Inc., Illinois, U.S.A.

Trichothecenes and Other Mycotoxins
Edited by J. Lacey

CHAPTER 8

Ecology of the *Aspergillus flavus* group and aflatoxin formation in maize and groundnut

R. A. Hill, D. M. Wilson, W. W. McMillian, N. W. Widstrom,
R. J. Cole, T. H. Sanders, and P. D. Blankenship

INTRODUCTION

Maize (*Zea mays* L.)

Contamination of maize ears by micro-organisms may occur at any time during development or harvesting, and in storage. In warm, humid, subtropical and tropical climates, *Aspergillus* and *Penicillium* species colonize grain before harvest in greater numbers than in grain grown in cooler, temperate regions of the world. Most of the existing information on maize microflora, e.g. Hesseltine and Bothast (1967), was obtained in temperate regions. Differences between temperate regions and regions with higher temperatures were especially evident when the field occurrence of the *A. flavus* group and aflatoxin contamination of maize in a state like Indiana was compared with that of Georgia. In Indiana, Rambo *et al.* (1974) found little *A. flavus* and aflatoxin in 1971 and 1972 in pre-harvest maize. By contrast, in Georgia, species of the *A. flavus* group were co-dominant among the fungi in mature pre-harvest maize in 1975, 1976, 1977, 1978 and 1979 (Wilson *et al.*, 1981). Maize grown in southern states was more likely to have aflatoxin contamination than that grown in the corn belt (Shotwell, 1977). Since the region where maize was grown affected the incidence of the *A. flavus* group, other components of the fungal flora may also be affected by the location.

Toxic fungal metabolites that contaminate food and feed have major implications in human and animal health. Not only can fungi cause deterioration in our food and feed, making it less nutritious; some fungi also produce toxic by-products, such as the aflatoxins and other mycotoxins, that have major economic effects on agriculture by destroying markets or by causing feed problems that affect the health of humans and farm animals (CAST

Report, 1979). Aflatoxin contamination of maize has caused important economic losses and is now a marketing consideration in the south-eastern United States (Nichols, 1983). An understanding of how, when and why aflatoxin contamination occurs is necessary in order to develop effective control measures. No previous study on maize has attempted to define the successions and interactions of the mycoflora and associated toxic metabolites of maize from silking to harvest. There is insufficient information on the factors that influence the incidence, distribution, population levels and infectivity of the *A. flavus* group in maize.

The specific objectives of the experiments on maize reported here were to assess the microflora of maize, grown under field plot conditions, from silking to harvest with emphasis on the ecology and taxonomy of *Aspergillus* and *Penicillium* species, and to relate the microflora to aflatoxins and damage.

Groundnuts (*Arachis hypogaea* L.)

An association between drought stress in groundnuts and increased aflatoxin contamination was observed in studies in South Africa (Sellschop, 1954), Nigeria (McDonald and Harkness, 1967) and in several states in the U.S.A.: Georgia (Sanders *et al.*, 1981), North Carolina (Dickens *et al.*, 1973), Alabama (Diener and Davis, 1977) and Texas (Pettit *et al.*, 1971). However, equivocal results were reported by Wilson and Stansell (1983) with increased aflatoxin contamination only in 2 of 4 years. *In vitro* studies on the effects of moisture and temperature on the growth of *Aspergillus flavus* and *A. parasiticus* and on aflatoxin production (Panasenko, 1967; Ayerst, 1969; Northolt *et al.*, 1976; Diener and Davis, 1977) have defined limiting and optimum conditions (Figure 1). Studies to determine precisely the effects of temperature and moisture availability in groundnuts (using six unique experimental plots with independent control of soil moisture and temperature) on kernel colonization by the *A. flavus* group and aflatoxin contamination were started in 1980 at the USDA National Peanut Research Laboratory, Dawson, GA, and results to date are reported here.

MATERIALS AND METHODS

Maize

Field studies

Randomized complete block field experiments were used in 1978 (with three planting dates) and in 1979 (with two planting dates) (Table 1). There were six replications of three treatments: insect control by insecticides, insect infested and untreated control. Five ears from each replicate of each treatment were harvested at random on 0, 15, 30, 45 and 60 days after the full

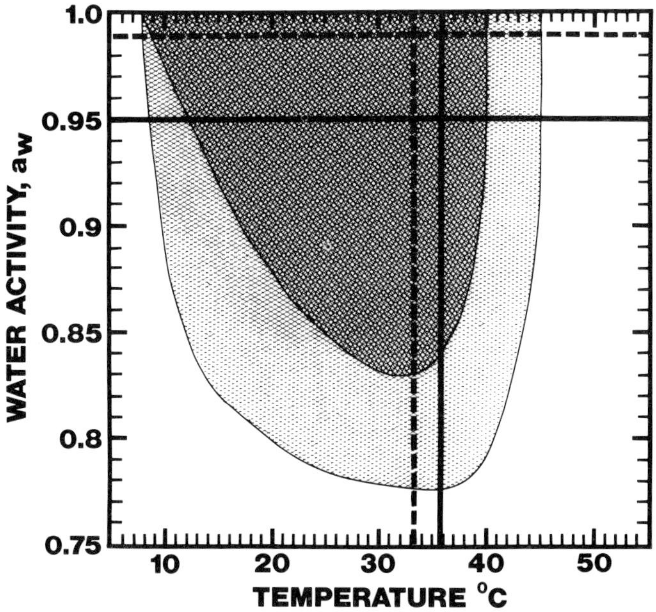

Figure 1 Effect of temperature and water activity on growth (optima, solid lines; range, whole shaded area) and aflatoxin production (optima, broken lines; range, darkly shaded area) *in vitro* by *Aspergillus flavus* Link and *A. parasiticus* Speare. (Based on results published by Panasenko, 1967; Ayerst, 1969; Diener and Davis, 1977, and Northolt *et al.*, 1976)

silk stage for assessment of the microflora. Commercial maize hybrids were grown at the University of Georgia, Coastal Plain Experiment Station, Tifton, using standard farm production practices. Samples at 45 and 60 days after full silk were analysed chemically for aflatoxins (Thean *et al.*, 1980). The presence or absence of damage on each ear (tip, middle and base) was recorded.

Microbiology

The microflora was evaluated by incubating whole untreated (1978 and 1979) and surface disinfected (1979) (0.5% NaOCl, 5 min, rinsed distilled water) kernels on 10% malt salt agar and a medium selective for the *A. flavus* group (Griffin and Garren, 1976). Kernels were taken from the tip, middle and base of each ear. Petri dishes were incubated at 26 and 35 °C. The major emphasis was on *Aspergillus* and *Penicillium* species, but other fungi were identified, as far as possible, to genus and species. Actinomycetes and bacteria were recorded but not identified further.

Penicillium species were classified with the aid of Raper and Thom (1949) as modified by Kulik (1968) and Samson *et al.* (1976). *Aspergillus* species were classified after Raper and Fennell (1965) and Samson (1979). Each species, genus or type of micro-organism was scored for each kernel of maize incubated. Results for 1978 were based on the assessment of over 60,000 kernels and in 1979 of approximately 80,000 kernels.

Statistical analysis

An analysis of variance and covariance analysis for the variables 'category of micro-organism', 'damage' and 'aflatoxin contamination' was performed using SAS programmes.

Groundnuts

The unique experimental plot facility, with independent control of soil temperature and moisture, has been described by Blankenship *et al.* (1980) and the methods used to assess the microflora and for aflatotoxin analysis by Hill *et al.* (1983).

RESULTS AND DISCUSSION

Maize

Weather

In 1978 for the period from 0 to 60 days after the full silk stage, temperature and evaporation were similar for the three different crop planting dates. Rainfall was greatest for the second planting date and least for the first planting date, but the differences between planting dates were small (Table 1). The same period for both planting dates in 1979 was cooler than in 1978, with similar rainfall and less evaporation. However, the mean temperature was 1.2 °C warmer for the later than the earlier planting date, in 1979 (Table 1). In peanuts, even such an apparently small difference in mean temperature (in the geocarposphere) affected both kernel colonization by the *Aspergillus flavus* group and aflatoxin contamination (Sanders *et al.*, 1981; Cole *et al.*, 1982; Hill *et al.*, 1983).

Incidence of the Aspergillus flavus *group on maize kernels in 1978*

Most *A. flavus* group isolates were identified as *A. flavus* Link (92% colonies) and a much smaller proportion as *A. parasiticus* Speare (8% colonies). Various other species of the *A. flavus* group were isolated erratically and in

Table 1 Summary of climatological data[a] for the period 0 to 60 days after full silk stage in Georgia field corn in 1978 and 1979

	Air temp (°C) 1.52 m above soil				
	Mean max.	Mean min.	Overall mean	Precipitation total (mm)	Evaporation total (mm)
1978 crop year					
Planting date I (March 29) Full silk stage, June 13	32.5	21.5	27.0	198	365
Planting date II (April 17) Full silk stage, June 25	33.0	22.0	27.5	227	357
Planting date III (May 8) Full silk stage, July 4	32.9	21.8	27.3	209	363
1979 crop year					
Planting date I (April 2) Full silk stage, June 8	31.7	19.3	25.6	217	341
Planting date II (April 30) Full silk stage, July 2	32.5	21.2	26.8	228	326

[a] Data collected at University of Georgia, Coastal Plain Experiment Station, Tifton.

small numbers. The incidence of the *A. flavus* group on maize kernels in 1978 increased from none to very few kernels colonized at the full silk stage, to between half and three-quarters of the kernels colonized at maturity (60 days after full silk) (Figure 2). Position of kernels on the ear (tip, middle or base), planting date (Table 1) and treatment (untreated, insecticide sprayed or insect infested) had little effect on kernel colonization by the *A. flavus* group, but insecticidal sprays significantly reduced colonization by the *A. flavus* group occasionally (Figure 2).

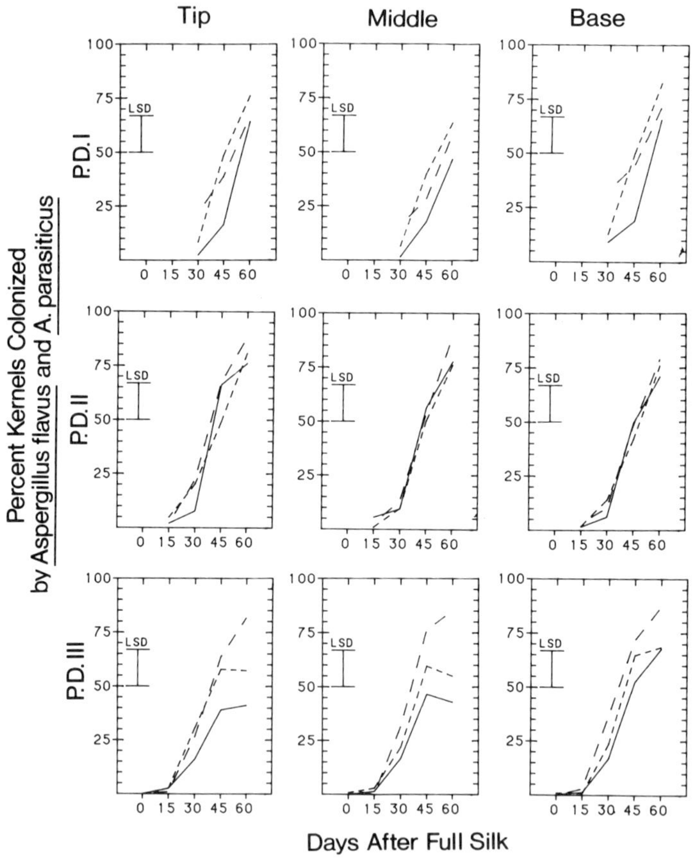

Figure 2 Incidence of *Aspergillus flavus* Link and *A. parasiticus* Speare on kernels (not surface disinfected) on the top, middle and base of ears of Georgia field corn (maize) in 1978, before harvest (P.D. = planting date; solid line = insecticide treated; long broken line = insect infested; short broken line = control).

Incidence of the A. flavus *group on maize kernels in 1979*

A. flavus again accounted for a much greater proportion (>90%) of the *A. flavus* group isolates identified than did *A. parasiticus*. The pattern of maize kernel colonization in 1979 was similar to that in 1978 but the incidence of the *A. flavus* group was decreased, especially for the early planting date with the coolest mean temperature during the ripening period (Figures 3 and 4; Table 1). Surface disinfection of kernels, on the whole, did not affect the recorded incidence of the *A. flavus* group (Figure 4) compared with non-disinfected kernels (Figure 3). However, at 45 days after full silk for kernels from the later planting date (Table 1) there was less *A. flavus* group after treatment with insecticides compared with kernels from insect infested and control treatments (Figure 4). The distribution of the *A. flavus* group on maize ears was not associated only with damage, since most damage, caused primarily by insects, occurred near the tip and base of maize ears grown in Georgia and there was a similar incidence of the *A. flavus* group in kernels irrespective of position on the ear.

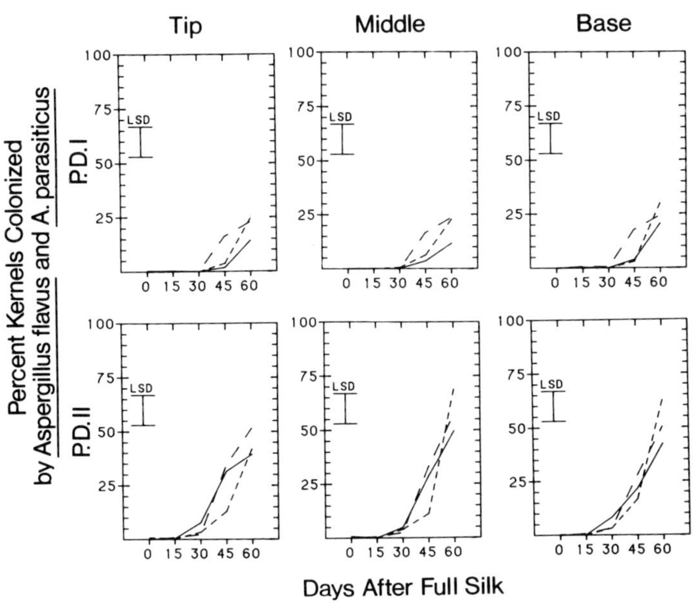

Figure 3 Incidence of *Aspergillus flavus* Link and *A. parasiticus* Speare on kernels (not surface disinfected) on the top, middle and base of ears of Georgia field corn (maize) in 1979, before harvest (P.D. = planting date; solid line = insecticide treated; long broken line = insect infested; short broken line = control).

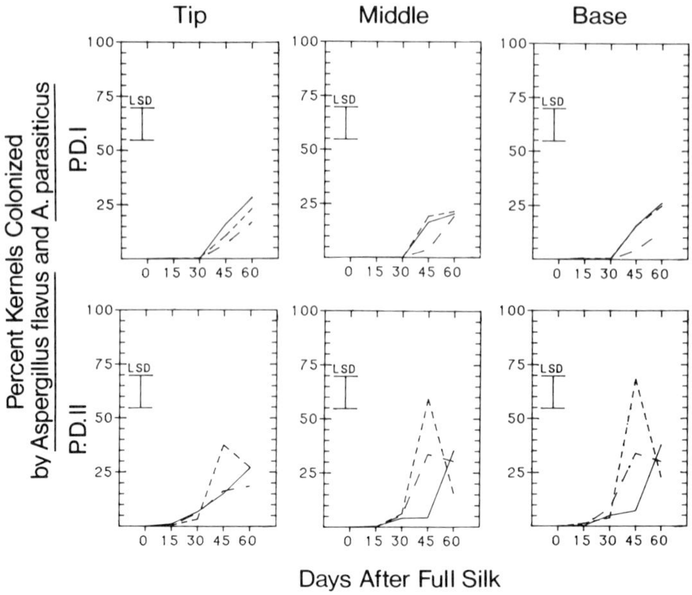

Figure 4 Incidence of *Aspergillus flavus* Link and *A. parasiticus* Speare on kernels (surface disinfected) on the top, middle and base of ears of Georgia field corn (maize) in 1979, before harvest (P.D. = planting date; solid line = insecticide treated; long broken line = insect infested; short broken line = control).

Correlations between the A. flavus *group, associated maize kernel fungi, actinomycetes and bacteria, aflatoxin contamination and damage*

Positive and negative correlations occurred in 1978 and 1979 between the *A. flavus* group and other micro-organisms on and within maize kernels, with the occurrence of aflatoxins and with damage (Tables 2, 3 and 4). *Aspergillus* species, other than *A. flavus* and *A. parasiticus*, that were isolated and identified from maize kernels in 1978 and 1979 were *A. candidus* Link, *A. clavatus* Desmazieres, *A. niger* V. Tieghem, *A. ochraceus* Wilhelm, *A. oryzae* (Ahlb.) Cohn, *A. sulphureus* (Fres.) Thom and Church, *A. tamarii* Kita and *A. terreus* Thom. In addition, three species of the ascosporic genus *Eurotium*, with *Aspergillus* anamorphs, *E. amstelodami* Mangin, *E. chevalieri* Mangin and *E. herbariorum* (Wiggers) Link ex Gray, were found. These *Aspergillus* and *Eurotium* species occurred erratically and infrequently and were grouped as 'other aspergilli' for statistical analysis. Because of the small incidence of 'other aspergilli' and their irregular distribution on maize ears there were no consistent, significant correlations between the *A. flavus* group and 'other aspergilli' (A.B in Tables 2, 3 and 4).

Table 2 Correlation coefficients (significant at $p > 0.01$) from covariance analysis of the *Aspergillus flavus* group with associated micro-organisms, aflatoxin contamination and damage on maize kernels (not surface disinfected) in 1978

Interaction								Variable												
Of planting date(s)[a]			With sampling date(s)[b]					A.B.	A.C	A.D	A.E	A.F	A.G	A.H	A.I	A.J	A.AFLA	A.DAM	AFLA.DAM
I	II	III	0	15	30	45	60	0.04	0.38	0.10	−0.60	−0.23	0.03	−0.24	−0.08	−0.22	0.12	0.36	0.18
I			0	15	30	45	60		0.34	−0.10	−0.10	−0.30	−0.19	−0.26	−0.07	−0.29	−0.29	0.31	0.24
	II		0	15	30	45	60		0.40			0.27	0.23	−0.34	−0.10	−0.28	0.13	0.17	
		III	0	15	30	45	60	0.06	0.38	0.30	−0.06	−0.19	0.40	−0.14	−0.05	−0.14		0.54	0.16
I	II	III		15				0.19											
I	II	III			30							−0.20	0.11	−0.15		−0.18		0.33	
I	II	III				45				0.30			0.15	−0.22		−0.10	0.20	0.33	0.28
I	II	III					60		−0.14	−0.31		−0.09	0.34	−0.21	−0.09			0.23	0.09
	II							0.27										−0.12	
		III		15															
I					30				0.15			−0.25				−0.28		0.38	
	II					30													
		III			30							−0.12		−0.16		−0.11		0.25	
I						45				−0.26		−0.15	0.14	−0.24		−0.13	0.41		0.34
	II					45				−0.32			0.16	−0.17			0.22		
		III				45			−0.15	−0.28		−0.13	0.20	−0.17				0.39	0.32
I							60		−0.23	−0.27			0.31	−0.17			0.23	0.22	0.15
	II						60		−0.22	−0.45			0.24	−0.24				0.14	
		III					60			−0.23			0.49	−0.25				0.39	

[a] Planting date I (March 29); II (April 17); III (May 8).
[b] Sampling date (days after full silk stage) I (June 13); II (June 25); III (July 4).
[c] A, *Aspergillus flavus* Link and *A. parasiticus* Speare; B, 'other *Aspergillus* species'; C, *Penicillium* species; D, *Fusarium moniliforme* Sheldon; E, 'other *Fusarium* species'; F, *Cladosporium* species; G, 'total yeasts'; H, '*Alternaria* and *Helminthosporium* species'; I, Mucorales; J, other fungi, actinomycetes and bacteria; AFLA, aflatoxin contamination; DAM, visible damage.

Table 3 Correlation coefficients (significant at $p > 0.01$) from covariance analysis of the *Aspergillus flavus* group with associated micro-organisms, aflatoxin contamination and damage on maize kernels (not surface disinfected) in 1979

Interaction							Variable[c]											
Of planting date(s)[a]		With sampling date(s)[b]					A.B	A.C	A.D	A.E	A.F	A.G	A.H	A.I	A.J	A.AFLA	A.DAM	AFLA.DAM
I	II	0	15	30	45	60		0.46	0.29		0.05	0.28	−0.11	−0.06	−0.01	0.27	0.12	0.15
I		0	15	30	45	60		0.42	0.31			0.28	−0.10		−0.08	0.17	0.15	0.15
	II	0	15	30	45	60		0.47	0.28		0.14	0.27	−0.10	−0.12	−0.10	0.20	0.06	0.12
I	II	0					0.24				0.09		0.09		0.07			
I	II			30							−0.07		−0.08		−0.09			
I	II				45			0.10	−0.27		0.16	−0.08	−0.14			0.25		−0.19
I	II					60	−0.11	0.13					−0.09	−0.12		0.27	0.14	0.24
	II	0					0.36	0.14					0.15		0.11			
	II		15								0.13				0.11			
	II			30					−0.10			−0.13					−0.13	
I					45			0.19	−0.24				−0.12	0.22			0.18	
	II				45				−0.32		0.13		−0.11			0.14	−0.10	−0.32
I						60	−0.10	0.18			−0.10			−0.14		0.13		
	II					60			0.18			0.14				0.18		0.23

[a] Planting date I (March 29); II (April 17).
[b] Sampling date (days after full silk stage) I (June 13); II (June 25).
[c] A, *Aspergillus flavus* Link and *A. parasiticus* Speare; B, 'other *Aspergillus* species'; C, *Penicillium* species; D, *Fusarium moniliforme* Sheldon; E, 'other *Fusarium* species'; F, *Cladosporium* species; G, 'total yeasts'; H, '*Alternaria* and *Helminthosporium* species'; I, Mucorales; J, other fungi, actinomycetes and bacteria; AFLA, aflatoxin contamination; DAM, visible damage.

Table 4 Correlation coefficients (significant at $p > 0.01$) from covariance analysis of the *Aspergillus flavus* group with associated micro-organisms, aflatoxin contamination and damage on maize kernels (surface disinfected) in 1979

Interaction							Variable[c]											
Of planting date(s)[a]		With sampling date(s)[b]					A.B	A.C	A.D	A.E	A.F	A.G	A.H	A.I	A.J	A.AFLA	A.DAM	AFLA.DAM
I	II	0	15	30	45	60		0.13	0.24		0.05	0.28	0.04	−0.04			0.08	0.06
I		0	15	30	45	60		0.12	0.20		0.10	0.28	0.14					
	II	0	15	30	45	60		0.12	0.25			0.27		−0.07			0.07	
I	II	0																
I	II		15				0.1		0.08			0.09	0.11		0.08			
I	II			30														
I	II				45			−0.11				0.26					0.09	0.19
I	II					60		−0.27	−0.10		0.17	0.17		0.11			0.10	
I		0																
	II	0																
I			15					0.17										
	II		15				0.13				0.12		0.16		0.12			
I				30														
	II			30					−0.15									
I					45				−0.28			−0.15				−0.20		0.13
	II				45			−0.18				0.40	−0.11					0.20
I						60		−0.29	−0.19		0.31	0.28	0.17					
	II					60		−0.32	−0.12					0.12			0.12	−0.11

[a] Planting date I (March 29); II (April 17).
[b] Sampling date (days after full silk stage) I (June 13); II (June 25).
[c] A, *Aspergillus flavus* Link and *A. parasiticus* Speare; B, 'other *Aspergillus* species'; C, *Penicillium* species; D, *Fusarium moniliforme* Sheldon; E, 'other *Fusarium* species'; F, *Cladosporium* species; G, 'total yeasts'; H, '*Alternaria* and *Helminthosporium* species'; I, Mucorales; J, other fungi, actinomycetes and bacteria; AFLA, aflatoxin contamination; DAM, visible damage.

Penicillium species comprised a major component of the maize kernel microflora in both years. *P. funiculosum* Thom. *P. citrinum* Thom, *P. purpurogenum* Stoll and an unidentified species with asymmetric penicillia were isolated frequently. *P. rugulosum* Thom, *P. tardum* Thom, P. *variabile* Sopp and *Paecilomyces lilacinus* (Thom) Samson (*P. lilacinum* Thom) occurred infrequently and in small numbers. Consistent and significant ($p > 0.01$) positive correlations between the *A. flavus* group and *Penicillium* species occurred in 1978 and 1979 for the interactions of all planting dates and each planting date with all sampling dates. However, there were some significant negative correlations between individual planting dates and sampling times (especially 45 and 60 days after full silk). In addition, the positive correlations of the *A. flavus* group with *Penicillium* species were smaller for surface disinfected kernels than for untreated kernels (1979, A.C in Table 4) and the negative correlations were greater (A.C in Tables 2, 3 and 4). These differences probably reflect different interactions between individual *Penicillium* species and the *A. flavus* group and differences between *Penicillium* species with superficial growth and those that had penetrated the kernels. A positive correlation between the *A. flavus* group and *P. funiculosum* on peanuts, has been reported (Joffe, 1969).

Fusarium moniliforme Sheldon occurred more abundantly on maize kernels in 1978 and 1979 even than the *A. flavus* group. Negative correlations between the *A. flavus* group and *F. moniliforme* occurred consistently for individual planting dates with sampling dates of 45 and 60 days after full silk. This was true both for 1978 and 1979 (with and without surface disinfection) (A.D in Tables 2, 3 and 4). Other *Fusarium* species occurred infrequently and in small numbers and there were no consistent correlations between these and the *A. flavus* group.

For practically all significant ($p > 0.01$) correlations between the *A. flavus* group and *Cladosporium* spp. in 1978 (A.F in Table 2), there was a negative correlation. In 1979 the observed incidence of *Cladosporium* spp. was greater than that in 1978 on kernels that were not surface disinfected, but less on surface disinfected kernels (which indicated predominantly superficial growth of *Cladosporium*) and correlations between the *A. flavus* group and *Cladosporium* spp. were mostly small and inconsistent (A.F in Tables 3 and 4).

Various yeasts were abundant in both years. Although yeasts were not identified to species, different entities were separated on the basis of colony colour and texture. The *A. flavus* group was most often positively correlated with 'total yeasts'. Significant negative correlations were small (A.G in Tables 2, 3 and 4).

In 1978 all significant correlations between the *A. flavus* group and '*Alternaria* and *Helminthosporium* species' were negative, as were most correlations for maize without surface disinfection of the kernels in 1979. However, with surface disinfection in 1979 there were some small positive correlations with the *A. flavus* group. These apparent differences may be attributable to superficial growth of *Alternaria* and internal mycelium of *Helminthosporium* (A.H in Tables 2, 3 and 4).

Fungi in the order Mucorales were rarely abundant on maize kernels in 1978 or 1979 and were irregularly distributed. Most often there was a small, negative correlation between the *A. flavus* group and the Mucorales (A.I in Tables 2, 3 and 4).

Bacteria, actinomycetes and unspecified 'other fungi' were usually negatively correlated with the *A. flavus* group in 1978 and 1979. Since most bacteria and many other micro-organisms in this group were present on maize kernels superficially, surface disinfection eliminated this negative correlation with the *A. flavus* group (A.J in Tables 2, 3 and 4).

Although there was colonization of maize kernels by the *A. flavus* group in the absence of visible damage, there was usually a positive correlation between damage and colonization by the *A. flavus* group (A.DAM in Tables 2, 3 and 4). This was most apparent in 1978 when the *A. flavus* group was more abundant than in 1979. Aflatoxin contamination and the *A. flavus* group (A.AFLA) and aflatoxin contamination and damage (AFLA.DAM) also were most often positively correlated, especially in 1978 (Tables 2, 3 and 4) despite the small incidence of aflatoxin in both years.

Groundnuts

Detailed results from studies of the effects of temperature and moisture conditions in the geocarposphere on kernel colonization by the *Aspergillus flavus* group and aflatoxin contamination of groundnuts for crops grown in 1980, 1981 and 1982 have been published (Sanders *et al.*, 1981; Cole *et al.*, 1982; Hill *et al.*, 1983). The experimental plot facility at Dawson was described by Blankenship *et al.* (1980). Figure 5 shows the relationship between mean geocarposphere temperature during the last 30–50 days before harvest and kernel colonization and aflatoxin contamination in undamaged, edible grades of groundnut grown with drought stress. There was no aflatoxin contamination of kernels grown with prolonged drought stress before harvest when the mean geocarposphere temperature during the drought-stress period was <25 °C or >32 °C. Aflatoxin concentration in edible-grade kernels increased with increased geocarposphere temperature to a maximum of nearly 500 ng/g (mean temperature, 30.5 °C). Groundnut kernel colonization by the *A. flavus* group also was maximal at 30.5 °C (95.7% kernels colonized at harvest) with a smaller proportion of kernels colonized at lower and higher temperatures. There was greater variability in the relationship between kernel colonization and mean geocarposphere temperature than that of aflatoxin concentration and mean geocarposphere temperature. However, the general pattern was similar with greater kernel colonization and aflatoxin

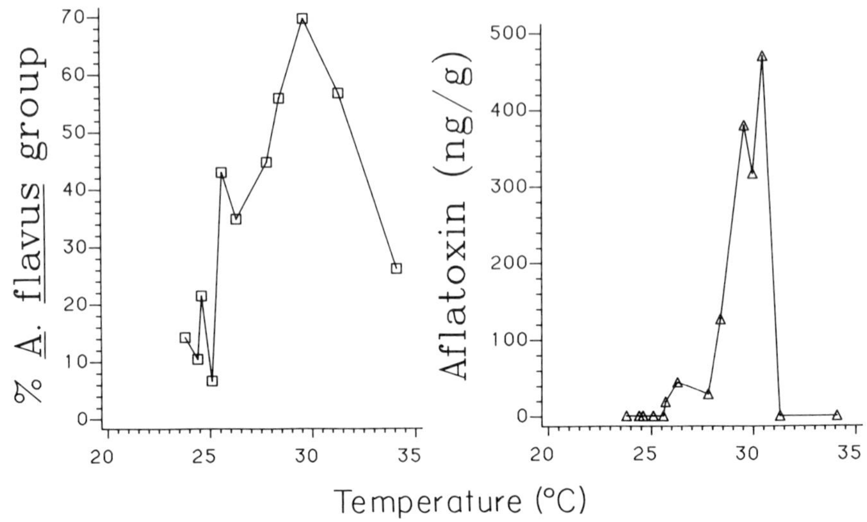

Figure 5 Percentage of edible-grade, drought-stressed groundnut kernels colonized by *Aspergillus flavus* Link and *A. parasiticus* Speare and aflatoxin concentration, in relation to geocarposphere temperature (during the drought-stress period) in 1980–83 crops grown in experimental plots at the National Peanut Research Laboratory, Dawson, Georgia.

concentration with increased temperature until the optimal temperature, with a decrease both in kernel colonization and aflatoxin concentration at supraoptimal temperatures. The effects of temperature and moisture availability on aflatoxin concentration and growth of *A. flavus*, based on results from *in vitro* studies (Figure 1) were similar to those from these studies (Figure 5), since in both instances aflatoxin production occurred over a more limited range of temperatures than did growth of the *A. flavus* group. Also, the optimum temperature for aflatoxin production in culture was similar to that observed for maximum contamination of groundnuts in the plot study.

Edible grades of groundnut kernels from irrigated crops, with unmodified and increased soil temperatures, contained no aflatoxin. Any kind of damage to kernels or pods increased the incidence of the *A. flavus* group and aflatoxin contamination of groundnut kernels. The greatest concentration of aflatoxin (>20,000 ng/g) was found in inedible grades of kernels (especially damaged) after drought stress, when the mean geocarposphere temperature was optimal for kernel colonization by the *A. flavus* group and aflatoxin production (ca. 30 °C).

Aflatoxin formation by *A. flavus* and *A. parasiticus*

Most *A. flavus* group isolates from maize were identified as *A. flavus* Link and a smaller proportion (<10%) of isolates as *A. parasiticus* Speare. In groundnuts, also, the majority of *A. flavus* group isolates were *A. flavus*. The proportion of *A. parasiticus* isolates varied with year but was greater than in maize (from 10 to 30%). Virtually all aflatoxigenic isolates of *A. flavus* tested at the National Peanut Research Laboratory produced only aflatoxins B_1 and sometimes B_2 in liquid culture. By contrast, nearly all isolates identified as *A. parasiticus* produced aflatoxins B_1, B_2, G_1 and G_2 under the same conditions. In addition, cyclopiazonic acid was produced only by some *A. flavus* isolates and not by any isolate identified as *A. parasiticus* (Dorner *et al.*, 1984). An analysis of over 4000 aflatoxin-positive maize and groundnut samples, tested by the Mycotoxin Analysis Research Center, University of Georgia Coastal Plain Experiment Station, showed that the majority of aflatoxin-positive and groundnut samples contained only B aflatoxins, but that the proportion of positive samples containing G aflatoxins was greater for groundnuts than for maize (Table 5). These findings are in accord with the incidence of the two species on each crop, reported here.

Table 5 Summary of B and G aflatoxins found in naturally contaminated field samples of maize and groundnuts in 1977–82[a]

Crop	Aflatoxins	Number of positive samples	Percentage	Ratio
Maize	Bs alone	3559	76.9	67:
	B + G	1015	21.9	19:
	Gs alone	53	1.1	1
Groundnuts	Bs alone	648	61.8	25:
	B + G	374	35.7	14:
	Gs alone	26	2.5	1

[a] Samples analysed at the Mycotoxin Analysis Research Center, University of Georgia, Coastal Plain Experiment Station, Tifton.

CONCLUSION

Temperature, of the air (maize) and in the geocarposphere (groundnuts), during seed development was the single most important determinant of colonization by the *Aspergillus flavus* group and subsequent aflatoxin contamination. Other factors, especially the availability of water to plant and fungus, interactions between temperature and the availability of water, and the effects of these variables on the *A. flavus* group and associated micro-organisms also were important. In groundnuts there were differences between edible and inedible grades of kernels and within U.S. commercial edible grades (Jumbo, Medium, No. 1 and other edible) both for *A. flavus* colonization and aflatoxin contamination. Any kind of kernel damage, irrespective of the crop growth conditions, could enhance aflatoxin contamination. No aflatoxin contamination was found in undamaged groundnut kernels from crops grown with adequate irrigation (irrespective of geocarposphere temperature) or from drought-stressed crops when the mean geocarposphere temperature during the last 45 days before harvest was <25 °C or >32 °C. in maize, kernel colonization by the *A. flavus* group and aflatoxin contamination were both maximal during the growing season when air temperatures were around 30 °C. Damaged ears also were more often colonized by the *A. flavus* group and more likely to be contaminated with aflatoxins than undamaged ears.

REFERENCES

Ayerst, G. (1969). 'The effect of moisture and temperature on growth and spore germination in some fungi', *J. Stored Prod. Res.* **5**, 127–141.

Blankenship, P. D., Cole, R. J., Sanders, T. H. (1980). Rainfall control plot facility at National Peanut Research Laboratory', *Proc. Am. Peanut Res. Educ. Soc.*, **12**, 46.

CAST Report (1979). 'Aflatoxin and other mycotoxins: an agriculture perspective'. Council Agric. Sci. Technol. (CAST) Report No. 80, 1979.

Cole, R. J., Hill, R. A., Blankenship, P. D., Sanders, T. H., and Garren, K. H. (1982). 'Influence of irrigation and drought stress on invasion by *Aspergillus flavus* of corn kernels and peanut pods'. *Develop. Indust. Microbiol.*, **23**, 229–236.

Dickens, J. W., Satterwhite, J. B., and Sneed, R. E. (1973). 'Aflatoxin-contaminated peanuts produced on North Carolina farms in 1968'. *J. Am. Peanut Res. Educ. Soc.*, **5**, 48–58.

Diener, U. L., and Davis, N. D. (1977). 'Aflatoxin formation in peanuts by *Aspergillus flavus*'. *Agric. Exp. Sta. Bull.* No. 493, U.S. Agricultural Experiment Station, Auburn, Ala.

Dorner, J. W., Cole, R. J., and Diener, U. L. (1984). 'The relationship of *Aspergillus flavus* and *Aspergillus parasiticus* with reference to production of aflatoxins and cyclopiazonic acid'. *Mycopathologia* **87**, 13–15.

Griffin, G. J., and Garren, K. H. (1976). 'Colonization of rye green manure and peanut fruit debris by *Aspergillus flavus* and *Aspergillus niger* group in field soils'. *Appl. Environ. Microbiol.*, **32**, 28–32.

Hesseltine, C. W., and Bothast, R. J. (1967). 'Mold development in ears of corn from tasseling to harvest'. *Mycologia*, **69** (2), 328–340.

Hill, R. A., Blankenship, P. D., Cole, R. J.,, and Sanders, T. H. (1983). 'Effects of soil moisture and temperature on preharvest invasion of peanuts by the *Aspergillus flavus* group and subsequent aflatoxin development'. *Appl. Environ. Microbiol.*, **45** (2), 628–633.

Joffe, A. Z. (1969). 'Relationships between *Aspergillus flavus, A. niger* and some other fungi in the mycoflora of groundnut kernels'. *Plant and Soil*, **31**(1), 57–64.

Kulik, M. M. (1968). A Compilation of Descriptions of New *Penicillium* Species. USDA Agriculture Handbook No. 351.

McDonald, D., and Harkness, C. (1967). 'Aflatoxin in the groundnut crop at harvest in northern Nigeria'. *Trop Sci.*, **9**, 148–161.

Nichols, T. E. (1983). 'Economic impact of aflatoxin in corn'. In *Aflatoxin and Aspergillus flavus in Corn*, pp. 67–71. (eds U. L. Diener, R. L. Asquith, and J. W. Dickens). Southern Cooperative Series Bulletin 279. Auburn University, Alabama 36849.

Northolt, M. D., Verhulsdonk, C. A. H., Soentoro, P. S. S., and Paulsch, W. E. (1976). 'Effect of water activity and temperature on aflatoxin production by *Aspergillus parasiticus. J. Milk Food Technol.*, **39**, 170–174.

Panasenko, V. T. (1967). 'Ecology of micro-fungi'. *The Botanical Review*, **33**(3), 189–215.

Pettit, R. E., Taber, R. A., Schroeder, H. W., and Harrison, A. L. (1971). 'Influence of fungicides and irrigation practice on aflatoxin in peanuts before digging'. *Appl. Microbiol.*, **22**, 629–634.

Rambo, B. W., Tuite, J., and Caldwell, R. W. (1974). '*Aspergillus flavus* and aflatoxin in preharvest corn from Indiana in 1971 and 1972', *Cereal Chem.* **51**(595), 600–604. Erratum 848–853.

Raper, K. B., and Fennell, D. I. (1965). *The Genus Aspergillus*. The Williams and Wilkins Co., Baltimore, MD, U.S.A.

Raper, K. B., and Thom, C. (1949). *A Manual of the Penicillia*. The Williams and Wilkins Co., Baltimore, MD, U.S.A.

Samson, R. A. (1979). A Compilation of the Aspergilli Described Since 1965. *Studies in Mycology*, No. 18. Centraalbureau voor Schimmelcultures, Baarn, The Netherlands (38 pp.).

Samson, R. A., Stolk, A. C., and Hadlok, R. (1976). Revision of the Subsection Fasciculata of *Penicillium* and Some Allied Species. *Studies in Mycology*, No. 11. Centraalbureau voor Schimmelcultures, Baarn, The Netherlands (47 pp.).

Sanders, T. H., Hill, R. A., Cole, R. J., and Blankenship, P. D. (1981). 'Effect of drought on occurrence of *Aspergillus flavus* in maturing peanuts'. *J. Am. Oil Chem. Soc.* **58**, 966A–970A.

Sellschop, J. P. R. (1954). Symp. *Mycotoxins in Foodstuff, Agric. Aspects*. Pretoria, South Africa, p. 47.

Shotwell, O. L. (1977). 'Aflatoxin in corn', *J. Am. Oil Chem. Soc.*, **54**, 216A–224A.

Thean, J. E., Lorenz, D. R., Wilson, D. M., Rodgers, K., and Gueldner, R. C. (1980). 'Extraction, cleanup and quantitative determination of aflatoxins in corn', *J. Assoc. Off. Anal. Chem.*, **63**, 631–633.

Wilson, D. M. and Stansell, J. R. (1983). 'Effects of irrigation regimes on aflatoxin contamination in peanut pods', *Peanut Sci.*, **10**, 54–46.

Wilson, D. M., Widstrom, N. W., Marti, L. R., and Evans, B. D. (1981). '*Aspergillus flavus* group, aflatoxin and bright greenish-yellow fluorescence in insect-damaged corn in Georgia', *Cereal Chem.*, **58**, 40–42.

Trichothecenes and Other Mycotoxins
Edited by J. Lacey

CHAPTER 9

Mycotoxins in crops grown in different climatic regions of Queensland

Barry J. Blaney

INTRODUCTION

Agriculture in Queensland differs from that in other eastern Australian states. The warmer climate supports production of such crops as sugar cane and tropical fruits, while the comparatively moist summer conditions have encouraged summer cropping of sorghum, maize, peanuts, sunflowers, navy beans and soybeans. While winter-grown wheat and barley may contribute up to half of the total Queensland grain production, these crops depend largely on soil moisture retained during the summer fallow. In 1980–81 sorghum production was 1,050,000 t; maize 123,000 t; peanuts 42,000 t; wheat 485,000 t; barley 170,000 t; which was 87%, 71%, 98%, 4% and 6% respectively, of the Australian total in that year.

Many of Queensland's cropping regions (Figure 1) feature rainfall which is both unreliable and poorly distributed. Droughts are therefore common and yields vary widely from season to season. This is particularly so for wheat and barley production, although all crops are affected to some extent (Figure 2).

Within Queensland, summer rainfall is more reliable in northern regions. Maize and peanuts are grown in both the Burnett in southern Queensland and on the tablelands in Far North Queensland. Temperatures are similar in these regions because the tablelands are elevated at about 750 m. While the Burnett receives irregular rainfall spread over both summer and winter, rainfall in Far North Queensland occurs almost exclusively in summer, is regular, and can be several times greater than in the Burnett region.

Mycotoxin contamination of crops is affected by fungal populations in the crop and by climatic influences during growth and maturation. In this chapter I will discuss the manner in which the climate affects the incidence of mycotoxins in Queensland crops.

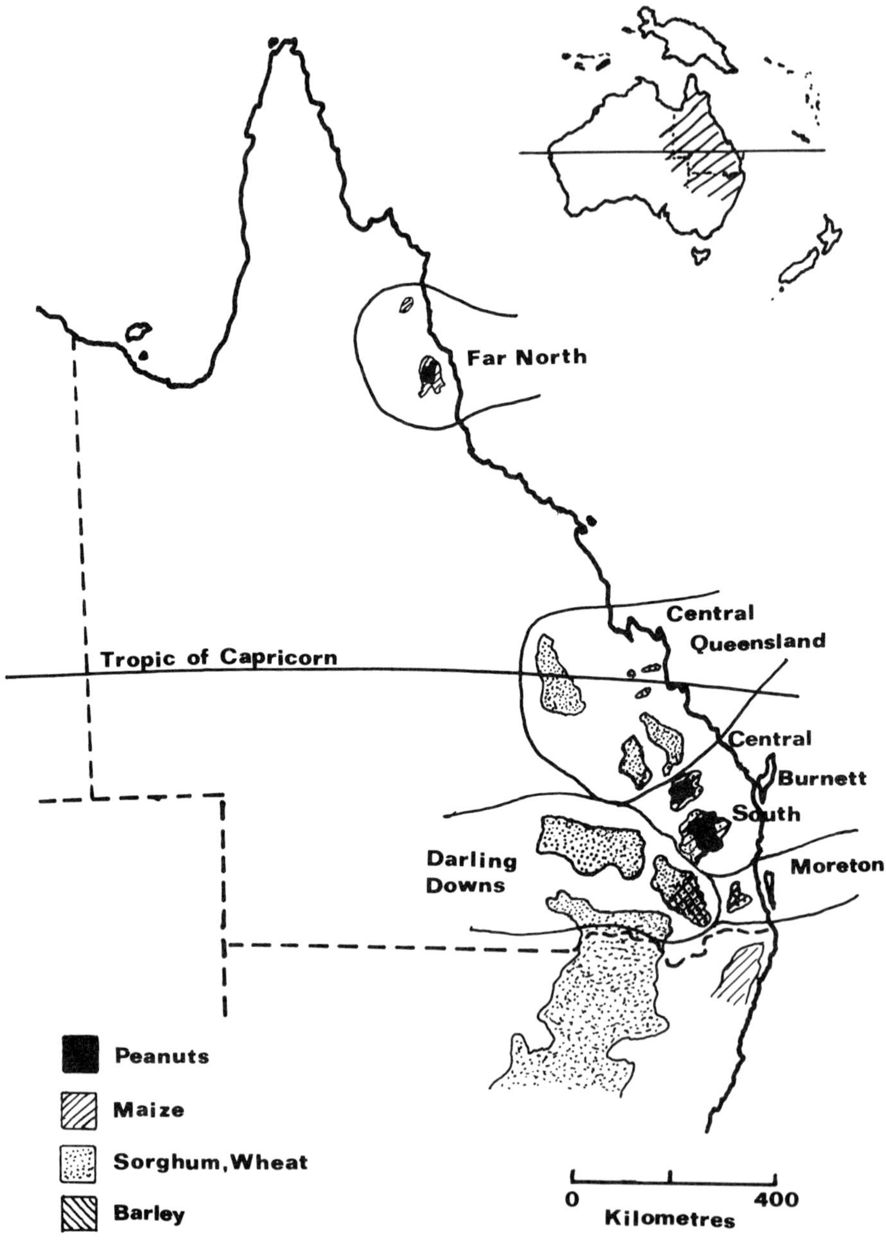

Figure 1 Grain and peanut production regions of Queensland.

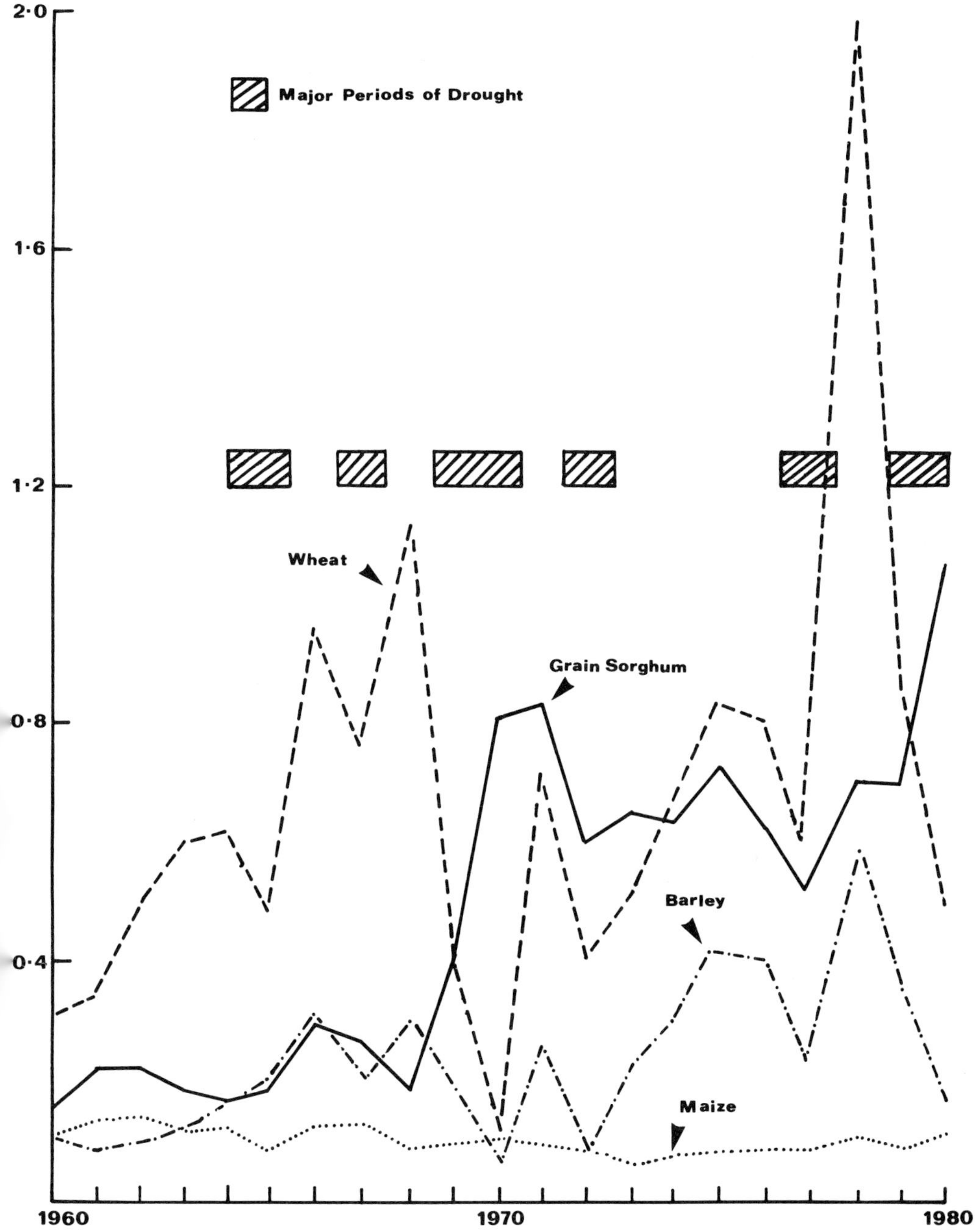

Figure 2 Grain production in Queensland (million tonnes).

PEANUTS

The three major peanut production regions within Queensland are the southern Burnett, the central Burnett and the far northern tablelands (Figure 1), which together produce over 90% of Australia's crop. Of this, about 70% is stored and marketed by the Peanut Marketing Board using intake depots in Kingaroy, Gayndah and Tolga respectively, and the remainder by independent shellers. Quantities of peanuts handled by the Board at these localities are given in Table 1.

Aflatoxin contamination of peanuts has been reported in many parts of the world. It is also a problem in Australia, where most *Aspergillus flavus* invasion and much aflatoxin production occurs before harvest (Graham, 1982a). Peanuts are screened by the Peanut Marketing Board using the minicolumn test (Holaday and Lansden, 1975) to ensure that none with significant concentrations of aflatoxin are used for human food (Telford, 1982). Peanuts with 0.015 mg total aflatoxin kg^{-1} or more are segregated and processed separately. Testing started in 1971, but has been intensive only since 1976. Aflatoxin contamination varies greatly, depending on weather during the growing season. In the Burnett, following drought in 1977, approximately 50% of the crop was used only for oil production because it contained more than 0.015 mg aflatoxins kg^{-1}. By comparison, only 15% of the crop was consigned to oil production in 1976, which had adequate rainfall (Blaney, 1982). Consequently, the Peanut Marketing Board installed a roasting and blanching plant at Kingaroy to aid removal of contaminated kernels from segregated deliveries. This enabled much of the 1980 drought-affected crop to be salvaged with only about 20% consigned to oil production.

Table 1 Quantities of peanuts handled by the Peanut Marketing Board at three localities in Queensland ('000 tonnes)

	Southern Burnett (Kingaroy)	Central Burnett (Gayndah)	Far North (Tolga)
1980	18.4	3.7	9.6
1981	16.4	4.3	10.3
1982	23.0	9.3	11.5

Pre-harvest aflatoxin contamination

Either irregular or inadequate rainfall is probably the largest single factor leading to aflatoxin contamination in the Burnett, especially on soils with lower water-holding capacities (Graham, 1982a). This situation seems similar to that in major peanut-growing regions of the U.S.A., where high temperatures combined with drought stress are major factors leading to contamination (Diener and Davis, 1977; Hill *et al.*, 1983). Table 2 shows the proportions of deliveries placed in segregation on the basis of aflatoxin testing at intake by the Peanut Marketing Board, as a guide to the effect of year and region on frequency of contamination (M. J. Read, personal communication). Contamination has been greater in the drier Burnett regions and the effect of drought in 1980 can also be seen.

Mechanical damage increases the susceptibility of peanuts to *A. flavus* invasion. The Red Spanish variety is more severely affected than Virginia Bunch, probably because of more pod splitting which can occur in the Burnett when rainfall follows a dry period (Graham, 1982b). In addition, some insects have caused increased damage during dry periods (Graham, 1982a). Rainfall at the stage of maturity of the damaged crop may then increase its aflatoxin content. However, besides *A. flavus*, the most commonly isolated fungus from Burnett-grown peanuts is *A. niger* (Graham, 1982a). The negative interaction between these fungi, as described by Joffe (1969) and Hill *et al.* (1983), may be important in the season-to-season aflatoxin incidence in the Burnett region, since *A. flavus* only predominates in hot, dry conditions. Less aflatoxin contamination in Far North Queensland may be a consequence of

Table 2 Percentage of deliveries of Queensland peanuts affected by *A. flavus* by region, year and variety

	Southern Burnett (Kingaroy)	Central Burnett (Gayndah)	Far North (Tolga)
Virginia Bunch			
1980	72	82	13
1981	8	6	5
1982	3	19	4
Red Spanish			
1980	58	96	23
1981	20	25	31
1982	22	5	11

more regular rainfall limiting pod damage, or of increased activity of micro-organisms antagonistic to *A. flavus* in the soil around the pods.

Aflatoxicosis from ingestion of peanut products

In Australia, the risk to man of liver cancer from consumption of peanut products appears to be low (Pitt, 1982). However, feeding of contaminated crop-residues and peanut by-products occasionally leads to aflatoxicosis of livestock. Cases occur more often in drought seasons, since not only are feeds more often contaminated, but pastures for grazing stock are depleted so that poor-quality supplements may be used. Stock are also in poorer condition and less able to resist the effects of toxins. Aflatoxicosis occurred in calves fed peanut hay containing a proportion of nut-in-shell (McKenzie *et al.*, 1981) and in pigs fed peanut waste (Ketterer *et al.*, 1982) in southern Queensland during the drought year of 1980. Peanut meals left after oil extraction and high-protein meals prepared from peanut shell and screenings present another hazard to livestock. However, these materials are sold through normal commercial channels and their aflatoxin content is regulated at point of sale.

MAIZE

Many mycotoxins have contaminated maize worldwide, including those produced by *A. flavus*, *Fusarium moniliforme*, *F. graminearum*, *F. tricinctum*, *F. oxysporum*, *F. equiseti*, *F. culmorum* and *Diplodia maydis*. All of these fungi have been isolated from Australian maize. However, only *F. moniliforme*, *F. graminearum* and *D. maydis* are important pathogens of the crop, although the prevalence of disease varies from season to season and between regions. Most Queensland maize production occurs in the Burnett and Moreton regions, on the Darling Downs, and in Far North Queensland (Figure 1).

Pre-harvest mycotoxin contamination

In the Burnett region, where maize is commonly rotated with peanuts, surveys for aflatoxin were carried out in 1978 and annually since 1980. In 1978, 140 samples of maize, each containing five or more bright greenish-yellow (BGY) fluorescent particles in the 2 kg samples, were selected from 805 deliveries in the region. Aflatoxins were detected in 15 of these samples. Average concentration in the 15 samples was 0.045 mg total aflatoxins kg^{-1}, but only eight exceeded 0.02 mg total aflatoxins kg^{-1}. Aflatoxins were detected on only two of 70 samples selected at random from the remaining 665 (0.005, 0.008 mg aflatoxins kg^{-1}) (Blaney, 1981). In the drought year of

1980 there was a similar survey of 768 samples by the Queensland Grain Growers Association. These came mainly from the Burnett, but included a few samples from the Darling Downs. Of these, 41 samples contained five or more BGY particles per 2 kg sample and 28 of these contained more than 0.02 mg total aflatoxins kg^{-1} (J. G. Twyford, personal communication). These and subsequent surveys have revealed a relatively low incidence of aflatoxin contamination in maize from these regions. Nevertheless, the Queensland Grain Growers Association has continued regular surveys which are intensified when contamination is considered more likely.

Most of the maize produced on the Darling Downs and in the Moreton regions has not been included in mycotoxin surveys. On the Darling Downs, *D. maydis* is one of the most common maize stalk rot pathogens (Francis and Burgess, 1975). Maize was heavily damaged by *D. maydis* and *F. moniliforme* cob rot during 1979 and 1980. In one investigation no adverse effects were found in either broiler chickens, laying hens or cattle fed the mouldy grain (Blaney *et al.*, 1981). However, cultures of Australian isolates of *D. maydis* have produced similar effects in young cockerels as toxic strains from southern Africa where diplodiosis occurs in cattle, sheep and horses (H. McFadden, personal communication). It therefore appears that there is a potential for toxicity to livestock.

The remaining maize-growing region is on the tablelands of Far North Queensland. This differs from the other regions in that it can have too much rain rather than too little, decreasing yields through lodging (Crosthwaite, 1983). Prolonged periods of drizzle with little sunshine late in the growth period increase the prevalence of *Gibberella* cob rots (Pont, 1963). In 1982, approximately 300 samples of freshly harvested maize were assayed for aflatoxins, ochratoxin A, sterigmatocystin and the trichothecene T-2 with negative results (B. J. Blaney, C. J. Moore and A. L. Tyler, unpublished data). However, zearalenone was detected in 85% of the samples, the concentrations correlating well with the proportion of kernels in each sample having a reddish-purple discolouration due to *Gibberella* cob rot. In four samples the concentration of zearalenone exceeded 1 mg kg^{-1}, enough to produce oestrogenic effects in pigs, but the mean in all samples was only 0.17 mg kg^{-1}. Concentrations were highest in districts that received the most rainfall during the growth period.

Zearalenone intoxication from ingestion of maize

By legislation, all maize produced on the far northern tablelands must be delivered to the Atherton Tableland Maize Marketing Board which dries it to 13.5–14% moisture before bulking and storage. However, zearalenone intoxication of pigs fed maize that had been stored with a high water content has recently been detected in the region (Blaney *et al.*, 1984). The maximum concentration of zearalenone in the feeds was 6.4 mg kg^{-1}. Most prepubertal gilts in the herd displayed enlarged teats and signs of oestrus such as red, swollen vulvas, and 25 pigs died as a direct result of rectal and vaginal prolapses. There was no evidence of mycotoxins other than zearalenone in the feeds.

GRAIN SORGHUM

The main sorghum-producing regions in Queensland are the Darling Downs and central Queensland (Figure 1) but little is known of the extent of mycotoxin contamination of the sorghum crop. The major diseases of sorghum in Queensland are seedling blight, root rot and stalk rot, all attributed to *F. moniliforme*. *F. moniliforme* var. *subglutinans* and *F. graminearum* Group 2 have been isolated occasionally from sorghum plants (Burgess *et al.*, 1981). Also, a wide range of fungi have been isolated from the stored grain including *A. flavus, A. niger, A. ochraceus, A. glaucus, Alternaria* spp., *Phoma* spp., *Penicillium* spp., *F. equiseti, F. oxysporum, F. solani* and *F. semitectum* (M. D. Connole, personal communication).

Climate differs between the two main sorghum-growing regions. Rainfall is more reliable in Central Queensland. On the other hand, the crop is harvested comparatively later than on the Darling Downs, when temperatures are lower and less favourable for drying, which leads to more problems in storage.

Pre-harvest mycotoxin contamination

Exceptional climatic conditions during the 1982–83 summer season offered an opportunity to study pre-harvest mycotoxin contamination of sorghum. One of the worst droughts on record extended well into the normal growing season for summer crops in most parts of Queensland, decimating many crops and leading to severe damage of others through drought stress, insects and disease. Heavy rain began just before harvesting and continued for 3–4 months in most regions. This provided a constantly humid environment which, following the drought stress and insect damage, provided ideal conditions for fungal growth. Floods submerged some crops, increasing water content, soil contamination and perhaps mechanical damage. Rain also delayed the harvest, allowing more time for fungal growth and leading to a high moisture content in the harvested grain.

On the Darling Downs the sorghum harvest was about half that originally anticipated, but the crops in Central Queensland received the most flooding, and original crop estimates of 600,000 t were reduced to 170,000 t. Samples obtained at harvest from all regions were badly affected by mould. Investigation of mycotoxins in these samples is incomplete but so far only 10 of 80 samples have contained aflatoxins, none with more than 0.02 mg aflatoxin B_1 kg^{-1}. In addition, small and probably biologically insignificant concentrations of zearalenone were detected in a few samples (B. J. Blaney, unpublished data). There is, as yet, no evidence of other toxins in the drought- and flood-affected sorghum at harvest.

Post-harvest mycotoxin contamination and mycotoxicoses

Moulding of sorghum grain stored with a high water content has been implicated in several accounts of mycotoxicosis of farm animals. In a small piggery in the Burnett region, poor appetite, depression, diarrhoea, vomiting and some mortalities followed introduction of a batch of sorghum into the feed. The sorghum was unripe when harvested at 16% moisture content and then heated in store. A sample contained 9.6 mg total aflatoxins (B_1, B_2, G_1 and G_2) kg^{-1} plus 0.1 mg ochratoxin A kg^{-1} (Ketterer *et al.*, 1982). Connole and Hill (1970) also reported cases of suspected aflatoxicosis in the Burnett region from feeding mouldy sorghum grain to pigs.

In a 200-sow breeding and fattening unit on the Darling Downs, poor appetite, depression and a decreased growth rate followed the introduction of a new batch of sorghum in the feed. This grain was also harvested before full ripening and although it was stored for less than a month, patches of black mould had developed. A sample contained 0.28 mg aflatoxins B_1 and B_2 kg^{-1} (Ketterer *et al.*, 1982).

In contrast to these cases of aflatoxicosis in southern Queensland, zearalenone intoxication of pigs fed mouldy sorghum has occurred on the tableland regions of Far North Queensland (Blaney *et al.*, 1984). In this instance up to 8 mg zearalenone kg^{-1} was detected in stored samples from three different crops, two of which were reported to have been affected by rain before harvest.

OTHER SUMMER CROPS

Insufficient information is available to determine whether or not safflower, sunflower, cottonseed, navy bean and soybean crops in Queensland are prone to mycotoxin contamination. There have been no systematic surveys. However, these crops have seldom been implicated in mycotoxicoses (Connole *et al.*, 1981). Although mould damage can occur as in the 1983 season, only low concentrations of these crops or their by-products are normally included in animal feeds.

WHEAT AND BARLEY

There is limited information on the occurrence of mycotoxins in Australian wheat and barley (Burgess, chapter 2 in this volume). The most common

Fusarium pathogens of wheat are isolates of *F. graminearum* Groups 1 and 2 (Francis and Burgess, 1977). While *F. graminearum* Group 1 (the cause of crown rot) is regarded as the most important *Fusarium* disease on wheat, only isolates of *F. graminearum* Group 2 (causing head blight) are known to produce mycotoxins such as deoxynivalenol, T-2 and other trichothecenes. Head blight is not usually significant except in wet seasons in higher rainfall areas such as the Burnett region where little wheat is grown (Burgess *et al.*, 1981). Wheat has never been surveyed at harvest for mycotoxins, while none were found in a single survey of the Queensland barley crop in 1979 (C. J. Moore, personal communication).

Queensland wheat and barley are mainly produced on the Darling Downs and in Central Queensland. Crops are usually harvested before the onset of summer rainfall at low moisture contents, and consequently deteriorate little in storage. However, when wetted they are as prone as other commodities to moulding, although unprocessed wheat appears to be a poorer substrate for aflatoxin production by some isolates of *A. flavus* than most other grains (Connole *et al.*, 1981). Bryden *et al.* (1975) reported the presence of aflatoxins in mouldy wheat and barley.

Because of irregular wheat yields, large crops exceed the capacity of existing silos and temporary bulk storages are used. Recently, we analysed samples from temporary stores of wheat from the 1981 and 1982 Queensland crops (B. J. Blaney, unpublished data). These had sustained water damage and portions were extensively moulded and rotten following about 18 months storage. Obviously moulded wheat was separated and evaluated as stock feed. Of 152 mouldy samples representing 12 stores of the 1981 crop, aflatoxins B_1 and B_2 were detected in 21 (14%) but no other mycotoxins. The mean concentration in the positive samples was 0.016 mg kg^{-1}, and they ranged from 0.002 to 0.060 mg aflatoxin B_1 kg^{-1}. Aflatoxins B_1 and B_2 were also present in 10 out of 13 mouldy samples (77%) from the only temporary store of wheat from the small 1982 crop. The mean concentration was 0.064 mg kg^{-1} and range 0.003 to 0.500 mg aflatoxin B_1 kg^{-1}. By contrast, two out of 65 samples of apparently sound grain (3%) from these same stores, contained 0.003 and 0.004 mg aflatoxin B_1 kg^{-1}. This latter result is reassuring, considering that wheat is the major grain used for human food in Australia.

Aflatoxicosis from mouldy bread

Although few cases are known of mouldy wheat being implicated in mycotoxicosis in Queensland, mouldy bread has caused aflatoxicosis on several occasions in dogs and pigs (Ketterer *et al.*, 1975, 1982).

CONCLUSIONS

Australia has many different climatic zones, and no generalizations can be made about the susceptibility of particular crops to mycotoxin contamination without consideration of both regional and seasonal factors. Queensland has a history of recurring drought, and drought stress is an important factor in pre-harvest aflatoxin contamination of the peanut and maize crop, particularly in the drier Burnett region. Conversely, while drought depresses yields, dry conditions usually limit the extent of fungal damage to wheat, barley and sorghum. However, the 1983 season was exceptional in that persistent heavy rain produced extensive mould damage to all crops.

Mycotoxins produced by *Fusarium* spp. have never been implicated in widespread mycotoxicosis in Queensland, although they may contaminate maize and sorghum in regions with high rainfall during the growing season, such as on the tablelands in Far North Queensland.

Mycotoxicosis of man seems extremely unlikely to occur in Queensland, but mycotoxicosis of livestock occurs occasionally. It is usually associated with feeding materials moulded from being stored damp. In addition, cases are likely to occur in drought seasons when feeds are not only scarce, but also likely to be of poorer quality.

REFERENCES

Blaney, B. J. (1981). 'Aflatoxin survey of maize from the 1978 crop in the South Burnett region of Queensland', *Qld. J. Agric. Anim. Sci.*, **38**, 7–11.

Blaney, B. J. (1982). 'Mycotoxin surveillance in Australia'. *J. App. Toxicol.*, **2**, 83–87.

Blaney, B. J., Bloomfield, R. C., and Moore, C. J. (1984). 'Zearalenone intoxication of pigs', *Aust. Vet. J..*, **61**, 24–27.

Blaney, B. J., Mannion, P. F., Tudor, G. D., and McKenzie, R. A. (1981). 'Examination of *Diplodia maydis* infected maize for toxicity to chickens and cattle', *Aust. Vet. J.*, **57**, 196.

Bryden, W. L., Rajion, M. A., Lloyd, A. B., and Cumming, R. B. (1975). 'Surveys of Australian feedstuffs for toxigenic strains of *Aspergillus flavus* and for aflatoxin', *Aust. Vet. J.*, **51**, 491–493.

Burgess, L. W., Dodman, R. L., Pont, W., and Mayers, P. (1981). 'Fusarium diseases of wheat, maize and grain sorghum in eastern Australia', in *Fusarium: Diseases, Biology, and Taxonomy* (eds P. E. Nelson, T. A. Toussoun and R. J. Cook), pp. 64–76. The Pennsylvania State University Press, University Park and London.

Connole, M. D., and Hill, M. W. M. (1970). '*Aspergillus flavus* contaminated sorghum grain as a possible cause of aflatoxicosis in pigs', *Aust. Vet. J.*, **46**, 503–505.

Connole, M. D., Blaney, B. J., and McEwan, T. (1981). 'Mycotoxins in animal feeds and toxic fungi in Queensland 1971–80', *Aust. Vet. J.*, **57**, 314–318.

Crosthwaite, I. C. (1983). 'Maize growing on the Atherton Tableland', *Qld. Agric. J.*, **109**, 40–46.

Diener, U L., and Davis, N. D. (1977). *Aflatoxin formation in peanuts by* Aspergillus flavus. Agricultural Experiment Station Bulletin No. 493. U.S. Agricultural Experiment Station, Auburn, Ala.

Francis, R. G., and Burgess, L. W. (1975). 'Surveys of *Fusaria* and other fungi associated with stalk rot of maize in eastern Australia', *Aust. J. Agric. Res.*, **26**, 802–807.

Francis, R. C., and Burgess, L. W. (1977). 'Characteristics of two populations of *Fusarium roseum* "Graminearum" in eastern Australia', *Trans. Br. Mycol. Soc.*, **68**, 421–427.

Graham, J. C. (1982a). *Preharvest Aflatoxin Contamination in Australian Peanuts.* Bulletin QB82001, Queensland Department of Primary Industries, Brisbane.

Graham, J. C. (1982b). 'The occurrence of aflatoxin in peanuts in relation to soil type and pod splitting', *Food Technol. Aust.*, **34**, 208–212.

Hill, R. A., Blankenship, P. D., Cole, R. J., and Sanders, T. H. (1983). 'Effects of soil moisture and temperature on preharvest invasion of peanuts by the *Aspergillus flavus* group and subsequent aflatoxin development', *Appl. Environ. Microbiol.*, **45**, 628–633.

Holaday, C. E., and Lansden, J. (1975). 'Rapid screening method for aflatoxin in a number of products', *Agric. Food. Chem.*, **23**, 1134–1136.

Joffe, A. A. (1969). 'Relationships between *Aspergillus flavus*, *A. niger*, and some other fungi in the mycoflora of groundnut kernels'. *Plant Soil*, **31**, 57–64.

Ketterer, P. J., Blaney, B. J., Moore, C. J., McInnes, I. S., and Cook, P. W. (1982). 'Field cases of aflatoxicosis in pigs', *Aust. Vet. J.*, **59**, 113–117.

Ketterer, P. J., Williams, E. S., Blaney, B. J., and Connole, M. D. (1975). 'Canine aflatoxicosis', *Aust. Vet. J.* **51**, 355–357.

McKenzie, R. A., Blaney, B. J., Connole, M. D., and Fitzpatrick, L. A. (1981). 'Acute aflatoxicosis in calves fed peanut hay', *Aust. Vet. J.*, **57**, 284–286.

Pitt, J. I. (1982). 'On regulatory standards and risk assessment for aflatoxins in Australia', *Food Technol. Aust.*, **34**, 232–234.

Pont, W. (1963). 'Maize diseases are common in north Queensland', *Qld. Agric. J.*, **89**, 357–365.

Telford, A. P. (1982). 'Control of aflatoxins in peanuts by segregation and sorting', *Food Technol. Aust.*, **34**, 214–215.

Trichothecenes and Other Mycotoxins
Edited by J. Lacey

CHAPTER 10

Mycoflora and aflatoxin in pre-harvest maize kernels of varied endosperm type

Raul G. Cuero, Eivind B. Lillehoj, William F. Kwolek and Marcus S. Zuber

INTRODUCTION

Although there is much information on environmental factors influencing aflatoxin production by *Aspergillus flavus* Link ex Fr (Mislivec and Tuite, 1970; Misra *et al.*, 1982), on pre- and post-harvest contamination of crops (Lillehoj *et al.*, 1976) and on predisposing factors such as insect damage (Lillehoj *et al.*, 1976), there is little on the effects of genotype. Variation in the susceptibility of plant genotypes to invasion by toxigenic fungi and to toxin formation has been observed (King, 1980; Loesch *et al.*, 1976; Zuber *et al.*, 1976, 1978), while Zuber *et al.* (1976) found the occurrence of *A. flavus* and aflatoxin in maize hybrids pre-harvest to vary between different regions in the U.S.A. and from year to year.

The present investigation was carried out to determine the interaction effects of other fungi on aflatoxin production by *A. flavus* on maize kernels of different endosperm types.

MATERIAL AND METHODS

Field experiments

Nineteen hybrids and one synthetic population were grown in 1980 and 1981 at eight sites: Gainesville, Florida (FL); Tifton, Georgia (GA); Raleigh, North Carolina (NC); Florence, South Carolina (SC); Mississippi States, Mississippi (MS); Lafayette, Indiana (IN); Wooster, Ohio (OH); and Knoxville, Tennessee (TN). Two replicates of each entry, each of 20 plants were grown at each site. Entries (Table 1) were selected on the basis of specific characteristics that might influence susceptibility to *A. flavus* infection of developing kernels and aflatoxin accumulation before harvest.

Table 1 Characteristics of test hybrids

Entry	Source	Hybrid	Characteristics
1	Custom seed	85588	High amylose, class 5 (ae—50% amylose)
2	Custom Seed	85473	High amylose, class 7 (ad—70% amylose)
3	Custom Seed	85070	Waxy (wx—0% amylose)
4	Custom Seed	Experimental	Normal counterpart, dent (27% amylose)
5	Pioneer	L3369	High lysine (0–2)
6	Pioneer	3369A	Normal counterpart
7	Experimental	SC54 *o-2* X SC76 *o-2*	High lysine (0–2)
8	Experimental	SC54 X SC76	Normal counterpart
9	Experimental	SC54 *fl-2* X SC76 *fl-2*	Floury endosperm (fl-2)
10	Pfister	HO	High oil (6.5% oil)
11	Experimental	Mo17 X B73	Near normal counterpart (4.5% oil)
12	Cargill	Hyb. I	Flint
13	Cargill	Hyb. I	Dent
14	Cargill	Hyb. II	Flint
15	Cargill	Hyb. II	Dent
16	Cargill	Hyb. III	Flint
17	Cargill	Hyb. III	Dent
18	Experimental	Mo17 X H60	'Aflatoxin-resistant' (Zuber *et al.*, 1978)
19	Experimental	Mo5 X Oh545	'Aflatoxin-susceptible' (Zuber *et al.*, 1978)
20	Experimental	MoSQA(H)C6	University of Missouri synthetic

Sampling procedure

At maturity, 10 ears from each plot were harvested at random and dried to a final kernel moisture content below 14% in a 60 °C forced-draught dryer for 5–7 days. The dried ears were shelled and a sample of kernels was taken for mycological studies. The remaining sample was ground in a Raymond hammer mill with a screen containing 3.2 mm perforations. The ground sample was blended for 15–30 min in a Patterson Kelley twin shell blender.

Aflatoxin analysis

Ground and blended corn samples were assayed for aflatoxin by the Official First Action Method (Association of Official Analytical Chemists, 1980).

Mycological examination

Mycological analysis was carried out by plating five kernels, previously surface-sterilized in 2% sodium hypochlorite, on 3% malt extract agar in each of 10 plates. Incubation was at 25 °C. Thus a total of 50 kernels was plated from each replicate of each entry.

Statistical analysis

The mean occurrence of fungi on mean ten kernels was used for statistical analysis. Correlation coefficients were computed to examine the interrelationships between different factors and the presence of *A. flavus* and aflatoxin levels for each combination of year and state (Steele *et al.*, 1980). Variability in the occurrence of fungi in kernels between states was examined by chi-square tests and variation between entry groups by analysis of variance.

RESULTS

Aflatoxin levels and *A. flavus* occurrence were relatively high in pre-harvest maize from the southern U.S. but sporadic and low in kernels from Midwest locations (Table 2). Toxin levels also varied between entries, kernels of entries, 3, 5 and 11 tended to have the highest toxin levels and samples of entries 8 and 20 the lowest. For instance in MS, samples of entries 3 and 11 contained 1417 and 1846 μg/kg aflatoxin respectively, while entry 20 contained only 31 μg/kg. In most other states the differences were smaller. There were also broad patterns in the occurrence of *A. flavus* in different regions (Table 2). Levels of aflatoxin were lowest in the Midwestern states (Indiana, Ohio, and Tennessee) and similarly there was little *A. flavus* in any entry. By contrast, samples from southern states had a high incidence of *A. flavus* in kernels. Although samples of entries from NC, SC and MS differed in the incidence of *A. flavus* and in toxin content, the pattern of variation was not consistent between the three locations.

Although overall incidence of *A. flavus* and levels of aflatoxin were similar in 1980 and 1981, incidence of *Fusarium moniliforme* was much less in 1981 than in 1980 (Table 3). This difference was largely the result of a high incidence of the fungus in Tennessee in 1980. Differences between states in the incidence of *A. flavus* were significant in 1980 and *F. moniliforme* in 1980 and 1981, although the incidence of both species in 1981 was about half that in 1980. However, the incidence of *F. moniliforme* was usually greater than that

Table 2 Mean aflatoxin B_1 and B_2 concentration and *A. flavus* occurrence (percentage) in maize for 1980 and 1981[a]

Entry	FL	GA	NC	SC	MS (1981 only)[b]	IN	OH	TN
1	14 (5.0)	223 (0.0)	153 (10.0)	168 (15.0)	331 (10.0)	9 (10.0)	0 (0.0)	16 (0.0)
2	24 (15.0)	224 (10.0)	92 (15.0)	94 (30.0)	398 (10.0)	13 (15.0)	0 (0.0)	4 (0.0)
3	194 (5.0)	563 (30.0)	76 (15.0)	96 (10.0)	1417 (15.0)	0 (5.0)	4 (5.0)	2 (0.0)
4	70 (0.0)	321 (40.0)	108 (5.0)	98 (10.0)	646 (0.0)	14 (20.0)	0 (20.0)	9 (5.0)
5	36 (10.0)	539 (20.0)	350 (25.0)	438 (50.0)	561 (0.0)	5 (0.0)	0 (0.0)	5 (0.0)
6	58 (10.0)	187 (10.0)	128 (0.0)	258 (20.0)	757 (0.0)	3 (10.0)	1 (0.0)	3 (0.0)
7	38 (20.0)	300 (5.0)	82 (0.0)	95 (25.0)	56 (5.0)	0 (0.0)	0 (0.0)	7 (0.0)
8	40 (0.0)	74 (0.0)	10 (0.0)	85 (15.0)	25 (0.0)	0 (0.0)	0 (15.0)	9 (15.0)
9	47 (5.0)	167 (10.0)	44 (0.0)	226 (30.0)	172 (0.0)	0 (0.0)	1 (0.0)	0 (0.0)
10	170 (10.0)	380 (5.0)	240 (10.0)	250 (25.0)	786 (0.0)	0 (5.0)	4 (0.0)	4 (0.0)
11	52 (15.0)	271 (15.0)	240 (0.0)	478 (65.0)	1846 (5.0)	0 (5.0)	7 (10.0)	0 (5.0)
12	21 (10.0)	237 (30.0)	59 (5.0)	168 (15.0)	489 (0.0)	0 (0.0)	7 (0.0)	12 (0.0)
13	29 (5.0)	147 (10.0)	644 (15.0)	36 (20.0)	466 (10.0)	0 (0.0)	0 (0.0)	0 (0.0)
14	12 (5.0)	241 (5.0)	112 (20.0)	79 (40.0)	852 (10.0)	0 (0.0)	2 (0.0)	5 (5.0)
15	31 (10.0)	350 (10.0)	51 (10.0)	46 (35.0)	368 (0.0)	0 (0.0)	0 (5.0)	5 (5.0)
16	25 (5.0)	264 (5.0)	152 (20.0)	302 (25.0)	652 (0.0)	0 (0.0)	0 (0.0)	4 (10.0)
17	36 (0.0)	134 (0.0)	128 (0.0)	252 (30.0)	571 (10.0)	0 (0.0)	0 (0.0)	2 (2.0)
18	15 (30.0)	231 (30.0)	70 (15.0)	159 (30.0)	107 (10.0)	0 (0.0)	0 (0.0)	3 (0.0)
19	16 (0.0)	332 (5.0)	57 (0.0)	117 (15.0)	844 (5.0)	1 (50.0)	0 (0.0)	1 (0.0)
20	12 (10.0)	222 (5.0)	2 (0.0)	177 (15.0)	31 (0.0)	0 (30.0)	0 (5.0)	0 (0.0)
LSD	16.7	5.0	12.2	5.0	62.2			

[a] Aflatoxin concentrations are geometric means and expressed as μg/kg based on log (x + 1) transformation.
[b] MS values on basis of 1 year (1981) data.
[c] Values in parentheses are numbers of *A. flavus* isolated from 100 surface-sterilized kernels. LSD = Least significant difference (p = 0.05).

Table 3 Mean microbial count and aflatoxin levels by state and year

	1980			1981		
State	*A. fl*[a]	*F. mon*[a]	B_1+B_2 (μg/kg)	*A. fl.*	*F. mon.*	B_1+B_2 (μg/kg)
FL	1.5	2.8	79	0.2	0.3	43
GA	1.4	1.7	359	1.1	2.2	289
NC	0.7	6.7	421	1.0	1.7	234
SC	4.0	4.7	134	1.3	4.4	181
MS	—	—	—	0.9	4.3	569
IN	1.0	4.3	8	0.6	0.7	0
OH	0	0.1	1	0.6	0	1
TN	0.4	8.3	7	0	0.9	3
Mean	1.3	4.1	144	0.7	1.8	136
χ^{2b}	181.1[c]	403.1[c]		5.07[ns]	248.7[c]	

A. fl = *Aspergillus flavus; F. mon* = *Fusarium moniliforme.*
[a] Mean occurrence per 10 kernels from each of 20 entries. See Table 1.
[b] Chi-square test of variation among states.
[c] $p < 0.01$, ns is not significant at the 0.05 level.

of *A. flavus* by a factor of about three in both years. There was no consistent relationship between microbial counts and aflatoxin levels, e.g. the state of Mississippi in 1981 and Indiana in 1980; although they differed in climate, they had almost the same incidence of *A. flavus* but extremes in aflatoxin levels, i.e. 569 and 8 μg/kg respectively. Year-to-year variation in aflatoxin levels was observed in some southern states, although the overall mean differed little.

Correlations between aflatoxin levels and occurrence of *A. flavus* and *F. moniliforme* were determined for 1980 and 1981 maize kernel samples from four locations in the southern U.S. (Table 4). There was a positive correlation between *A. flavus* and aflatoxin in four of the eight observations but only a single positive association was noted between *F. moniliforme* incidence and toxin concentrations.

Although there were significant correlations of *A. flavus* count with aflatoxin levels in the four states with common occurrence of aflatoxin (Table 4) there were striking entry differences in the same locations for fungal incidence (Table 5). In total, kernels of entries 1, 2, 3, 5, 11, 14, 18 and 20 carried most fungi, and entries 6, 8, 9 and 13 fewest. From the aggregate data from the four northern states in both years, large differences in the occurrence of *A. flavus* and *F. moniliforme* can be seen.

Table 4 Correlation (*r* coefficients) of the relationship between microbial count and aflatoxin level

	A. flavus vs aflatoxin $B_1 + B_2$ (μg/kg)		*F. moniliforme* vs aflatoxin $B_1 + B_2$ (μg/kg)	
State	1980	1981	1980	1981
FL	0.52[a]	−0.05	0.34	−0.15
GA	0.56[a]	0.40	0.35	0.12
NC	−0.04	0.67[b]	−0.09	0.50[a]
SC	0.59[b]	−0.26	−0.10	0

[a] Significant at 0.05 level.
[b] Significant at 0.01 level.

Although there were no consistent differences in the responses of different endosperm types to either fungal growth or aflatoxin production, entries 5, 11 and 18 tended to have more *A. flavus* and aflatoxin and entries, 1, 2, 11 and 20 more *F. moniliforme*, but with no change in toxin levels.

Few fungi other than *A. flavus* and *F. moniliforme* were found in kernel samples from any state in either year with the exception of Mississippi (MS) in 1981 where 29 out of 200 kernels carried *A. niger*. This association may suggest a possible antagonistic relation between the two species. The remaining data were insufficient for analysis, although some of the other species identified are known to be toxigenic.

DISCUSSION

This study shows that *A. flavus* and *F. moniliforme* frequently occur together before harvest on kernels of different endosperm types from different geographical areas but often associated with insect damage. Similar associations have been found in Colombia (Cuero, 1980; Cuero *et al.*, 1982) and the United States (Fennell *et al.*, 1973). However, previous studies have not recognized any correlation between incidence of *A. flavus* and *F. moniliforme* and aflatoxin production, and even the relationship between *A. flavus* incidence and aflatoxin production has not been consistent.

F. moniliforme was always isolated from about three times more kernels than *A. flavus*, but there was no indication of antagonism with *A. flavus* or of any effect on the levels of aflatoxin found. On the contrary there could be synergism between the two species favouring aflatoxin synthesis. This could

Table 5 Entry means of microbial occurrence in maize kernels from four states (FL, GA, NC, SC) in 1980 and 1981[a]

Entry	*A. flavus*	*F. moniliforme*	*A. flavus* + *F. moniliforme*
1	0.75	4.37	5.12
2	1.75	4.87	6.62
3	1.50	3.87	5.37
4	1.37	2.25	3.62
5	2.62	2.50	5.12
6	1.00	1.62	2.62
7	1.25	3.12	4.37
8	0.37	1.87	2.25
9	1.12	1.62	2.75
10	1.25	3.62	4.87
11	2.37	4.37	6.75
12	1.50	2.37	3.87
13	1.25	1.00	2.25
14	1.75	3.75	5.50
15	1.62	3.37	5.00
16	1.37	2.25	3.62
17	0.75	2.37	3.12
18	2.62	3.25	5.87
19	0.50	3.87	4.37
20	0.75	4.37	5.12
LSD =	1.29	1.88	2.45
F ratio for entry variation	$p<0.05$	$p<0.01$	$p<0.01$

[a] Mean per 10 kernels based on 16 samples (four states, two replications, 2 years) (see Table 1).

be the result of metabolites (alcohols, acids, secondary metabolites, etc.) of *F. moniliforme* which stimulate the metabolism of *A. flavus* increasing aflatoxin production.

By contrast, the strong suggestion of a negative correlation between *A. flavus* and *A. niger* suggested possible antagonism between these two species. The extent of such interaction on the grain is unknown despite numerous attempts using mixed cultures of a whole range of fungi *in vitro* and on grain (Boller and Schroeder, 1972, 1973; King, 1980; Lillehoj *et al.*, 1976; Loesch *et al.*, 1976; Reiss, 1973; Wells *et al.*, 1972; Wicklow *et al.*, 1980; Zuber *et al.*, 1978).

ACKNOWLEDGEMENT

We are indebted to G. E. Scott (USDA, State College, MS), E. S. Horner (University of Florida, Gainesville, FL), N. W. Widstrom (USDA, Tifton, GA), D. L. Thompson (NC State University, Raleigh, NC), A. Manwiller (Clemson University, Florence, SC), W. R. Findley (USDA, Wooster, OH), H. Warren (Purdue University, Lafayette, IN), D. West (University of Tennessee, Knoxville, TN) for planting, tending and harvesting the sample maize.

REFERENCES

Association of Official Analytical Chemists (1980). *Official Methods of Analysis*, 11th edn, p. 436, sect. 26.

Boller, R. A., and Schroeder, H. W. (1972). 'Influence of *Aspergillus chevalieri* in production of aflatoxin in rice by *Aspergillus parasiticus*', *Phytopathology*, **63**, 1507–1510.

Boller, R. A., and Schroeder, H. W. (1973). 'Influence of *Aspergillus candidus* on production of aflatoxin in rice by *Aspergillus parasiticus*', *Phytopathology*, **64**, 121–123.

Cuero, R. G. (1980). 'Ecological distribution of *Fusarium solani* and its opportunistic action related to mycotic keratitis in Cali, Colombia', *J. Clin. Microbiol.*, **12**(3), 445–461.

Cuero, R. G., Hernandez, I., Cardenas, H., and Osorio, E. (1982). Mycotoxin contamination of corn harvested from the Highlands and Lowlands in Colombia (unpublished).

Fennell, D. I., Bothast, R. J., Lillehoj, E. B., and Peterson, R. E. (1973). Bright greenish-yellow fluorescence and associated fungi in white corn naturally contaminated with aflatoxin. *Cereal Chem.*, **50**, 404–414.

King, S. B. (1980). 'Time of infection of maize kernels by *Fusarium moniliforme* and *Cephalosporium acremonium*', *Phytopathology*, **71**(8), 796–799.

Lillehoj, E. B., Fennell, D. I., and Kwolek, W. F. (1976). '*Aspergillus flavus* and aflatoxin in Iowa corn before harvest', *Science,* **193**, 495–496.

Loesch, P. J., Foley, D. C., and Cox, D. F. (1976). 'Comparative resistance of opaque-2 and normal inbred lines of maize to ear-rotting pathogens', *Crop Sci.* **16**, 841–842.

Mislivec, P. B., and Tuite, J.(1970). 'Species of *Penicillium* occurring in freshly harvested and in stored dent corn kernels', *Mycologia*, **62**, 67–88.

Misra, R. S., Sinha, K. K., and Singh, P. (1981). 'Aflatoxin production by *Aspergillus parasiticus* (NRRL-3240) on maize seeds in competitive environment', *Nat. Acad. Sci. Letters*, **4**(3).

Reiss, J. (1973). 'Influence of the mycotoxins aflatoxin B_1, rubratoxin B, patulin and diacetoxyscirpenol on the fermentation activity of baker's yeast', *Mycopathologia et Mycologia Applicata*, **51**, 337–345.

Steele, R. G. D., and Torrie, J. H. (1980). *Principles and Procedures of Statistics*. McGraw-Hill, New York.

Wells, T. R., Kreutzer, W. A., and Lindsey, D. L. (1972). 'Colonisation of gnotobiotically grown peanuts by *Aspergillus flavus* and selected interacting fungi', *Phytopathology*, **62**, 1238–1242.

Wicklow, D. T., Hesseltine, C. W., Shotwell, O. L., and Adams, G. L. (1980). 'Interference competition and aflatoxin levels in corn', *Phytopathology*, **70**, 761–764.

Zuber, M. S., Calvert, O. H., Lillehoj, E. B., and Kwolek, W. F. (1976). 'Preharvest development of aflatoxin B_1 in corn in the United States', *Phytopathology*, **66**, 1120–1121.

Zuber, M. S., Calvert, O. H., Kwolek, W. F., Lillehoj, E. B., and Kang, M. S. (1978). 'Aflatoxin B_1 production in eight-line diallel of *Zea mays* infected with *Aspergillus flavus*, *Phytopathology*, **68**, 1346–1349.

Trichothecenes and Other Mycotoxins
Edited by J. Lacey

CHAPTER 11

Toxigenic *Pithomyces* and *Myrothecium* species in pasture

Margaret E. di Menna and Peter H. Mortimer

INTRODUCTION

Pithomyces chartarum (Berk. & Curt.) M. B. Ellis and the three toxigenic *Myrothecium* spp., *M. roridum* Tode ex Fr., *M. verrucaria* (Alb. & Schw.) Ditm. ex Fr and M. *leucotrichum* (Peck) Tulloch, are cosmopolitan, have been isolated from a variety of substrates and can be found on pasture leaves, particularly those of warm temperate regions. Only *M. roridum* is a plant pathogen. It is plurivorous and has caused losses in coffee and cotton but can also grow as a saprophyte (Fitton and Holliday, 1970). *P. chartarum* causes seasonal outbreaks of toxicosis in pasture-fed sheep, cattle and deer in New Zealand and less frequent incidents in other parts of the world, but myrothecia have rarely been implicated in cases of disease. The reasons for this difference in effectiveness as toxic agents can be looked for in optimal growth conditions for the fungi, production and mode of action of their toxins and access by animals to the toxins formed.

PITHOMYCES CHARTARUM

Growth conditions

Optimal growth conditions for *P. chartarum* have been well defined (Brook, 1963). Spores germinate and are produced only at RH values close to 100%, although mycelial growth persists to values as low as 60%. Spores are formed at temperatures from 5–30 °C but sporulation is most rapid at 24–26 °C. Maximum growth temperature is 30–33 °C. In the field accelerated sporulation with a spore-to-spore generation time as short as 2 days occurs at the base of pasture when at least 4 mm of rain is followed immediately by two or more nights when grass minimum temperatures do not drop below 12 °C (Mitchell *et al.*, 1959). In New Zealand these danger periods occur intermittently from

December to May as warm moist air streams cross the country. However, they are not always followed by sufficient increases in spore numbers to make pastures toxic because the values given are the least rain and the lowest temperature known to precede outbreaks of clinical disease.

Toxin production

Sporidesmin A, the major toxic metabolite of *P. chartarum*, is formed when the fungus spores, and in amounts proportional to the number of spores produced (di Menna *et al.*, 1970). Strains vary in rate and amount of spore and sporidesmin production but the relationship between the two has held for all New Zealand strains examined given that the incubation temperature is constant. The optimal temperature for toxin production is *ca.* 20 °C, which is lower than that for sporulation (Ross and Thornton, 1962; Davison and Marbrook, 1965). A number of strains which spore freely but produce no sporidesmin have been isolated in South Africa and North America (Marasas, 1974, private communication; Halder *et al.*, 1981), but spores in New Zealand pastures are taken to indicate that sporidesmin has been formed, although not necessarily that it is still present there. Sporidesmin is unstable in water solution and rapidly destroyed by light (Clare and Mortimer, 1964). Measurement of sporidesmin in pasture samples has only become possible with the availability of HPLC and the small amount of evidence so far obtained suggests that sporidesmin levels rise in parallel with spore numbers but drop more rapidly, spores persisting in pasture when dead and leached of toxin (Jury, 1982).

Disease

Sporidesmin, given orally to sheep or cattle at rates comparable to those taken in with toxic pasture, inflames the epithelium of the bile duct. This leads to a chain of events taking 10–20 days: obstruction of the duct, retention of bile, liver damage, spill of phylloerythrin into circulating blood and consequent photosensitization of areas of hairless and unpigmented skin which has given the disease the trivial and misleading name of facial eczema (Mortimer *et al.*, 1978). Oral doses totalling 0.4–0.7 mg sporidesmin/kg, given to sheep over 3 days, cause clinical disease with some mortality. When such doses are divided further and given over 48 days the effects are similar. Doses too small to produce clinical disease cause weight losses and also potentiate the effects of any later doses (Mortimer *et al.*, 1978). Some individual animals are genetically less susceptible to the toxin than others (Jury, 1982) but in all the effects are cumulative and exposure does not confer immunity.

Facial eczema is limited to warm temperate zones where stock are fed exclusively on pasture and particularly where they are grazed intensively, the most efficient use of pasture meaning maximum access to its base where most spores are produced. In New Zealand the high-risk areas are the lowlands of the North Island, the high country and most of the South Island being too cool. For the same reason outbreaks occur in Victoria, New South Wales and Western Australia but not in Tasmania (Hore, 1960; Edwards *et al.*, 1983). Low rainfall restricts its incidence in South Africa to one relatively wet district (Marasas *et al.*, 1972). It has been reported from some South American countries (Andrade *et al.*, 1978; Regalado, 1981). It is only possible to suggest that facial eczema is unknown in tropical countries such as Fiji because temperatures there are higher than is favourable for sporidesmin production. Its absence from temperate parts of the Northern Hemisphere may be the consequence of a combination of continental rather than oceanic climates, and of farming systems based on cropping or extensive grazing. The role played by non-sporidesmin producing strains of *P. chartarum* can only be speculated upon.

In New Zealand control measures for facial eczema are based on detection of dangerous pasture by spore counts made by the farmers themselves. Putting a figure to a dangerous spore count becomes more difficult as stocking rates rise and evidence mounts for the cumulative effects of small doses of sporidesmin. It has been possible to avoid outbreaks, although not cases, of clinical disease by treating pastures as toxic when counts reach 100,000 spores/g, wet weight, of pasture leaves but it is obvious that this figure has a large inbuilt safety factor because spore production is seasonal and intermittent and sporidesmin is more ephemeral than spores. It may become necessary to lower the dangerous spore count with each succeeding spore rise of a season.

When toxic pasture is detected, control measures are to avoid using it, to reduce existing *P. chartarum* populations by applying fungicides (Parle and di Menna, 1972), or to give prophylactic doses of zinc to animals grazing it (Smith *et al.*, 1977).

MYROTHECIUM spp.

Growth conditions

Myrothecia appear to prefer conditions similar to those favourable for *P. chartarum* but their growth patterns in pasture are harder to establish. The diagnostic spores of *P. chartarum* can be counted directly in leaf washings or on trap slides but *Myrothecium* spores are unicellular and impossible to distinguish from those of many other groups. Estimation of populations is made tediously by quantitative culture of leaf washings and identification of isolates recovered. In New Zealand such counts have been higher in summer and autumn than in winter and spring, in irrigated pasture than in unirrigated and in material from North Island pastures than from the cooler South Island (di Menna *et al.*, 1973). Only four species were isolated, the three toxin producers together with the non-toxic *Myrothecium cinctum* (Corda) Sacc.

Toxin production

The characterized *Myrothecium* toxins are trichothecenes but highly toxic non-crystalline compounds of unknown structure are also formed by all three toxigenic species (White *et al.*, 1977). *In vitro* toxin production varies with strain; most New Zealand isolates of *M. verrucaria*, some of *M. roridum* and few of *M. leucotrichum* produced detectable mammalian toxins but only 12% of the strains recovered from five series of pasture samples were toxigenic (di Menna *et al.*, 1973). Toxin production may not be related to sporulation, for Bowden and Schantz (1955) enhanced yields with forced aeration or stirring, which usually inhibits spore production.

Disease

The purified *Myrothecium* toxins, roridin A and verrucarin A, caused death of sheep in 2 days from haemorrhagic gastroenteritis when dosed orally at a rate of 2 mg/kg (Mortimer *et al.*, 1971). The least lethal dose for sheep of broth culture of the most toxic New Zealand strain was 0.5 mg/kg (di Menna and Mortimer, 1971). Toxic cultures given in sub-lethal doses caused weight losses and reduction in numbers of circulating leucocytes (di Menna and Mortimer, 1971). All *Myrothecium* toxins cause severe contact dermatitis.

Only one instance of naturally occurring myrotheciotoxicosis has been reported, in sheep on rye stubble in the Soviet Union (Vertinskii *et al.*, 1967). Di Menna and Mortimer (1971) thought that *Myrothecium* toxins from pasture could be responsible for outbreaks of acute disease with high mortality within 48 h, as in kikuyu poisoning, and for subclinical disease 'indistinguishable from that part of the ill-thrift syndrome remaining when causes such as parasitism and trace element deficiencies have been eliminated'. However, examination of kikuyu grass (*Pennisetum clandestinum*) pastures associated with sudden death of grazing cattle in New Zealand (di Menna *et al.*, 1973), Australia (Gabbedy *et al.*, 1974) and Natal (Bryson and Newsholme, 1978) yielded some *Myrothecium* isolates but insufficient to account for the toxicity of the pastures. *Myrothecium* levels in the New Zealand material, for instance, were no higher than those in non-toxic mixed-species pastures and only one toxic strain was isolated from 23 kikuyu grass samples.

Sheep and cattle in New Zealand often fail to make expected weight gains in summer and autumn but *P. chartarum* is common at this time as well as myrothecia, and it was not possible to separate any effects of the two groups (di Menna and Wyatt, 1975).

Failure to associate myrothecia with disease in pasture-fed animals may be because toxic strains are in the minority and stock become habituated to small doses of *Myrothecium* toxins. Pasture may not be the favoured habitat of the toxigenic species. The instance of myrotheciotoxicosis cited was in sheep on rye stubble. *Dendrodochium toxicum* Pidopl. & Bil., reported to grow on straw, is likely to be a *Myrothecium* sp. for it is green-spored and produces roridin A and verrucarin A (Tulloch, 1972; White *et al.*, 1977). Regarded as the most toxic fungus known, it has caused deaths of horses and pigs in the Soviet Union (Sarkisov, 1954).

Nevertheless, myrothecia in pasture should not be dismissed as innocuous. The distribution of diseases is often related to the distribution of researchers, the areas covered by mycologists are restricted and it is usual to arrive at the sites of outbreaks too late. Workers looking for the cause of facial eczema were badly handicapped by the 2–3-week interval between intake of toxic pasture and appearance of external symptoms in affected animals. Farming practices change, background information accumulates, technology advances and problems can rarely be neatly filed as finally solved or totally insoluble.

REFERENCES

Andrade, S. O., Lopes, H. O., da S., Barros, M. de A., Leite, G. G., Dias, S. M. C., Saueressig, M., Nobre, D., and Temperini, J. A. (1978). 'Aspectos da fotossensibilizacao em bovinos em pastagens de *Brachiaria decumbens* Stapf infestadas por *Pithomyces chartarum* (Berk. & Curt.) M. B. Ellis', *Arq. Inst. Biol.*, **45**, 117–136.

Bowden, J. P., and Schantz, E. J. (1955). 'The isolation and characterization of dermatitic compounds produced by *Myrothecium verrucaria*', *J. Biol. Chem.*, **214**, 365–372.

Brook, P. J. (1963). 'Ecology of the fungus *Pithomyces chartarum* (Berk. & Curt.) M. B. Ellis in pasture in relation to facial eczema disease of sheep', *N.Z. J. Agric. Res.*, **6**, 147–228.

Bryson, R. W., and Newsholme, S. J. (1978). 'Kikuyu grass poisoning of cattle in Natal', *J. S. Afr. Vet. Assoc.*, **49**, 19–21.

Clare, N. T., and Mortimer, P. H. (1964). 'The effect of mercury arc radiation and sunlight on the toxicity of water solutions of sporidesmin to rabbits', *N.Z. J. Agric. Res.*, **7**, 258–263.

Davison, S., and Marbrook, J. (1965). 'The effect of temperature on the toxicity of spores of *Pithomyces chartarum* (Berk. & Curt.) M. B. Ellis', *N.Z. J. Agric. Res.*, **8**, 126–130.

Edwards, J. R., Richards, R. B., Love, R. A., Harrison, M. R., and Gwynn, R. V. R. (1983). 'An abattoir survey of the prevalence of facial eczema of sheep in Western Austraiia', *Aust. Vet. J..*, **60**, 157.

Fitton, M., and Holliday, P. (1970). '*Myrothecium roridum*', *C.M.I. Desc. Path. Fungi Bact.*, No. 253.

Gabbedy, B. J., Gwynn, R., Hopkinson, W. I., Kay, B. E., and Wood, P. McR. (1974). 'Kikuyu poisoning of cattle in Western Australia', *Aust. Vet. J..*, **50**, 369.

Halder, C. A., Taber, R. A., and Camp, B. J. (1981). 'Absence of sporidesmin production by twelve Texas isolates of *Pithomyces* spp.', *Appl. Env. Microbiol.*, **41**, 212–215.

Hore, D. E. (1960). 'Facial eczema', *Aust. Vet. J.*, **36**, 172–176.

Jury, K. E. (1982). 'Developments in facial eczema', *Proc. Ruakura Farm. Conf.*, 65–69.

Marasas, W. F. O., Adelaar, T. F., Kellerman, T. S., Minne, J. A., van Rensburg, I. B. J., and Burroughs, G. W. (1972). 'First report of facial eczema in sheep in South Africa', *Onderstepoort J. Vet. Res.*, **39**, 107–112.

di Menna, M. E., and Mortimer, P. H. (1971). 'Experimental myrotheciotoxicosis in sheep and calves', *N.Z. Vet. J.*, **19**, 246–248.

di Menna, M. E., and Wyatt, P. A. (1975). 'Examination of animal fodders for toxin-producing fungi', *Ann. Rep. Res. Div. N.Z. Min. Agric. Fish.* 118–119.

di Menna, M. E., Campbell, J., and Mortimer, P. H. (1970). 'Sporidesmin production and sporulation in *Pithomyces chartarum*', *J. Gen. Microbiol.*, **61**, 87–96.

di Menna, M. E., Mortimer, P. H., Smith, B. L., and Tulloch, M. (1973). 'The incidence of the genus *Myrothecium* in New Zealand pastures and its relation to animal disease', *J. Gen. Microbiol.*, **79**, 81–87.

Mitchell, K. J., Walshe, T. O., and Robertson, N. G. (1959). 'Weather conditions associated with outbreaks of facial eczema', *N.Z. J. Agric. Res.*, **2**, 584–604.

Mortimer, P. H., Campbell, J., di Menna, M. E., and White, E. P. (1971). 'Experimental myrotheciotoxicosis and poisoning in ruminants by verrucarin A and roridin A', *Res. Vet. Sci.*, **12**, 508–515.

Mortim,er P. H., White, E. P., and di Menna, M. E. (1978). 'Pithomycotoxicosis "facial eczema" in sheep', in *Mycotoxic Fungi, Mycotoxins, Mycotoxicoses*, Vol. 2 (eds T. D. Wyllie and L. G. Morehouse), pp. 195–203. Marcel Dekker, New York and Basel.

Parle, J. N., and di Menna, M. E. (1972). 'Fungicides and the control of *Pithomyces chartarum*. II. Field trials', *N.Z. J. Agric. Res.*, **15**, 54–63.

Regalado, A. M. (1981). 'Intoxication por *Pithomyces chartarum* en bovinos en el partido de Daireaux, Buenos Aires', *Gaceta Vet.*, **43**, 471–473.

Ross, D. J., and Thornton, R. H. (1962). 'A study of the physiology of *Pithomyces chartarum* (Berk. & Curt.) M. B. Ellis. 3. Production of toxin on some laboratory media', *N.Z. J. Sci.*, **5**, 165–183.

Sarkisov, A. H. (1954). *Mycotoxicoses (Fungal Poisonings)*, Publisher of Agricultural Literature, Moscow.

Smith, B. L., Embling P. P., Towers, N. R., Wright, D. E., and Payne, E. (1977). 'The protective effect of zinc sulphate in experimental sporidesmin poisoning of sheep', *N.Z. Vet. J.*, **25**, 124–127.

Tulloch, M. (1972). *The Genus* Myrothecium *Tode ex Fr.* C.M.I., Kew.

Vertinskii, K. I., Dzhilavyan, Kh. A., and Koroleva, V. P. (1967). 'A disease of sheep caused by the fungus *Myrothecium verrucaria', Byull. Vses. Inst. Eksp. Vet.*, No. 2, 86–90.

White, E. P., Mortimer, P. H., and di Menna, M. E. (1977). 'Chemistry of the *Myrothecium* toxins', in *Mycotoxic Fungi, Mycotoxins, Mycotoxicoses*, Vol. 1, (eds T. D. Wyllie and L. G. Morehouse), pp. 465–469. Marcel Dekker, New York and Basel.

Trichothecenes and Other Mycotoxins
Edited by J. Lacey

CHAPTER 12

The endophyte of tall fescue (*Festuca arundinacea* Schreb.)

M. C. Johnson, and M. R. Siegel and L. P. Bush

INTRODUCTION

Tall fescue (*Festuca arundinacea* Schreb.) has often been associated with poor cattle performance, despite evidence that the grass has a chemical composition indicative of a high-quality forage (Bush *et al.*, 1979; Miles *et al.*, 1964; Phillips *et al.*, 1954; Sullivan *et al.*, 1956). This poor animal performance (fescue toxicity syndrome) is characterized by reduced weight gain, decreased milk production, lower feed intake, rough haircoat, increased respiration rate, and high rectal temperature, and primarily occurs during the summer months. Other animal disorders associated with tall fescue and possibly related are fescue foot (Garner and Cornell, 1978) and fat necrosis (Stuedemann *et al.*, 1975) in cattle, and agalactia in horses (Garrett *et al.*, 1980).

Over the past 20 years several fungal pathogens of tall fescue have been considered as being involved in fescue toxicity syndrome. These include species of *Aspergillus* (Futrell *et al.*, 1974), *Claviceps* (Jacobson *et al.*, 1963; Riggs, 1966), *Helminthosporium* (Flores, 1966), and *Rhizoctonia* (Gentry, 1968). None of these, however, has been successfully implicated in the syndrome. More recently, Bacon *et al.* (1977) demonstrated that tall fescue, in pastures on which cattle showed signs of the disorder, was infected with an endophytic fungus. This association between poor animal performance and the presence of a fungal endophyte has been corroborated by additional lines of investigation (Jones, 1981; Schmidt *et al.*, 1982). The endophyte was identified as *Epichloë typhina* (Pers.) Tul. (asexual state = *Sphacelia typhina* Sacc.) (Bacon *et al.*, 1977), but is also referred to as *Acremonium coenophialum* Morgan-Jones & Gams (Morgan-Jones and Gams, 1982). Transmission of the endophyte in nature appears to be limited to seed produced on infected plants (Bacon *et al.*, 1977; Sampson, 1933).

The role of the endophyte in fescue toxicity syndrome may be direct, or indirect through manipulation of the toxic chemistry of the host plant. *E. typhina* produces several ergot alkaloids *in vitro*, including agroclavine, chanoclavine, elymoclavine, festuclavine, penniclavine, ergovaline, and ergovalinine (Porter *et al.*, 1979, 1981). It has not been determined if these alkaloids are produced *in vivo* in endophyte-infected tall fescue (Porter *et al.*, 1981).

The presence of the endophyte in tall fescue plants has been closely linked to pyrrolizidine alkaloid accumulation in seed and forage (Bush *et al.*, 1982). It has been postulated that an interaction between the endophyte and plant tissue is necessary for stimulation of pyrrolizidine alkaloid synthesis. As is the case with the ergot-type alkaloids, it has not been established if the pyrrolizidine alkaloids contribute to the animal symptoms of fescue toxicity syndrome.

The recognition that many of the outward animal symptoms are related to ambient temperature (Hemken *et al.*, 1981) has led to the use of high-temperature rooms to study the syndrome (Boling *et al.*, 1983). In addition, the finding that serum prolactin levels are depressed (Hurley *et al.*, 1981; Bolt *et al.*, 1982) after only 3 days of feeding endophyte-infected tall fescue (Jackson *et al.*, 1982) provides a technique to study the toxic entity (or entities). By monitoring prolactin secretions attempts can be made to isolate and identify toxin(s) that may be present in various extracts of endophyte-infected fescue.

DETECTION OF THE ENDOPHYTE

Because endophyte-infected tall fescue plants exhibit no external symptoms, detection of the fungus in tall fescue has often proved laborious and challenging. Several methods have been used, however, to ascertain whether plants and/or seed contain the endophyte. These include microscopic examination (Neill, 1940; Bacon *et al.*, 1977), isolation in callus culture (Conger and McDaniel, 1983), pyrrolizidine alkaloid analysis (Bush *et al.*, 1982; Robbins *et al.*, 1972), and enzyme-linked immunosorbent assay (ELISA) (Johnson *et al.*, 1982, 1983).

Our own work has been greatly facilitated by the use of ELISA. The technique is a serological assay that has a high degree of specificity and sensitivity. With ELISA we could detect as little as 100 ng of freeze-dried *E. typhina* mycelium per millilitre (Johnson *et al.*, 1982). Furthermore, antigens of the fungus were detected in samples consisting of only one endophyte-infected seed extracted together with as many as 19 endophyte-free seeds (Johnson *et al.*, 1983).

Viable seed containing dead endophyte still tested positive in our ELISA system (Johnson *et al.*, 1983). (Death of the endophyte in the seed is most likely due to the temperatures infected seed is subjected to and/or length of time in storage.) Thus, ELISA testing of seed directly often indicates a higher endophyte content than the testing of the same seed indirectly (by germinating seeds and assaying the resultant seedlings 4–8 weeks later with ELISA) (Table 1).

Table 1 shows results of ELISA testing and pyrrolizidine alkaloid analysis for four (A, B, C, and D) tall fescue seed lots. All four lots were tested with ELISA on an individual seed basis. In addition, seed from each lot was sown in 7.5 cm plastic containers in the greenhouse, seedlings thinned to one per container, and all containers numbered. Either 25 or 30 seedlings were assayed from each lot at 8 weeks after seeding, allowed to regenerate, and then assayed again at 20 weeks. As determined by ELISA, the same seedlings that were infected at 8 weeks were infected at 20 weeks. Seedlings in which antigens of the endophyte were not detected at 8 weeks were the same seedlings that appeared to be free of the endophyte at 20 weeks. Alkaloid determinations at 8 weeks were made on pooled forage samples from all seedlings of each seed lot. The concentration of alkaloids at 20 weeks were determined in eight individual seedlings from each of lots A–C. (Four of the seedlings were endophyte-free, whereas the other four were endophyte-infected as indicated by ELISA). Four endophyte-free seedlings from lot D were also analysed to determine alkaloid content. Alkaloids, *N*-acetyl and *N*-formyl loline, were detected only in those samples known to contain seedlings that tested positive for endophyte presence using ELISA.

Table 1 Enzyme-linked immunosorbent assay (ELISA) and loline analysis of tall fescue seeds and seedlings

	ELISA test result[a]			*N*-acetyl + *N*-formyl loline (seedlings)[b] (μg/kg)		
		Seedlings			20 weeks	
Lot	Seed	8 weeks	20 weeks	8 weeks	+	−
A	55/55[c] (100%)	26/30 (87%)	26/30 (87%)	494	1033	ND[d]
B	63/65 (97%)	10/25 (40%)	10/25 (40%)	118	1307	ND
C	44/60 (73%)	12/30 (40%)	12/30 (40%)	138	1482	ND
D	0/30 (0%)	0/30 (0%)	—[e]	ND		ND

[a] Seeds and seedlings ELISA tested on an individual basis.
[b] Loline determinations at 8 weeks of pooled forage samples from all seedlings of each lot. Loline determinations at 20 weeks represent means of four endophyte-infected (+) seedlings or four endophyte-free (−) seedlings as identified by ELISA.
[c] Number of positive samples/number tested.
[d] ND = Not detected.
[e] Not tested.

CONTROL OF THE ENDOPHYTE

Attempts to eradicate the endophyte from established tall fescue pastures with the use of fungicides have not been successful (Williams and Backman, 1983; Siegel *et al.*, 1983). However, a temporary suppression of growth of the endophyte with an apparent alleviation in animal symptoms was obtained on pastures following treatment with sterol inhibitor fungicides (Williams and Backman, 1983). Elimination of the endophyte from mature tall fescue plants under greenhouse conditions has been accomplished with repeated foliar applications of fungicides, i.e. triadimefon, propiconazole (Williams and Backman, 1983), or with the fungicide, bitertanol, incorporated into the soil (Siegel *et al.*, 1983).

Because infected seed is the primary, if not the only, means of spread of the endophyte, relatively clean pastures may be established by planting seed that will produce a low percentage of infected plants. This requires certifying that a given seed source is relatively free of the endophyte initially, or treating infected seed to kill the endophyte.

A two-stage sampling procedure was developed for incorporation into a certification programme that tests for endophyte content in tall fescue seed lots by means of ELISA (Johnson *et al.*, 1983). This sampling procedure had a high probability ($\alpha = 0.97$) of accepting lots with an endophyte content of 5% (acceptable level for certification), and a low probability ($\alpha = 0.02$) of accepting lots with an endophyte content of 10% (unacceptable level for certification).

The three methods that have been tried to kill the endophyte in tall fescue seed and still retain acceptable levels of seed germination and seedling vigour are chemical treatments (fungicides), i.e. triadimefon, triadimenol (Williams *et al.*, 1982; Siegel *et al.*, 1983), therapeutic treatments (high temperature) (Williams and Backman, 1983), and long-term storage (Neill, 1941).

An experiment was conducted to determine the longevity of the endophyte in tall fescue seed under several different storage conditions. A 100% endophyte-infected seed lot was identified using ELISA. The seed was divided into five sublots, and each sublot placed in a specific storage condition. The five storage conditions were: freezer (−20 °C), refrigerator (6 °C), germplasm maintenance facility (GPM) (10 °C), room temperature (21 °C), and the seed house which ranged from 21 to 35 °C. Beginning after 3 months storage and then at 4-month intervals, seed from each storage environment was monitored for germination and percentage of seedlings infected with the endophyte as determined by ELISA. After only 11 months of storage, seed maintained at room temperature and in the seed house produced endophyte-free seedlings (Table 2). In contrast, seed stored for 15 months in the freezer gave rise to only endophyte-infected seedlings (Table 2). There was no appreciable decline in percentage germination under any of the storage conditions during the 15-month period.

Table 2 Percentage of tall fescue seed[a] containing viable endophyte as related to storage environment and duration of storage

Storage environment	Temperature (°C)	Months in storage 3	7	11	15
Freezer	−20	100[b]	100	100	100
Refrigerator	6	100	90	85	90
GPM	10	90	100	80	75
Room temperature	21	95	55	0	0
Seed house	21–35	95	60	0	0

[a] Moisture content of seed was determined to be 9.9%.
[b] Percentage of viable endophyte was determined indirectly (grow-out test). Each number represents the ELISA testing of 20 seedlings.

RELATIONSHIP TO PERENNIAL RYEGRASS ENDOPHYTE

Similarities between endophytic fungi present in *Festuca* species and *Lolium* species were noted as long as 50 years ago (Peyronel, 1930; Sampson, 1933, 1937; Neill, 1941). In 1981 Fletcher and Harvey in New Zealand associated a nervous disease of sheep, known as ryegrass staggers, with the presence of an endophytic fungus in perennial ryegrass (*Lolium perenne* L.) plants from staggers-producing pastures. This endophyte had previously been discounted as being involved in ryegrass staggers based partly on a report that fungal cultures and endophyte-infected perennial ryegrass seed were non-toxic when fed to birds, mice, rats and sheep (Cunningham, 1958).

Perennial ryegrass seed from a cultivar considered to have a high level of endophyte (Ellett RG 8201) and a cultivar considered to have no endophyte (Grasslands Ruanui) were supplied in February 1982 by Dr I. C. Harvey, Lincoln, New Zealand, and Dr G. C. M. Latch, Palmerston North, New Zealand. Seed from each cultivar was tested on a single-seed basis in our ELISA system (antiserum prepared to a Kentucky isolate of the endophyte from tall fescue). Additional seed from each cultivar was sown in 7.5 cm plastic containers in the greenhouse and the resultant seedlings tested with ELISA after 11 weeks. Of the Ellett RG 8201 seeds, 14 of 15 elicited high absorbances in ELISA, whereas none of the Ruanui seeds elicited absorbances above background levels (Table 3). The absorbances elicited by the Ellett RG 8201 seeds were at least as high as those elicited by endophyte-infected tall fescue seeds. Vegetative stem tissue from all the 11-week-old seedlings of Ellett RG 8201 elicited high absorbances in the ELISA test and were considered positive. Stem sections from the Ruanui seedlings were all negative in our ELISA system (Table 3). The ease with which antiserum

Table 3 Detection of the perennial ryegrass endophyte by means of ELISA using an antiserum prepared to the tall fescue endophyte

	ELISA test result[a]	
Perennial ryegrass cultivar	Seed	Seedlings
Ellett RG 8201	14/15[b]	15/15
Ruanui	0/15	0/ 5

[a] Seeds and seedlings tested on an individual basis.
[b] Number of positive samples/number tested.

prepared to the tall fescue endophyte was able to detect the perennial ryegrass endophyte in the context of ELISA, suggests that the two endophytes may be identical or closely related. However, in contrast to endophyte-infected tall fescue, endophyte-infected perennial ryegrass seed and forage did not contain *N*-acetyl or *N*-formyl loline.

OBSERVATIONS

The exact nature of the relationship between the endophyte of tall fescue and its plant host has not been determined. Infected plants appear to grow and reproduce in the same manner as non-infected plants. Detrimental effects of endophyte presence on the physiology of the tall fescue plant have not been observed.

From the perspective of the endophyte, the host plant provides nutrients and a well-insulated home. Its apparent reliance on seed transmission for dissemination renders it less dependent on exacting environmental conditions, i.e. wind, rain, and long periods of free moisture, usually necessary for spread of fungal spores and the infection process.

From the perspective of the host plant there is some evidence that endophyte infection may even be of some benefit. Because reduced intake is one of the animal symptoms of the syndrome, especially during hot weather, there is less of a tendency for endophyte-infected tall fescue pastures to be overgrazed. Small mammals have also been shown to consume considerably less endophyte-infected tall fescue seed as compared to endophyte-free seed (Neal and Schmidt, 1982). Possibly the most significant factor as regards any selective advantage endophyte-infected tall fescue might have over endophyte-free tall fescue would be that of enhanced insect resistance as found in perennial ryegrass cultivars with high levels of endophyte (Funk *et al.*, 1983; Prestidge *et al.*, 1982; Mortimer and di Menna, chapter 15 in this volume).

Thus the nature of the relationship between tall fescue and its endophyte may indeed be a mutualistic one, or at the least a commensal one in which tall fescue is neither harmed nor helped by the endophyte.

REFERENCES

Bacon, C. W., Porter, J. K., Robbins, J. D., and Luttrell, E. S. (1977). '*Epichloë typhina* from toxic tall fescue grasses', *Appl. Environ. Microbiol.*, **34**, 576–581.

Boling, J. A., Hemken, R. W., Bush, L. P., Buckner, R. C., Jackson, J. A., Jr., and Yates, S. C. (1983). 'Role of alkaloids and toxic compound(s) in the utilization of tall fescue by ruminants', in *Proc. 14th Int. Grassl. Congr. Lexington, Ky. 15–24 June 1981* (eds J. A. Smith and V. W. Hays), pp. 722–725. Westview Press, Boulder, Colo.

Bolt, D. J., Bond, J., Lynch, G. P., and Elsasser, T. (1982). 'Concentrations of PRL, LH, FSH, GH and TSH in plasma and pituitary of ewes grazing tall fescue and orchardgrass pastures', *J. Anim. Sci.*, **55**, (Suppl. 1), 4.

Bush, L. P., Boling, J. A., and Yates, S. G. (1979). 'Animal disorders', in *Tall Fescue Monograph* (eds R. C. Buckner and L. P. Bush) pp. 247–292. Amer. Agron. Soc., Madison, Wisconsin.

Bush, L. P., Cornelius, P. L., Buckner, R. C., Varney, D. R., Chapman, R. A., Burrus, P. B., Kennedy, C. W., Jones, T. A., and Saunders, M. J. (1982). 'Association of *N*-acetyl loline and *N*-formyl loline with *Epichloë typhina* in tall fescue', *Crop Sci.*, **22**, 941–943.

Conger, B. V., and McDaniel, J. K. (1983). 'Use of callus cultures to screen tall fescue samples for *Acremonium coenophialum*', *Crop Sci.*, **23**, 172–174.

Cunningham, I. J. (1958). 'Non-toxicity to animals of ryegrass endophyte and other endophytic fungi of New Zealand grasses', *N.Z. J. Agric. Res.*, **1**, 489–497.

Fletcher, L. R., and Harvey, I. C. (1981). 'An association of a *Lolium* endophyte with ryegrass staggers', *N.Z. Vet. J.*, **29**, 185–186.

Flores, J. M. (1966). '*Helminthosporium dictyoides* Drechs and *H. vagans* Drechs: sporulation, infectivity, and effect of alkaloid development on tall fescue (*Festuca arundinacea* Schreb.)', M.S. thesis, Univ. Kentucky.

Funk, C. R., Halisky, P. M., Johnson, M. C., Siegel, M. R., Stewart, A. V., Ahmad, S., Hurley, R. H., and Harvey, I. C. (1983). 'An endophytic fungus and resistance to sod webworms: association in *Lolium perenne* L.', *Bio/Technology*, **1**, 189–191.

Futrell, M. C., Farnell, D. R., Poe, W. E., Watson, V. H., and Coats, R. E. (1974). 'Fungal populations in the rumen associated with fescue toxicosis', *J. Environ. Quality*, **3**, 140–143.

Garner, G. B. and Cornell, C. N. (1978). 'Fescue foot in cattle', in *Mycotoxic Fungi, Mycotoxins, Mycotoxicosis: An Encyclopedic Handbook*, Vol. 2 (eds T. D. Wyllie and L. G. Morehouse), pp. 45–62. Marcel Dekker, New York.

Garrett, L. W., Heiman, E. D., Pfander, W. H., and Wilson, L. L. (1980). 'Reproductive problems of pregnant mares grazing fescue pastures', *J. Anim. Sci.*, **51** (Suppl. 1), 237.

Gentry, C. E. (1968). 'Interrelationship of *Rhizoctonia solani* Kuhn, environment, and genotype on the alkaloid content of tall fescue, *Festuca arundinacea* Schreb.', Ph.D. thesis, Univ. Kentucky. Univ. Microfilms, Ann Arbor, MI (*Diss. Abstr.*, **30**, 1446B).

Hemken, R. W., Boling, J. A., Bull, L. S., Hatton, R. H., Buckner, R. C., and Bush, L. P. (1981). 'Interaction of environmental temperature and anti-quality factors on the severity of summer fescue toxicosis', *J. Anim. Sci.*, **52**, 710–714.

Hurley, W. L., Convey, E. M., Leung, K., Edgerton, L. A., and Hemken, R. W. (1981). 'Bovine prolactin, TSH, T_4 and T_3 concentrations as affected by tall fescue summer toxicosis and temperature', *J. Anim. Sci.*, **51**, 374–379.

Jackson, J. A., Jr., Hemken, R. W., Boling, J. A., Harmon, R. J., Buckner, R. C., and Bush, L. P. (1982). 'Effect of toxic tall fescue on prolactin levels in dairy cattle', *J. Dairy Sci.*, **65** (Suppl. 1), 172.

Jacobson, D. R., Miller, W. M., Seath, D. M., Yates, S. G., Tookey, H. L., and Wolff, I. A. (1963). 'Nature of fescue toxicity and progress toward identification of the toxic entity'. *J. Dairy Sci.*, **46**, 416–422.

Johnson, M. C., Anderson, R. L., Kryscio, R. J., and Siegel, M. R. (1983). 'Sampling procedures for determining endophyte content in tall fescue seed lots by ELISA', *Phytopathology*, **73**, 1406–1409.

Johnson, M. C., Pirone, T. P., Siegel, M. R., and Varney, D. R. (1982). 'Detection of *Epichloë typhina* in tall fescue by means of enzyme-linked immunosorbent assay', *Phytopathology*, **72**, 647–650.

Jones, T. A. (1981). 'The association of an endophytic fungus *Sphacelia typhina* (Pers.) Sacc. with the production of pyrrolizidine alkaloids in tall fescue', M.S. thesis, Univ. Kentucky.

Miles, D. G., Griffith, G., and Walters, R. J. K. (1964). 'Variation in the chemical composition of four grasses', *Annu. Rep. Welsh Plant Breeding Stn., 1963* (Aberystwyth), pp. 110–114.

Morgan-Jones, G., and Gams, W. (1982). 'Notes on hyphomycetes. XLI. An endophyte of *Festuca arundinacea* and the anamorph of *Epichloë typhina*, new taxa in one of two new sections of *Acremonium*', *Mycotaxon*, **15**, 311–318.

Neal, W. D., and Schmidt, S. P. (1982). 'Effect of the "fescue toxicity" fungus on the growth of weanling rats', *J. Anim. Sci.*, **55** (Suppl. 1), 317.

Neill, J. C. (1940). 'The endophyte of rye-grass (*Lolium perenne*)', *N.Z. J. Sci. Technol.*, Sect. A 21, 280–291.

Neill, J. C. (1941). 'The endophytes of *Lolium* and *Festuca*', *N.Z. J. Sci. Technol.*, Sect. A 23, 185–193.

Peyronel, B. (1930). 'Simbiosi fungina tipo "*Lolium*" in alcune graminacee del genre "*Festuca*"', *Nuovo Giorn. Bot. Ital.*, N.S. **37**, 643–648.

Phillips, T. G., Sullivan, J. T., Loughlin, M. E., and Sprague, V. G. (1954). 'Chemical composition of some forage grasses I. Changes with plant maturity', *Agron. J.*., **46**, 361–369.

Porter, J. K., Bacon, C. W., and Robbins, J. D. (1979). 'Ergosine, ergosinine, and chanoclavine I from *Epichloë typhina*', *J. Agric. Food Chem.*, **27**, 595–598.

Porter, J. K., Bacon, C. W., Robbins, J. D., and Betowski, D. (1981). 'Ergot alkaloid identification in clavicipitaceae systemic fungi of pasture grasses', *J. Agric. Food Chem.*, **29**, 653–657.

Prestidge, R. A., Pottinger, R. P., and Barker, G. M. (1982). 'An association of *Lolium* endophyte with ryegrass resistance to Argentine stem weevil', *Proc. 35th New Zealand Weed and Pest Control Conference*, pp. 119–122.

Riggs, R. K. (1966). 'The interrelationship of tall fescue and *Claviceps purpurea* upon the development of alkaloids in each', Ph.D. thesis, Univ. Kentucky. Univ. Microfilms, Ann Arbor, MI (*Diss. Abstr.*., **31**, 1643B).

Robbins, J. D., Sweeny, J. G., Wilkinson, S. R., and Burdick, D. (1972). 'Volatile alkaloids of Kentucky 31 tall fescue seed (*Festuca arundinacea* Schreb.)', *J. Agric. Food Chem.*, **20**, 1040–1043.

Sampson, K. (1933). 'The systemic infection of grasses by *Epichloë typhina* (Pers.) Tul.', *Trans. Brit. Mycol. Soc.*, **18**, 30–47.

Sampson, K. (1937). 'Further observations on the systemic infection of *Lolium*', *Trans. Brit. Mycol. Soc.*, **21**, 84–97.

Schmidt, S. P., Hoveland, C. S., Clark, E. M., Davis, N. D., Smith, L. A., Grimes, H. W., and Holliman, J. L. (1982). 'Association of an endophytic fungus with fescue toxicity in steers fed Kentucky 31 tall fescue seed or hay', *J. Anim. Sci.*, **55**, 1259–1263.

Siegel, M. R., Varney, D., Johnson, M., and Nesmith, W. (1983). 'Tall fescue endophyte: distribution in Kentucky and chemical control', *Phytopathology*, **73**, 506.

Stuedemann, J. A., Wilkinson, S. R., Williams, D. J., Ciordia, H., Ernst, J. W., Jackson, W. A., and Jones, J. B., Jr. (1975). 'Long-term broiler litter fertilization of tall fescue pastures and health and performance of beef cows', in *Managing livestock wastes. Proc. 3rd Int. Symp. on Livestock Wastes*, pp. 264–268. Am. Soc. Agric. Eng., St. Joseph, Mich.

Sullivan, J. T., Phillips, T. G., Loughlin, M. E., and Sprague, V. G. (1956). 'Chemical composition of some forage grasses. II. Successive cuttings during the growing season', *Agron. J..*, 48, 11–14.

Williams, M. J., and Backman, P. A. (1983). 'Control of the fungal endophyte *Acremonium coenophialum* in seed and established plants of tall fescue', in *Proceedings Tall Fescue Toxicosis Workshop, March 17–18, 1983 Atlanta, GA*, pp. 52–58. Cooperative Extension Service, Univ. of Georgia, Athens.

Williams, M. J., Backman, P. A., and Crawford, M. A. (1982). 'Fungicidal control of a fungal endophyte in seed and established plants of tall fescue', *Phytopathology*, **72**, 971.

Trichothecenes and Other Mycotoxins
Edited by J. Lacey

CHAPTER 13

Endophytes and ryegrass staggers

G. C. M. Latch

INTRODUCTION

Animals which graze ryegrass dominant pastures in New Zealand sometimes suffer from a neuromuscular disorder known as ryegrass staggers (Cunningham and Hartley, 1959). The disorder usually occurs in summer and early autumn when pastures are drought-stressed and the weather is hot. Severe symptoms often occur several days after light rain or heavy dews. Affected animals appear normal until they are disturbed or chased, whereupon they run with a stiff gait, stagger, fall and suffer muscular spasms. In mild cases they will shortly rise to their feet again; in very severe cases they remain down until they die. However, deaths directly caused by ryegrass staggers are not common and the most important effect of the disorder to farmers is that animals gain weight more slowly, or even lose weight, and cannot be readily moved to non-toxic pastures.

CAUSE OF TOXICITY

Neill (1939) appears to be the first to suggest that a fungus could be responsible for the toxicity in ryegrass. Neill (1940) described a seedborne endophytic fungus which was present in the leaves of many perennial ryegrass (*Lolium perenne*) plants in New Zealand. However, because of its general distribution in certified ryegrass at all stages of growth, he now thought that it was unlikely to cause stock troubles associated with particular seasons and localities. Subsequently, Neill (1941) reported the absence of effects after feeding mice, rats, chickens and one sheep with endophyte-infected seed. Cultures of the fungus grown on ryecorn, fed to mice, also proved to be non-toxic (Cunningham, 1958). Studies to link other fungi in pastures with the disorder have been fruitless (Latch *et al.*, 1976; di Menna *et al.*, 1976; Gallagher *et al.*, 1977).

The association of an endophytic fungus in tall fescue (*Festuca arundinacea*) with the incidence of fescue toxicity was reported by Bacon *et al.*

(1977). Fletcher and Harvey (1981) have now shown that a similar type of association occurs with ryegrass staggers and endophyte-infected ryegrass (see also Harvey, chapter 14 in this volume). In a replicated grazing trial involving three different cultivars of ryegrass, severe ryegrass staggers developed in sheep grazing one of the cultivars, mild symptoms occurred in sheep on the second cultivar, and none developed in those sheep on the third cultivar. When the three ryegrasses were examined for endophyte mycelium there was an excellent correlation between the percentage of plants in the sward infected with fungus and the incidence and severity of ryegrass staggers. This correlation was confirmed in a subsequent trial by Mortimer *et al.* (1982).

Two neurotoxins have been found in toxic ryegrass and named lolitrem A and B (Gallagher *et al.*, 1981). When these were fed to mice the onset and duration of the induced tremors was closely allied to that observed in sheep with ryegrass staggers. Two further neurotoxins were also isolated from endophyte-infected ryegrass seed (Gallagher *et al.*, 1982). Fed to sheep, this seed produced symptoms similar to ryegrass staggers. Lolitrems have not been detected in pure cultures of the commonly found endophyte in ryegrass (Gallagher, personal communication). Although proof that endophytes cause ryegrass staggers is not yet irrefutable, the association is very strong, and it seems likely that toxic compounds are produced by the plant in response to the presence of the endophyte mycelium.

RYEGRASS ENDOPHYTES AND THEIR CONTROL

Two endophytic fungi have been found in ryegrasses: species of *Acremonium* and of *Gliocladium* (Latch, 1983). It is likely that *Acremonium* is involved with ryegrass staggers because it is by far the most common, but *Gliocladium* may also be involved. Both fungi are present in all above-ground parts of infected plants including the seed, and *Gliocladium* is also present as mycelium in the roots. Endophyte mycelium is best detected in the leaf sheath where it is most concentrated. If the adaxial epidermis is peeled off and stained with lactophenol cotton blue the intercellular mycelium can be clearly seen. *Gliocladium* produces a profuse quantity of highly branched mycelium, whereas *Acremonium* has relatively few strands of mycelium that run longitudinally in the leaf and have few branches. Musgrave (1984) has shown, with ELISA, that 75% of the total *Acremonium* mycelium is in the lowest 2 cm of the plant from ground level.

Both fungi produce small numbers of conidia in culture but sporulation has not been seen in leaves or leaf sheaths. Hence it is likely that the only way the fungi can spread in nature is through infected seed. If, as seems likely, endophytes are involved with ryegrass staggers, the disorder can be overcome by growing endophyte-free ryegrass. The incidence of endophytes in seed of New Zealand ryegrasses was determined by planting seed of 36 freshly harvested lines of perennial, Italian and hybrid ryegrasses. Microscopic examination of 20 plants from each line 3 weeks after sowing showed that the percentage infection varied greatly, and that a high percentage of seed of many lines of perennial ryegrass was endophyte-infected (Table 1).

If seed of most perennial ryegrass contains endophyte mycelium what is the best way to kill this mycelium without harming seed viability? Neill (1940) found that endophyte mycelium within ryegrass seed died if the seed was stored for 18–24 months at room temperature. Endophyte-free pastures could result from sowing this stored seed. However, seed which has been stored commercially for 2½ years has also been found with viable endophyte, so that storage conditions may be important. Thus Latch (1983) grew perennial ryegrass seed which had been stored for 15 years at 0–5 °C and at low humidity, and found viable endophyte in the seedlings. Endophyte mycelium can also be killed by soaking infected seed in water at 57 °C for 15 min (Latch and Christensen, 1982). However, seedling viability was decreased by almost 50%. Treating seed with fungicides appears the best way of killing mycelium within seed. The systemic fungicides prochloraz or propiconazole, applied at rates of 1 g/kg infected seed, resulted in endophyte-free seedlings (Latch and Christensen, 1982). However, these fungicides are slightly phytotoxic and it would be more satisfactory if seed was harvested from disease-free plants.

Endophyte mycelium can be killed in ryegrass plants by drenching benomyl into the root zone (Latch and Christensen, 1982). Plants were potted into a sand/peat medium and an aqueous suspension of 0.1 g of benomyl per litre of potting medium was poured into the sealed pots. Plants were free of infection within 8 weeks of treatment. Thus, endophyte can be removed from parent plants of ryegrass cultivars and health seed produced from them. The parent

Table 1 Incidence of endophytes in 20 seedlings grown from each of 36 freshly harvested lines of ryegrass seed

Ryegrass type	Cultivar	Number of seed lines tested	Endophyte infection (%) Range	Endophyte infection (%) Average
Perennial	Ellett	7	75–95	87
	Nui	11	15–95	44
	Ruanui	3	0–10	4
	Uncertified	5	45–80	64
Hybrid	Ariki	3	0	0
	Manawa	4	0–5	1
Italian	Paroa	3	0	0

plants of most New Zealand ryegrass cultivars have been examined for endophyte infection and when present the mycelium was killed by treatment with benomyl. Observations on endophyte-free mown plots and experimental swards indicated that infection of healthy plants with endophytes probably does not occur in the field, for they have remained healthy for a period of 5 years. We have sequentially planted endophyte-infected and healthy ryegrasses at 5 cm spacings into field plots so that each healthy ryegrass is surrounded by four infected plants. After 11 months all the original healthy plants were still free of endophyte. Thus, ryegrass staggers may be overcome by establishing endophyte-free pastures.

ENDOPHYTES AND INSECT RESISTANCE

Unfortunately control of ryegrass staggers was far from solved. Mortimer *et al.* (1982) reported that their endophyte-free ryegrass plots were severely attacked by Argentine stem weevil (*Listronotus bonariensis*) and many plants died (see also Mortimer and di Menna, chapter 15 of this volume). It has now been shown (Gaynor *et al.*, 1983) that *Acremonium* endophyte-infected ryegrass produces compounds which deter the weevil from feeding on the plant, whereas infection with *Gliocladium* does not have this effect. Other insects have also been reported as being deterred from feeding on endophyte-infected ryegrass (Funk *et al.*, 1983; Ahmad and Funk, 1983).

CONTROL OF RYEGRASS STAGGERS

The present situation is that New Zealand farmers have the alternatives of establishing endophyte-free pastures and risk losses from weevil attack or establishing high endophyte pastures and risk ryegrass staggers. Breeders are seeking true plant resistance to the Argentine stem weevil. If this is achieved endophyte-free seed will be available to farmers. In the U.S.A. no harmful insects have been found attacking endophyte-free tall fescue, so farmers can plant endophyte-free seed to avoid fescue toxicity problems.

In addition to breeding endophyte-free ryegrass plants with resistance to the weevil it may be possible to breed animals with resistance to ryegrass staggers. One New Zealand farmer whose stock suffer from ryegrass staggers each year has reported that two of his rams were not noticeably affected by the disorder. The majority of their progeny also appeared to be unaffected (Hewett, 1983).

NEW ROLES FOR ENDOPHYTES

Endophytes have been reported from many species of grass (Mühle, 1971; Bacon *et al.*, 1983; Johnson *et al.*, chapter 12 in this volume) and it is now possible to transfer some from one species to another. Thus, protection against attacks by insects and other fungi may be conferred to previously susceptible species. This may be satisfactory for turf grasses but there is the risk in pasture grasses that toxins, harmful to animals, may also be produced by these new grass/endophyte combinations. Clearly any introduction of endophytes into new grass hosts must be done with caution.

REFERENCES

Ahmad, S., and Funk, C. R. (1983). 'Bluegrass billbug tolerance of ryegrass cultivars and selections', *J. Econ. Ent.* **76**, 414–416.

Bacon, C. W., Robbins, J. D., and Porter, J. K. (1983). 'Toxins from fungal endophytes of pasture weed grass species as additional sources of grass toxicities', *Proc. Forage and Turfgrass Endophyte Workshop*, Corvallis Oregon, p. 35.

Bacon, C. W., Porter, J. K., Robbins, J. D., and Luttrell, E. S. (1977). '*Epichloë typhina* from toxic tall fescue grasses', *Appl. Environ. Microbiol.*, **35**, 576–581.

Cunningham, I. J. (1958). 'Non-toxicity to animals of ryegrass endophyte and other endophytic fungi of New Zealand grasses', *N.Z. J. Agric. Res.*, **1**, 489–497.

Cunningham, I. J., and Hartley, W. J. (1959). 'Ryegrass staggers', *N.Z. Vet. J.*, **7**, 1–7.

Fletcher, L. R., and Harvey, I. C. (1981). 'An association of a *Lolium* endophyte with ryegrass staggers', *N.Z. Vet. J.*, **29**, 185–186.

Funk, C. R., Halisky, P. M., Johnson, M. C., Siegel, M. R., Stewart, A. V., Ahmad, S., Hurley, R. H., and Harvey, I. C. (1983). 'An endophytic fungus and resistance to sod webworms: association in *Lolium perenne* L.', *Bio/Technology*, **1**, 189–191.

Gallagher, R. T., Campbell, A. G., Hawkes, A. D., Holland, P. T., McGaveston, D. A., Pansier, E. A., and Harvey, I. C. (1982). 'Ryegrass staggers: the presence of lolitrem neurotoxins in perennial ryegrass seed', *N.Z. Vet. J..*, **30**, 183–184.

Gallagher, R. T., White, E. P., and Mortimer, P. H. (1981). 'Ryegrass staggers: isolation of potent neurotoxins lolitrem A and lolitrem B from staggers-producing pastures', *N.Z. Vet. J.*, **29**, 189–190.

Gallagher, R. T., Keogh, R. G., Latch, G. C. M., and Reid, C. S. W. (1977). 'The role of fungal tremorgens in ryegrass staggers', *N.Z. J. Agric. Res.* **20**, 431–440.

Gaynor, D. L., Rowan, D. D., Latch, G. C. M., and Pilkington, S. (1983). 'Preliminary results on the biochemical relationship between adult Argentine stem weevil and two endophytes in ryegrasses', *Proc. N.Z. Weed and Pest Control Conf.*, **36**, 220–224.

Hewett, R. L. (1983). 'Breeding for resistance to ryegrass staggers, the inflation-proof long-term solution?', *Proc. N.Z. Grasslands Assoc.*, **44**, 244–247.

Latch, G. C. M. (1983). 'Incidence of endophytes in seed lines and their control with fungicides', *Proc. N.Z. Grasslands Assoc.*, **44**, 251–253.

Latch, G. C. M., and Christensen, M. J. (1982). 'Ryegrass endophyte, incidence, and control'. *N.Z. J. Agric. Res.*, **25**, 443–448.

Latch, G. C. M., Falloon, R. E., and Christensen, M. J. (1976). 'Fungi and ryegrass staggers', *N.Z. J. Agric. Res.*, **19**, 233–242.

di Menna, M. E., Mantle, P. G., and Mortimer, P. H. (1976). 'Experimental production of a staggers syndrome in ruminants by a tremorgenic *Penicillium* in soil', *Res. Vet. Sci.*, **24**, 45–46.

Mortimer, P. H., Fletcher, L. R., di Menna, M. E., Harvey, I. C., Smith, G. S., Barker, G. M., Gallagher, R. T., and White, E. P. (1982). 'Recent advances in ryegrass staggers', *Proc. Ruakura Farmers' Conf.*, pp. 71–74.

Mühle, E. (1971). *Krankheiten und Schädlinge der Futtergräser*, pp. 147–152. S. Hirzel Verlag, Leipzig.

Musgrave, D. R. (1984). 'Detection of the endophytic fungus of *Lolium perenne* by enzyme-linked immunosorbent assay (ELISA)', *N.Z. J. Agric. Res.*, **27**, 283–288.

Neill, J. C. (1939). DSIR Internal Report, N.Z.

Neill, J. C. (1940). 'The endophyte of ryegrass (*Lolium perenne*)', *N.Z. J. Sci. Technol.*, **A21**, 280–291.

Neill, J. C. (1941). 'The endophytes of *Lolium* and *Festuca*', *N.Z. J. Sci. Technol.*, **A23**, 185–193.

Trichothecenes and Other Mycotoxins
Edited by J. Lacey

CHAPTER 14

The *Lolium* endophyte and ryegrass staggers in New Zealand

Ian C. Harvey

INTRODUCTION

Until recently, ryegrass staggers (RGS) of cattle, sheep and deer in New Zealand was considered possibly to have been caused by *Penicillium* spp., either by the ingestion of spores in close-grazed toxic pastures (di Menna *et al.*, 1976; di Menna and Mantle, 1978) or by the ingestion of translocated tremorgenic compounds in the sward, produced by the fungus or fungi in the rhizosphere (White *et al.*, 1980). Study of the cause of ryegrass staggers has recently been intensified following the discovery of a relationship between the condition and a common endophytic fungus found in many perennial ryegrass lines, and known simply as the *Lolium* endophyte (Fletcher and Harvey, 1981). A chain of almost fortuitous discoveries has resulted in a plethora of publications on a fungus that had been the subject of only one paper since 1941 (Lloyd, 1959). Concurrent studies in Kentucky on the cause of tall fescue toxicosis in cattle focused on a very closely related fungus, now known as *Acremonium coenophialum* Morgan Jones and W. Gams, endophytic in tall fescue (Bacon *et al.*, 1977), and followed a similar pattern to the investigations in New Zealand.

INITIAL FINDINGS

Fletcher and Harvey (1981) gave convincing evidence that the link between the *Lolium* endophyte and RGS should be further investigated. In the same journal, Gallagher *et al.* (1981b) described potent neurotoxins (lolitrems) from toxic pastures, which were unrelated to the tremorgens produced by *Penicillium* spp. These two reports provided new perspectives on the aetiology of RGS. The *Lolium* endophyte had previously been dismissed as the cause of either RGS or facial eczema on the basis of three arguments. Firstly, Neill (quoted in Cunningham and Hartley, 1959) considered that 'a widely distributed fungus such as the *Lolium* endophyte could not be responsible for localized outbreaks of ryegrass staggers'. Secondly, when seed containing the *Lolium* endophyte was fed to a range of laboratory animals and to one sheep, no toxicity was noted (Neill, 1941). Thirdly, when pure cultures of the fungus were fed to sparrows and mice, no ill effects were noted (Cunningham, 1958). However Gallagher *et al.* (1981a) have recently found that not all sheep fed a diet composed largely of endophyte-rich seed reacted with RGS-like symptoms, perhaps explaining why Neill (1941) was unsuccessful using only one sheep.

Fungi other than the penicillia investigated have included *Fusarium* spp. (Latch *et al.*, 1976) and *Ascochyta* (McKenzie and Byford, unpublished data), but none of these could be verified as the aetiological agent of RGS.

The investigations of Fletcher and Harvey (1981) were aided by the proximity of a randomized perennial ryegrass cultivar grazing trial at Lincoln which demonstrated dramatic differences of RGS scores (0–5 scale after Keogh, 1973) among the cultivars on offer to the trial sheep.

The cultivar Grasslands 'Ruanui' produced no staggers in grazing sheep, Grasslands 'Nui' was awarded an intermediate RGS score of 2.4, and a North Canterbury ecotype was scored at 4.3. Investigation of the *Lolium* endophyte level of these cultivars (0, 42 and 95% respectively) showed a near-perfect correlation with the RGS scores. Fletcher and Harvey (1981) recognized that the *Lolium* endophyte had been largely overlooked in recent investigations due to its initial requirement for anaerobic conditions and its very slow growth rate in culture. In addition is is relatively resistant to a wide range of fungicides, many of which had been tested on toxic pastures as a remedial measure (Latch *et al.*, 1976).

The fungus is found only in certain parts of the plant. It is largely absent from roots and is in only small concentrations in leaf blades. The leaf sheath (Figure 1), stems and seeds contain the largest concentrations of the fungus. Both Keogh (1973) and Mortimer (1978) had noted that it was 'the base of ... pastures' which appeared to contain the factor(s) that caused RGS.

VERIFICATION AND A DILEMMA

The findings from the Lincoln trial were verified by Mortimer *et al.* (1982) using a single cultivar with and without endophyte, but the trial revealed a disturbing fact: although endophyte-free pastures produced no RGS, they were susceptible to the predation of Argentine stem weevil, *Listronotus bonariensis* (Kuschel), causing rapid late-summer pasture failure and subsequent lack of autumn regrowth (see also Mortimer and di Menna, chapter 15 in this volume). Almost all summer persisting perennial ryegrass cultivars and selections in New Zealand have now been found to contain *Lolium* endophyte (Harvey,

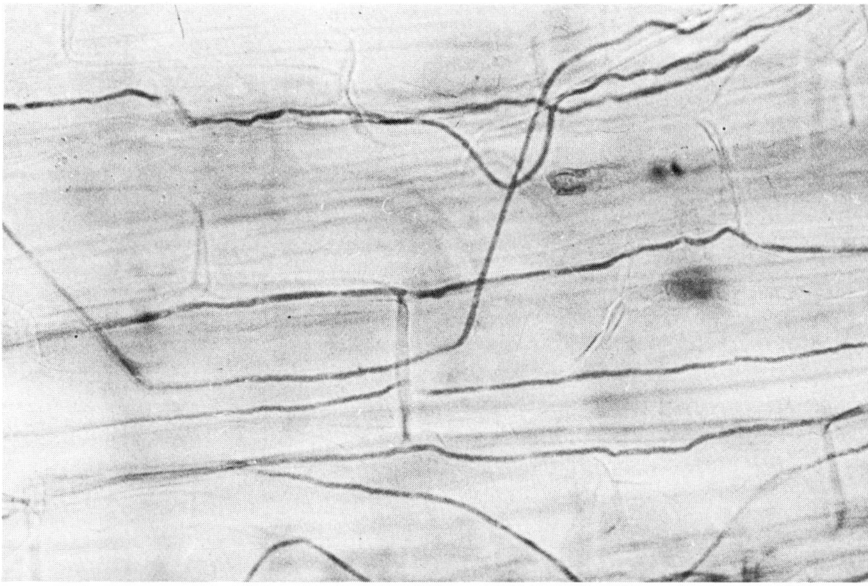

Figure 1 Photomicrograph of the *Lolium* endophyte in the leaf sheath of a ryegrass seedling.

unpublished data; A. V. Stewart, Plant Breeders (NZ) Ltd, Christchurch, personal communication). This suggests that natural selection in older pastures increases the percentage of endophyte-infected plants, and that breeders' selections have been basically for endophyte-infected phenotypes. Moreover, endophyte almost invariably dies in seed stored for more than 2 years at ambient temperatures (Neill, 1940), and these factors explain why ryegrass breeding results, especially those relating to persistence and herbage regrowth, have often been spurious and non-repeatable. It has long been recognized, however, that persistence and Argentine stem weevil resistance were conferred only through the maternal parent; the one passing on the endophyte infection via the seed.

The paradox, that a propensity to cause RGS is the penalty for good pasture persistence, has caused much debate, and it is unfortunate that areas with a major Argentine stem weevil problem also have poor ryegrass persistence, especially coupled with drought, and considerable RGS incidence. It may be some time before this dilemma is resolved either through breeding for genetic, non-endophyte-related Argentine stem weevil resistance, development of remedial or prophylactic pharmaceuticals for RGS, or the breeding of resistant livestock (Hewett, 1983).

FUNGICIDES AND THE *LOLIUM* ENDOPHYTE

Once the strong link between RGS and the *Lolium* endophyte was established, it appeared that ridding seed of the fungus before sowing would be a simple way of reducing RGS, albeit in new sowings only. Alternatively, seed could be stored at ambient temperature for 1–2 years, in which time the fungus largely dies (Neill, 1940). Independently, two groups simultaneously discovered that prochloraz (BFC Chemicals 'Sportak') gave very good control of the fungus in the seedborne phase (Latch and Christensen, 1982; Harvey *et al.*, 1982). The endophyte could also be destroyed in endophyte-infected plants growing in pots containing high concentrations of benomyl (Latch and Christensen, 1982) or decreased by propiconazole (Ciba Geigy 'Tilt') applied to foliage at normal label concentration to run-off point or to seeds (Harvey *et al.*, 1982). *In vitro*, the *Lolium* endophyte was more sensitive to prochloraz than to propiconazole and imazalil (Janssen 'Fungaflor'). Many fungicides had little effect on the growth of the fungus in culture, reflecting the difficulty of freeing plants from infection or reducing the incidence of RGS with pasture-applied fungicides (Harvey *et al.*, 1982).

With the discovery that the *Lolium* endophyte affected resistance to the major pasture pest, Argentine stem weevil, the impetus for freeing lines of perennial ryegrass from endophyte infection was lost, and breeders decided that at this time endophyte was best left in the seed lines and farmers should be encouraged to moderate the effect of ryegrass staggers on stock by suitable management techniques (Keogh, 1983). However, the effect of the endophyte on agriculture has yet to be fully elucidated; the closely related endophyte (*Acremonium coenophialum*) in tall fescue can produce dramatic reductions in cattle live weight gain (Hoveland *et al.*, 1983) in the U.S.A., and some evidence of a similar effect on lamb growth by the *Lolium* endophyte has been described from New Zealand by Fletcher (1983).

FINE STRUCTURE OF THE *LOLIUM* ENDOPHYTE

The ultrastructure of the endophyte in perennial ryegrass leaf sheaths, stems, seeds and growing in axenic culture has been studied (Fineran *et al.*, 1983; Harvey and Fineran, unpublished) and graphically confirm the benign nature of the endophyte. The fungus is illustrated totally confined to an extracellular niche, causing no disruption of the host cytoplasm and surviving on apolplastic nutrient leakage through the host cell walls (Figure 2). This contrasts with the report by Bacon (1983) that *A. coenophialum* growing in tall fescue 'is not happy', and describing fungus cytoplasm disruption, suggesting that the two are not compatible. Comparatively massive crystalloids and aggregates of tubules have been found in the cytoplasm of the *Lolium* endophyte growing in leaf sheaths (Fineran *et al.*, 1983). Endophyte cytoplasm in flowering stems

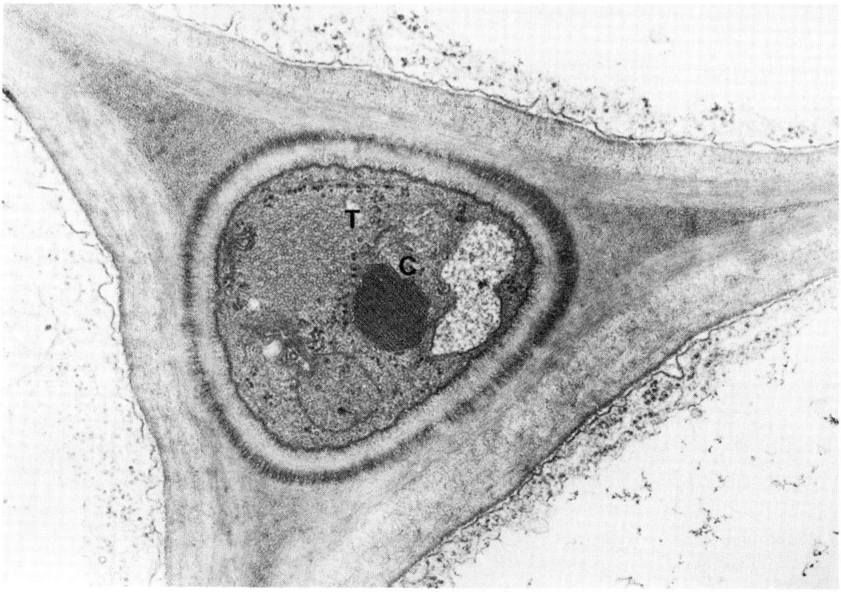

Figure 2 An electron micrograph showing the transverse section of a hyphae of the *Lolium* endophyte occupying the intercellular space between three host cells. Note the fibrillar wall interface between host and endophyte, crystalloids (C) and aggregates of tubules (T) on the hyphal cytoplasm. (Photo by M. Ingerfeld, Canterbury University, New Zealand.)

contained only aggregates of tubules, and in maturing seeds only crystalloids, while mycelium from culture contained neither (Harvey and Fineran, unpublished).

Neurotoxic lolitrems have been found in leaf sheaths and seeds containing the *Lolium* endophyte, but not in axenic cultures of the fungus (Gallagher *et al.*, 1981a). One may thus speculate that possibly either the crystalloids or aggregates of tubules, or both, might be the site of production or accumulation of the lolitrems linked to RGS (Gallagher *et al.*, 1981b) or to the Argentine stem weevil anti-feeding compound reported by Gaynor *et al.* (1983). Alternatively, it has been suggested (Gallagher *et al.*, 1982; Harvey, 1983; Mortimer, 1983) that the fungus might instead stimulate the host to produce the lolitrems either by its physical presence or by diffusates that elicit a 'phytoalexin'-type response in the host. Gallagher *et al.* (1982) suggest that cold-stressed endophyte-rich plants produce seven times the amount of lolitrem B than do non-stressed plants, but the endophyte levels were not measured after stressing. Thus, the increase may have been due to an increased rate of colonization by the endophyte in the host in response to stimulation.

ENDOPHYTE DETECTION

Johnson *et al.* (1982 and chapter 12 in this volume) reported that the fescue endophyte could be readily detected using enzyme-linked immunosorbent assay (ELISA), and this method was modified to determine levels of *Lolium* endophyte in pasture samples and seed lines of perennial ryegrass in New Zealand (Musgrave, 1984). The Plant Health Station of the Ministry of Agriculture and Fisheries at Lincoln utilized this method, with occasional confirmatory light microscope evaluation of infection (Harvey *et al.*, 1982), to determine levels of infection in breeders' selection seed lines (Funk *et al.*, 1983), in pasture samples where RGS outbreaks were suspected (Harvey, 1983) and in seed lines produced from toxic pastures (Table 1). The results have allowed breeders to select seed lines with endophyte levels that confer resistance to insect attack, and farmers to recognize permanent pastures that are high in endophyte and which require special grazing management to minimise the effects of RGS.

Table 1 *Lolium* endophyte levels in perennial ryegrass seed line through one generation

	Test method			
	ELISA		Light microscope	
Test material	μg/endophyte /mg host	Infection index (0–3[a])	Percentage of plants infected	Infection index (0–3[a])
Original seed line	5	2.5	100	3
Resultant pasture (8 months old)	17	3	100	3
Harvested seed (11 months after sowing)	200	3	90	2.7
Resultant seedlings in glasshouse (6 weeks old)	—	—	80	1.3

[a] 0 = No infection; 3 = heavy infection.

REFERENCES

Bacon, C. W. (1983). 'The fungal endophyte of tall fescue', *Proc. Tall Fescue Toxicosis Workshop*, University of Georgia, College of Agriculture, Athens, Georgia, pp. 34–47.

Bacon, C W., Porter, J. K., Robbins, J. D., and Luttrell, E. S. (1977). '*Epichloë typhina* from toxic tall fescue grasses', *Appl. Environ. Microbiol.*, **34**, 576–581.

Cunningham, I. J. (1958). 'Non-toxicity to animals of ryegrass endophyte and other endophytic fungi of New Zealand grasses, *N.Z. J. Agric. Res.*, **1**, 489–497.

Cunningham, I. J., and Hartley, W. J. (1959). 'Ryegrass staggers', *N.Z. Vet. J.*, **7**, 1–7.

Fineran, B. A., Harvey, I. C., and Ingerfeld, M. (1983). 'Unusual crystalloids and aggregates of tubules in the *Lolium* endophyte of ryegrass leaf sheaths', *Protoplasma*, **117**, 17–23.

Fletcher, L. R. (1983). 'Effects of presence of *Lolium* endophyte on growth rates of weaned lambs, growing on to hoggets, on various ryegrasses', *Proc. N. Z. Grasslands Assoc.*, **44**, 237–239.

Fletcher, L. R., and Harvey, I. C. (1981). 'An association of a *Lolium* endophyte with ryegrass staggers', *N.Z. Vet. J.*, **29**, 185–186.

Funk, C. R., Halisky, P. M., Johnson, M. C., Siegel, M. R., Stewart, A. V., Ahmad, S., Hurley, R. H., and Harvey, I. C. (1983). 'An endophytic fungus and resistance to sod webworms: association in *Lolium perenne* L.', *Bio/Technology*, **1**, 189–191.

Gallagher, R. T., Campbell, A. G., Hawker, A. D., Holland, P. T., McGavesten, D. A., Pansier, E. A., Harvey, I. C. (1981a). 'Ryegrass staggers: the presence of lolitrem neurotoxins in perennial ryegrass seed', *N.Z. Vet. J.*, **30**, 183–184.

Gallagher, R. T., Smith, G. S., di Menna, M. E., Young, P. W. (1982). 'Some observations on the neurotoxic production in perennial ryegrass', *N.Z. Vet. J.*, **30**, 203–204.

Gallagher, R. T., White, E. P., and Mortimer, P. H. (1981b). 'Ryegrass staggers. Isolation of potent neurotoxins Lolitrem A and Lolitrem B from stagger-producing pastures', *N.Z. Vet. J.*, **29**, 189–190.

Gaynor, D.L., Rowan, D.D., Latch, G. C. M., and Pilkington, S.(1983). 'Preliminary results on the biochemical relationship between adult argentine stem weevil and two endophytes in ryegrass', *Proc. 36th N.Z. Weed and Pest Control Conference*, pp. 220–224.

Harvey, I. C. (1983). 'The *Lolium* endophyte: return from anonymity', *Proc. N.Z. Grasslands Assoc.*, **44**, 234–236.

Harvey, I. C., Fletcher, L. R., and Emms, L. M. (1982). 'Effects of several fungicides on the *Lolium* endophyte in ryegrass plants, seeds and in culture', *N.Z. J. Agric. Res.*, **25**, 601–606.

Hewett, R. L. (1983). 'Breeding for resistance to ryegrass staggers, the inflation-proof long-term solution?', *Proc. N.Z. Grasslands Assoc.*, **44**, 244–247.

Hoveland, C. S., Schmidt, S. P., King, C. C., and Clark, E. M. (1983). 'Summer syndrome on tall fescue'. *Proc. Tall Fescue Toxicosis Workshop*, University of Georgia, College of Agriculture, Athens, Georgia, pp. 5–10.

Johnson, M. C., Pirone, T. P., Siegel, M. R., and Varney, D. R. (1982). 'Detection of *Epichloë typhina* in tall fescue by means of enzyme-linked immunosorbent assay', *Phytopathology*, **72**, 647–650.

Keogh, R. G. (1973). 'Induction and prevention of ryegrass staggers in grazing sheep', *N.Z. J. Exp. Agric.*, **1**, 55–57.

Keogh, R. G. (1983). 'Ryegrass staggers: Management and control', *Proc. N.Z. Grasslands Assoc.*, **44**, 248–250.

Latch, G. C. M., and Christensen, M. J. (1982). 'Ryegrass endophyte, incidence, and control', *N.Z. J. Agric. Res.*, **25**, 443–448.

Latch, G. C. M., Falloon, R. E., and Christensen, R. E. (1976). 'Fungi and ryegrass staggers'. *N.Z. J. Agric. Res.*, **19**, 233–242.

Lloyd, A. B. (1959). 'The endophytic fungus of perennial ryegrass', *N.Z. J. Agric. Res.*, **2**, 1187–1194.

di Menna, M. E., and Mantle, P. G. (1978). 'The role of penicillia in ryegrass staggers', *Res. Vet. Sci.*, **24**, 347–351.

di Menna, M. E., Mantle, P. G., and Mortimer, P. H. (1976). 'Experimental production of a staggers syndrome in ruminants by a tremorgenic *Penicillium* from soil', *N.Z. Vet. J.*, **24**, 45–46.

Mortimer, P. H. (1978). 'Perennial ryegrass staggers in New Zealand', in *Effects of Poisonous Plants on Livestock*, (eds P. F. Keeler, K. R. van Kempen and L. F. James), pp. 353–361. Academic Press, New York.

Mortimer, P. H. (1983). 'Ryegrass staggers: clinical, pathological and aetiological aspects', *Proc. N.Z. Grasslands Assoc.*, **44**, 230–233.

Mortimer, P. H., Fletcher, L. R., di Menna, M. E., Harvey, I. C., Smith, G. S., Barker, G. M., Gallagher, R. T., and White, E. P. (1982). 'Recent advances in ryegrass staggers', *Proc. Ruakura Farmers' Conference* 1982, pp. 71–74.

Musgrave, D. R. (1984). 'Detection of an endophytic fungus of *Lolium perenne* using enzyme-linked immunosorbent assay (ELISA)', *N.Z. J. Agric. Res.*, **27**, 283–288.

Neill, J. C. (1940). 'The endophyte of ryegrass (*Lolium perenne*)', *N.Z. J. Sci. Technol.*, **21A**, 280–291.

Neill, J. C. (1941). 'The endophytes of *Lolium* and *Festuca*', *N.Z. J. Sci. Technol.*, **23A**, 185–193.

White, E. P., Smith, G. S., di Menna, M. E., and Mortimer, P. . (1980). 'Absorption and translocation of *Penicillium* by ryegrass plants', *N.Z. Vet. J.*, **28**, 123–124.

Trichothecenes and Other Mycotoxins
Edited by J. Lacey

CHAPTER 15

Interactions of *Lolium* endophyte on pasture production and perennial ryegrass staggers disease

Peter H. Mortimer and Margaret E. di Menna

INTRODUCTION

A sequence of field observations and research has led to discoveries which have implications for pasture and animal production based on grazing systems used in New Zealand and elsewhere. These emerged from investigations into the aetiology of perennial ryegrass staggers (PRGS), a neurological disorder of livestock in New Zealand and other countries (Cunningham and Hartley, 1959; Mortimer, 1978, 1983; Mantle and Penny, 1981).

PRGS occurs in summer and autumn, is commonly associated with warm conditions and with grazing of short pasture swards. A mycotoxic aetiology for PRGS has been suspected for over 20 years (see Mantle and Penny, 1981) but significant progress has been recent (Mortimer *et al.*, 1982). Perhaps the most perceptive observations on PRGS were those of Keogh (1973, 1978) and Byford (1978), who asserted that severe PRGS outbreaks occurred only when stock overgrazed pasture in which perennial ryegrass (*Lolium perenne* L.) was the dominant component and that the toxic agent responsible was closely associated with perennial ryegrass.

PERENNIAL RYEGRASS STAGGERS

In New Zealand the disease commonly affects sheep, cattle and horses, and has recently been identified in several species of farmed deer. It is believed that both incidence and severity have markedly increased during the past decade. The susceptibility of individual animals to PRGS varies widely within flocks and herds, and there is field evidence that susceptibility has a strong genetic basis (Hewett, 1983).

The earliest clinical signs are fine head tremors and fasciculation of neck and limb muscle groups. As severity increases mild postural inco-ordination develops and is seen as head nodding, and swaying while standing. If animals are made to move, jerky, overcompensated, inco-ordinated movements are produced. Forced exercise exacerbates the signs and severely affected animals may become stiff-legged and prance on fore, hind or all four limbs, sometimes for a few steps, often for many metres. Such a display usually ends in collapse to the ground, the animal seized in rigid tetanic spasms. Usually these effects clear quickly, the animal regaining a standing position and walking away appearing little affected. This gradation of clinical signs, seen when stock are driven, is used for scoring clinical severity in PRGS field trials (Keogh, 1973).

When stock are removed from toxic pasture recovery is usually complete in 1–3 weeks. However, deaths from falls or drowning are common in severe outbreaks. Immobility in severely affected individuals restricts or prevents foraging and watering, and stock losses may be particularly high in drought seasons.

ASPECTS OF THE AETIOLOGY OF PRGS

The isolation of lolitrems from toxic ryegrass

The recent isolation and study of increasing numbers of tremorgenic mycotoxins from cultures of *Penicillium* and *Aspergillus* spp. (Mantle and Penny, 1981) suggested a possible cause of PRGS. Cultures of strains of several species, found to contain known tremorgens, were dosed to sheep and cattle and produced clinical effects very similar to those of PRGS (di Menna *et al.*, 1976). However, fractions from extracts of toxic ryegrass leaf material obtained from field outbreaks of PRGS repeatedly yielded no evidence for the presence of known tremorgenic mycotoxins. Extracts from 25 g samples of ryegrass leaves collected from many PRGS outbreaks in New Zealand were concentrated (× 30,000) and purified. These, injected into mice, produced strong tremors which developed more slowly and persisted 2–3 days longer than the tremors obtained with injection of known tremorgens (Mortimer and White, 1980). These different tremorgenic effects in mice raised doubts that the *Penicillium* and *Aspergillus* tremorgens being investigated were the neurotoxins responsible for PRGS. Further purification of fractions from toxic ryegrass samples led to the isolation of two new tremorgens, lolitrems A and B (Gallagher *et al.*, 1981). These tremorgens have also been extracted from ryegrass seed together with two other minor fractions, lolitrems C and D (Table 1). Samples from the same seed batches, pelleted and fed to sheep, produced 'severe PRGS-like symptoms' (Gallagher *et al.*, 1982). The structure of lolitrem B, the major neurotoxin associated with ryegrass staggers, has since been determined (Gallagher *et al.*, 1984).

Table 1 The lolitrems, novel tremorgens isolated from PRGS-producing leaf material and seed of perennial ryegrass (*Lolium perenne*)

Tremorgen	Molecular weight	Formula
Lolitrem A	701	$C_{42}H_{55}NO_8$
Lolitrem B	685	$C_{42}H_{55}NO_7$
Lolitrem C	687	
Lolitrem D	703	

The association between PRGS, lolitrems and *Lolium* endophyte

Chance played a major role in discovering the agency by which lolitrems were produced in perennial ryegrass pasture. A severe outbreak of PRGS occurred in a trial in which lambs grazed recently sown, replicated and controlled plots of pure ryegrass during summer drought (Fletcher, 1982). Three ryegrass cultivars were used in the trial. The incidence of PRGS was 100% in animals on all six plots sown with one cultivar, no PRGS emerged in lambs grazing plots sown with a second, and there was a moderate incidence in lambs on plots sown with the third. Extracts of leaf samples from the PRGS-producing plots contained large amounts of lolitrems while little was found in leaf samples from the PRGS-free plots.

It seemed possible that a seedborne infection might be responsible for the differences in PRGS incidence in animals grazing the different cultivars, and attention was focused on the endophyte of perennial ryegrass. A strong association was found between (1) the presence of *Lolium* endophyte, (2) lolitrems in ryegrass leaves, and (3) the occurrence of PRGS (Fletcher and Harvey, 1981; Fletcher, 1982; Mortimer *et al.*, 1982). Two further trials were mounted to substantiate the association.

In the first of these (1982 Ruakura trial) four plots were sown in spring with two perennial ryegrass seedlines with a high (>95%) incidence of viable endophyte infection and four with two lines with a low (<3%) incidence, all lines being of the same cultivar. Seed lines with a high endophyte incidence were freshly harvested, those with a low incidence had been stored for 2 and 4 years respectively. The incidence of PRGS in the 39 set-stocked lambs used to close-graze the high-endophyte plots in January–February 1982 reached 100% after 24 days, whereas there was no PRGS in 39 lambs identically grazed on the low-endophyte plots (Table 2; Mortimer *et al.*, 1982).

Table 2 Summary of the results of the Ruakura 1982 PRGS trial which confirmed the association of *Lolium* endophyte with the occurrence of PRGS in grazing lambs

	Seed		Sheep (n = 39) Clinical ryegrass staggers (%)		
Line[a]	Year harvested	Endophyte present (%)	Day 12	Day 19	Day 26
A	1979	<3	0	0	0
B	1977				
C	1981	>90	15	61	100
D	1981				

[a] Two replicates each.

In the next year (1983 Rukuhia trial) plots were sown in autumn 1982 with the same line of newly harvested perennial ryegrass seed. Four plots were sown with untreated seed and the endophyte infection rate in plants in the established swards was >90%. Another four plots were sown with the same line but after treatment of the seed with prochloraz (BFC Chemicals, 'Sportak'), a fungicide previously shown to control endophyte infection in ryegrass seed (Harvey *et al.*, 1982; Latch and Christensen, 1982). Endophyte infection in plots sown with fungicide-treated seed was <25%. When the plots were set-stock grazed during January–March 1983 all 40 lambs on the high-endophyte plots developed clinical PRGS, half of them being scored 'most severely' affected. On the low-endophyte plots, identically grazed, only three of 40 lambs showed clinical signs of PRGS and these were scored 'mildly' affected (Table 3; Mortimer *et al.*, 1984).

The design of the 1983 Rukuhia trial restricted variables to one, the amount of endophyte infection in the pasture, and the significance of the results from it needs no statistical evaluation. From the series of trials it is clear that without endophyte in ryegrass there will be no occurrence of PRGS. How the lolitrems relate to *Lolium* endophyte remains largely undetermined.

FIELD OBSERVATIONS OF THE *LOLIUM* ENDOPHYTE

The history and taxonomy of the *Lolium* endophyte have been described in this volume by Latch (chapter 13) and Harvey (chapter 14). Endophyte can be detected in ryegrass tissue by microscopic examination of material stained

Table 3 Summary of PRGS trial results (Rukuhia 1983) showing that prochloraz fungicide treatment of seed reduces *Lolium* endophyte infection in emergent swards and controls PRGS in lambs grazing the swards

Seed treatment	Percentage of endophyte in sward	Ryegrass staggers clinical score				Percentage of sheep affected
		Nil	Mild	Mod.	Sev.	
No treatment[a]	>93	0	9	11	20	100
Prochloraz fungicide[a]	<25	37	3	0	0	8

[a] Four replicates each—40 sheep per treatment. Very similar results at two other trial locations.

by heating in lacto-phenol cotton blue. A comparative assessment of the amount of endophyte in infected plants from the 1982 Ruakura and 1983 Rukuhia trials was made by counting the number of hyphal strands/mm of breadth of leaf sheath, a complete traverse being made of the breadth at a point as close as possible to the point of attachment to the tiller. A standardized system of staining and mounting was adopted and the same microscope used throughout, the depth to which hyphae can be detected within tissue being dependent on the resolving power of the instrument. As counts varied from plant to plant the mean was taken of counts from a sheath from one tiller from each of 40 plants/sample.

Mean counts made at the same time on the two trials were similar and there were similar seasonal trends on both. Counts rose from *ca.* 20 strands/mm in October to a peak of 40/mm in February and fell to 10 in July, closely paralleling mean monthly temperatures for the region (Figure 1).

LOLIUM ENDOPHYTE INFECTIONS AND INSECT PREDATION OF PASTURE

The 1982 Ruakura trial demonstrated another biological association, previously unsuspected, of considerable agronomic importance. The eight plots, all equally grazed under dry conditions, were much overgrazed after being stocked for 28 days and animals were removed to allow pasture regrowth. Approximately 100 mm of rain fell following destocking, producing optimal physical conditions for regrowth. Differences in regrowth pattern between the high- and low-endophyte plots became noticeable within 7 days of destocking and became more apparent daily. Regrowth was expectedly good

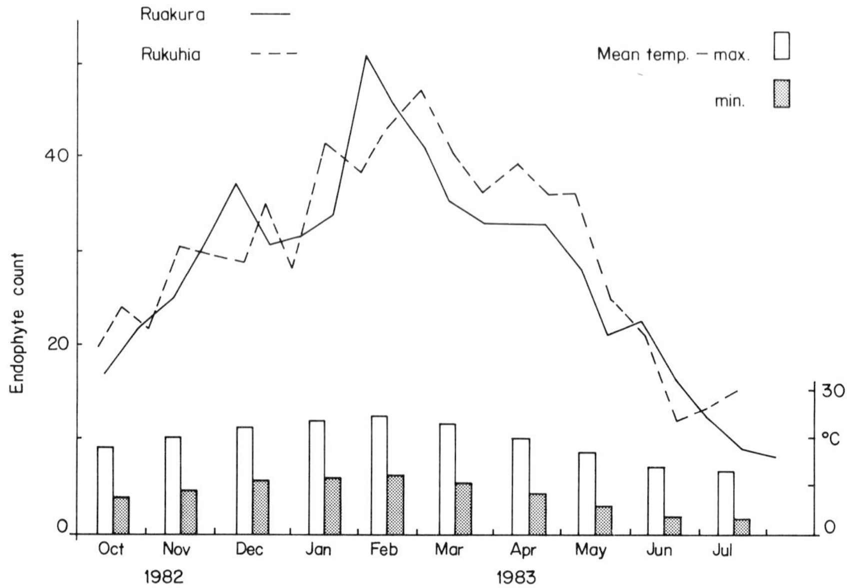

Figure 1 Mean counts of *Lolium* endophyte in leaf sheath of perennial ryegrass samples from two (Ruakura and Rukuhia) PRGS trial sites—October 1982 to July 1983 (graphs) and mean monthly maximum and minimum air temperatures at Ruakura—40-year averages (columns).

on all four high-endophyte plots but poor on those with low endophyte. In these plots, ryegrass plants showed severe residual injury and extensive tiller death. The same injury, caused by larvae burrowing into the tillers, was not prominent in the high-endophyte plots. Severe predation by Argentine stem weevil was suspected and later confirmed. Counts of adult weevil populations on the plots and estimates of dry matter (DM) production during regrowth are given in Table 4 (Mortimer *et al.*, 1982; Mortimer and di Menna, 1983). Pasture in the low-endophyte plots was almost totally destroyed by weevil damage superimposed on the earlier dry conditions so that autumn reseeding of these plots was necessary.

DISCUSSIONS AND CONCLUSIONS

The discovery of the likely role of lolitrems and of *Lolium* endophyte in the aetiology of PRGS has been a considerable advance in our knowledge. Chemical characterization of the lolitrems may assist in determining whether they originate as fungal or plant metabolites, or whatever. At present there is insufficient evidence to make sensible speculation on this matter. So far

Table 4 Results of 1982 PRGS trial showing the association of *Lolium* endophyte in ryegrass with stem weevil insect populations and their predatory feeding effects on pasture dry matter (DM) production

<table>
<tr><th colspan="3">Seed</th><th rowspan="2">Weevil populations (adults/m²)</th><th colspan="2">Pasture regrowth (DM kg/ha)</th></tr>
<tr><th>Line[a]</th><th>Year harvested</th><th>Endophyte present (%)</th><th>@ 4 weeks</th><th>@ 10 weeks</th></tr>
<tr><td>A</td><td>1979</td><td rowspan="2"><3</td><td>184</td><td>113</td><td>677</td></tr>
<tr><td>B</td><td>1977</td><td>229</td><td>321</td><td>1021</td></tr>
<tr><td>C</td><td rowspan="2">1981</td><td rowspan="2">>90</td><td>25</td><td>526</td><td>2196</td></tr>
<tr><td>D</td><td>40</td><td>885</td><td>2853</td></tr>
</table>

[a] Two replicates of four seed lines—*all* same cultivar—plots sown 1981.

cultures of *Lolium* endophyte have neither been tremorgenic to mice nor yielded lolitrems. Extrapolation from a similar situation is tempting. *Acremonium coenophialum* Morgan-Jones and Gams (previously referred to as *Epichloë typhina* (Fr.) Tulasne), an endophyte of *Festuca arundinacea* Schreb., which has been associated with tall fescue toxicity, produces *in vitro* the classes of alkaloids which produce both fescue foot and fescue toxicity (Porter *et al.*, 1979). However, evidence that the same alkaloids are present in toxic fescue is incomplete (Porter and Russell, 1983).

The association of perennial ryegrass and *Lolium* endophyte is a stable one, ryegrass if infected remains infected, but the quantity of fungus in infected plants varies with season (Figure 1). We do not know if amounts of lolitrems present in pasture are a reflection of the amount of endophyte present or of its growth activity. Again, the strictly seasonal occurrence of PRGS and its often sudden onset may suggest that lolitrem production is triggered by environmental changes which affect the endophyte–ryegrass association.

Results from our grazing trials in three consecutive years suggest that PRGS outbreaks can be induced in most, if not all, summers provided that pure perennial ryegrass stands with high endophyte infection are managed to give continuous and adequate grazing of the basal parts of the sward. This allows a more intensive study of the disease, its pathogenesis, treatment and control.

Discovery of the associationn of *Lolium* endophyte infection with resistance of perennial ryegrass to Argentine stem weevil predation came as a bonus in PRGS investigations, and research now reflects the agronomic interest aroused by the finding. Presence of endophyte has explained both the insect resistance and the increased pasture productivity previously recognized in some ryegrass seedlines (Kain *et al.*, 1982). Regrettably, this finding now raises doubts over the interpretation of many ryegrass varietal trials involving pasture and animal production, since endophyte infection of emergent swards was not determined.

Prestidge *et al.* (1982) reported that Argentine stem weevil showed a feeding preference for ryegrass tillers free of endophyte, and Gaynor *et al.* (1983) extracted a feeding deterrent from infected leaves but could not detect it in endophyte cultures. The deterrent proved to be highly polar and hydrophylic, dialysable and therefore of low molecular weight. These properties suggest that the deterrent factor is distinct from the lipophilic lolitrems associated with PRGS (Gallagher *et al.*, 1981).

When weevil resistance was first detected it was suggested that the resistance might apply to other insect pests of pasture (Mortimer *et al.*, 1982). Funk *et al.* (1983), examining seeds from a collection of ryegrass lines susceptible or resistant to sod webworm (*Crambus* spp.) feeding, found little endophyte in susceptible lines and a high percentage of infection in the resistant ones. They also suggested that endophytes might enhance resistance to other insect pests in other plant species.

With optimal pasture production and control of animal diseases as research objectives it may be unimportant whether or not *Lolium* and *Festuca* endophytes themselves elaborate materials toxic to animals or deterrent to insect pests of pasture. Whether these materials are phytoalexins or mycotoxins is not established, but a fungus is involved, so that the technological disciplines, expertise and input are the same.

REFERENCES

Byford, M. J. (1978). 'Ryegrass staggers in sheep and cattle', *Proc. N.Z. Soc. Anim. Prod.*, **38**, 42–46.

Cunningham, I. J., and Hartley, W. J. (1959). 'Ryegrass staggers', *N.Z. Vet. J.*, **7**, 1–7.

Fletcher, L. R. (1982). 'Observations of ryegrass staggers in weaned lambs grazing different ryegrass pastures'. *N.Z. J. Exp. Agric.*, **10**, 203–207.

Fletcher, L. R., and Harvey, I. C. (1981). 'An association of *Lolium* endophyte with ryegrass staggers, *N.Z. Vet. J.*, **29**, 185–186.

Funk, C. R., Halisky, P. M., Johnson, M. C., Siegel, M. R., Stewart, A. V., Ahmad, S., and Hurley, R. H. (1983). 'Association of an endophytic fungus with resistance to sod webworms in perennial ryegrass', *Biotechnology*, **1**, 189–191.

Gallagher, R. T., Campbell, A. G., Hawkes, A. D., Holland, P. T., McGaveston, D. A., Pansier, E. A., and Harvey, I. C. (1982). 'Ryegrass staggers: the presence of lolitrem neurotoxins in perennial ryegrass seed', *N.Z. Vet. J.*, **30**, 183–184.

Gallagher, R. T., Hawkes, A. D., Steyn, P. S. and Vleggaar, R. (1984). 'Tremorgenic neurotoxins from perennial ryegrass causing ryegrass staggers disorder in livestock: Structural elucidation of lolitrem B', *J. Chem. Soc., Chem Commun.*, 614–616.

Gallagher, R. T., White, E. P., and Mortimer, P. H. (1981). 'Ryegrass staggers: isolation of potent neurotoxins lolitrem A and lolitrem B from staggers-producing pastures', *N.Z. Vet. J.*, **29**, 189–190.

Gaynor, D. L., Rowan, D. D., Latch, G. C. M., and Pilkington, S. (1983). 'Preliminary results on the biochemical relationship between adult Argentine stem weevil and two endophytes in ryegrass'. *Proc. 36th N.Z. Weed and Pest Control Conf.* pp. 220–224.

Harvey, I. C., Fletcher, L. R., and Emms, L. M. (1982). 'Effects of several fungicides on the *Lolium* endophyte in ryegrass plants, seeds, and in culture'. *N.Z. J. Agric. Res.*, **25**, 601–606.

Hewett, R. L. (1983). 'Breeding for resistance to ryegrass staggers, the inflation-proof longterm solution?', *Proc. N.Z. Grasslands Assoc.*, **44**, 244–247.

Kain, W. M., Wyeth, T. K., Gaynor, D. L., and Slay, M. W. (1982). 'Argentine stem weevil (*Hyperodes bonariensis* Kuschell) resistance in perennial and hybrid ryegrasses', *N.Z. J. Agric. Res.*, **25**, 255–259.

Keogh, R. G. (1973). 'Induction and prevention of ryegrass staggers in grazing sheep', *N.Z. J. Exp. Agric.*, **1**, 55–57.

Keogh, R. G. (1978). 'Development and control of ryegrass staggers in sheep', *Proc. N.Z. Soc. Anim. Prod.*, **38**, 47–52.

Latch, G. C. M., and Christensen, M. J. (1982). 'Ryegrass endophyte, incidence, and control', *N.Z. J. Agric. Res.*, **25**, 443–448.

Mantle, P. G., and Penny, R. H. C. (1981). 'Tremorgenic mycotoxins and neurological disorders—a review', in *The Veterinary Annual*, 21st issue. (eds C. G. S. Grunsell and F. W. G. Hill), pp. 51–62. John Wright and Sons, Bristol.

di Menna, Margaret E., Mantle, P. G. and Mortimer, P. H. (1976). 'Experimental production of a staggers syndrome in ruminants by a tremorgenic Penicillium from soil', *N.Z. Vet. J.*, **24**, 45–46.

Mortimer, P. H. (1978). 'Perennial ryegrass staggers in New Zealand'. In "*Effects of Poisonous Plants on Livestock*". Eds R. F. Keeler, K. R. Van Kempen and L. F. James), pp. 353–361, Academic Press, New York.

Mortimer, P. H. (1983). 'Ryegrass staggers: clinical, pathological and aetiological aspects', *Proc. N.Z. Grassld Ass.*, **44**, 230–233.

Mortimer, P. H., and di Menna, M. E. (1983). 'Ryegrass staggers; further substantiation of a *Lolium* endophyte aetiology and the discovery of weevil resistance of rye-grass pastures infected with *Lolium* endophyte', *Proc. N.Z. Grassld Ass.*, **44**, 240–243.

Mortimer, P. H., and White, E. P. (1980). 'Tremorgen concentration from pasture and comparisons of tremorgenic effects', *Agric. Res., N.Z. Min. Ag. Fish.; A. Rep., Res. Div.*, **1979–80**, 92–93.

Mortimer, P. H., Fletcher, L. R., di Menna, M. E., Harvey, I. C., Smith, G. S., Barker, G. M., Gallagher, R. T., and White, E. P. (1982). 'Recent advances in ryegrass staggers'. *Proc. Ruakura Fmrs' Conf.*, **34**, 71–74.

Mortimer, P. H., Young, P. W. and di Menna, M. E. (1984). 'Perennial ryegrass staggers research – an overview. *Proc. N.Z. Soc. Anim. Prod.*, **44**, 181–184.

Neill, J. C. (1940). 'The endophyte of ryegrass (*Lolium perenne*)'. *N.Z. J. Sci. Tech.*, **A21**, 280–291.

Porter, J. K., Bacon, C. W., and Robbins, J. D. (1979). 'Ergosine, ergosinine and chanoclavine I from *Epichloë typhina*', *J. Agric. Food Chem.*, **27**, 595–598.

Porter, J. K., and Russell, R. B. (1983). 'Chemical constituents of tall fescue associated with toxicity', In *Proceedings—Tall Fescue Toxicosis Workshop*, pp. 27–33. Co-operative Extension Service, University of Georgia College of Agriculture, Athens.

Prestidge, R. A., Pottinger, R. P., and Barker, G. M. (1982). 'An association of *Lolium* endophyte with ryegrass resistance to Argentine stem weevil', *Proc. 35th N.Z. Weed and Pest Control Conf.*, pp. 119–122.

Trichothecenes and Other Mycotoxins
Edited by J. Lacey

CHAPTER 16

Annual ryegrass toxicity

C. C. J. Culvenor and M. V. Jago

THE DISEASE

Annual ryegrass toxicity (ARGT) is a disease of sheep and cattle in the 14–22 inch rainfall regions of South and Western Australia (Figure 1), where the land use is predominantly wheat–sheep farming (McIntosh *et al.*, 1967; Stynes and Wise, 1980). Annual ryegrass (*Lolium rigidum*) is well adapted to these regions and is a productive, volunteer species valued for the pasture it provides in non-cropping years. In affected districts annual ryegrass seedheads are infected with a nematode, *Anguina agrostis*, and the bacterium *Corynebacterium rathayi*. Infection was first found at Black Springs, South Australia. From there it initially spread slowly but then more rapidly, particularly in Western Australia where it was recognized 12–14 years ago. It is now endemic in an area of several million hectares in the two states and is thought likely to appear soon in Victoria and New South Wales.

Infected seedheads become toxic as the grass dries off in October–November. Affected sheep show signs of ataxia, fall behind the mob when driven, develop a characteristic, stiff-legged gait and collapse with convulsions. Death may occur 1 or 2 days after first signs. Mortality rates are variable and can approach 100%. The impact on affected properties is severe and if all paddocks are infected some farmers cease to run stock.

THE INFECTION

Infection of the plant is initiated by *A. agrostis*, which survives the summer as the anhydrobiotic second-stage larva. The opening autumn rains release the nematode larvae from dry galls and germinate unaffected ryegrass seeds. The larvae ascend the seedlings and eventually enter the primordia of the florets, inducing the formation of galls in which they develop to produce the new generation. The bacterium adheres to the cuticle of the nematode and so is carried into the developing floret where it multiplies rapidly, causes the death of the nematode and fills the nematode-induced gall. Yellow bacterial slime sometimes occurs on the outside of the seed heads. At plant maturity the

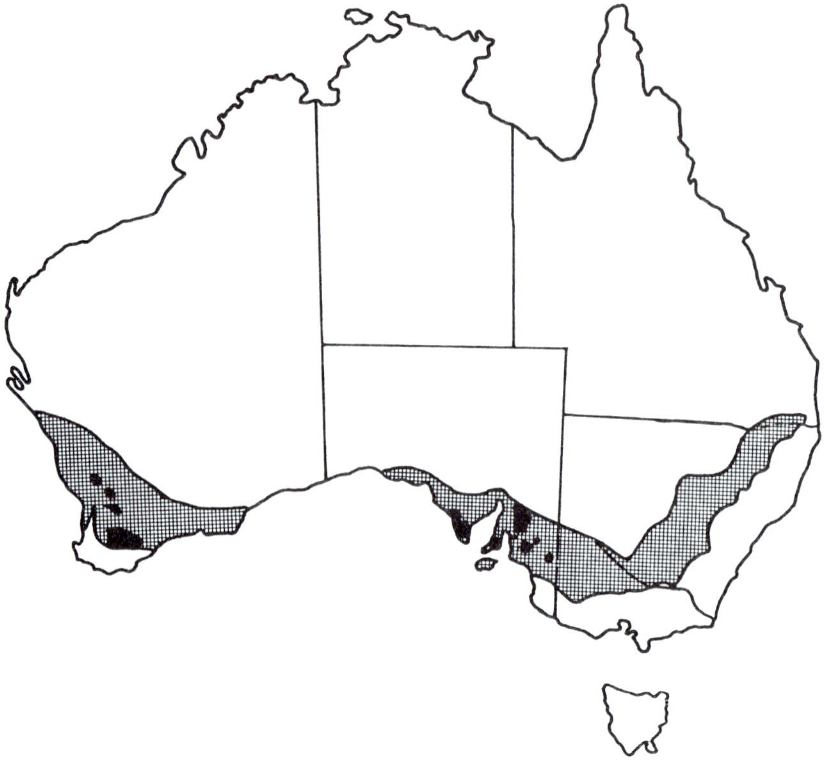

Figure 1 The distribution of annual ryegrass (*Lolium rigidum*) (hatched areas) and areas of toxicity (black areas) in Australia.

galled seedheads, containing nematodes or bacteria, dry out and drop to the ground.

Thus the mature grass carries normal seed, nematode-filled galls and bacteria-filled galls. Only the bacteria-filled galls are toxic (Lanigan *et al.*, 1976), the toxicity appearing at a late stage in gall development, usually about November. Despite considerable effort, only small amounts of toxin have been produced by *C. rathayi* in culture on callus tissue of *L. multiflorum* (Stynes and Petterson, 1980) or in liquid medium completely free of ryegrass cell material or extractives (Payne *et al.*, 1983). In the latter instance the toxin was characterized by its chemical and antiglycosylation properties (Figure 2).

THE TOXIN(S)

The toxic agent is a family of closely related glycolipids which have been called corynetoxins (Vogel *et al.*, 1981) whose structure resembles that of the tunicamycin group of antibiotics (Edgar *et al.*, 1982; Frahn *et al.*, 1983).

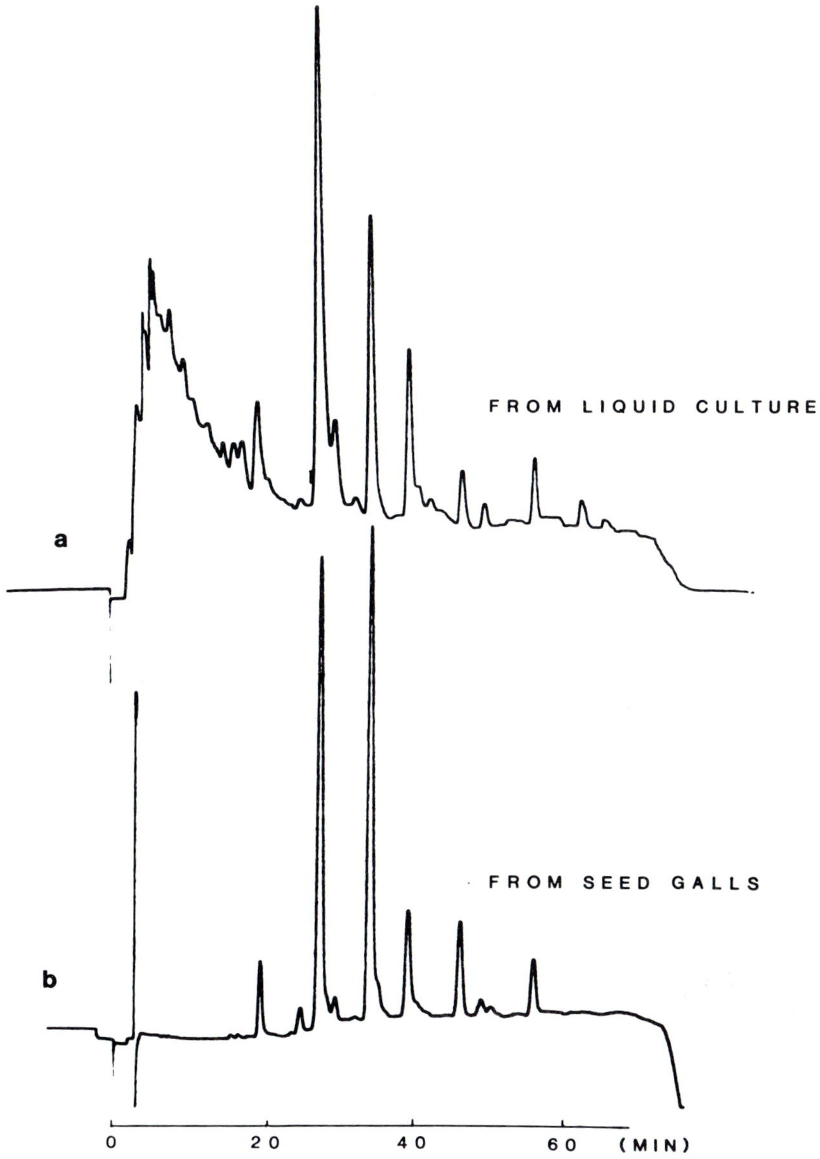

Figure 2 Corynetoxins isolated from *Corynebacterium rathayi* growing in liquid culture.

The tunicamycin molecule is a glycosidic nucleoside, acetyl-glucosaminyl-tunicaminyl-uracil (AGTU), bearing a long-chain fatty acid in amide linkage. Tunicamine is a unique C_{11}-aminosugar which is essentially composed of carbon-linked ribose and galactosamine units. The AGTU portion is constant in all members of the group, except in some members of the streptovirudins, in which the uracil is replaced by dihydrouracil (Tresselt *et al.*, 1983). Variations in the fatty acid create the groups of related compounds. The tunicamycin acids previously reported are C_{13} to C_{17}, either saturated or αβ-unsaturated, with normal or *iso* [i.e. $(CH_3)_2CH–CH_2–$] chain terminations. The corynetoxin acids are C_{15} to C_{19}, saturated, αβ-unsaturated or β-hydroxy-, and normal, *iso* or *anteiso* [$CH_3CH_2CH(CH_3)CH_2–$] chains.

The β-hydroxy acids are new to the tunicamycin group. The *anteiso* chain endings, predominant in the corynetoxins, have been reported previously only in the streptovirudin series. In a comparative study of corynetoxins, tunicamycins and streptovirudins, *anteiso* acids have been identified in some individual tunicamycins (Cockrum and Edgar, 1983). The two main corynetoxins are H17a (β-hydroxy, C_{17}, *anteiso* acid) (Figure 3) and U16a (αβ-unsaturated, C_{16}, *anteiso* acid).

For some of these groups of compounds, identification of the main portion of the molecule as AGTU has not been fully accomplished. It was

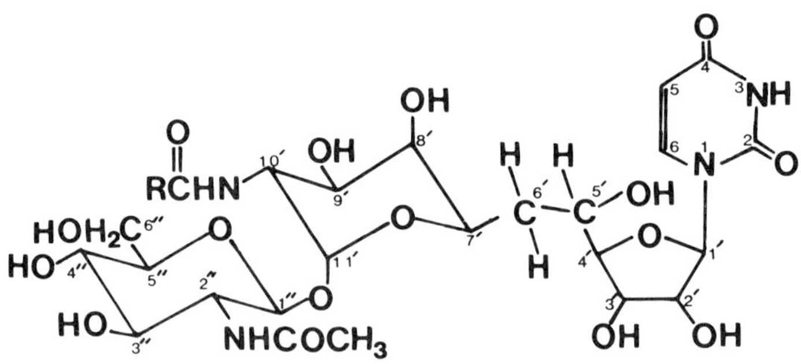

Corynetoxin U17a: R = $CH_3CH_2CH(CH_2)_{10}CH{=}CH–$ (with CH_3 on the branched CH)

Corynetoxin H17a: R = $CH_3CH_2CH(CH_2)_{10}CHOHCH_2–$ (with CH_3 on the branched CH)

Figure 3 The chemical structure of the corynetoxins.

achieved for the corynetoxins by converting both them and the tunicamycins into di-trifluoroacetyl-glucosaminyl-tunicaminyl-uracil by heating with trifluoroacetic anhydride and trifluoroacetic acid (50 : 1) (Frahn *et al.*, 1983). The stereochemistry of tunicamine and the glycosidic linkages was also not previously known with certainty. In an NMR (nuclear magnetic resonance) study on corynetoxin H17a, we have now resolved the sugar resonances sufficiently to identify nearly all the vicinal coupling constants relating to the CH protons of the sugar chains (Frahn *et al.*, 1983). In this way we have verified that tunicamine is carbon linked, tail-to-tail ribofuranose and galactosamine, and we have shown that the head-to-head glycosidic linkage with acetyl glucosamine is α-galactosamine–β–glucosamine rather than β–galactosamine–α–glucosamine as previously thought (Ito *et al.*, 1980). This changed stereochemistry has been incorporated in Figure 3.

TOXICITY OF CORYNETOXINS

Corynetoxins, extracts of the toxic seedheads and toxic ryegrass, have been found lethal to all animal species tested to date, including sheep, cattle, guinea pigs, rats, mice and chickens (McIntosh *et al.*, 1967; Berry and Wise, 1975; Berry *et al.*, 1976, 1980a,b; Lanigan *et al.*, 1976; Peterson and Jago, 1977; Schneider, 1981). The characteristic neurological disturbances of staggering gait, inco-ordination and convulsions have been reported in all affected species except mice and chickens. The earliest clinical sign of annual ryegrass toxicity is loss of appetite leading to anorexia and loss of body weight. Sometimes there is diarrhoea.

Reports of gross pathology indicate some variation among the species from extensive lesions in sheep to lack of obvious damage in chickens. The vascular system is often affected, particularly in the brain and spinal cord with increased vascular permeability and haemorrhages. Haemorrhages have also been reported in the heart, lungs, lymph nodes, bladder and gastrointestinal tract. The liver is sometimes yellow and friable, the lungs oedematous and severely congested and the kidneys pale. Studies in nursling rats poisoned with extracts of the toxic ryegrass suggest a long-lasting vasoconstrictive action. Widespread lesions of focal necrosis in the brain are consistent with anoxia. Evidence of restricted blood flow can be seen in the kidneys, lungs and brain and the peripheral circulation is reduced to such an extent that gangrene of the tail or hind limb sometimes occurs.

Very little quantitative data on the toxicity of the corynetoxins is available. The acute LD_{50} for nursling rats is about 130 μg/kg given by subcutaneous injection. Single sheep (young crossbred wethers) have been killed by 50 μg/kg given either as a single subcutaneous injection or as repeated daily doses of 12.5 μg/kg.

MECHANISM OF TOXICITY

As was expected from the structural relationship between the corynetoxins and the tunicamycins, both showed similar inhibitory activity against certain viruses, bacteria and mammalian cells in culture (Jago *et al.*, 1983a). This activity is due to the specific inhibition of lipid-linked *N*-glycosylation of glycoproteins (Struck and Lennarz, 1980) and of an analogous process in peptido-glycan (cell wall) synthesis in bacteria (Bettinger and Young, 1975). In eucaryotic cells the enzyme inhibited is UDP-Glc*N*Ac:dolichol–P Glc*N*Ac–1–P transferase which affects the first step in the assembly of the oligosaccharide on the lipid carrier. The toxins are considered to bind to the active site because of similarity in structure to the substrate–product transition complex (Heifetz *et al.*, 1979). Because of the highly specific nature of this inhibition, tunicamycin has been used extensively in examination of the role of the carbohydrate side chains in glycoproteins. Almost all of this work has been with tissue culture systems. Annual ryegrass toxicity (ARGT), the first disease of animals known to be caused by tunicamycin-related compounds, is probably the result, either directly or indirectly, of the depletion or impaired synthesis of essential *N*-glycosylated glycoproteins.

The carbohydrate components are believed to act as modulators in establishing the conformation of glycoproteins (Schwarz and Datema, 1982). They stabilize the conformation and protect the molecule against denaturing conditions of temperature, pH and proteolytic attack. Current evidence suggests that the carbohydrate side chains are mainly situated in the β-turns of the polypeptide backbone, loci which become vulnerable if aglycosylated. In some, but not all, glycoproteins, the carbohydrate component is essential for activity.

The *N*-glycosylated glycoproteins are widely distributed in mammalian systems performing a number of essential functions. They include enzymes, hormones, structural components of the plasma membrane and membrane receptors. The membrane receptors include those for acetylcholine, insulin, low-density lipoproteins, transferrin, epidermal growth factor, asialoglycoproteins and for the phosphomannosyl residues of lysosomal enzymes (Bhargava and Makman, 1980; Ronnett and Lane, 1981; Reed *et al.*, 1981; Sahagian *et al.*, 1981; Prives and Bar-Sagi, 1983; Brown *et al.*, 1983). In the presence of tunicamycin the numbers of functional receptors at the cell surfaces are reduced either by accumulation of inactive, aglycosylated receptors or by enhanced proteolytic degradation of underglycosylated receptors. Other gylcoproteins, the glycosylation of which has been shown to be inhibited by tunicamycin, include pro-opiomelanocortin (the precursor for ACTH, endorphin and other peptide hormones) (Budarf and Herbert, 1982), transferrin and the apoprotein B chain of very low-density lipoproteins (Struck *et al.*, 1978), the immunoglobulins IgA and IgE (Hickman *et al.*,

1977), glycosidases (Speake *et al.*, 1981) and fibronectin (Olden *et al.*, 1978), the multifunctional glycoprotein of the cell surface and of the blood. Fibronectin is of particular interest. It is present on the surface of normal cells, in the connective tissue matrix and basement membranes of blood vessels and, in a soluble form, in the blood. It is an adhesive glycoprotein involved in cell-to-cell and cell-to-substrate interactions, and as such is thought to modulate microvascular integrity and vascular permeability (Saba and Jaffe, 1980). Reduced fibronectin levels may thus be a cause of the breached blood–brain barrier observed in corynetoxin poisoned animals (Peterson and Jago, 1977; Berry *et al.*, 1980b). Plasma fibronectin also plays an essential role in the functioning of the reticuloendothelial system (RES). Depression of this system, which is partly if not largely a consequence of depletion of blood fibronectin, leads to impaired cardiovascular function including decreased peripheral circulation and oxygen utilization (Saba, 1970; Saba and Jaffe, 1980; Niehaus *et al.*, 1980; Dillon *et al.*, 1982). Some of the observations made in the study of corynetoxin poisoning in nursling rats (Peterson and Jago, 1977) strongly suggest RES depression. This aspect is currently under investigation and preliminary studies indicate that fibronectin levels are depressed in the poisoned rats (Jago *et al.*, 1983b).

PROTECTIVE MEASURES AND CURRENT STUDIES

Eradication of the nematode is considered impractical in Australia and animal losses are now minimized by a range of measures designed to decrease exposure to seeding ryegrass. These include the use of herbicides to reduce growth of annual ryegrass in crops, use of haying-off chemicals to prevent seeding, and daily inspection of flocks in the danger period so that animals may be moved if toxicity becomes evident. These methods are costly and time-consuming and substantial losses may still occur. New methods are urgently needed, that will allow unrestricted utilization of ryegrass pastures.

Protective immunization of animals against corynetoxins is one possibility being investigated by the Western Australian Department of Agriculture, and also in our laboratory. There are many problems, and progress has been slow.

Another possibility is to control corynetoxin production by the bacteria in ryegrass seedheads. Further information is being sought on the factors which allow *C. rathayi* to produce toxin in the field but which are lacking in laboratory cultures. Analogy with *C. diphtheriae* and the production of antibiotics by other bacteria suggests that a plasmid or bacteriophage may be necessary for corynetoxin production. Particles of 20–30 nm diameter have been observed in the capsule layer of bacteria from toxic galls (Bird *et al.*, 1980) and their nature and relationship to toxicity are currently being studied.

The clarification of the principal mechanisms involved in animal poisoning by corynetoxins raises the further possibility of fibronectin replacement therapy or prophylactic induction of the RES, to which biochemical studies are also now being addressed.

POTENTIAL FOR SPREADING

Infection of *L. rigidum* by *A. agrostis* and *C. rathayi* has created a serious problem in parts of South and Western Australia, and has great potential for spreading throughout the annual ryegrass habitat in Australia and elsewhere. It has recently appeared in South Africa (Schneider, 1981). There is also a risk that it will spread to other grasses since *A. agrostis* and a co-infecting bacterium, probably a *Corynebacterium*, have infected *Festuca rubra commutata* in Oregon and the resulting seedheads have induced signs resembling ARGT in animals (Galloway, 1961). In Western Australia, natural infection by *C. rathayi* has been observed on *Phalaris paradoxa*, *P. minor*, *Avena sativa*, *A. fatua* and *Danthonia caespitosa*, growing with annual ryegrass; extracts of infected plants were capable of inducing in nursling rats the characteristic signs of ARGT (D. L. Chatel and D. S. Petterson, personal communication). However, there was no evidence of a persistent infection in pastures free of annual ryegrass. Nor has infection been reported in wheat, *Triticum aestivum*, although the related organisms, *Anguina tritici* *C. tritici*, can infect this crop under appropriate conditions, as found in India (Bird, 1981). If infection reaches Victoria, there is the possibility of spread to the large areas of perennial ryegrass, *L. perenne*, which merge with the annual ryegrass region.

REFERENCES

Berry, P. H., and Wise, J. L. (1975). 'Wimmera ryegrass toxicity in Western Australia', *Aust. Vet. J.*, **51**, 525–530.

Berry, P. H., Howell, J. McC., and Cook, R. D. (1980a). 'Morphological changes in the central nervous system of sheep affected with experimental annual ryegrass (*Lolium rigidum*) toxicity', *J. Comp. Pathol.*, **90**, 603–617.

Berry, P. H., Cook, R. D., Howell, J. McC., White, R. R., and Purcell, D. A. (1976). 'Lesions in sheep and guinea pigs pen fed parasitised annual ryegrass (*Lolium rigidum*)', *Aust. Vet. J..*, **52**, 540–541.

Berry, P. H., Howell, J. McC., Cook, R. D., Richards, R. B., and Peet, R. L. (1980b). 'Central nervous system changes in sheep and cattle affected with natural or experimental annual ryegrass toxicity', *Aust. Vet. J.*, **56**, 402–403.

Bettinger, G. E., and Young, F. E. (1975). 'Tunicamycin, an inhibitor of *Bacillus* peptidoglycan synthesis: a new site of inhibition'. *Biochem. Biophys. Res. Commun.*, **67**, 16–21.

Bhargava, G., and Makman, M. H. (1980). 'Effect of tunicamycin on the turnover of epidermal growth factor receptors in cultured calf aorta smooth muscle cells', *Biochim. Biophys. Acta*, **629**, 107–112.

Bird, A. F. (1981). 'The *Anguina–Corynebacterium* association', in *Plant Parasitic Nematodes*, Vol. 3 (eds B. M. Zuckerman and R. A. Rohde), pp. 303–323. Academic Press, New York.

Bird, A. F., Stynes, B. A., and Thomson, W. W. (1980). 'A comparison of nematode and bacteria-colonized galls induced by *Anguina agrostis* in *Lolium rigidum*', *Phytopathology*, **70**, 1104–1109.

Brown, M. S., Anderson, R. G. W., and Goldstein, J. L. (1983). 'Recycling receptors: the round-trip itinerary of migrant membrane proteins', *Cell*, **32**, 663–667.

Budarf, M. L., and Herbert, E. (1982). 'Effect of tunicamycin on the synthesis, processing and secretion of pro-opiomelanocortin peptides in mouse pituitary cells', *J. Biol. Chem.*, **257**, 10128–10135.

Cockrum, P. A., and Edgar, J. A. (1983). 'HPLC comparison of tunicaminyluracil-based antibiotics—corynetoxins, tunicamycins, streptovirudins and antibiotic MM 19290', *J. Chromatogr.*, **268**, 245–254.

Dillon, B. C., Saba, T. M., Cho, E., and Lewis, E. (1982). 'Opsonic fibronectin deficiency in the etiology of starvation-induced reticuloendothelial phagocytic dysfunction', *Exp. Mol. Pathol.*, **36**, 177–192.

Edgar, J. A., Frahn, J. L., Cockrum, P. A., Anderton, N., Jago, M. V., Culvenor, C. C. J., Jones, A. J., Murray, K., and Shaw, K. J. (1982). 'Corynetoxins, causative agents of annual ryegrass toxicity; their identification as tunicamycin group antibiotics', *J. Chem. Soc. Chem. Commun.*, 222–224.

Frahn, J. L., Edgar, J. A., Jones, A. J., Cockrum, P. A., Anderton, N., and Culvenor, C. C. J. (1983). 'Structure of the corynetoxins, metabolites of *Corynebacterium rathayi* responsible for toxicity of annual ryegrass (*Lolium rigidum*) pastures', *Aust. J. Chem.* **37**, 165–182.

Galloway, J. H. (1961). 'Grass seed nematode poisoning in livestock', *J. Am. Vet. Med. Assoc.*, **139**, 1212–1214.

Heifetz, A., Keenan, R. W., and Elbein, A. D. (1979). 'Mechanism of action of tunicamycin on the UDP-GlcNAc: dolichyl-phosphate Glc–NAc–1–phosphate transferase', *Biochemistry*, **18**, 2186–2192.

Hickman, S., Kulczycki, A., Jr., Lynch, R. G., and Kornfeld, S. (1977). 'Studies of the mechanism of tunicamycin inhibition of IgA and IgE secretion by plasma cells', *J. Biol. Chem.*, **252**, 4402–4408.

Ito, T., Takatsuki, A., Kawamura, K., Sato, K., and Tamura, G.(1980). 'Isolation and structures of components of tunicamycin', *Agric. Biol. Chem.*, **44**, 695–698.

Jago, M. V., Payne, A. L., Peterson, J. E., and Bagust, T. J. (1983a). 'Inhibition of glycosylation by corynetoxin, the causative agent of annual ryegrass toxicity: a comparison with tuncamycin', *Chem. Biol. Interact.* (in press).

Jago, M. V., Payne, A. L., Peterson, J. E., and Bagust, T. J. (1983b). 'Mechanism of action of corynetoxins the causative agents of annual ryegrass toxicity', *Toxicon*, Suppl. 3, 195–197.

Lanigan, G. W., Payne, A. L., and Frahn, J. L. (1976). 'Origin of toxicity in parasitised annual ryegrass (*Lolium rigidum*)', *Aust. Vet. J.*, **52**, 244–246.

McIntosh, G. H., Rac, R., and Thomas, M. R. (1967). 'Toxicity of parasitised Wimmera ryegrass, *Lolium rigidum*, for sheep and cattle', *Aust. Vet. J.*, **43**, 349–353.

Niehaus, G. D., Schumacker, P. R., and Saba, T. M. (1980). 'Reticuloendothelial clearance of blood-borne particulates. Relevance to experimental lung microembolism and vascular injury', *Ann. Surg.*, **191**, 479–487.

Olden, K., Pratt, R. M., and Yamada, K. M. (1978). 'Role of carbohydrates in protein secretion and turnover: effects of tunicamycin on the major cell surface glycoprotein of chick embryo fibroblasts', *Cell*, **13**, 461–473.

Payne, A. L., Cockrum, P. A., Edgar, J. A., and Jago, M. V. (1983). 'Production of corynetoxins by *Corynebacterium rathayi* in liquid cultures', *Toxicon*, Suppl. 3, 345–348.

Peterson, J. E., and Jago, M. V. (1977). 'Brain damage by extracts of parasitised annual ryegrass (*Lolium rigidum*) in nursling rats', *Aust. J. Exp. Biol. Med. Sci.*, **55**, 233–244.

Prives, J., and Bar-Sagi, D. (1983). 'Effect of tunicamycin, an inhibitor of protein glycosylation, on the biological properties of acetylcholine receptor in cultured muscle cells', *J. Biol. Chem.*, **258**, 1775–1780.

Reed, B. C., Ronnett, G. V., and Lane, M. D. (1981). 'Role of glycosylation and protein synthesis in insulin receptor metabolism by 3T3–L1 mouse adipocytes', *Proc. Natl Acad. Sci. U.S.A.*, **78**, 2908–2912.

Ronnett, G. V., and Lane, M. D. (1981). 'Post-translational glycosylation-induced activation of aglycoinsulin receptor accumulated during tunicamycin treatment', *J. Biol. Chem.*, **256**, 4704–4707.

Saba, T. M. (1970). 'Physiology and physiopathology of the reticuloendothelial system', *Arch. Inten. Med.*, **126**, 1031–1052.

Saba, T. M., and Jaffe, E. (1980). 'Plasma fibronectin (opsonic glycoprotein): its synthesis by vascular endothelial cells and role in cardiopulmonary integrity after trauma as related to reticuloendothelial function', *Am. J. Med.*, **68**, 577–594.

Sahagian, G. G., Distler, J., and Jourdian, G. W. (1981). 'Characterization of a membrane-associated receptor from bovine liver that binds phosphomannosyl residues of bovine testicular β-galactosidase', *Proc. Natl. Acad. Sci. U.S.A.*, **78**, 4289–4293.

Schneider, D. J. (1981). 'First report of annual ryegrass toxicity in the Republic of South Africa', *Onderstepoort J. Vet. Res.*, **48**, 251–255.

Schwarz, R. T., and Datema, R. (1982). 'The lipid pathway of protein glycosylation and its inhibitors: the biological significance of protein-bound carbohydrates', *Adv. Carbohydr. Chem. Biochem.*, **40**, 287–379.

Speake, B. K., Malley, D. J., and Hemming, F. W. (1981). 'The effect of tunicamycin on secreted glycosidases of *Aspergillus niger*', *Arch. Biochem. Biophys.*, **210**, 110–117.

Struck, D. K., and Lennarz, W. J. (1980). 'The function of saccharide-lipids in synthesis of glycoproteins', in *The Biochemistry of Glycoproteins and Proteoglycans* (ed. W. J. Lennarz), pp. 35–83. Plenum Press, New York.

Struck, D. K., Suita, P. B., Lane, M. D., and Lennarz, W. J. (1978). 'Effect of tunicamycin on the secretion of serum proteins by primary cultures of rat and chick hepatocytes', *J. Biol. Chem.*, **253**, 5332–5337.

Stynes, B. A., and Petterson, D. S. (1980). 'Production of a neurotoxin in a cell culture of endosperm from *Lolium multiflorum* colonised by *Corynebacterium rathayi*', *Physiol. Plant Pathol.*, **16**, 163–167.

Stynes, B. A., and Wise, J. L. (1980). 'The distribution and importance of annual ryegrass toxicity in Western Australia and its occurrence in relation to cropping rotations and cultural practices', *Aust. J. Agric. Res.*, **31**, 557–569.

Tresselt, D., Eckhardt, K., Ihn, W., and Krebs, D. (1983). 'Zur chemischen Struktur der Streptovirudine', *J. Nat. Prod. (Lloydia)*, **46**, 483–489.

Vogel, P., Petterson, D. S., Berry, P. H., Frahn, J. L., Anderton, N., Cockrum, P. A., Edgar, J. A., Jago, M. V., Lanigan, G. W., Payne, A. L., and Culvenor, C. C. J. (1981). 'Isolation of a group of glycolipid toxins from seedheads of annual ryegrass (*Lolium rigidum* Gaud.) infected by *Corynebacterium rathayi*', *Aust. J. Exp. Biol. Med. Sci.*, **59**, 455–467.

MYCOTOXINS IN FOOD AND FEED

Trichothecenes and Other Mycotoxins
Edited by J. Lacey

CHAPTER 17

Fusarium mycotoxins and the malting of barley

Brian Flannigan, John G. Morton and Richard J. Naylor

INTRODUCTION

The objective in producing malt from barley for the brewing and distilling industries is to allow the grain to germinate and grow sufficiently for the starchy endosperm to be well modified and for enough enzymes to be synthesized for later conversion of the remaining polysaccharide to fermentable sugars. Initially, the malting process involves raising the moisture content (m.c.) to 43–47% by steeping in water, usually each of a series of two or more wet periods being preceded by an air-rest to accelerate subsequent germination (after draining) at an ambient temperature of 13–16.5 °C. The green malt (*ca.* 42% m.c.) so produced is then kilned in a stream of air at 45–60 °C for up to 24 h in order to reduce the moisture content to 5–6%. A final curing at 70–80 °C reduces it finally to 3–4% before the culms and rootlets are removed by screening (Briggs *et al.*, 1981).

Barley purchased for malting is normally of high quality, especially for the brewing and distilling industries, but although malt enters into a wide range of human foods as malt extract, the grain which is malted is not always so. For example, faulty storage may lead to deterioration of part or all of a bulk retained for malting. Culms and rootlets are used in compounded animal feeds, and spent grains remaining after aqueous extraction (mashing) prior to fermentation are also fed to cattle.

The aim of the present investigation was to establish the fate of zearalenone and T-2 toxin during the malting of artificially contaminated (spiked) barley. We also examined the effect of T-2 toxin on the germination and growth of barley and on characteristics of the finished malt, and of both zearalenone and T-2 toxin on the growth of brewing yeast.

MATERIALS AND METHODS

Preparation of malt from barley contaminated with toxin

Barley (cv. Golden Promise) was spiked with toxin dissolved in chloroform by continuously tumbling 125 g lots of grain mixed with 7 ml toxin solution in the

250 ml conical flask of a rotary evaporator for 5 min. The solvent was then removed by rotary evaporation under reduced pressure at 25–27 °C. After removal of the grain, the flask was washed with 1 ml chloroform and the washings were added to the spiked grain in a polyethylene bag. After further manual mixing, the grain was held at room temperature for 24 h to allow the chloroform to evaporate.

Aliquots (25 g) of spiked barley were malted at 16 °C for 7 days in polyvinyl tubes (80 × 40 mm bore) fitted with a 2 mm nylon-mesh base. The tubes were steeped in tap water (60 ml) for 16 h, drained to give an air rest of 8 h, and then steeped for a further 16 h (60 ml tap water) before being drained. The grain was then allowed to germinate for the remainder of the period with daily mixing to prevent entanglement of rootlets. The green malt so produced was then dried in the kiln of a micro-malting apparatus (Seeger, Fellbach, W. Germany) for 2.5 h at 50 °C, 3.5 h at 55 °C, 3.5 h at 65 °C and 2 h at 80 °C to produce a pale malt.

Extraction of steep liquor and malt for toxins

Aliquots (50 ml) of steep liquor were extracted by shaking with 50 ml chloroform for 1 h (after acidification with 1 ml 1 N HCl in the case of those containing zearalenone). The aqeuous phase was twice re-extracted with 20 ml chloroform, and the three chloroform fractions were pooled and evaporated to dryness at 50 °C under vacuum. Samples of dried malt (10.0 g), previously comminuted for 30 s in a Waring Blendor, were extracted with chloroform as above. To separate zearalenone from unwanted lipid, the combined chloroform fractions were extracted by shaking with 50 ml 1 N NaOH for 5 min and then discarded. The alkaline aqueous phase containing the zearalenone was acidified with 60 ml 1 N HCl, and then extracted twice with 50 ml chloroform for 5 min. Lipid was removed from chloroform extracts of dried malt containing T-2 toxin by shaking, partitioning into methanol and for 5 min with 50 ml petroleum ether (BP 70 °C), which fraction was then discarded, and re-extracting into chloroform. In both cases, chloroform extracts were then evaporated to dryness as above (Morton, 1983).

Analysis of samples for toxins

Zearalenone in dried films was extracted into methanol : water (75 : 25, v/v) and aliquots chromatographed in this solvent on a Waters Associates HPLC instrument using a 30 cm reverse phase C_{18}-coated silica gel column and a flow rate of 1 ml min^{-1}. The UV detector was set at a wavelength of 274 nm and the amount of toxin determined from the integrated area under the appropriate peak (Morton, 1983).

Dried samples of T-2 toxin were silylated with a 5 : 1 (v/v) mixture of N,O-*bis*-(trimethylsilyl) trifluoracetamide and trimethylchlorosilane for 1 h at

60 °C and excess reagent was removed *in vacuo* at 60 °C. Derivatized samples were dissolved in chloroform for gas–liquid chromatography (GLC) and aliquots chromatographed in a Packard 928 instrument fitted with a 1 m × 2 mm column of 3% OV17 on Chromasorb WAW (Phase Separations Ltd, Queensferry, Clwyd, U.K.), the oven temperature being 260 °C and the nitrogen flow-rate 30 ml min^{-1}. Silyl T-2 emerged at 3.9 ± 0.14 min and was measured in terms of peak height from the output of the flame ionization detector (Morton, 1983).

Germination and growth of barley contaminated with T-2 toxin

The percentage germination of spiked grain was determined after 2 days at 18 °C by the 4 ml test for germinative energy recommended by the Institute of Brewing (Institute of Brewing, 1977). Rootlet and coleoptile growth of 2.5 g samples of spiked grain, which had been steeped in 6 ml tap water according to the regime described previously, were examined after incubation for 5 days at 16 °C in Petri dishes set up as in the test for germinative energy.

Characteristics of malt prepared from barley contaminated with toxin

To test for α-amylase activity, 50-kernel samples of malt were ground in a domestic coffee grinder (Moulinex MC3) for 30 s along with 20 g solid carbon dioxide (crushed). After thawing, 0.6 g of ground malt was extracted for 1 h with 20 ml 50 mM sodium acetate buffer (pH 5.7) containing 100 mM sodium chloride and 10 mM calcium chloride. After centrifuging, the supernatant was used as the source of enzyme (Freeman, 1982). Activity was assayed by a modification of the Institute of Brewing (1977) recommended adaptation of the iodine-staining method of Briggs (1961). Purified β-limit dextrin prepared in this laboratory was used as the substrate and α-amylase-free β-amylase (Clodor, Manchester, U.K.) was included in reaction mixtures. For assay, 0.5 ml of a 0.3% (w/v) solution of β-limit dextrin in 20 mM sodium acetate buffer containing 10 mM calcium chloride and 0.2% (w/v) α-amylase-free β-amylase was incubated with 0.5 ml enzyme extract at 25 °C for 10 min. Dilute iodine solution (10 ml) was then added to the reaction mixture and the absorbance at 565 nm read on a Pye Unicam SP600 spectrophotometer (Freeman, 1982). Enzyme extracts were initially diluted in assay buffer to give a decrease of 0.3–0.4 in absorbance relative to controls (containing no enzyme extract) with an absorbance of 1.0. Activity is calculated as iodine-dextrin-colour-units, where

$$\begin{aligned} A_{565nm} \text{ units} &= A_{565nm} \text{ control} - A_{565nm} \text{ test} \\ &= 1.0 - A_{565nm} \text{ test,} \end{aligned}$$

and is expressed as units per kernel (Baxter *et al.*, 1974). The α-amino nitrogen content of wort prepared by hot-water extraction of coarsely ground samples of malt, which gives a measure of amino acids released from protein by combined action of endopeptidases, dipeptidases and carboxypeptidases, was determined by the standard methods (Institute of Brewing, 1977).

Growth of yeast in the presence of toxin

Log phase cultures of *Saccharomyces cerevisiae* NCYC 1108 in Sabouraud broth were diluted to give an absorbance of 0.10–0.12, as measured on a Pye-Unicam SP550 spectrophotometer, when added to toxin containing broth. Growth was observed for up to three generations and the dry weight of the culture calculated by reference to a standard graph of dry weight against absorbance at 570 nm (Morton, 1983).

RESULTS

Presence of toxin in steep liquor and malt

Complete recovery of zearalenone was achieved from barley spiked with 200 μg g^{-1} (Table 1), but only 57% was recovered with 20 μg g^{-1}. The efficiency of extraction from first-steep liquor (four replicates, spiked and left overnight before extraction) was examined and the relationship between toxin applied and recovered is shown in Figure 1. Because the first steep liquor contained the highest levels of extraneous solvent-extractable material, this graph was used to correct the observed zearalenone content of steep liquors and kilned malts to the absolute quantity. When used for malt it gave the same values as a 'spot' correction for which malt had been spiked with toxin as for barley (Table 2).

The results of a malting experiment (four replicates) with spiked barley are shown in Tables 2 and 3. With 200 μg zearalenone g^{-1} barley (Table 2), 4%

Table 1 Recovery of zearalenone (F-2) from barley

F-2 applied (μg)	2000[a]	200[b]
F-2 recovered (μg ± SD)	2067 ± 228	114 ± 2
Percentage recovery ± SD	103 ± 12	57 ± 1

[a] 200 μg toxin g^{-1} barley
[b] 20 μg toxin g^{-1} barley.

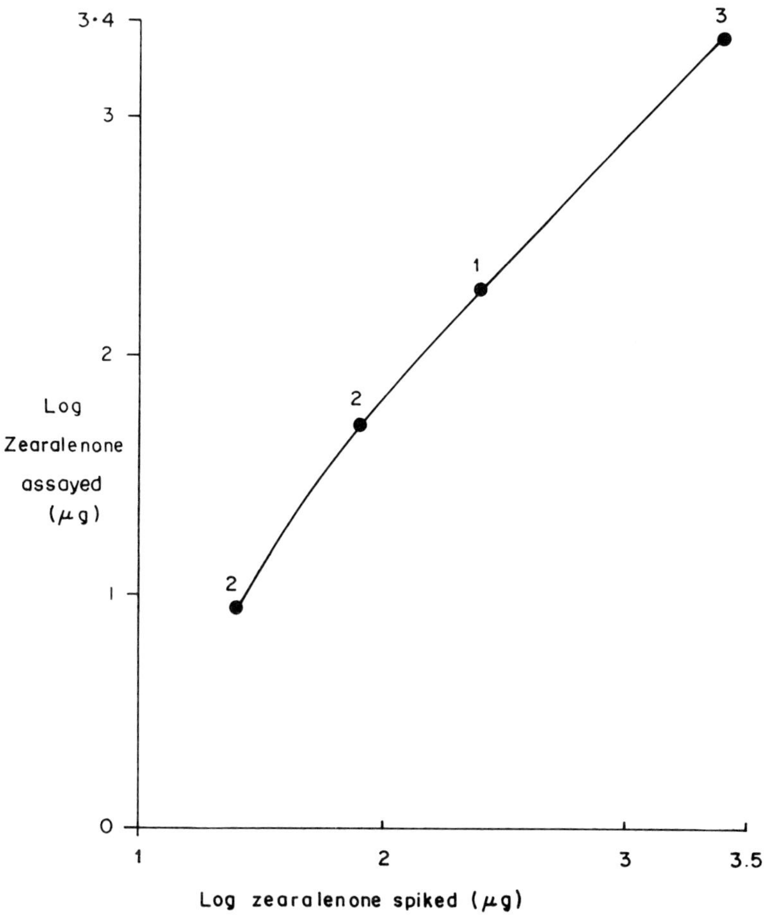

Figure 1 Recovery of zearalenone from spiked steep liquor. Percentage errors are shown above points.

was recovered in each steep and approx. 13% from the malts; with 20 μg toxin g^{-1} (Table 3), about 11% was recovered in the first steep, approx. 7% in the second and approx. 9% in the malt. Although the two methods of correction for extraction efficiency gave different values for zearalenone recovered with the lower rate of spiking, the difference was not significant.

Table 4 shows the results of a malting experiment (two replicates) in which barley spiked with 10 μg T-2 toxin g^{-1} was malted. The quantities of T-2 recovered have been corrected to absolute values on the basis that the recovery of toxin from first steep liquor spiked with 50 μg T-2 was 11 ± 1 μg

Table 2 Fate during malting of zearalenone (F-2) applied to barley (200 μg g^{-1})

	Barley	Steep 1	Steep 2	Malt	
Aliquot size	25 g	50 ml	50 ml	18.5 ± 0.5g	
F-2 per aliquot (μg ± SD)	5000	128.9 ± 9.5	138.5 ± 3.5	264.7 ± 21.1[a]	
				(*a*)	(*b*)
Corrected F-2 per aliquot (μg ± SD)	—	178.3 ± 12.3	190.6 ± 4.4	653.3 ± 69.7[b]	652.5 ± 70.0[b]
Percentage recovery of F-2 (± SD)	—	3.6 ± 0.3	3.8 ± 0.1	13.1 ± 1.4	13.1 ± 1.4

(*a*) Corrected using spot extraction efficiency test (75 ± 1% recovered of 300 μg spiked).
(*b*) Corrected using liquid sample extraction efficiency plot (Figure 1).
[a] Per 10 g aliquot.
[b] Per 18.5 ± 0.5 g malt.

Table 3 Fate during malting of zearalenone (F-2) applied to barley (20 μg g^{-1})

	Barley	Steep 1	Steep 2	Malt	
Aliquot size	25 g	50 ml	50 ml	18.1 ± 0.5 g	
F-2 per aliquot (μ ± SD)	500	30.4 ± 2.9	16.2 ± 0.3	11.5 ± 1.6[a]	
Corrected F-2 per aliquot (μg ± SD)	—	54.4 ± 3.1	35.9 ± 0.4	(*a*) 44.5 ± 9.7[b]	(*b*) 49.3 ± 2.0[b]
Percentage recovery of F-2 (± SD)	—	10.9 ± 0.6	7.2 ± 0.1	8.9 ± 1.9	9.9 ± 0.4

(*a*) Corrected using spot extraction efficiency test (47.5 ± 2.5% recovered of 11 μg spiked).
(*b*) Corrected using liquid sample extraction efficiency plot (Figure 1).
[a] per 10 g aliquot
[b] per 18.1 ± 0.5 g malt.

Table 4 Fate of applied T-2 toxin (10 μg g^{-1}) during malting

	Barley	Steep 1	Steep 2	Malt
Aliquot size	25 g	50 ml	50 ml	10 g
T-2 per aliquot (μg)	250	19, 21	4, 6	Trace
Corrected T-2 per aliquot (μg)	—	79–105	17–30	Trace
Percentage recovery of T-2 toxin	—	32–42	7–12	Trace

(20–24%). This value fell between the quantities found in the two steeps and was therefore applicable, although it introduced an overestimate of the absolute quantity of T-2 recovered from the first steep, and an underestimate of that in the second steep, when used as a correction factor. The low extraction efficiency means that on worst-case analysis the comparatively small errors are considerably amplified in the corrected values. The recovery from the first steep was 32–42% of the T-2 toxin applied to the barley, and from the second steep 7–12%, but only non-measurable traces were detected in the malt.

Effect of T-2 toxin on germination and growth of barley and on malt characteristics

Like that of control grain, percentage germination of barley spiked with 2 μg or 10 μg T-2 toxin g^{-1} barley was 95–97%. When the growth of rootlets and coleoptiles of grain which had been similarly spiked was examined, again the toxin had no effect on the number of kernels germinating (chitting), i.e. with the coleorhiza emerging. However, the numbers of kernels which chitted without further growth, and those in which the coleoptile did not emerge, rose with increasing toxin concentration (Table 5). Although the mean number of rootlets was not significantly altered, both the total rootlet and the coleoptile lengths (measured from the point of emergence from below the husk) were markedly reduced.

The α-amylase activity (six replicates) of malt prepared from barley spiked with 10 μg toxin g^{-1} was significantly reduced (*t* test, $p<0.001$), the reduction produced by 2 μg g^{-1} being approximately one-third of this and not significantly different from the control (Table 6). The α–amino nitrogen content of worts (three replicates) prepared from the malts fell by 5% at the lower rate of spiking and 9% at the higher (Table 6), but these reductions

Table 5 Effect of T-2 toxin on growth of rootlets and coleoptiles of malting barley

T-2 spiked ($\mu g\ g^{-1}$)	Total no. kernels	Kernels chitted but not grown (%)	Rootlets		Coleoptile	
			Mean no. per kernel (± SD)	Mean total length per kernel (mm ± SD)	Kernels lacking coleoptile (%)	Mean length per kernel (mm ± SD)
10	61	23.0	4.7 ± 1.2	28.6 ± 14.2	30.0	10.5 ± 6.7
	64	37.5	4.4 ± 1.1	16.2 ± 10.4	55.0	4.7 ± 2.9
2	66	27.3	5.5 ± 0.8	38.4 ± 15.2	10.0	18.3 ± 5.8
	64	18.8	5.3 ± 0.8	39.8 ± 16.6	12.5	15.1 ± 7.0
Solvent control	56	14.3	5.0 ±1.2	49.5 ± 16.2	0	20.3 ± 6.0
	64	16.7	5.4 ± 0.8	48.4 ± 15.7	2.5	21.2 ± 7.9

Table 6 Effect of T-2 toxin on malt characteristics

T-2 spiked ($\mu g\ g^{-1}$ barley)	α-Amylase activity[a] of kilned malt (units $kernel^{-1} \pm SD$)	α-Amino nitrogen content of wort ($mg\ l^{-1} \pm SD$)
10	12.3 ± 0.8	168.1 ± 3.3
2	14.9 ± 1.7	173.5 ± 1.3
Solvent control	15.9 ± 1.7	182.9 ± 12.6

[a] See Materials and Methods, for derivation of units.

were not significant, and the SG of the worts from malts prepared using spiked barley did not differ from that of the control, all being 1.0280–1.0283.

Growth of yeast in the presence of toxin

The effect of zearalenone on the dry weight of log-phase cultures of *S. cerevisiae* is shown in Figure 2. Graphs A and B, derived from the same experiment, demonstrate the inverse linear relationship between growth and increasing toxin concentration over the range 5–50 μg toxin ml^{-1} broth culture, irrespective of the duration of the assay. However, the linear relationship was not consistently exhibited at lower concentrations of toxin (Graph C). When cultures were allowed to grow on into stationary phase, those exposed to 10 μg toxin ml^{-1} showed higher yields than controls, but at higher concentrations of zearalenone dry weights were markedly reduced (Figure 3). An inverse linear relationship was observed in cultures exposed to T-2 toxin over the range 0.1–1.0 μg toxin ml^{-1} broth culture (Figure 4).

DISCUSSION

Although analysis of T-2 toxin would have benefited from more refined methodology, the results indicate that approximately 73–79% of zearalenone applied to the barley, and 46–61% of T-2 toxin, was apparently lost during the malting process. Relatively little toxin was recovered from the kilned malt, only 9–13% of zearalenone and unmeasurable traces of T-2 being detected. Although Krogh *et al.* (1974) made no mention of extraction efficiency, they reported that ochratoxin A in moderately contaminated barley (0.4–0.8 $\mu g\ g^{-1}$) was 'entirely degraded' during malting, but when heavily contaminated grain (1–5 $\mu g\ g^{-1}$) was malted 11–19% of the toxin appeared in the wort prepared from the malt. Although the apparent loss of zearalenone and T2

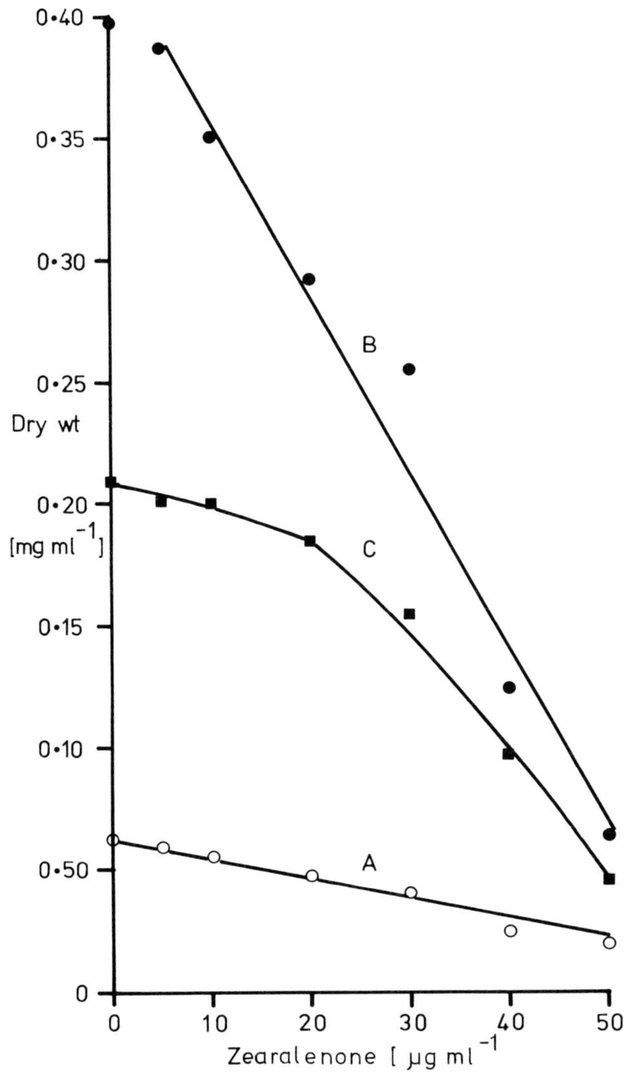

Figure 2 Effects of zearalenone on growth of log-phase cultures of *Saccharomyces cerevisiae*. Graphs A and B refer to one experiment when samples for measurement of dry weight were taken after 3 h (A) and 7.5 h (B). Graph C refers to a separate experiment when samples were taken after 4 h.

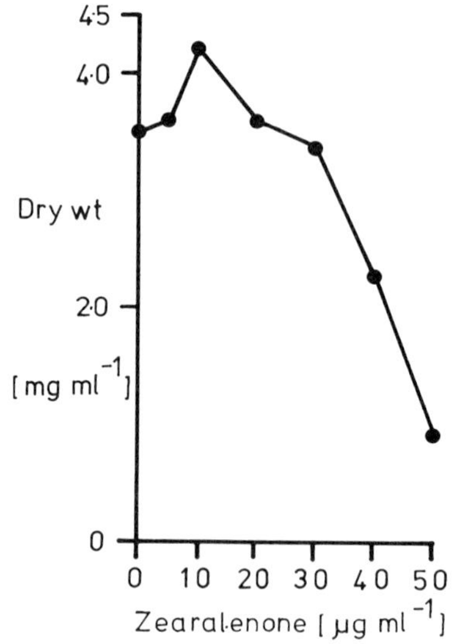

Figure 3 Effect of zearalenone on maximum growth yield of *Saccharomyces cerevisiae* in broth culture.

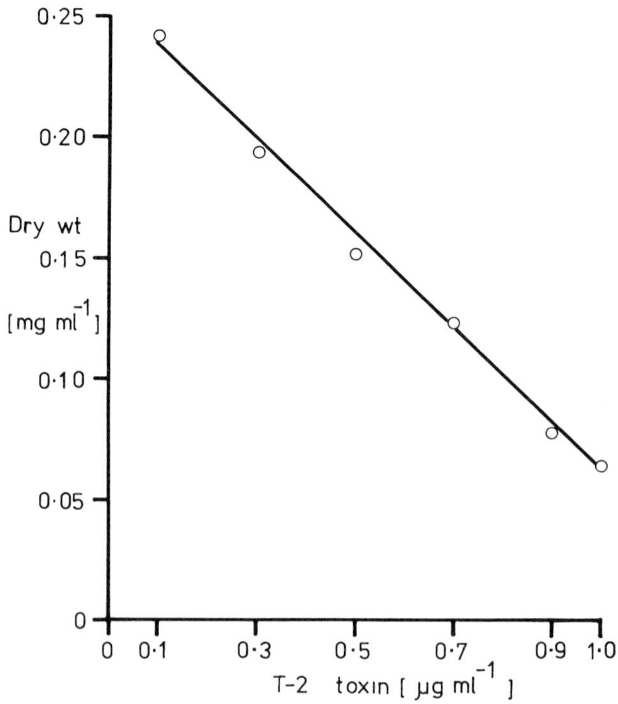

Figure 4 Effect of T-2 toxin on growth of log-phase cultures of *Saccharomyces cerevisiae*.

during malting might be attributable to degradation, the relatively low efficiency of extraction both of toxins from spiked steep liquor and of zearalenone from spiked malt might indicate binding of toxin to substances extracted from barley into steep liquor and present in kilned malt. It is also possible that the toxins may be bound to or degraded by micro-organisms associated with barley and malt. Both the effects observed on the growth of *S. cerevisiae* in this investigation and the recovery of ochratoxin A in yeast after brewing (Nip *et al.*, 1975) illustrate uptake of toxins by yeast. The possibility also exists that during kilning hydroxyl groups on zearalenone and T-2 may react to form esters with carboxyl groups of proteins, amino acids and oxidative or other degradation products of malt sugars.

The report of Ylimaki *et al.* (1979) that *Fusarium*-inoculated barley containing 20 μg T-2 toxin kg^{-1} had a germinative capacity (percentage of living kernels) of only 74% contrasts with the fact that the much higher concentrations used in the present investigation did not reduce germinative energy (percentage germination), but it is not clear whether the effects noted by Ylimaki *et al.* (1979) were entirely due to the toxin or to germ damage caused by other factors associated with the presence of fusaria on the grain. The effects on rootlet and coleoptile growth (Table 5) are in agreement with the findings of Nummi *et al.* (1975), who used toxin solutions of unspecified concentrations. However, these workers appear to have used a greater maximum concentration of toxin than in the present study, as they recorded complete inhibition of rootlet and coleoptile growth and a reduction of 90% in α-amylase activity in malt, compared to a reduction of only 23% in this investigation.

Since α-amylase is synthesized *de novo* in germinating barley (Filner and Varner, 1967), the reduced α-amylase activity observed in malts prepared from spiked barley could indicate inhibition of synthesis of this enzyme by T-2. It is well known that trichothecene mycotoxins variously inhibit protein synthesis (Wyllie and Morehouse, 1977). There also appears to have been some reduction in proteolytic activity, as judged by the reduction in α-amino nitrogen in the worts (Table 6). Some proteolytic enzymes are synthesized *de novo* (Jacobsen and Varner, 1967) but it is possible that the reduced activity both of these enzymes and α-amylase may reflect inhibition of enzyme action rather than repression of *de novo* synthesis. However, the depression of α-amylase and protease activity appears to have had little effect on the modification of the starchy endosperm during malting, since the SG of the worts prepared by mashing of the malts was not reduced relative to the controls.

In addition to the presence of toxins in finished malt and the depression by T-2 of growth and enzyme synthesis of barley during malting, the inhibitory effects of both toxins on the growth of yeast may have implications for the fermentation industry. T-2 toxin has also been noted to cause 58–80% reduction in the rate of fermentation of *S. cerevisiae* var. *carlsbergensis*

(Lafont *et al.*, 1981). The results obtained with T-2 (Figure 4) also suggest that *S. cerevisiae* could be a useful organism for the bioassay of purified toxin. Burmeister and Hesseltine (1970) demonstrated that T-2 solutions in the same range of concentrations as used in this investigation reduced the percentage germination of peas. However, the zone inhibition test of these authors, who used *Rhodotorula rubra*, showed no detectable inhibition of growth of this yeast by discs bearing less than 4 μg T-2 toxin.

REFERENCES

Baxter, E. D., Booer, C. D., and Palmer, G. H. (1974). 'Improved malting performance of freshly harvested barley after abrasion', *J. Inst. Brewing*, **80**, 549–555.

Briggs, D. E. (1961). 'Modification of the Sandstedt, Kneen and Blish assay of α-amylase', *J. Inst. Brewing*, **67**, 427–431.

Briggs, D. E., Hough, J. S., Stevens, R., and Young, T. W.(1981). *Malting and Brewing Science*, Vol. 1. Chapman and Hall, London.

Burmeister, H. R., and Hesseltine, C. W. (1970). 'Biological assays for two mycotoxins produced by *Fusarium tricinctum*', *Appl. Microbiol.*, **20**, 437–440.

Filner, P., and Varner, J. E. (1967). 'A test for *de novo* synthesis of enzymes: density labelling with $H_2{}^{18}O$ for barley α-amylase induced by gibberellic acid' *Proc. Nat. Acad. Sci.*, **58**, 1520–1526.

Freeman, P. L. (1982). 'Physiological aspects of enzyme modification in malting barley' Ph.D. thesis, Heriot-Watt University, Edinburgh.

Institute of Brewing (1977). *Recommended Methods of Analysis*, Institute of Brewing, London.

Jacobsen, J. V., and Varner, J. E. (1967). 'Gibberellic acid-induced synthesis of proteases by isolated aleurone layers of barley', *Pl. Physiol.*, **42**, 1596–1600.

Krogh, P., Hald, B., Gjertsen, P., and Myken, F. (1974). 'Fate of ochratoxin and citrinin during malting and brewing experiments', *Appl. Microbiol.*, **28**, 31–34.

Lafont, J., Romand, A., and Lafont, P. (1981). 'Effets de mycotoxines sur l'activité fermentaire de *Saccharomyces cerevisiae*', *Mycopathologia*, **74**, 119–123.

Morton, J. G. (1983). 'Fusarium mycotoxins and the malting of barley', M.Sc. thesis, Heriot-Watt University, Edinburgh.

Nip, W. K., Chang, F. C., Chu, F. S., and Prentice, N. (1975). 'Fate of ochratoxin in brewing', *Appl. Microbiol.*, **30**, 1048–1049.

Nummi, M. Niku-Paavola, M.-L., and Enari, T.-M. (1975). 'Der Einfluss eines *Fusarium*-Toxin auf die Gersten-Vermälzung, *Brauwissenschaft*, **28**, 130–133.

Wyllie, T. D., and Morehouse, L. G. (1977). *Mycotoxic Fungi, Mycotoxins, Mycotoxicoses*, Vol. 1. Dekker, New York.

Ylimaki, A., Koponen, H., Hintikka, E.-L., Nummi, M., Niku-Paavola, M.-L., Ilus, T., and Enari, T.-M. (1979). *Mycoflora and Occurrence of Fusarium Toxins in Finnish Grain*. Materials and Processing Technology, Publication 21, Technical Research Centre of Finland, Espoo.

Trichothecenes and Other Mycotoxins
Edited by J. Lacey

CHAPTER 18

Effect of processing on aflatoxin in contaminated maize

Luzviminda P. Celino and Lilia S.P. Madamba

INTRODUCTION

Aflatoxin, a secondary metabolite produced by *Aspergillus flavus* Link ex Fries and *A. parasiticus* Speare was first found in peanut meal that caused the death of many turkey poults in England in 1960 (Sargent *et al.*, 1961). Subsequently it has been found in other agricultural commodities, including maize (Slocum, 1963).

Maize is important in the Philippines both as a food crop and an industrial raw material. White maize is converted to grits, processed in a like manner to rice and eaten in areas where maize is the staple food. It is also used to manufacture starch, oil, syrup, dextrins and glucose which are utilized in a wide variety of products both commercially and in the home. Some maize is popped, roasted, and sold as salted popped maize or sweetened crackers. Yellow maize is used mainly for livestock and poultry feeds.

A high incidence of aflatoxin in maize and its by-products in the Philippines has been established (Santamaria *et al.*, 1972). The ubiquity of the fungi and carcinogenicity of the toxins to various animal species (Wogan, 1966) have led to concern over the risks from toxin contamination of food products. As a result of this concern, attempts have been made to remove or inactivate aflatoxin contaminants. All the aflatoxins are stable to heat and resistant to destruction during the heat processing of foods (Carnaghan, 1965). It is not removed by ordinary washing procedures (Cucullu *et al.*, 1966). Detoxification of aflatoxins in contaminated oilseeds has been attempted using solvent extraction of aflatoxin (Rayner and Dollear, 1968; Rayner *et al.*, 1970) and using hydrogen peroxide (Sreenivasamurthy *et al.*, 1967) but ammoniation seems to be the most promising treatment (Lee *et al.*, 1974).

Maize has been used in most studies on the effects of processing on aflatoxins. The processing of oil by alkali refining, washing and bleaching destroys any aflatoxin extracted from germ (Parker and Melnick, 1966). After milling, most aflatoxin from maize artificially contaminated with large amounts of aflatoxin was found in the steepwater (40%) and fiber (30–38%), with the remainder in the gluten and germ. The starch contained only 1% of the original aflatoxin (Ruark and Watson, 1967). Ensiling high moisture shelled maize contaminated with aflatoxin caused little change in aflatoxin B_1 content. The lactic fermentation was insufficiently acid to convert the aflatoxin into the non-toxic derivative aflatoxin B_{2a} (Linderfelser and Ciegler, 1970).

The main objective of our study was to determine the effect of home processing on the aflatoxin content of contaminated maize.

MATERIALS AND METHODS

Preparation of samples

A 5-kg sample of naturally contaminated maize grits was used to determine the effects of washing and cooking on aflatoxin content. Fifty gram samples of pop maize and glutinous maize varieties were artificially inoculated by adding dropwise 5 ml of an aqueous spore suspension containing 4.8×10^6 spores of *A. flavus* per ml (peanut isolate); incubated for 1 week at room temperature and stored in the freezer until use. The inoculated pop maize and glutinous maize sample varieties were used in the popping and roasting experiments respectively.

Home processing methods

Steam cooking

One set of samples (50 g) was washed twice with 150 ml tap water, then cooked with 150 ml tap water using an automatic rice cooker or an ordinary aluminium cooking pan. Another set of samples was cooked without prior washing.

Popping procedure

An aluminium cooking pan (8 l) hitched to an iron wire and heated by a Humboldt Fisher burner was used in this popping experiment. Cooking medium (coconut oil and margarine) (10 g) were preheated in the cooking pan to 50–60 °C. Artificially contaminated maize (50 g) was popped for 3.5 min at 130 ± 5 °C. The pan was continuously shaken throughout the duration of the popping process to ensure even heat distribution.

Roasting procedure

Cooking oil (100 g) was preheated to 130 °C in a small cast-iron frying pan. Aflatoxin-contaminated samples (50 g) were roasted for 15 min with constant stirring at 130–150 °C. Actual roasting temperature was measured with a thermometer in the cooking oil.

For dry roasting, the frying pan was preheated for 2–3 minutes. Contaminated samples (50 g) were roasted for 30 min with constant stirring at 60 °C.

'Binatog' preparation

'Binatog' is a maize food preparation common in the Southern Luzon region. It is usually eaten as a snack with grated coconut.

In this experiment, 2 g lime and 100 ml tap water were added to 50 g of inoculated maize and the mixture heated to boiling for 10 minutes. After cooling, the samples were washed twice with 200 ml tap water and filtered in a wire screen. The washings were pooled and the total volume noted. The lime-treated samples were cooked with 500 ml tap water in a small cast-iron cooking utensil for 2.5 h to produce a soft sticky product.

Analytical procedure

After each cooking experiment, 50 g each of raw and cooked samples were weighed and stored in the freezer. All samples were brought to room temperature and extracted following the method of Alexander and Baur (1977). The wash-water and lime washings were extracted following the method of Lee (1965). The extraction process of Parker and Melnick (1966) was adopted for oil samples.

Thin-layer chromatography

Five, 10 and 15 μl of samples and known quantities of aflatoxin standards were spotted on Adsorbosil I (Applied Science Lab, State College, Pa.) precoated, activated plates. Aflatoxin levels were adjusted so that intensities were similar to those of the unknown samples. The plates were developed in a non-equilibrated tank containing 0.5 cm of 9 : 1 v/v chloroform–acetone, for 30 min at 25 °C. After development, the plates were dried and examined under a high-intensity (365 nm) long-wave ultraviolet light. Aflatoxin was quantified by visually estimating the intensity of unknown spots against known aflatoxin spots and the concentration of aflatoxins in the original sample calculated using appropriate dilution factors.

RESULTS AND DISCUSSION

Steam cooking

Although aflatoxin is insoluble in water, washing of naturally contaminated maize grits lessened the aflatoxin B_1 content (Table 1). For instance, when 50 g sample containing 9.4 μg aflatoxin B_1 (198 μg/kg) were washed twice with 150 ml water, an average of 3.1 μg were recovered in the wash water. This represents about 33% of the original amount. Possibly some of the aflatoxin was bound to water-soluble components and was extracted with these substances. A similar observation was noted by Feuell (1966) in his solvent extraction studies of aflatoxins.

After cooking, the same sample still contained 5.7 μg aflatoxin B_1 (34 μg/kg) or 61% of the initial amount. The total loss was 39% and as 33% was lost during washing, 6% could be attributed to the effect of heat in the rice cooker. When cooked in an ordinary pan the loss was 0.6 μg or 7% of the initial amount. The difference in the reduction of aflatoxin during cooking in the two cooking vessels, however, was not significant, although cooking time in the rice cooker was about 5 min longer than the ordinary pan.

In both cases the amount of aflatoxin in the cooked samples (34 μg/kg) was above the 20 μg/kg level established as the safe level by U.S. Food and Drug Administration.

Cooking maize grits without prior washing caused a reduction in aflatoxin content of 9% regardless of cooking vessel. It appears that more aflatoxin is lost by washing than by cooking. Repeated washing may cause more reduction but is not practical due to significant losses of soluble nutrients.

Table 1 Effect of washing and cooking on the aflatoxin B_1 content of naturally contaminated maize grits

Cooking method	Stage	Toxin Content (μg/50 g)	Toxin reduction (%)
Rice cooker	Initial sample	9.4 ± 0.2	—
	Washed and cooked	5.7 ± 0.4	39 ± 2
	Washings	3.2 ± 0.4	—
	Unwashed and cooked	8.5 ± 0.5	9 ± 2
Cooking pan	Initial sample	9.3 ± 0.2	—
	Washed and cooked	5.8 ± 1.4	37 ± 4
	Washings	2.9 ± 1.0	—
	Unwashed and cooked	8.4 ± 0.2	9 ± 1

Popping experiment

The effect of popping artificially contaminated maize using two different media is shown in Table 2. When 50 g kernel of a local maize variety containing 349 μg aflatoxin B_1 (6980 μg/kg) and 551 μg (11,000 μg/kg) aflatoxin G_1 were popped in margarine, a large reduction in aflatoxin content was noted. The popped samples contained 92 μg aflatoxin B_1 (1730 μg/kg) and 195 μg aflatoxin G_1 (3670 μg/kg), a 73% and 64% reduction of aflatoxin B_1 and G_1, respectively. Using cooking oil as the popping medium, reduction was 87% for aflatoxin B_1 and 84% for aflatoxin G_1. This suggests that cooking oil is a better medium for popping the local variety of maize than margarine.

The reverse was true with the imported variety of pop maize (TNT brand). A 78% reduction in aflatoxin B_1 was found using oil as the medium but 88% with margarine. With aflatoxin G_1, reductions obtained were 95% and 97% respectively, for samples popped in oil and margarine. Aflatoxin G_1 was more effectively reduced than B_1 regardless of popping medium. Investigations at the U.S. Food and Drug Administration (Roegner, 1967) showed that aflatoxin B_1 is heat-stable, while the less toxic aflatoxin G_1 is not.

Similar studies by Escher *et al.* (1973) on roasting pecan halves with oil or margarine showed little difference between the two in toxin degradation. Reductions were, respectively, 65% (seedling) and 60% (Stuart) in oil and 60% (seedling) and 59% (Stuart) in margarine. Although the margarine temperature (107 °C) was much lower than that of the oil bath (191 °C), this was compensated for by a much longer treatment time (60 min vs 6 min).

In our study, although popping was always done at 130 ± 5 °C for 3 min, significant differences in aflatoxin reduction were noted between popping media. No explanation can be offered for this difference, but it is possible that the nature of the oil and margarine may contribute to this effect.

Table 2 Effect of popping on the aflatoxin content of artificially contamined maize kernels

		Toxin content (μg/50 g)		Toxin reduction (%)	
Source	Stage	B_1	G_1	B_1	G_1
Local variety	Initial toxin content	349 ± 36	551 ± 51	—	—
	Popped in margarine	92 ± 12	195 ± 6	73 ± 4	64 ± 4
	Popped in oil	43 ± 18	85 ± 6	87 ± 8	84 ± 4
Imported variety	Initial toxin content	349 ± 32	717 ± 1	—	—
	Popped in margarine	41 ± 12	21 ± 11	88 ± 8	97 ± 8
	Popped in oil	76 ± 23	30 ± 5	78 ± 8	95 ± 4

Roasting experiment

Fifty gram maize kernels artificially contaminated within 38 μg aflatoxin B_1 (780 μg/kg) and 64 μg aflatoxin G_1 (1280 μg/kg) were dry roasted. The resulting product contained only 2 μg aflatoxin B_1 (46 μg/kg) and 7 μg aflatoxin G_1 (161 μg/kg), decreases of 94% and 89% respectively (Table 3). Oil roasting was less effective, perhaps because of the shorter heating time.

The oil remaining after roasting contained some of the aflatoxin from the sample but this only accounted for 2% of aflatoxin B_1 and 3% of the G_1. Thus re-use of such oil could be a source of aflatoxin contamination of other food (Parpia and Sreenivasamurthy, 1971).

The loss of aflatoxins B_1 and G_1 from artificially contaminated maize was more than previously found with peanuts. Naturally contaminated single peanut halves dry roasted at 150 °C for 30 min lost 80% and 60% respectively

Table 3 Effect of roasting on the aflatoxin content of artificially contaminated maize kernel

	Toxin content (μg/50 g)		Toxin reduction (%)	
	B_1	G_1	B_1	G_1
Initial toxin content	38 ± 2	64 ± 4	—	—
Dry roasted	2 ± 0.8	7 ± 2	94 ± 8	89 ± 8
Oil roasted	5 ± 1	4 ± 1	86 ± 4	93 ± 8
Oil	0.6 ± 0.4	1.7 ± 1	—	—

of aflatoxins B_1 and B_2 while inoculated peanuts roasted at 163 °C for 7 min or 174 °C for 3 min lost 65% of aflatoxins B_1 and G_1 (Lee *et al.*, 1968). Dry roasting time used in our study was the same as that of Lee *et al.*, (1968), but their oil roasting temperature was higher (174 °C) although we used a longer treatment time.

‘Binatog’ preparation

Lime washing during ‘binatog’ preparation removed 68% of both aflatoxin B_1 and G_1, most of which was recovered in the lime washings (Table 4). This suggests some aflatoxin is deep within the maize kernel as found in peanuts (Cucullu *et al.*, 1966).

Table 4 Effect of lime washing on the aflatoxin content of artificially contaminated maize kernels for 'binatog' preparation

	Toxin content (μg/50 g)		Toxin reduction (%)	
	B_1	G_1	B_1	G_1
Initial toxin content	38 ± 2	64 ± 4	—	—
Lime washed/uncooked	12 ± 3	20 ± 1	68 ± 6	68 ± 2
Lime washings	21 ± 1	22 ± 1	—	—
Lime washed/cooked	2 ± 0.7	6 ± 2	94 ± 8	90 ± 8

No aflatoxin could be recovered from the lime washings unless these were acidified with 1 N HCl. This is probably a consequence of the lactone ring structure of aflatoxin B_1 or G_1, opened or hydrolysed in the aqueous lime being restored by the acid.

The ease with which aflatoxin was converted to the alkali salt derivative and back again following acidification, using acid of the type and concentration found in the stomach, supports Feuell's (1966) observation that a non-fluorescent form of aflatoxin obtained by mild alkaline treatment may still become toxic. However, we found that thorough washing of the lime-treated maize with tap water, until it gave a neutral reaction, removed most of the salt derivative.

About 13% of the original aflatoxin B_1 and 34% of G_1, was unaccounted for after totalling recoveries from the maize and lime washings. These differences could be attributed to low recoveries of the aflatoxins or to incomplete conversion of calcium salt to original aflatoxin by acid treatment.

After cooking, 2 μg aflatoxin B_1 (10 μg/kg) and 6 μg aflatoxin G_1 (40 μg/kg) remained, giving a loss of about 23–27% due to cooking and total losses of aflatoxins B_1 and G_1 of 94% and 90% respectively. Loss due to lime washing was 68% for both aflatoxin B_1 and G_1. The total aflatoxin found in the cooked sample (50 μg/kg) was still above the 20 μg/kg level established as the safe level by U.S. Food and Drug Administration.

REFERENCES

Alexander, R. J., and Baur, M. C. (1977). 'Note on a two-dimensional TLC procedure for determining aflatoxins in corn', *Cereal Chem.*, **54**(3), 699–704.

Carnaghan, R. B. A. (1965). 'Hepatic tumors in ducks fed a low level of toxic groundnut meal', *Nature*, **208**, 308.

Cucullu, A. F., Lee, L. S., Mayne, R. Y., and Goldblatt, L. A. (1966). 'Determination of aflatoxins in individual peanuts and peanut sections', *J. Am. Oil Chem. Soc.*, **43**, 89–92.

Escher, F. E., Koehler, P. E., and Ayres, J. C.(1973). 'Effect of roasting on aflatoxin content of artificially contaminated pecans', *J. Food Sci.*, **38**, 889–892.

Feuell, A. J. (1966). 'Aflatoxin in groundnuts. Part IX—Problems of detoxification', *Tropical Sci.*, **7**, 60–70.

Lee, W. V. (1965). 'Quantitative determination of aflatoxin in groundnut products, *Analyst*, **90**, 305–307.

Lee, L. S., Cucullu, A. F., and Goldblatt, L. A. (1968). 'Appearance and aflatoxin content of raw and dry roasted peanut kernels', *Food Technol.* **22**, 1131–1134.

Lee, L. S., Stanley, J. B., Cucullu, A. F., Pons, W. A. Jr., and Goldblatt, L. A. (1974). 'Ammoniation of aflatoxin B_1: Isolation and identification of the major product', *J. Assoc. Off. Anal. Chem.*, **57**, 408–410.

Linderfelser, L. A. and Ciegler, A.(1970). 'Studies on aflatoxin detoxification in shelled corn by ensiling', *J. Agric. Food Chem.*, **18**, 640–648.

Parker, W. A., and Melnick, D. (1966). 'Absence of aflatoxin from refined vegetable oils', *J. Am. Oil Chem. Soc.*, **43**, 635–638.

Parpia, H. A. B., and Sreenivasamurthy, V. (1971). 'Importance of aflatoxin in foods with reference to India', 3rd Intl. Congress Food Science and Tech. Proceedings, p. 701 (ed. Stewart G. F. and Willey, C. L.). Inst. Food Technologists, Chicago, Ill.

Rayner, E. T., and Dollear, F. G. (1968). 'Removal of aflatoxins from oil seed meals by extraction with aqueous isopropanol', *J. Am. Oil Chem. Soc.*, **45**, 622–624.

Rayner, E. T., Dollear, F. G., and Codifer, L. P. Jr. (1970). 'Extraction of aflatoxins from cotton seed and peanut meals with ethanol', *J. Am. Oil Chem. Soc.*, **47**, 26.

Roegner, F. R. (1967). 'The FDA looks at aflatoxins', 38th Ann. Tech. Conf. Grain Elevator Processing Superintendents Assoc., Chicago, Illinois, 28 Feb. 1967.

Ruark, R. G., and Watson, S. A. (1967). 'Survey of mycotoxin potentiality in the wet-milling industry', p. 11. In Summary Proc. 1967 *Mycotoxin Research Seminar*, USDA, Washington, D.C.

Santamaria, P. A., Pizarro, A. C., and Jackson, C. R. (1972). 'Aflatoxin contamination of raw agricultural crops and their by-products in the Philippines', *Phil. Phytopathol.*, **8**, 12–20.

Sargent, K. A., Sheridan, J. O'Kelly, and Carnaghan, R. B. A. (1961). 'Toxicity associated with certain samples of groundnuts', *Nature*, **192**, 1096–1097.

Slocum, G. G. (1963). 'Let's look at some microbiological problems of cereal goods', *Cereal Sci. Today*, **8**, 313–314.

Sreenivasamurthy, V., Parpia, H. A. B., Srikanta, S., and Murti, A. S. (1967). 'Detoxification of aflatoxin in peanut meal by hydrogen peroxide', *J. Assoc. Off. Anal. Chem.*, **50**, 350–354.

Wogan, G. N. (1966). 'Chemical nature and biological effects of the aflatoxins', *Bacteriol. Rev.*, **30**, 460–470.

Trichothecenes and Other Mycotoxins
Edited by J. Lacey

CHAPTER 19

Occurrence and significance of penicillia and other fungi in mould-spoiled foods

B. Jarvis, A. P. Williams, Ailsa D. Hocking, W. B. Chapman, Sara J. Cooper, Gillian M. Wood, and J. I. Pitt

INTRODUCTION

In addition to their beneficial role in the production of fermented foods such as 'blue cheese', members of the genus *Penicillium*, as well as many other fungi, are responsible for extensive spoilage of food materials. Statistics on the occurrence of such spoilage are sparse and those which are published probably only represent the tip of an iceberg. Data published annually by the Association of Environmental Health Officers in the U.K. indicate a continuing level of complaints of mould in foods running at about 4000–6000 per year. However, these data refer only to those public complaints made to enforcement officers and exclude foods thrown away by the consumer and those returned direct to the manufacturer, either by the consumer or the retailer. About two-thirds of the complaints each year relate to bread, flour confectionery, cheese and dairy products, and meats and meat products. No comparable figures exist for Australia. The State Health Departments do not issue figures on consumer complaints about mould-spoiled foods. The CSIRO Division of Food Research deals with approximately 20–30 problems each year, concerning mould spoilage, but these originate at the manufacturer level, rather than the consumer. The Division receives many more telephone enquiries regarding mouldy foods and mycotoxins which do not require any laboratory investigations.

In a survey undertaken in the early 1970s, Jarvis and Rhodes (unpublished) observed a predominance of penicillia in the spoilage mycoflora of meats and meat products stored under refrigeration. Similar observations were made by Torrey and Marth (1977), who found that although penicillia were the dominant spoilage fungi in refrigerated foods, a higher incidence of aspergilli occurred in non-refrigerated mould-spoiled products.

In common with fungi of other genera, many penicillia are capable of producing mycotoxins in foods, feedstuffs and culture media. The nature of the mycotoxins is diverse and ranges from citrinin, penicillic acid and ochratoxins to mycophenolic acid, PR toxin and tremorgenic toxins. Further details of the types of mycotoxin and the species reported to produce them are to be found in Moreau (1979) and other reviews. Although surveys of mycotoxigenic penicillia in cereal and other agricultural products have been detailed, those of mould-spoiled food products have been limited and generally restricted to mouldy cheese. Bullerman and Olivigini (1974), Bullerman (1976, 1980), Lafont *et al.* (1979), Northolt *et al.* (1980), Josefsson (1981) and Jarvis (1983) all reported the isolation of potentially mycotoxigenic penicillia from mouldy cheese.

The objectives of studies in the U.K. by some of us, of which this chapter describes but one part, are to assess the nature of the fungi commonly isolated from spoiled manufactured foods; to assess the extent to which the isolated fungi may be potentially mycotoxigenic; and to investigate alternative, and preferably rapid, methods for their identification.

No such systematic studies have been, or are being, done in Australia. Thus the work described has, for the most part, been undertaken at Leatherhead Food RA and most of the data presented are derived from these studies.

EXPERIMENTAL

Food samples

Samples of U.K. mould-spoiled food of many types were obtained from three sources; some were provided by industrial members of the Leatherhead Food RA and could be considered to be typical of foods spoiled but not necessarily offered for sale to the public; others were provided by colleagues and represent foods spoiled during domestic storage; the remainder originated from enforcement agencies and were the subject of public complaint under the U.K. Food and Drugs Act (Anon, 1955).

Australian samples, examined at the CSIRO Division of Food Research, mostly originated from food manufacturers, either as finished product or suspect raw materials. A few samples of mouldy foods were brought in by colleagues.

Isolation of fungi

At Leatherhead Food RA, fungi were picked from the spoiled foods and plated onto Rose Bengal chloramphenicol agar (Oxoid modification of the medium of Jarvis, 1973), Czapek–Dox agar and malt extract agar. Isolated colonies were picked onto a range of media to check purity, and identification was done according to the following methods: for penicillia according to Raper and Thom (1949), Pitt (1979a), Samson *et al.* (1976) and Frisvad (1981); for aspergilli according to Raper and Fennell (1965); for other genera according to Onions *et al.* (1981).

Similar methods were used at CSIRO Food Research except that the primary isolation medium was Dichloran Rose Bengal chloramphenicol agar (DRBC; King *et al.*, 1979) and penicillia were identified according to Pitt (1979a).

Detection of mycotoxins in mouldy foods

Samples of mouldy foods were, in some instances, examined for the presence of a range of mycotoxins using a modification of the extraction method described by Roberts and Patterson (1975) and HPLC or reverse-phase HPLC analysis of the extract as described by Jarvis *et al.* (1982).

Mycotoxin assays were not carried out on Australian food samples.

Detection of biological activity in culture extracts

Representative fungal isolates were cultured for 7 days at 25 °C in Czapek solution and in media containing yeast extract (20 g/l) and sucrose (40 or 200 g/l). The cultures (comprising both mycelium and culture medium) were extracted sequentially using hexane, chloroform and ethylacetate. The extracts were concentrated *in vacuo* or under nitrogen, pooled and tested against baby hamster kidney cells, brine shrimps and germinating pea seedlings (Burmeister and Hesseltine, 1970; Jarvis *et al.*, 1982). All bioassays were done in parallel with both positive and negative controls.

RESULTS

Identification of fungal isolates

Penicillia were isolated from over 50% of 215 mouldy food samples examined at the Leatherhead Food RA, representing six major classes of foods: meat products, cheese, bakery products, fruit and vegetable products, nuts and miscellaneous. In total, 413 different fungal isolates were obtained; of these, 219 (53%) were penicillia, 56 (14%) were aspergilli, 65 (16%) were zygomycetes and 36 (9%) were cladosporia. Table 1 illustrates the foods from which the various isolates were obtained. The highest relative proportion of penicillia was associated with meat products, cheeses and bakery products.

A total of 106 different *Penicillium* isolates from mouldy food samples was examined at CSIRO Food Research. The food samples were not classified and a breakdown of other fungal species isolated from these samples is not available.

Table 1 Frequency of isolation of fungi from spoiled U.K. foods—numbers (and percentages)[a] of fungi

Genus	Meat products (n=51)	Cheese (n=32)	Bakery products (n=46)	Fruit and vegetable products (n=50)	Nuts (n=11)	Misc. (n=25)	Total (n=215)
Penicillium	52 (52)	51 (89)	58 (56)	33 (44)	6 (24)	19 (37)	219 (53)
Aspergillus	2 (2)	2 (4)	29 (28)	10 (13)	7 (28)	6 (12)	56 (14)
Cladosporium	25 (25)	1 (2)	2 (2)	3 (4)	0 (0)	5 (10)	36 (9)
Mucor	13 (13)	1 (2)	3 (2)	14 (18)	5 (20)	6 (12)	42 (10)
Rhizopus	7 (7)	1 (2)	3 (3)	3 (4)	4 (16)	5 (10)	23 (6)
Other	1 (1)	1 (2)	9 (9)	12 (16)	3 (12)	11 (21)	37 (9)
Total	100	57	104	75	25	52	413

[a] Rounded to nearest whole number.

Identification of *Penicillium* isolates

The majority of *Penicillium* isolates from both English and Australian samples belonged in the subgenus *Penicillium*, the subgenus which contains the largest percentage of reported toxigenic isolates. Most of these isolates were in series *Viridicata* and series *Expansa* (Pitt, 1979a), although a few representatives of other series were also isolated (see Table 2). Of the U.K. isolates the greatest number belonged in *P. aurantiogriseum* (20–25%), followed by *P. crustosum* (16%) and *P. chrysogenum* (11%). The percentages of other species isolated are shown in Table 3. Of the 211 U.K. isolates, only one belonged in the subgenus *Biverticillium*.

A similar pattern was seen in the 106 Australian isolates of penicillia from foods (Table 3). Over 50% of the isolates were placed in subgenus *Penicillium*, most in series *Viridicata* and series *Expansa*. *P. crustosum* was the most commonly isolated species (15%), followed by *P. chrysogenum* (9%), then *P. aurantiogriseum*, *P. expansum* and *P. roquefortii* (each 6%). It may be significant that P. *viridicatum*, which is reported to produce ochratoxin, has not been isolated from a sample of food at CSIRO Food Research, although the closely related *P. verrucosum* has. Unlike the U.K. isolates, a substantial proportion of the Australian isolates were from subgenera other than subgenus *Penicillium*. There were 14% of isolates from subgenus *Biverticillium*; *P. islandicum* from this subgenus is known to be able to produce mycotoxins such as islanditoxin and luteoskyrin. A further 14% were placed in subgenus *Furcatum* which contains some mycotoxigenic species such as *P. citrinum* (citrinin) and *P. janthinellum* (tremorgenic toxins).

This difference in distribution of *Penicillium* isolates from the U.K. and Australia may reflect the different environment or climatic conditions in these two parts of the world. For example, *Biverticillum* are generally able to grow at higher temperatures than those in subgenus *Penicillium*, many of which grow strongly at refrigeration temperatures. It may, however, simply be a reflection of the different methods of sample collection; samples examined at Leatherhead Food RA were part of a systematic survey, whereas samples examined at CSIRO Food Research were acquired on an *ad hoc* basis.

Comparative identification of selected isolates from subgenus *Penicillium* using the methods of Pitt (1979a), Raper and Thom (1949), Samson *et al.* (1976) and Frisvad (1981) illustrates the taxonomic problems facing the food mycologist. For example, isolates of P. *chrysogenum* (sensu Pitt) include the Raper and Thom taxonomic groups of *P. chrysogenum*, *P. notatum* and *P. meleagrinum*. Similarly, *P. viridicatum* (*sensu* Pitt) incorporates five species from Raper and Thom, three species from Samson *et al.* and parts of two species from Frisvad. These and other comparisons are illustrated in Table 4.

Table 2 Sources of identified U.K. isolates belonging to genus *Penicillium* according to series

Series	Meat products	Cheese	Bakery products	Fruit and vegetable products	Nuts	Misc.	Total
Expansa	8	6	14	8	3	4	43 (20)[a]
Viridicata	33	43	31	14	2	8	131 (62)
Urticicola	4	0	7	6	0	1	18 (8)
Cylindrospora	1	0	0	1	0	1	3 (1)
Furcatum	0	0	4	0	0	1	5 (2)
Divaricatum	0	2	0	0	0	1	3 (1)
Aspergilloides	1	0	1	3	0	3	8 (4)
Biverticillium	0	0	0	1	0	0	1 (1)
Eupenicillium	0	0	0	0	0	0	0 (0)

[a] Figures in parentheses indicate percentage of total penicillia to nearest whole number.

Table 3 Number and percentage of *Penicillia* isolated from foods

Subgenus	Species (after Pitt, 1979a)	No. and (percentage) U.K. isolates	Australian isolates
Penicillium	*P. aurantiogriseum*	42 (20)	6 (6)
	Near *P. aurantiogriseum*	11 (5)	0 (0)
	P. crustosum	35 (16)	16 (15)
	P. chrysogenum	24 (11)	10 (9)
	P. expansum	19 (9)	6 (6)
	P. roquefortii	19 (9)	6 (6)
	P. brevicompactum	13 (6)	3 (3)
	P. echinulatum	10 (5)	4 (4)
	P. hirsutum	7 (3)	0 (0)
	P. verrucosum	5 (2)	5 (5)
	P. viridicatum	4 (2)	0 (0)
	P. italicum	2 (1)	1 (1)
	*P. palitans**	2 (1)	0 (0)
	P. camembertii	1 (0.5)	0 (0)
	P. digitatum	1 (0.5)	0 (0)
	P. olsonii	0 (0)	1 (1)
	unplaceables	2 (1)	0 (0)
Furcatum	*P. corylophilum*	4 (2)	7 (7)
	P. griseoroseum	2 (1)	1 (1)
	P. citrinum	1 (0.5)	6 (6)
	P. janthinellum	0 (0)	2 (2)
	P. miczynskii	1 (0.5)	2 (2)
	P. canescens	0 (0)	1 (1)
Aspergilloides	*P. glabrum*	0 (0)	3 (3)
	P. thomii	0 (0)	1 (1)
	P. implicatum	0 (0)	1 (1)
	P. spinulosum	8 (4)	4 (4)
Biverticillium	*P. minioluteum*	1 (0.5)	2 (2)
	P. pinophilum	0 (0)	4 (4)
	P. islandicum	0 (0)	2 (2)
	P. purpurogenum	0 (0)	1 (1)
Eupenicillium	*E. ludwigii*	0 (0)	3 (3)
	E. erubescens	0 (0)	4 (4)

* Identified according to Raper and Thom (1949).

Table 4 Identification of isolated penicillia according to different taxonomic schemes

Name (Pitt, 1979a)	Name (Raper and Thom, 1949)	Name (Samson *et al*., 1976)	Name (Frisvad, 1981)
P. expansum	*P. expansum*	*P. expansum*	*P. expansum*
P. chrysogenum	*P. chrysogenum*	*P. chrysogenum*	*P. chrysogenum*
	P. notatum		
	P. meleagrinum		
P. atramentosum	*P. atramentosum*		
P. viridicatum	*P. viridicatum*	*P. verrucosum* var *verrucosum*	
	P. palitans	*P. verrucosum* var *melanochlorum* (*sensu* SSH)	'*P. cyclopium* p' pro parte
	P. olivino-viride	*P. verrucosum* var *cyclopium*	
	P. lanoso-viride	*P. verrucosum* var *verrucosum*	'*P. viridicatum* o-c' pro parte
	P. psittacinum	*P. verrucosum* var *verrucosum*	
P. crustosum	*P. crustosum*	*P. verrucosum* var *melanochlorum* (*sensu* Pitt)	'*P. crustosum* p.A.'
	P. terrestre	*P. verrucosum* var *cyclopium*	'*P. melanochlorum*'
P. roquefortii	*P. roqueforti*	*P roqueforti*	*P. roquefortii*
	P. casei	*P. verrucosum* var *cyclopium*	
P. hirsutum	*P. corymbiferum*	*P. verrucosum* var *corymbiferum*	
		P. hordei	*P. hirsutum*
	P. carneo-lutescens (sensu Pitt)	*P. verrucosum* var *ochraceum* (*sensu* Pitt)	

P. echinulatum	*P. cyclopium* var *echinulatum*	*P. echinulatum*	
P. aurantiogriseum	*P. aurantio-candidum*	*P. verrucosum* var *cyclopium*	
	P. solitum	*P. verrucosum* var *cyclopium*	'*P. cyclopium* p' pro parte
	P. cyclopium	*P. verrucosum* var *cyclopium*	
	P. martensii	*P. verrucosum* var *cyclopium*	
	P. aurantio-virens		
P. camembertii	*P. camemberti*	*P. camemberti*	
	P. caseicolum		
	P. biforme	*P. verrucosum* var *cyclopium*	
P. griseofulvum	*P. urticae*	*P. griseofulvum*	*P. patulum*
P. puberulum	*P. commune*	*P. commune*	'*P. cyclopium* p.'
	P. lanosum	*P. lanosum*	
	P. lanoso-griseum	*P. verrucosum* var *cyclopium*	
	P. puberulum		
P. olivicolor	*P. ochraceum*	*P. verrucosum* var *ochraceum*	
P. brevicompactum	*P. stoloniferum*		*P. brevicompactum*
	P. brevicompactum		
P. granulatum	*P. granulatum*	*P. granulatum*	
P. verrucosum	*P. viridicatum*	*P. verrucosum* var *verrucosum*	'*P. viridicatum* o-c' pro parte

Mycotoxins

Initially, some 60 mouldy food samples were examined for the presence of a range of mycotoxins which were detected at trace levels or greater in 35 samples; more than one toxin was isolated from some samples. The predominant toxins were ochratoxin (14 samples), patulin (8 samples), citrinin (16 samples) and aflatoxin (6 samples). Trace amounts of zearalenone and tricothecenes were detected in some samples. Fungi potentially capable of producing the specific mycotoxins detected were isolated from seven samples of food (Table 5) with a further 11 possibles, i.e. the toxin is known to be produced by a species closely related to the identified isolate. In total, therefore, 18 out of 35 samples contained detectable levels of one or more toxins which could reasonably be supposed to have been produced by one or more of the isolates obtained. Of the remaining samples, toxins detected could not be related back to the fungal isolates. However, in many of these (nut and cereal products particularly) it is realistic to suppose that toxin production could have occurred in the raw material used for the manufactured product, and that it was not associated with the subsequent mould spoilage.

Table 5 Association of fungal isolates with specific toxins detected in various foods

Isolate	Toxin	Food
P. roquefortii	Patulin	Cheese
A. flavus	Aflatoxin	Bread
P. citrinum	Citrinin	Cornish pasty
P. expansum	Citrinin	Cheese (three samples)
P. viridicatum	Ochratoxin	Cheese

Bioassay is being done on extracts of broth cultures of a number of the fungal isolates which were selected to represent approximately 10% of each species isolated. To date, only 20 isolates have been tested in the full bioassay system which comprises tests against *Artemia salina* (brine shrimp), baby hamster kidney cells (BHK) and germinating pea seedling. Of the isolates tested (Table 6), all 10 penicillia exerted toxic effects in one or more assay systems and five were toxic to all three bioassay systems. Only one of the 10 non-penicillia isolates tested was unable to cause a toxic effect in any test system used—this was a *Cladosporium* sp.

Table 6 Toxicity of some fungal isolates in various test systems

Genus	No. of isolates tested	No. causing positive effect in			
		Pea seedlings	Brine shrimp	BHK cells	One or more systems
Penicillium	10	9	5	8	10
Aspergillus	3	2	2	3	3
Cladosporium	3	2	0	2	2
Alternaria	1	1	1	1	1
Zygomycetes	3	2	0	3	3

DISCUSSION

The high incidence of penicillia isolated from the spoiled foods was not unexpected—especially from the high proportion of chilled foods examined. These results, therefore, merely confirm the earlier observations of other workers including ourselves. The occurrence of specific mycotoxins in some foods correlated well with isolation of known mycotoxigenic species, especially producers of ochratoxin, citrinin and patulin, for all except cereal and nut-based products. In these latter foods the mycotoxins probably originated from the raw materials (e.g. peanuts, wheat, corn etc.). What is surprising is the very high proportion of isolates which exerted a toxic effect in one or more bioassay systems. Since the isolates were deliberately selected to be representative of those obtained, bias in relation to mycotoxin potential is unlikely. A fuller assessment must await the outcome of further investigation.

Undoubtedly, many problems remain in relating the occurrence of specific toxins to fungal isolates, resulting from difficulties in the identification and taxonomy of penicillia and other genera. Different workers have identified isolates using a variety of systems with different names to describe the same species. The species boundaries are often unclear, especially for closely related taxa. Although the taxonomic studies of Pitt (1979a) have broadened the taxonomic bands and clarified the boundaries, data on the mycotoxin association of particular series and species is still sparse. Ciegler *et al.* (1973) have made some useful contributions in relation to production of ochratoxin and citrinin by *P. viridicatum* and Pitt (1979b) has clarified the taxonomy of the species producing tremorgenic toxins, particularly with reference to *P. crustosum* and the production of Penitrem A.

In the longer term it may be helpful to relate toxin production to substrate utilization and metabolic recognition tests, as has been done for bacteria. The methods described by Frisvad (1981) and in a further paper, soon to be published, provide an early essay in the art of relating distinct metabolic groups in the subgenus *Penicillium* to toxin formation. One of us (A.P.W.) has successfully repeated Frisvad's work with several groups of penicillia and has extended it to many more 'metabolic intermediates' than those suggested by Frisvad, especially in *P. chrysogenum*. However, much work remains to be done in the genus *Penicillium* to elucidate the species–toxin relationships.

Although the taxonomy and nomenclature in *Penicillium* is less confused than for some other mycotoxigenic genera such as *Fusarium* and *Trichoderma*, we still know only a small proportion of the specific relationships between species and the toxins that they produce. Until we have this information, it will be difficult to assess the significance of a particular *Penicillium* species isolated from a food sample, or to predict which toxins may be present in that food.

ACKNOWLEDGEMENTS

The U.K. work described was funded by the U.K. Ministry of Agriculture, Fisheries and Food, to whom we are indebted for permission to publish. We are also indebted to Miss Debbie Head and Messrs Peter Mann and Martin Page for excellent assistance.

REFERENCES

Anon (1955). *Food and Drugs Act*. HMSO, London.

Bullerman, L. B. (1976). 'Examination of Swiss cheese for incidence of mycotoxin producing moulds', *J. Food Sci.*, **41**, 26–28.

Bullerman, L. B. (1980). 'Incidence of mycotoxic moulds in domestic and imported cheese', *J. Food Safety*, **2**, 47–58.

Bullerman, L. B., and Olivigni, F. J.(1974). 'Mycotoxin producing potential of moulds isolated from cheddar cheese', *J. Food Sci.*, **39**, 1166–1168.

Burmeister, H. R., and Hesseltine, C. W. (1970) 'Biological assays for two mycotoxins produced by *Fusarium tricinctum*' *Appl. Microbiol.* **20**, 437–440.

Ciegler, A., Fennell, D. I., Sansing, G. A., Detroy, R. W., and Bennett, G. A. (1973). 'Mycotoxin producing strains of *Penicillium viridicatum*: classification into subgroups', *Appl. Microbiol.*, **26**, 271–278.

Frisvad, J. C. (1981). 'Physiological criteria and mycotoxin production as aids in the identification of common asymmetric penicillia', *J. Appl. Environ. Microbiol.*, **41**, 569–578.

Jarvis, B. (1973). 'Comparison of an improved Rose Bengal chlortetracycline agar with other media for the selective isolation and enumeration of moulds and yeasts in foods', *J. Appl. Bact.*, **36**, 723–727.

Jarvis, B. (1983). 'Moulds and mycotoxins in mouldy cheeses', *Microbiologie Aliments Nutrition*, **1**, 187–191.

Jarvis, B., Chapman, W. B., Williams, A. P., Norton, D.M., and Toule, G.M. (1982). 'Methods for the detection and identification of selected mycotoxins', in *Isolation and Identification Methods for Food Poisoning Organisms* (eds J. E. L. Corry, D. Roberts and F. A. Skinner), pp. 367–92. *SAB Technical Series* No. 17. Academic Press, London.

Joseffson, E. (1981). 'Mycotoxiner i ost', *Var Foda*, **33**, 237–48.

King, A. D., Hocking, Ailsa, D., and Pitt, J. I. (1979). 'Dichloran Rose Bengal medium for enumeration and isolation of moulds from foods', *Appl. Environ. Microbiol.*, **37**, 959–964.

Lafont, P., Siriwardana, M. G., and Lafont, J. (1979). 'Contamination de fromages par des metabolites fongiques', *Med. Nutr.*, **15**, 257–262.

Moreau, C. (1979). *Moulds, Toxins and Food* (ed. and translated by M. O. Moss). John Wiley, Chichester.

Northolt, M. D., van Egmond, H. P., Soentoro, P., and Deijll, E. (1980). Fungal growth and the presence of sterigmatocystin in hard cheese', *J. Assoc. Off. Anal. Chem.*, **63**, 115–119.

Onions, A. H. S., Allsopp, D., and Eggins, H. O. W. (1981). *Smith's Introduction to Industrial Mycology*, 7th edn. Edward Arnold, London.

Pitt,J. I. (1979a). *The Genus* Penicillium *and Its Teleomorphic States,* Eupenicillium *and* Talaromyces. Academic Press, London and New York.

Pitt, J. I. (1979b). *Penicillium crustosum* and *P. simplicissimum*, the correct names for two common species producing tremorgenic mycotoxins', *Mycologia*, **71**, 1166–1177.

Raper, K. B., and Fennell, D. I. (1965). *The Genus* Aspergillus. Krieger, Huntington, New York.

Raper, K. B., and Thom, C. T. (1949). *A Manual of the Penicillia*. Williams & Wilkins, Baltimore, USA.

Roberts, B. A., and Patterson, D. S. P. (1975). 'Detection of twelve mycotoxins in mixed animal feedstuffs using a novel membrane clean-up procedure', *J. Ass. Off. Analyt. Chem.*, **58**, 1178–1181.

Samson, R. A., Stolk, A. C., and Hadlock, R. (1976). 'Revision of the subsection Fasciculata of Penicillium and allied species', *Studies in Mycology. II.* Centraalbureau voor Schimmelcultures, Baarn.

Torrey, G. S., and Marth, E. H. (1977). 'Temperature in home refrigerators and mould growth at refrigeration temperatures', *J. Food Prot.*, **40**, 393–397.

Trichothecenes and Other Mycotoxins
Edited by J. Lacey

CHAPTER 20

Factors influencing activity of fungi and antifungal agents in poultry feed

Pat B. Hamilton

THE PROBLEM OF MOULDING IN FEEDS

The use of antifungal agents in poultry feeds assumes that fungal activity can occur in feeds. This is an assumption that many scientists find difficult to accept. It would, therefore, seem appropriate to present some data that indicate that fungal activity in feed is common and causes large economic losses to the animal industry. These data will indicate some of the problems to be faced in controlling fungal activity in feeds.

The first example is a survey (Table 1) conducted in a broiler chicken operation experiencing clinical aflatoxicosis. The frequency of contamination with aflatoxin, which was moderate in the incoming ingredients, increased to 52% during manufacture of the feed and to 91% by the time the feed reached the birds. Thus, aflatoxin was forming before, during and after manufacture of feed.

Further evidence of harmful fungal activity in feed was obtained from a broiler chicken operation with two mills to manufacture feed from the same ingredients, using the same dietary formulas (Hamilton, 1978). The birds fed from one of the mills, the problem mill, displayed a 1% increase in feed conversion ratio and a 2% decrease in body weight compared to birds receiving feed from the other, normal mill, but otherwise receiving the same management. This incident involved over 10 million birds before it was solved, and cost $400,000 in decreased productivity. It was associated with the occurrence of aflatoxin (> 1 μg/kg) in 30% of the feed samples from the problem mill as compared to only 2% in feed samples from the normal mill. The aflatoxin concentrations ranged from 0–30 μg/kg in the problem mill as opposed to 0–6 μg/kg in the normal mill. When better cleaning procedures and the use of an antifungal agent were introduced, the aflatoxin frequency and level of contamination returned to those in the normal mill, as did the

Table 1 Survey of feed and ingredients for aflatoxin during an outbreak of aflatoxicosis in chickens

Material	Aflatoxin frequency[a] (%)
Soybean meal	5
Maize	30
Other ingredients	0
Feed from mill	52
Feed from troughs	91

[a] Frequency is percentage of samples of each material that contained aflatoxin. Limits of detection were 100 to 10,000 μg/kg.

feed conversion ratios and the body weights in the flocks. The absence of infectious diseases was attested by diagnosticians from five states of the United States and three other countries who inspected these flocks. It seems obvious that low levels of fungal activity and aflatoxin were associated with severe economic losses.

Perhaps even more persuasive were the results of a survey of five independent broiler chicken operations in North Carolina in the U.S.A. (Jones *et al.*, 1982). Growers in each operation were classified as good, mediocre or poor using an objective index based on their productivity for the previous year. During an experimental period of about 2 months the good producers marketed more birds of higher weight, with better feed conversion, with lower mortality and with fewer condemnations. The good producers received a higher payment ($0.1202/bird) than the mediocre ($0.1149/bird) and poor ($0.1086/bird) growers, while the cost of production approximated $0.1050/bird. Of particular interest, however, is the frequency of feed contamination with aflatoxin which was 18% for the good growers, 22% for the mediocre and 31% for the poor. Of the positive samples, those from the good growers averaged 6.1 μg/kg, those from the mediocre growers averaged 6.5 μg/kg and those from the poor growers averaged 14.0 μg/kg. The design of the survey ensured that all growers received equivalent feed of the same concentration and frequency of contamination with aflatoxin. The maize used to manufacture the feed contained a mean 1.2 μg/kg aflatoxin. The results of this survey indicate clearly that fungal activity and aflatoxin formation occur during feed manufacture and storage prior to consumption. They further demonstrate that levels of aflatoxin below those considered safe by regulatory agencies are associated with significant economic problems.

FUNGAL ACTIVITY IN FEED HANDLING EQUIPMENT

The question of where some of this fungal activity occurs can be answered by close inspection of the equipment which handles the feed prior to its consumption by birds. I have seen feed mills with holes in the roofs of bins and buildings large enough for rain, snow, wild birds and even inspectors to enter. Mills and feed delivery trucks are seldom inspected, maintained and cleaned sufficiently to satisfy a microbiologist. In areas of stagnant flow, accumulations of caked, mouldy feed held together by mould mycelia occur all too frequently, and feed flowing over the surface of such accumulations becomes contaminated with fungi and products of their activity.

The storage bins on the farm are further sites of fungal activity. The interiors of these bins should be bright, shiny metal, but all too frequently the metal cannot be seen because of a coating of caked, mouldy feed which is probably the result of moisture migration induced by temperature differentials between the feed and the exterior of the bin. Within poultry houses the feed distribution equipment is subject to similar contamination and accumulation. The potential magnitude of these problems is indicated by the presence of over 42 miles of feed lines in a single 500,000 laying hen facility. All of this fungal activity can result in loose, free-flowing chicken feed frequently becoming caked and clogging the feed distribution system. Quite frequently, spots of corrosion may be seen on the inside of feed handling equipment resulting from cathodic depolarization caused by microbial activity.

Evidence that fungal activity in the feed handling and distribution equipment is detrimental to animal productivity was obtained in an experiment (Table 2) in which the feed delivery trucks, storage bins, auger ways, and feeder pans and cones in some houses were cleaned down to bare metal and disinfected and compared with other houses on the same farms which were not cleaned. There was about a 6% improvement in body weight, about a 20% improvement in serum carotenoids (which correlate with carcass pigmentation and are an important economic parameter) and 22% improvement in carcass grades. This experiment was conducted on farms which had had

Table 2 Beneficial effect of cleaning feed distribution equipment

Variable	Control	Experimental
Body weight (g)	1529	1612
Serum carotenoids (μg/ml)	21.7	26.1
Carcass grade (weighted average)	1.84	1.43

mould problems and the beneficial effects of cleaning out the caked, mouldy material were obvious. However, the beneficial effect of cleaning lasted only for one flock and subsequent flocks in these houses showed no such improvement. While cleaning between flocks is strongly recommended as a management practice to prevent the inoculation and contamination of good feed, it has little effect on toxins already formed, and it is labour-intensive and expensive.

Another study (Good and Hamilton, 1981) of two identical, adjacent chicken houses with birds receiving identical feed delivered at the same time and identical care and management indicated the importance of fungal activity even during normal operation. Through 16 consecutive flocks, one house always yielded better results than the other. Assuming fungal colonization to be time-dependent, the mean residence time of feed in one house was decreased by delivering half as much feed per delivery as usual but twice as often. The other house received feed on a normal schedule. In the first trial, the house giving the poorer results had its delivery frequency increased with the result that productivity parameters improved to become better than those in the normally treated house. Treatments were reversed in the second trial and, as expected, productivity again became better than the first house. The results of the first trial were repeated in a third trial but the intended fourth trial was abandoned because the grower, after seeing the results of the first three trials, insisted that both houses be on a short delivery time.

THE EFFECT OF WATER ON FUNGAL ACTIVITY

It is commonly, but mistakenly, assumed that the moisture contents of grain and feed ingredients are too small to permit fungal activity except in rare and unusual circumstances. This mistaken assumption is probably based on the belief that U.S. grain standards are set by the U.S. regulatory agencies to protect consumers. However, U.S. grain standards are set by the U.S. grain traders for their benefit and not necessarily the benefit of the purchasers of the grain. The subterfuges and duplicity of the grain trade were highlighted by a headlined article in a prominent farm journal in the U.S. this past year. The article advised farmers to add water to grain containing less than the maximum permitted by U.S. grain standards. In a specific example, maize at 10.3% moisture was 'improved' by adjusting its moisture to 15.5% permitted in U.S. grade #2 maize. By this adulteration (a U.S. court has held that adding such water is an illegal adulteration), the farmer sells water for the price normally reserved for grain. One of the justifications offered by this journal was that the grain dealers would do it in any event. The article was well written and contained photographs of watering equipment in actual operation. Alas, the article contained tables whereby one could calculate the amount of water required to bring the moisture up to the desired level. Such adulteration is stupid from a microbiological viewpoint and it cannibalistically profits one segment of agriculture at the expense of another segment, the animal producer. Reform is necessary when farmers and grain dealers cannot make a reasonable profit without adulterating grain with water and trash. It is most disheartening that such practices in the international grain trade persist despite studies and supposed banning over 80 years ago (Christensen and Kaufmann, 1969; Shanahan *et al.*, 1910).

The overwhelming importance of water to the activity of fungi and antifungal agents was illustrated by a simple study. The water content of individual kernels of commercial maize equilibrated to a constant A_w (water activity, a measure of available water) was determined and found to follow a normal distribution with a range of about 5% around the mean (Hamilton, 1983; Christensen and Kaufmann, 1969). This raised the question of whether such differences influenced the growth and respiration of fungi. By determining the production of respiratory CO_2 from individual kernels and plotting CO_2 production against the moisture content of those kernels a linear relationship between water content and CO_2 production was established. Kernels with the higher moisture content were shown to be more susceptible to microbial attack with fungal fruiting bodies more abundant on kernels with the greater water content and higher respiratory CO_2 production. My conclusion is that at a given A_w, the wetter grain will deteriorate more rapidly than the drier. Also broken kernels, separated from the intact kernels in a mass of corn equilibrated to constant A_w, had about five times the microbial activity of intact kernels. Thus the fraction of commercial corn known as cracked kernels and foreign matter should be regarded with suspicion.

MEASUREMENT OF FUNGAL AND ANTIFUNGAL ACTIVITY

A traditional method of controlling fungal activity when water content cannot be controlled is the use of antifungal agents. There are many antifungal agents on the market and it would appear that one merely has to determine which is best and use it. Many methods have been used to measure fungal activity in feed, but the most common by far has been the use of fungal counts on agar plates. Nevertheless, I consider the use of mould counts for the evaluation of mould inhibitors in feed to be theoretically and practically unsound. An equation for fungal activity can be written as

$$(CH_2O)_n + n\ O_2 \rightarrow n\ CO_2 + n\ H_2O + \text{heat} + \text{biomass}$$

Mould count is an indicator of fungal biomass but because mould counts are heavily biased toward moulds that produce spores profligately the biomass may be erroneously estimated. Ergosterol (Seitz *et al.*, 1977) and glucosamine (Nandi, 1978), normal constituents of most fungi, are more direct measures of biomass. However, in my laboratory, their measurement is not sufficiently reproducible and sensitive to be used successfully with animal feeds except under limited experimental conditions.

Other components in the above equation could give more direct assessments of fungal activity than colony counts. Tabib *et al.* (1982a) reported that a commercial inhibitor, Mold-X, inhibited heat production, accumulation of water of respiration, oxygen consumption, CO_2 production and loss of dry matter when the inhibitor was used in feed. Fungal activity obviously had been inhibited; nevertheless, the mould count was increased not decreased. Thus, the inhibitor stimulated sporulation while inhibiting other measures of fungal activity. Reliance on mould count under such circumstances would lead to erroneous conclusions about efficacy of mould inhibitors.

Extensive testing has suggested that the measurement of fungal respiration or its indicators, especially CO_2 production, is much superior to mould counts for the evaluation of fungal activity and antifungal agents (Dixon and Hamilton, 1981a,b; Tabib *et al.*, 1982a). Representative samples of feed may be placed in ordinary culture tubes and capped with a rubber septum designed for use with serum bottles. Thus, each tube resembles a miniature feed storage bin with a headspace. The rubber septum prevents equilibration of the headspace gas with the ambient atmosphere. The headspace gas may be sampled simply with a disposable tuberculin syringe and needle, and injected into a gas chromatograph capable of separating and quantitating CO_2 and O_2. At best, such samples can be analyzed in less than 1 min (Dixon and Hamilton, 1981a). This technique has permitted several factors that limit the activity of mould inhibitors in animal feeds to be evaluated.

INFLUENCE OF pH ON ANTIFUNGAL ACTIVITY OF ORGANIC ACIDS

Some of the mould inhibitors more widely used in the feed industry are based on short-chain fatty acids such as propionic, acetic, sorbic, benzoic and formic acids. These are all weakly acidic with a pKa around pH 5 and are all more active in the free acid form than the salt form. Hence, the pH of the feed determines the antifungal activity of the mould inhibitor. Maize is a major constituent of most U.S.A. animal feeds. Its pH averages 6.5; the range shown by 125 varieties was an entire pH unit although the distribution curve of the pH was normal. Such variation may significantly affect the antifungal activity of organic acid mould inhibitors.

Other ingredients of animal feeds may affect this overall pH. For instance, calcium carbonate, as limestone, is a nutritional necessity in typical animal diets but it is also an excellent buffer and neutralizes free acids, converting them to their less active salts. When graded amounts of limestone were incorporated into corn meal, the activity of propionic acid in the mixture was a function of the concentration of limestone (Dixon and Hamilton, 1981b).

Similarly, the percentage inhibition of moulding obtained with varying concentrations of propionic acid was dependent upon the amount of soybean meal mixed with corn meal (Table 3). The greater the soybean meal concentration, the less active was the propionic acid. Fish meal, poultry by-product meal, and other high-protein supplements used in animal diets have similar effects but not corn gluten meal (Table 4). The reason for this difference was that corn gluten meal is deficient in dibasic amino acids compared to other protein supplements (Scott *et al.*, 1969) and so is less able to buffer and neutralize free organic acids.

Table 3 Effect of soybean meal on inhibition of mould activity in corn meal by propionic acid

Propionic acid (mg/g)	Soybean meal (percentage in corn meal)		
	0	10	25
0	0[a]	0	0
0.125	58	46	15
0.25	73	52	19
0.50	83	65	67
1.00	92	87	81

[a] Tabular values are percentage inhibition of CO_2 produced in 48 h at 29 °C from unsterilized corn meal (25% H_2O) containing stated amounts of propionic acid and soybean meal.

Table 4 Antifungal activity of propionic acid in mixtures of corn gluten meal and corn meal

Propionic acid (mg/g)	Corn gluten meal (percentage of corn meal)		
	0	10	25
0	0[a]	0	0
0.125	11	16	16
0.25	10	5	21
0.50	26	21	35
1.00	69	78	78

[a] Tabular values are percentage inhibition of CO_2 produced in 48 h at 29 °C from unsterilized corn meal (25% H_2O) containing the stated amounts of propionic acid and corn gluten meal.

ROLE OF PARTICLE SIZES OF SUBSTRATE AND INHIBITOR

Another factor influencing fungal and antifungal activity is the particle size of the substrate and of the mould inhibitor. In general, the smaller the particle size of feed the greater the amount of respiratory CO_2 produced by the resident fungi, presumably because of the greater surface area associated with the smaller particles (Dixon and Hamilton, 1981c). Fine material separated from pellets of feed gave almost double the CO_2 production of intact pellets weight for weight (Jones and Hamilton, 1983). Similarly, the activity of the mould inhibitor may depend on its own particle size (Dixon and Hamilton, 1981c). Both sorbic acid and gentian violet of small particle size were more inhibitory than identical amounts of large particle material. The particle size effect was independent of the moisture content of feed and occurs, presumably, because the inhibitor must be finely dispersed in order to come in contact with fungi on as many particles of feed as possible. For good inhibition, the inhibitor must dissolve and diffuse into the interior of the feed particle where fungi reside and grow.

Another important determinant of the activity of antifungal agents in maize-based diets is the nature of the maize itself (Tabib *et al.*, 1982b). Comparison of the dose–response curves for propionic, sorbic, benzoic, a commercial mixture of volatile fatty acids, and gentian violet enabled maize meal to be classified into sensitive and resistant batches, i.e. an inhibitor could have excellent antifungal activity or little. Why maize meal is heterogeneous is not known, and cannot be attributed to the sensitivity of micro-organisms isolated from the different batches nor to the pH of the maize meal or its ability to bind chemically and physically with the inhibitors.

INFLUENCE OF PELLETING

Pelleting is an important determinant of fungal activity in poultry feeds (Tabib *et al.*, 1984). Practically all poultry feed manufactured in the U.S.A. today is pelleted to ensure nutrient uniformity and to increase the digestibility of the feed. Even the feed of young chicks is pelleted to gain these advantages, but is then crumbled to give bite-sized material. In a survey, paired samples of mash and pellets from four independent feed mills were obtained, taken to the laboratory, adjusted to identical moisture concentrations, incubated, and the CO_2 production measured. It was found that the pelleting process in which live steam is injected into the mash prior to passing it through pellet dies delayed the onset of fungal activity. In another portion of this survey, paired samples of mash and pellets were taken from four mills which were using Mold-X, a commercial inhibitor. The onset of fungal activity in pellets containing the mould inhibitor was delayed beyond that of the pelleted untreated material or of the treated mash. The heating process and the propionic acid combined to improve the antifungal activity of the propionic acid. This effect was duplicated in laboratory experiments and in general confirms earlier observations of an interaction (activity greater than the sum of the individual activities) between organic acid inhibitors and temperature when yeasts were cultured in nutrient broth (Beuchat, 1981).

DOSE–RESPONSE CHARACTERISTICS

A lack of appreciation of the nature of dose–response curve often results in moves in the feed industry to economize by decreasing the amount of mould inhibitor used. It is assumed that half the concentration of mould inhibitor will only decrease its activity by half, but often all activity is lost (Table 5). There may be a threshold dose which must be exceeded before any antifungal activity is evident.

It is often not appreciated that the activity of antifungal agents in feed is time-dependent. The use of mould inhibitors often merely delays the onset of moulding (Table 6), the extent depending on the amount of antifungal agent used. Once the antifungal agent has been degraded, fungal activity continues at a normal rate. This effect can be seen with propionic, acetic, sorbic and benzoic acids, gentian violet and several commercial mould inhibitors. It is possible to pickle feed with an antifungal agent but the concentrations typically used are much smaller than those used in other materials, e.g. silage, hay. Consequently, only short periods of time are free from fungal activity. For economic reasons the quantities of antifungal agents used are the minimum necessary to give protection for the few days or weeks between preparation and feeding.

Table 5 Non-linear dependence of antifungal activity of propionic acid on its concentration in corn meal

Propionic acid (mg/g corn meal)	Inhibition (%)
0	0[a]
0.25	14
0.50	20
1.00	93

[a] Tabular values are percentage inhibition of CO_2 produced from corn meal (30% H_2O) in 48 h at 29 °C.

Table 6 Delay of onset of fungal activity in corn meal caused by propionic acid[a]

Time of incubation (days)	Weight loss (percentage of initial weight)	
	Control	Treated
0	0	0
1	0	0
2	5	0
3	11	0
4	—	0
5	17	6
7	—	10
9	30	18

[a] Weighed amounts of unsterilized corn meal (30% H_2O) with and without propionic acid (0.5 mg/g) were incubated at 29 °C in serum bottles sealed with a cotton plug. At the indicated times the dry weights of the contents of bottles were measured after heating (110 °C) overnight in a forced-draught oven.

FUNGAL ACTIVITY IN FEED OF LOW WATER CONTENT

Much emphasis has been placed on the relationship between fungal activity and the water content of feeds and their ingredients. This is further demonstrated by samples of feed, from three poultry companies, which were analysed for water content and production of CO_2. The water contents averaged 11.5, 12.5 and 13% respectively but feed from all three companies produced measurable CO_2 and the correlation between CO_2 and moisture content was highly significant (Jones and Hamilton, 1983). It is generally assumed that feed below 13% has an A_w low enough to inhibit microbial activity but this study showed that this assumption was clearly untrue.

Similarly samples of feed taken from chicken houses showed CO_2 production by feed with less than 13% water (Jones and Hamilton, 1983). Both feed moisture and CO_2 production increased with increasing time in the poultry houses. It was also found that crumbles (feed that has been pelleted and then reduced in particle size) were more susceptible to fungal attack than unpelleted mash. Apparently, pelleted feed is more easily digested than non-pelleted by fungi as well as by animals.

CORRELATION OF AFLATOXIN WITH ZINC CONTENT

Aspergillus flavus requires zinc for *in vitro* production of aflatoxin. A survey of feed samples from commercial chicken operations revealed a significant correlation between aflatoxin contamination and zinc content (Jones *et al.*, 1984). The zinc content of unamended feed (<20 μg/g) is supplemented normally with a mineral premix containing zinc to meet the nutrient requirements of chickens (40 μg Zn/g feed). The mean zinc concentration of the feed samples (117 μg/g) was about three-fold greater than the nutrient requirement and suggested that stricter control of zinc levels during manufacture could reduce aflatoxin of feed. The mineral premix added to poultry rations also contains Mn, Fe and Cu to correct nutrient deficiencies. This suggests that toxins requiring these minerals for their formation will be amplified if quality control during manufacture is inadequate.

CONCLUSION

The study of fungal activity and antifungal activity in feed and feed ingredients is a neglected area. Such studies are necessary to achieve control over fungal activity by using antifungal agents. Some of the factors which influence the activity of antifungal agents in feed can be manipulated now, while other factors appear inaccessible from the current state of knowledge. Even if complete mastery through scientific principles becomes possible, it appears that economics and politics will play a modulating if not commanding role in the application of such principles. Nevertheless, research on fungi and mycotoxins remains a fruitful area of endeavour.

REFERENCES

Beuchat, L. R. (1981). 'Synergistic effects of potassium sorbate and sodium benzoate on thermal inactivation of yeasts', *J. Food Sci.*, **46**, 881–777.

Christensen, C. M., and Kaufmann, H. H. (1969). *Grain Storage: the Role of Fungi in Quality Loss*. University of Minnesota Press, Minneapolis, MN.

Dixon, R. C., and Hamilton, P. B. (1981a). 'Evaluation of some organic acids as mold inhibitors by measuring CO_2 production from feed and ingredients', *Poultry Sci.*, **60**, 2182–2188.

Dixon, R. C., and Hamilton, P. B. (1981b). 'Effect of feed ingredients on the antifungal activity of propionic acid', *Poultry Sci.*, **60**, 2407–2411.

Dixon, R. C., and Hamilton, P. B. (1981c). 'Effect of particle sizes of corn meal and a mold inhibitor on mold inhibition', *Poultry Sci.*, **60**, 2412–2415.

Good, R. E., and Hamilton, P. B. (1981). 'Beneficial effect of reducing the feed residence time in a field problem of suspected moldy feed', *Poultry Sci.*, **60**, 1403–1405.

Hamilton, P. B. (1978). 'Fallacies in our understanding of mycotoxins', *J. Food Prot.*, **41**, 404–408.

Hamilton, P. B. (1983). 'Nonhomogeneity of microbial activity in corn kernels equilibrated to constant A_w', *Abst. Ann. Meet. Amer. Soc. Microbiol.*, **83**, 247.

Jones, F. T., and Hamilton, P. B. (1983). 'Evidence for fungal activity in poultry feed under typical field conditions', *Abst. Ann. Meet. Amer. Soc. Microbiol.*, **83**, 246.

Jones, F. T., Hagler, W. M., and Hamilton, P. B. (1982). 'Association of low levels of aflatoxin in feed with productivity losses in commercial broiler operations', *Poultry Sci.*, **61**, 861–868.

Jones, F. T., Hagler, W. M., and Hamilton, P. B. (1984). 'Correlation of aflatoxin with zinc content of chicken feed', *Appl. Environ. Microbiol.*, **47**, 478–800.

Nandi, B. (1978). 'Glucosamine analysis of fungus-infected wheat as a method to determine the effect of antifungal compounds in grain preservation', *Cereal Chem.*, **55**, 121–126.

Scott, M. L., Nesheim, M. C., and Young, R. J. (1969). *Nutrition of the Chicken*, W. F. Humphrey Press, Inc., Geneva, NY.

Seitz, L. M., Mohr, H. E., Burroughs, R., and Sauer, D. B. (1977). 'Ergosterol as an indicator of fungal invasion in grains', *Cereal Chem.*, **54**, 1207–1217.

Shanahan, J. E., Leighly, C. E., and Boerner, E. G. (1910). 'American export corn (Maize) in Europe', U.S.D.A. Bureau of Plant Industry Circular 55.

Tabib, Z., Hagler, W. M., and Hamilton, P. B. (1982a). 'Comparison of physiological parameters useful for assessment of activity of antifungal agents in feed and ingredients', *J. Food Prot.*, **45**, 1030–1037.

Tabib, Z., Hagler, W. M., and Hamilton, P. B. (1982b). 'Factors influencing activity of antifungal agents in animal feed and ingredients', *Abst. Ann. Meet. Amer. Soc. Microbiol.*, **1982**, 204.

Tabib, Z., Jones, F. T., and Hamilton, P. B. (1984). 'Effect of pelleting of poultry feed on the activity of mold and mold inhibitors', *Poultry Sci.*, **63**, 70–75.

MYCOTOXIN PRODUCTION IN CULTURE AND THE EFFECT OF ENVIRONMENTAL FACTORS

Trichothecenes and Other Mycotoxins
Edited by J. Lacey

CHAPTER 21

Mycotoxin production from *Myrothecium* species

Bruce B. Jarvis, Gowsala Pavanasasivam and George A. Bean

INTRODUCTION

The trichothecene complex of antibiotics (Ueno, 1983) can be divided into two groups: the simple trichothecenes (Figure 1), (e.g. diacetoxyscirpenol (**1**), T-2 toxin (**2**), verrucarol (**3**), trichodermol (**4**), etc.) and the macrocyclic trichothecenes (Figure 2) (e.g. roridins (**7**, **8**,) and verrucarins (**5**, **6**). Although the first group has received widespread interest because of their presence in feed grain (Ueno, 1980) and their alleged use as chemical warfare agents ('yellow rain') (Rosen and Rosen, 1982), the macrocyclic trichothecenes are more interesting because of the complexity of their structures.

Our interest in macrocyclic trichothecenes arose from the finding of a series of potent antileukemic macrocyclic trichothecenes (the baccharinoids; Figure 2, **9**) in the Brazilian shrub, *Baccharis megapotamica* (Kupchan *et al.*, 1977). Later, Jarvis *et al.*, (1981a) showed that the presence of baccharinoids in *B. megapotamica* probably results from a fungus producing roridins near the plant. The plant then takes up the roridins and enzymatically transforms them oxidatively to the baccharinoids. This is curious, because trichothecenes are potent phytotoxins (Brian *et al.*, 1961; Cole *et al.*, 1981). We have found a wide variety of plants (corn, beans, artichokes, peppers, tomatoes, tobacco and wheat), including a north American *Baccharis* species, *B. halmifolia*, to be very sensitive to the toxic effects of macrocyclic trichothecenes, including the baccharinoids (Cutler and Jarvis, 1985). However, recently, another Brazilian plant, *Baccharis coridifolia* has been found to contain roridins A (**7**) and E (**8**) (Busam and Habermehl, 1982).

The anticancer activity of trichothecenes (Loeffler *et al.*, 1965; Stähelin *et al.*, 1968) led to our interest in the development of macrocyclic trichothecenes as anticancer agents (Jarvis and Mazzola, 1982). Based on structure–activity relationships, we concluded that the presence of an oxygen atom in the A-ring was important for *in vivo* activity against P388 leukaemia in mice. Thus, the baccharinoids which possess either an 8β-hydroxyl group (e.g. **9**) or a

Diacetoxyscirpenol (**1**)

R_1 = OH; R_2 = Ac; R_3 = OAc; R_4 = H

T-2 Toxin (**2**)

R_1 = OH; R_2 = Ac; R_3 = OAc; R_4 = $\overset{O}{\|}{OC}CH_2CH(CH_3)_2$

Verrucarol (**3**)

R_1 = R_2 = R_4 = H; R_3 = OH

Trichodermol (**4**)

R_1 = R_2 = R_3 = R_4 = H

Figure 1

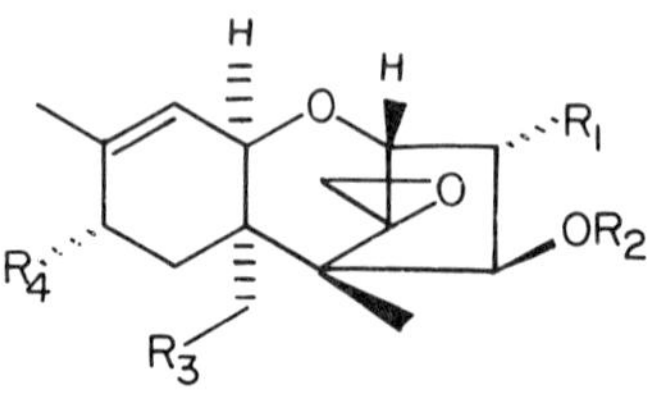

Verrucarin A (**5**)

R_1, R_2 = O; R_3 = $-\underset{(S)}{\overset{2'}{C}}HOH\underset{(R)}{\overset{3'}{C}}H\overset{12'}{C}H_3\overset{4'}{C}H_2-$; R_4 = H

Verrucarin J (**6**)

R_1, R_2 = O; R_3 = $\underset{(E)}{CH{=}CCH_3CH_2-}$; R_4 = H

Roridin A (**7**)

R_1 = R_4 = H; R_2 = $-\underset{(R)}{\overset{13'}{C}}HOH\overset{14'}{C}H_3$; R_3 = $-\underset{(S)}{\overset{2'}{C}}HOH\underset{(R)}{\overset{3'}{C}}H\overset{12'}{C}H_3\overset{4'}{C}H_2-$

Roridin E (**8**)

R_1 = R_4 = H; R_2 = $-\underset{(R)}{C}HOHCH_3$; R_3 = $-\underset{(E)}{CH{=}CCH_3CH_2}$

Baccharinoid B7 (**9**)

R_1 = H; R_2 = $-\underset{(R)}{C}HOHCH_3$; R_3 = $-\underset{(R)}{C}HOH\underset{(R)}{C}HCH_3CH_2-$; R_4 = OH

Figure 2

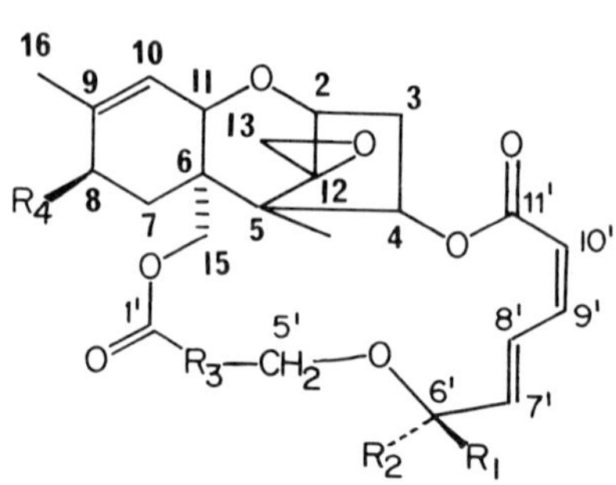

9β,10β-epoxide group, are active *in vivo* against P388 mouse leukemia, while the roridins and verrucarins which lack such functionality in the A-ring (see Figure 2) are inactive *in vivo* although highly active against other cell lines and tumours (Harri *et al*, 1962; Rüsch and Stähelin, 1965). We have increased *in vivo* activity of a number of roridins and verrucarins against P388 mouse leukemia by converting them into baccharinoid-like compounds (Jarvis *et al.*, 1980a; 1984).

In order to carry out these transformations, large quantities of the starting macrocyclic trichothecenes were required. Of the four genera reported to produce macrocyclic trichothecenes (*Myrothecium*, *Stachybotrys*, *Cyclindrocarpon*, and *Verticimonosporium*) *Myrothecium* was the best producer (Jarvis *et al.*, 1983a). Our major effort in trichothecene production has centred on the two species of *Myrothecium* reported to be producers of roridins and verrucarins *M. roridum* and *M. verrucaria* (Tamm, 1974).

TOXIN PRODUCTION

Precautions

People working with these compounds should be aware that they are acutely toxic. The principal problem is one of skin irritation. Certain areas of the body are especially sensitive, particularly around the nose, mouth and eyes. Small quantities of material transferred from the hands to the face can result in several days of misery. Gloves should be worn and if particles are likely to become airborne, also a surgical face mask.

Culture methods

Liquid cultures give the best results and large-scale fermentations (800–1350 l) with *Myrothecium* species are possible (Jarvis *et al.*, 1982; Bloem *et al.*, 1983). Toxin production is often low on solid media although some isolates do produce well on selected rice substrates (Lee and Mirocha, 1984). Since we have wanted to be able to scale up production, virtually all of our efforts have been directed toward liquid cultures. The seed medium needs to be inoculated with spores since only those species which sporulate well give good trichothecene production. Spores may be produced on solid media with Sabouraud agar, rice containing 0.5% added peptone, oats, and bee pollen is particularly useful. Toxin production is best when the fermentation is carried out in two stages. First, the spores are suspended in a rich seed medium (e.g. corn steep) where they grow for *ca.* 2 days. This medium is then used as a 5–10% inoculum for the production medium. The best production media are defined media which do not contain complex nutrients such as peptone, corn steep, and the like. If growth of the organism is either poor or too luxuriant, production of antibiotics is poor.

Analysis for macrocyclic trichothecenes

Thin-layer chromatography (TLC) is the most convenient method of analysis. All of the macrocyclic and trichoverroid (Jarvis *et al.*, 1982) trichothecenes possess a 7′, 8′, 9′, 10′-dienic lactone system which absorbs strongly in the UV ($\lambda_{max} \sim 260$ nm; $\epsilon \sim 20{,}000$). Thus, these compounds are easily visualized by UV light on the appropriate TLC plates. Furthermore, these compounds give a distinctive coloration with several spray reagents: (1) with vanillin reagents (Kupchan *et al.*, 1977), they turn a steel-blue colour when the plate is heated (roridins turn colour much faster than do the verrucarins); (2) with ceric sulphate reagent they turn an orange-brown when heated; (3) with 4-(*p*-nitrobenzyl) pyridine (NBP) reagent (Takitani *et al.*, 1979), they turn a light sky-blue colour. If appropriate standards are available, (high-performance liquid chromatography (HPLC) analysis (either on C-18 reversed phase, MeOH/H_2O, or on normal silica gel phase, EtOAc/hexane) is very useful for quantitative analysis. Mass spectroscopy (MS) has been of limited use for the macrocyclic trichothecenes because of the low volatility of these compounds (roridins generally melt at 200–250 °C, while verrucarins do not melt below 300 °C). Derivatization usually fails to yield better results. Electron impact MS often does not yield the parent ions; chemical ionization (CI), particularly negative ion CI is probably the method of choice for MS analysis. However, gas chromatography–mass spectroscopy (GC–MS) can be very useful following hydrolysis since most of the macrocyclic trichothecenes and trichoverroids give verrucarol (**3**) with small amounts of 8α-hydroxyverrucarol (Jarvis *et al.*, 1982), 8β-hydroxyverrucarol (Kupchan *et al.*, 1977) and 16-hydroxyverrucarol (Pavanasasivam and Jarvis, 1983, unpublished results; and Jarvis *et al.*, 1983b).

Bioproduction of trichothecenes from myrothecium species

The original work with *M. verrucaria* and *M. roridum* was carried out in the early 1960s by Professor Tamm and his group at Basel. In a series of very elegant papers (Tamm, 1974), they showed that liquid cultures of these fungi produce the highly toxic roridins (e.g. **7** and **8**) and verrucarins (e.g. **5** and **6**). Our earlier work was carried out with *M. verrucaria* (ATCC 24571) under conditions reported by Tamm (Harri *et al.*, 1962) with Czapek–Dox medium (glucose is the principal carbon source), followed later by a two-stage process involving medium A as a seed medium and medium B (see Table 1) as production medium (see below). This process was employed on an 800 l scale to yield gram quantities of previously characterized verrucarins and roridins (Jarvis *et al.*, 1982). Later, we found corn steep liquor-glucose (medium 1, Table 1) to be a better seed medium with this strain (Pavanasasivam and Jarvis, 1984, unpublished results). In addition to the known verrucarins and

Table 1 Composition of media used in liquid cultures

Medium	Composition (g/ℓ)
A	Glucose (20), malt extract (2), yeast extract (2), peptone (2), KH_2PO_4 (2), $MgSO_4 \cdot 7H_2O$ (2), $FeSO_4 \cdot 7H_2O$ (0.2), NH_4Cl (3).
Modified A	As above except no $FeSO_4 \cdot 7H_2O$ and KH_2PO_4 (1) and $MgSO_4 \cdot 7H_2O$ (0.5).
N-Z Amine	N-Z amine[a] (20), corn steep liquor (3), glucose (50)
B	Sucrose (40), glycerol (10), $NH_4H_2PO_4$ (1), K_2PHO_4 (3), $MgSO_4 \cdot 7H_2O$ (0.2), NaCl (5)
C	Corn steep liquor (10), glucose (15.6)
D	Sucrose (40), KH_2PO_4 (1), $MgSO_4 \cdot 7H_2O$ (0.5), $CaCl_2$ (0.5), KNO_3 (0.86), monobasic ammonium tartatrate (1), NaOAc (1), glycerol (10)

[a] Sheffield Chemicals.

roridins isolated from the above fermentation, we have isolated a number of new macrocyclic trichothecenes and a new class of trichothecenes, the trichoverroids (Jarvis *et al.*, 1982; Figure 3). These latter compounds are characterized by having the central trichothecene ring system and one or more portions of the macrocyclic ring which spans C4 to C15, but they lack the macrocyclic ring itself. Examples of these trichoverroids are trichodermadiene, (**10**) (Jarvis *et al.*, 1980b), trichodermadienediols A (**11a**) and B (**11b**), trichoverrols A (**12a**) and B (**12b**), trichoverrins A (**13a**) and B (**13b**) (Jarvis *et al*, 1981b), roridin L-2 (**14**) (Bloem *et al.*, 1983), and trichoverritone (**15**) (Jarvis *et al.*, 1983b). Trichoverrins are especially interesting because they appear to be the biosynthetic precursors to the roridins (Jarvis *et al.*, 1982).

We have examined over a dozen isolates of *M. verrucaria*, available from the American Type Culture Collection and U.S. Department of Agriculture in Beltsville, Maryland, and find little variation in the trichothecenes produced, although some isolates produce more than others; ATCC 24571 is particularly good. The only significant variation we have observed is that liquid cultures of *M. verrucaria*, ATCC 24571 in medium D (Table 1) produce 3-acetyl-4-hydroxymethylpyrrole (*ca.* 25 mg/l) (Pavanasasivam and Jarvis, 1984, unpublished results), a compound previously characterized by Professor Tamm, from a culture of *M. verrucaria* (Tamm, 1974). We have had good success scaling up our fermentation with *M. verrucaria* ATCC 24571 (Jarvis *et al.*, 1982) but although *M. verrucaria* ATCC 26146 produces good yields of roridin A (**7**) (*ca.* 50 mg/l) on a small scale (50 ml of medium/250 ml shake

Trichodermadiene (**10**), $R_1 = -\overset{(R)}{CH}-\overset{(R)}{CHCH_3}$ (epoxide, O bridging the two carbons); R_2 = H

Trichodermadienediol A (**11a**), R_1 = $\underset{(S)}{CHOH}\underset{(S)}{CHOH}CH_3$; R_2 = H

B (**11b**), R_1 = $-\underset{(S)}{CHOH}\underset{(R)}{CHOH}CH_3$; R_2 = H

Trichoverrol A (**12a**), R_1 = $-\underset{(S)}{CHOH}\underset{(S)}{CHOH}CH_3$; R_2 = OH

B (**12b**), R_1 = $-\underset{(S)}{CHOH}\underset{(R)}{CHOH}CH_3$; R_2 = OH

Trichoverrin A (**13a**), R_1 = $-\underset{(S)}{CHOH}\underset{(S)}{CHOH}CH_3$; R_2 = $O\overset{O}{\overset{\|}{C}}\underset{(E)}{CH{=}CCH_3}CH_2CH_2OH$

B (**13b**), R_1 = $-\underset{(S)}{CHOH}\underset{(R)}{CHOH}CH_3$; R_2 = $O\overset{O}{\overset{\|}{C}}\underset{(E)}{CH{=}CCH_3}CH_2CH_2OH$

Roridin L-2 (**14**), R_1 = $-CH(CHOHCH_3)OCH_2CH_2$ 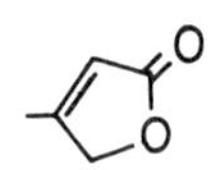 ; R_2 = OH

Trichoverritone (**15**), R_1 = $-CH(CHOHCH_3)OCH_2CH_2$ 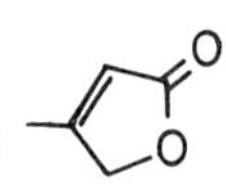

$R_2 = -\underset{\underset{O}{\|}}{OC}-\underset{(E)}{CH{=}CCH_3}CH_2CH_2OH$

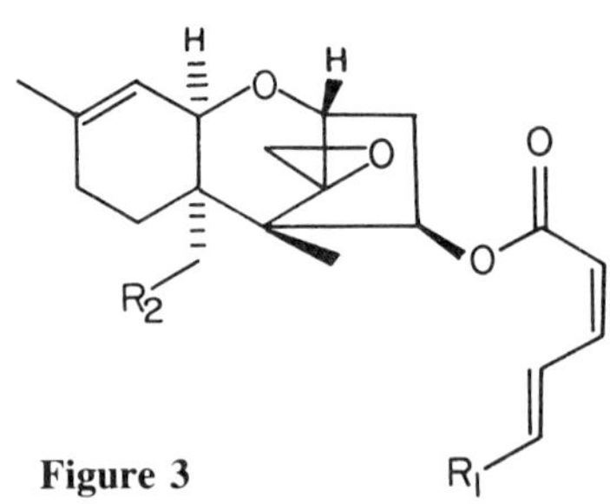

Figure 3

flask), it gives poor yields (≤ 10 mg/l) when scaled up to a 10 l fermentation (Pavanasasivam and Jarvis, 1984, unpublished results).

In recent times we have focused our attention on *M. roridum*. Our initial lack of success with *M. roridum* was due to our inability to obtain spores of this fungus. However, sporulation can be obtained on the media described above, although *M. roridum* usually does not sporulate as well as *M. verrucaria*. In general, a greater range of trichothecenes is produced by *M. roridum* than by *M. verrucaria* but the amounts are smaller. It is not known whether there is any connection between toxin production and pathogenicity but *M. roridum* is often found as a plant pathogen while *M. verrucaria* is less often reported to be pathogenic (Leath and Kendall, 1983).

Although yields of antibiotics produced by *M. verrucaria* vary with the seed and production media used (Table 1), the types of antibiotics produced vary little. However, liquid cultures of *M. roridum* on different media can give rise to quite different products. *M. roridum* strains often exhibit sensitivity to certain transition metal ions. With *M. roridum*, toxin production is usually inhibited by Fe^{2+} and somewhat increased with Zn^{2+}. Thus, use of modified medium A (Table 1) with *M. roridum* is often desirable. Toxin production by *M. verrucaria* varies little with the presence or absence of these metal ions. The spectrum of antibiotics produced by *M. roridum* varies with medium much more than with *M. verrucaria*. For example, *M. roridum* ATCC 24570 produces trichothecenes very poorly when corn steep is used as the seed medium and medium B (Table 1) as the production medium. If medium A is used as the seed medium with medium B, roridins A (**7**) and E (**8**) are produced in reasonably good yield (*ca*. 25 mg/l, each). However, when this strain is grown in liquid culture on rice infusion, no trichothecenes are produced, but instead significant quantities of (−)-epoformin (**16a**) (*ca*. 150 mg/l) and (−)-epiepoformin (**16**b) (*ca*. 20 mg/l) (Pavanasasivam and Jarvis, 1984, unpublished results). Earlier, the enantiomers of **16a** and **16b** were reported by Japanese workers to be produced by *Penicillium claviforme* (Yamamoto *et al*., 1973) and an unidentified pathogen (Nagasawa *et al*., 1978). The epoformins, like the macrocyclic trichothecenes, can be visualized on TLC plates by both UV light and the NBP reagent. However, trichothecenes give a light blue colour with the NBP reagent, while epoformins give an intense dark royal blue colour with this reagent.

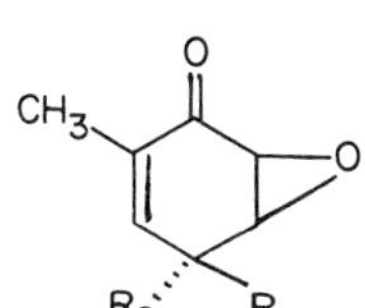

16a, R_1=H, R_2=OH
16b, R_1=OH; R_2=H

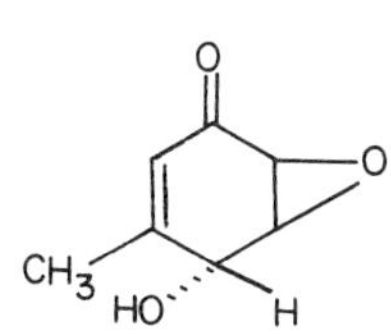

17

The most interesting strain we have studied was provided by the Parke-Davis Division of the Warner-Lambert Company, Ann Arbor, Michigan. This group had conducted a 1350 l fermentation with *M. roridum* CL-514 and isolated, in addition to roridins A (**7**), isoE and verrucarins A (**5**) and J (**6**), two new congeners, roridin L-2 (**14**) (Bloem *et al.*, 1983), isosatratoxin H (**18**) and 12′-hydroxyisoverrucarin J (Smitka *et al.*, 1984). From a fraction from this fermentation we isolated trichoverritone (**15**), the largest heretofore reported trichothecene, and 16-hydroxyroridin L-2 (Jarvis *et al.*, 1983b). *M. roridum* strain 81-131, a pathogen on *Aglaonema* sp., also produces trichoverritone (**15**) and roridin L-2 (**14**) (Jarvis *et al.*, 1983b). In addition, CL-514 produces 16-hydroxytrichodermadienediols A and B (see **11a** and **11b**), compounds also isolated in very low yield from a culture of *M. verrucaria* ATCC 24571 (Jarvis and Vrudhula, 1983). These are the first reported naturally occurring C-16 substituted trichothecenes, although this position has been hydroxylated by chemical (Jarvis *et al*, 1980a and 1984) and microbial (Pavanasasivam and Jarvis, 1983) methods.

When *M. roridum* CL-514 was first grown by us in liquid culture (seed medium: A; production medium: B, Table 1), the major compound produced was roritoxin A (**19**), a novel macrocyclic trichothecene closely related to vertisporin produced by *Verticimonosporium diffractum* (Minato *et al.*, 1975). Compound **19** has been fully characterized by ^{1}H- and ^{13}C-NMR (nuclear magnetic resonance) spectroscopy and high-resolution MS (Pavanasasivam and Jarvis, 1984, unpublished results). Unfortunately, all further attempts to obtain **19** in liquid culture have failed although recently we have obtained **19** when CL–514 was grown on rice substrate (Jarvis and Yatawara, 1985, unpublished). Instead in liquid culture, we obtain either quantities of trichothecenes previously reported (Bloem *et al.*, 1983) to be produced by CL-514 or the epoformin type compound, (−)-anti-5-hydroxy-4-methyl-7-oxabicyclo[4.1.0]hept-3-en-2-one (**17**) (Jarvis and Yatawara, 1984, unpublished results). This previously unreported compound is produced best in

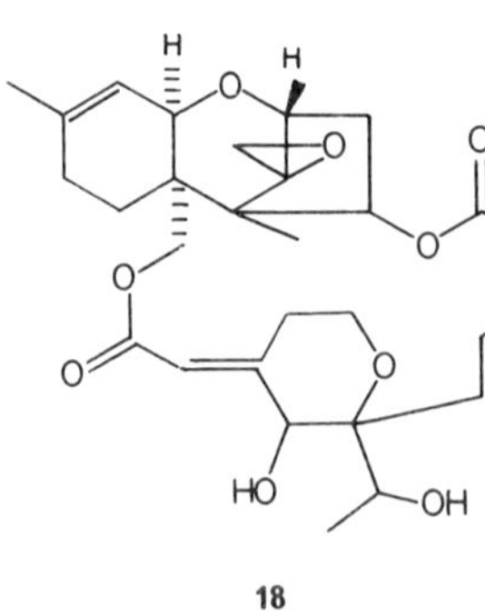

18

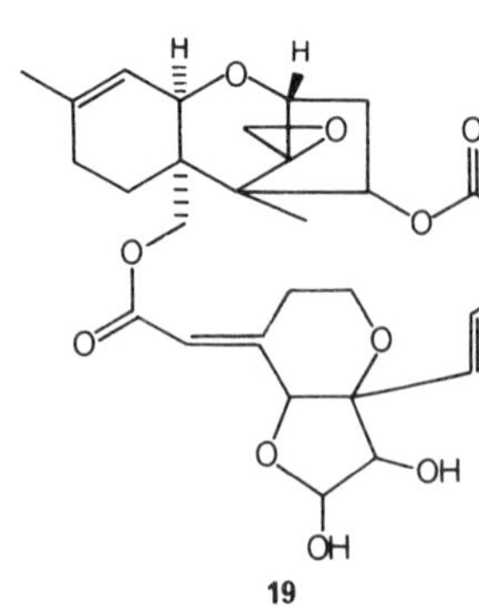

19

medium B when glycerol is omitted. It is interesting that **17** is produced by *M. roridum* CL-514 while (−)-epoformin (**16a**) is produced by *M. roridum* ATCC 24570.

We also have been working with a strain of *M. roridum* isolated from diseased muskmelons (Bruton, 1982). It is a highly pathogenic isolate that is capable of attacking all parts of the muskmelon including leaves, stems, mature and immature fruits. When grown in liquid media(medium: D, Table 1), the following compounds are produced: trichodermadienediols A (**11a**) and B (**11b**), roridin L-2 (**14**) and 16-hydroxyroridin L-2. Trichothecene production by this strain of *M. roridum* was also low compared to that from *M. verrucaria*. The level of trichodermadienediols A and B was approximately 10 mg/l, and the levels of roridin L-2 and 16-hydroxyroridin L-2 was approximately 5 mg/l. When muskmelon seedling roots were exposed to roridin A (**7**), 8β-hydroxyroridin A was detected in the leaves 24h later (Bean *et al.*, 1984). Plants exposed to roridin A showed a phytotoxic response with leaf tips becoming necrotic, indicating that trichothecene production by *M. roridum* could be involved in the establishment of disease by this fungus.

Recently, we have found that *M. roridum* strain 4582 in liquid culture (seed medium: C; production medium: B) gives no macrocyclic or trichoverroid trichothecenes. Instead, trichodermol (**4**) and 8β-hydroxy-trichodermol are produced (Jarvis and Yatawara, 1984, unpublished results). This appears to be a unique strain of *Myrothecium* in which the biosynthetic pathway to the trichoverroids and macrocyclic trichothecenes has been blocked. Careful study of this strain may lead to some important insights into the biosynthesis of this important class of compounds.

ACKNOWLEDGEMENT

These investigations were supported by Grant No. 259–67 awarded by the National Cancer Institute, DHHS and by Contract No. DMAD-17-82-C-2240 awarded by the U.S. Army.

REFERENCES

Bean, G. A., Fernando, T., Jarvis, B. B. and Bruton, B. (1984). 'The isolation and identification of trichothecene metabolites from a plant pathogenic strain of *Myrothecium roridum*', *J. Natl. Prods.*, **47**, 727–729.

Bloem, R. J., Smitka, T. A., Bunge, R. H., French, J. C., and Mazzola, E. P. (1983). 'Roridin L-2, a new trichothecene', *Tetrahedron Lett.*, **24**, 249–252.

Brian, P. W., Dawkins, A. W., Grove, J. F., Hemming, H. G., Lowe, D., and Norris, G. L. F. (1961). 'Phytotoxic compounds produced by *Fusarium equiseti*', *J. Exp. Bot.*, **12**, 1–20.

Bruton, B. D. (1982). '*Myrothecium roridum*, a potentially devastating pathogen of muskmelons in south Texas', (Abstr.) *Phytopathology*, **72**, 355.

Busam, L. and Habermehl, G. G. (1982). 'Accumulation of mycotoxins by *Baccharis coridifolia*: a reason for livestock poisoning', *Naturwissenschaften*, **69**, 392–393.

Cole, R J., Dorner, J. W., Cox, R. H., Cunfer, B. M., Cutler, H. G., and Stuart, B. P. (1981). 'The isolation and identification of several trichothecene mycotoxins from *Fusarium heterosporun', J. Natl. Prods..*, **44**, 324–330.

Cutler, H. C. and Jarvis, B. B. (1985). 'Preliminary observations on the effects of macrocyclic trichothecenes on plant growth', *Exper. Environ. Bot.*, **25**, 115–128.

Harri, E., Loeffler, W., Sigg, H. P., Stähelin, H., Stoll, Ch., Tamm, Ch., and Wiesinger, D. (1962). 'Über die verrucarine und roridine, eine gruppe von cytostatisch hochwirksamen antibiotica aus *Myrothecium*-Arten', *Helv. Chim. Acta*, **45**, 839–853.

Jarvis, B. B., and Mazzola, E. P. (1982). 'Macrocyclic and other novel trichothecenes: their structure, synthesis, and biological significance', *Acc. Chem. Res.*, **15**, 388–395.

Jarvis, B. B., and Vrudhula, V. M. (1983). 'New trichoverroids from *Myrothecium verrucaria*: 16-hydroxytrichodermadienediols', *J. Antibiotics*, **36**, 459–461.

Jarvis, B. B., Stahly, G. P., Pavanasasivam, G., and Mazzola, E. P. (1980a). 'Antileukemic compounds derived from chemical modification of macrocyclic trichothecenes. 1. derivatives of verrucarin A', *J. Med. Chem.*, **23**, 1054–1058.

Jarvis, B. B., Midiwo, J. O., Stahly, G. P., Pavanasasivam, G., and Mazzola, E. P. (1980b). 'Trichodermadiene: a new trichothecene', *Tetrahedron Lett.*, **21**, 787–788.

Jarvis, B. B., Midiwo, J. O., Tuthill, D., and Bean, G.A. (1981a). 'Interaction between the antibiotic trichothecenes and the higher plant *Baccharis megapotamica', Science* (Washington, D.C.), **214**, 460–462.

Jarvis, B. B., Pavanasasivam, G., Holmlund, C. E., DeSilva, T., Stahly, G. P., and Mazzola, E. P. (1981b). 'Biosynthetic intermediates to the macrocyclic trichothecenes', *J. Am. Chem. Soc.*, **103**, 472–474.

Jarvis, B. B., Stahly, G. P., Pavanasasivam, G., Midiwo, J. .O., DeSilva, T., Holmlund, C. E., Mazzola, E. P., and Geohegan, R. F., Jr. (1982). 'Isolation and characterization of the trichoverroids and new roridins and verrucarins', *J. Org. Chem.*, **47**, 1117–1124.

Jarvis, B. B., Eppley, R. M., and Mazzola, E. P. (1983a). 'Bioproduction of macrocyclic trichothecenes', in *Trichothecenes: Chemical, Biological, and Toxicological Aspects* (ed. Y. Ueno), pp. 20–38. Kodansha, Tokyo.

Jarvis, B. B., Vrudhula, V. M., and Pavanasasivam, G. (1983b). 'Trichoverritone and 16-hydroxyroridin L-2, new trichothecenes from *Myrothecium roridum*', *Tetrahedron Lett.*, 24, 3539–3542.

Jarvis, B. B., Midiwo, J. O., and Mazzola, E. P. (1984). 'Antileukemic compounds derived by chemical modification of macrocyclic trichothecenes. 2. derivatives of roridins A and H and verrucarins A and J', *J. Med. Chem.*, **27**, 239–244.

Kupchan, S. M., Streelman, D. R., Jarvis, B. B., Dailey, R. G., Jr., and Sneden, A. T. (1977). 'Isolation of potent new antileukemic trichothecenes from *Baccharis megapotamica*', *J. Org. Chem.*, **42**, 4221–4225.

Leath, K. T., and Kendall, W. A. (1983). '*Myrothecium roridum* and *Myrothecium verrucaria* pathogenic to roots of red clover and alfalfa', *Plant Disease*, **67**, 1154–1155.

Lee, Y.-W. and Mirocha, C. J. (1984). 'Production of nivalenol and fusarenone-X by *Fusarium roseum* Fn-2B on a rice substrate'. *Appl. Environ. Microbiol.*, **48**, 857–858.

Loeffler, W., Mouli, R., Rüsch, M. E., and Stähelin, H. (1965). 'Anguidin and derivatives, a new antibiotic and antitumoral product', Fr. 1,372,122', *Chem. Abst.* **62**, 5856d.

Minato, H., Katayama, T., and Tori, K. (1975). 'Vertisporin, a new antibiotic from *Verticimonosporium diffractum*', *Tetrahedron Lett.*, **16**, 2579–2582.

Nagasawa, H., Suzuki, A., Tamura, S. (1978). 'Isolation and structure of (+)-deoxyepiepoxydon and (+)-epiepoxydon, phytotoxic fungal metabolites', *Agric. Biol. Chem.*, **42**, 1303–1304.

Pavanasasivam, G., and Jarvis, B. B. (1983). 'Microbial transformation of macrocyclic trichothecenes', *Appl. Environ. Microbiol.*, **46**, 480–483.

Rosen, R. T., and Rosen, J. D. (1982). 'Presence of four *Fusarium* mycotoxins and synthetic material in 'Yellow Rain': evidence for the use of chemical weapons in Laos', *Biomed. Mass Spectr.* **9**, 443–450.

Rüsch, M. E., and Stähelin, H. (1965) 'Über einige biologische Wirkungen des Cytostaticum Verrucarin A', *Arzneim.-Forsch.* **15**, 893–897.

Smitka, T. A., Bunge, R. H., Bloem, R. J. and French, J. C. (1984). 'Two new trichothecenes, PD 113, 325 and PD 113, 326', *J. Antibiotics,* **37**, 823–828.

Stähelin, V. H., Kalberer-Rüsch, M. E., Signer, E., and Lazary, S. (1968). 'Über einige biologische Wirkungen des Cytostaticum Diacetoxyscirpenol', *Arzneim.-Forsch.*, **18**, 989–994.

Takitani, S., Asabe, Y., Kato, T., Suzuki, M., and Ueno, Y. (1979). 'Spectrodensitometric determination of trichothecene mycotoxins with 4-(ρ-nitrobenzyl) pyridine on silica gel thin-layer chromatograms', *J. Chromatogr.*, **172**, 335–339.

Tamm, Ch. (1974). 'The antibiotic complex of the verrucarins and roridins', *Fortschr. Chem. Org. Naturst.*, **31**, 63–117.

Ueno, Y. (1980). 'Trichothecene mycotoxins: mycology, chemistry, and toxicology', *Adv. Nutr. Sci.*, **3**, 301–353.

Ueno, Y. (ed.) (1983). *Trichothecenes: Chemical, Biological, and Toxicological Aspects*, Kodansha, Tokyo.

Yamamoto, I., Mizuta, E., Henmi, Y., Yamano, T., and Yamatodani, S. (1973). 'Epoformin, a new antibiotic produced by *Penicillium claviforme*', *Takeda Kenkyusho Ho*, **32**, 532–538.

Trichothecenes and Other Mycotoxins
Edited by J. Lacey

CHAPTER 22

Effects of controlled atmospheres on *Penicillium patulum* growth and patulin production*

N. Paster and N. Lisker

INTRODUCTION

The effects of controlled atmospheres in retarding deterioration of various food products have been studied extensively and a considerable amount of information is available on their effects on stored-products insects. A clear correlation between increases in CO_2 levels and insect mortality was found (Jay, 1980; Navarro and Calderon, 1980). Although complete mortality has not always been obtained, modified atmospheres for controlling storage insects have already been successful in field trials (Bailey and Banks, 1980). Navarro and Calderon (1980) suggested that controlled atmosphere treatments should be integrated with other physical means (e.g. cooling and aeration) in order to increase insect mortality in storage.

As species of the genera *Aspergillus* and *Penicillium* comprise most of the fungi involved in the deterioration of stored grains, the effects of controlled atmospheres on their mycotoxin production have been investigated, particularly with regard to aflatoxin (Diener and Davis, 1972; Epstein *et al.*, 1979; Landers *et al.*, 1967; Sanders *et al.*, 1968; Wilson and Jay, 1976). Much less has been done on the effects of modified atmospheres on the production of other mycotoxins. Lillehoj *et al.* (1972) described the effects of controlled atmospheres on penicillic acid production by *P. martensii* in corn. Paster *et al.* (1983) reported that concentrations of 30% CO_2 and above totally inhibited the production of ochratoxin by *A. ochraceus* grown on a synthetic medium at 26 °C. Although patulin is produced by several species of *Aspergillus* and *Penicillium* (Ciegler, 1977) and has carcinogenic, teratogenic and mutagenic properties, only one brief report was published on the effects of controlled atmospheres on patulin production by *P. expansum* (Beer and Woods, 1977).

* Contribution from the Agricultural Research Organization, The Volcani Center, Bet Dagan, Israel. No. 787-E, 1983 series.

The present study was therefore aimed at providing much broader information on the effects of controlled atmospheres on patulin production by *P. patulum*.

MATERIAL AND METHODS

General

The fungus used throughout these studies was *P. patulum* Bain. (NRRL 2159a). Mixtures of gases of the desired compositions were prepared according to the system described by Navarro and Donahaye (1972). Erlenmeyer flasks of 100 ml capacity, containing 20 ml Czapek agar (CA), served as test chambers and were connected to the system after inoculation. The chambers were flushed with the prepared gas composition and 100 μl gas samples were withdrawn from them periodically in order to test their composition with a gas chromatograph.

Inoculation

In preliminary studies a disc of CA (0.3 cm in diameter) covered with the fungal mycelium was placed in the centre of each flask containing 20 ml CA. The growth rate of the colony was very slow during the first 10 days of growth, unlike growth from spores. Therefore, mycelial discs were not used further and spores were used as a source of inoculum. Each Erlenmeyer was inoculated with 0.2 ml of a 0.1% agar suspension containing 30 or 60 spores and the suspension was gently distributed over the surface of the medium using a Drugalsky glass rod.

After connection to the system the flasks were observed daily until visible growth occurred and colonies could be counted. Unconnected Erlenmeyer flasks, placed alongside the test chambers, served as controls. All experiments were carried out at 26 ± 1 °C.

Patulin assay

Two flasks of each treatment were taken for patulin assay. The entire contents of both flasks were blended together (twice for 1 min) in a mixture of 200 ml of acetonitrile : 4% KCl : 20% H_2SO_4 (178 : 20 : 2, v/v/v). After filtration through Whatman No. 1 filter paper, 50 ml of distilled water was added to 100 ml of the filtrate in a separatory funnel. The whole mixture was then extracted twice with chloroform (50 and 30 ml, respectively). Both chloroform fractions were combined and evaporated first to near dryness under reduced pressure and then under a N_2 stream until dry. The residue was suspended in a 0.5 ml methanol and immediately loaded on precoated silica gel 60 thin-layer chromatography (TLC) plates with a F_{254} fluorescent indicator (E. Merck, catalog No. 5715). Plates were developed in benzene : methanol : acetic acid (85.0 : 7.5 : 7.5, v/v/v) and patulin concentration was estimated under a short-wave ultraviolet (UV) light (254 nm) where the intensities of the fluorescent spots were compared with patulin standards. An additional estimation of patulin concentration was made after spraying the plates with 0.5% 3-methyl-2-benzothiazolinone hydrazone (MBTH) hydrochloride monohydrate solution and heating for 15 min at 130 °C (AOAC, 1980). Patulin appears as a yellow spot under visible light. Mycelial dry weight of colonies in each test chamber was determined as follows: Erlenmeyer flasks were heated in a water bath until complete melting was attained. Mycelium was then removed from each flask, washed with distilled water to remove the remaining agar, and dried over a filter paper (Whatman No. 1) at room temperature, followed by overnight drying at 80 °C.

RESULTS

Effects of controlled atmospheres on the appearance of *P. patulum* colonies

In Erlenmeyer flasks not connected to the modified atmosphere system, as well as those in which air was flushed through flasks, colonies were first seen 2 days after inoculation. However, in most of the gas mixtures tested, the appearance of colonies was retarded 2–4 days. The delay before colonies were visible in the various combinations of atmospheric gases are given in Table 1. When spores were exposed to 60% CO_2/20% O_2, colonies were visible 6 days after inoculation, while under 0% CO_2/1% O_2, 0% CO_2/60% O_2, 0% CO_2/70% O_2 or 40% CO_2/20% O_2, they were visible after 5 days. The addition of 1% CO_2 to 5% or 60% O_2 cancelled out the inhibitory effects detected at these O_2 concentrations in the absence of CO_2. The numbers of colonies under all gas combinations tested were at least 70% of those in control flasks (Table 1), but no colonies were observed in 7 days under 100% CO_2 or 100% N_2. However, when the Erlenmeyers were subsequently disconnected from the pure CO_2 and N_2 and exposed to air, colonies could be seen after 3 days.

The effect of controlled atmospheres on mycelial dry weight and patulin production

Patulin production in flasks inoculated with 30 or 60 spores per flask was compared daily. More patulin was produced with 60 spores than with 30 only during the first 3 days after colonies became visible. Afterwards, similar amounts of patulin were produced by both levels of inoculum. Most patulin

Table 1 The effect of controlled atmospheres on patulin production and growth of *Penicillium patulum* in Czapek agar maintained at 26 °C

Controlled atmospheres CO_2/O_2 (%)	Colony appearance[a] (days)	Number of colonies (as percentage of control)	Days of incubation after colony appearance[b]			
			7		9	
			Patulin (mg/40 ml)	Mycelium dry weight (g/40 ml)	Patulin (mg/40 ml)	Mycelium dry weight (g/40 ml)
Control	2	100	45.0	0.26	40.0	0.24
0/1	5	74.5	1.0	0.10	3.0	0.12
0/5	4	81	14.0	0.15	19.0	0.20
0/10	4	80	45.0	0.20		
0/20	4	86	45.0	0.26		
0/40	4	72	45.0	0.28		
0/60	5	73	20.0	0.24	32.5	0.26
0/70	5	80	1.3	0.24	25.0	0.24
1/1	4	83	1.0	0.06	1.2	0.12
1/5	2	86	45.0	0.16		
1/60	2	95	45.0	0.24		
10/20	4	100	42.0	0.30		
20/20	4	90	25.0	0.26	25.0	0.22
25/20	4	90	25.0	0.26	25.0	0.23
30/10	4	100	16.0	0.26	12.5	0.22
30/20	4	76	10.0	0.28	10.0	0.26
30/40	4	86	12.5	0.28		
40/20	5	85	10.0	0.22	12.5	0.23
60/20	6	72	2.5	0.12	2.5	0.12

[a] Number of days before colonies were visible.
[b] Only those treatments showing a decrease in patulin production on the seventh day were examined on the ninth day.

was obtained from 5 to 9 days after colony appearance (45.0 mg/40 ml media). Therefore, production of patulin was estimated only on the seventh and ninth days after colony appearance. For example, when colonies appeared in test chambers kept under a certain CO_2/O_2 combination, 2 days after they appeared in controls, patulin estimation in these chambers was also carried out 2 days after the estimation of patulin in the controls. The amount of patulin produced by colonies grown for 7 days after their appearance under 1% or 5% O_2 but no CO_2, was less than that produced by the control; less mycelium was also produced (Table 1).

The amount of patulin and the mycelial dry weight of colonies growing under these conditions for a further 2 days were still lower than controls. Under 1% O_2 given without CO_2 and kept for an additional 5 days, the amount of patulin and mycelial dry weight were respectively only 15.0 mg/40 ml and 0.18 mg/40 ml. When 10%, 20%, or 40% O_2 was given without CO_2, production of both patulin and mycelial dry weight was similar to the controls, although with these gas combinations colonies appeared 2 days later. Further increase to 60% or 70% O_2 given without CO_2, decreased patulin production but not mycelial dry weight. Keeping the Erlenmeyers at the same conditions for an additional 2 days increased the amount of patulin, and 2 days later the amount of patulin was the same as in the control flasks.

Increasing CO_2 to 1% in combination with 1% O_2 gave the same amount of patulin as 1% O_2 alone. However, adding 1% CO_2 to 5% or 60% O_2 overcame the inhibitory effect on patulin production of these two O_2 concentrations in the absence of CO_2.

Patulin production was not decreased by up to 10% CO_2. However, it was decreased when CO_2 was increased to 20% or more while O_2 was maintained at 20% (close to normal atmosphere).

Mycelial dry weight was decreased only with 60% CO_2 and an additional 2 days incubation under the same conditions gave no further increase in patulin production or mycelial dry weight.

When spores were kept at 100% CO_2 or N_2 for 7 days, no colonies developed. However, when test chambers were removed from CO_2 or N_2 atmospheres and subsequently exposed to air, colonies appeared after 3 days and the amount of patulin detected 7 days later was the same as in controls (inoculated when test cultures were removed from CO_2 or N_2 atmospheres). No reduction in mycelial dry weight was recorded.

DISCUSSION

Colonies from spores kept under different combinations of atmospheric gases usually became visible 2–4 days after those exposed to air. Only when 1% CO_2 was given in combination with 5% or 60% O_2, did colonies appear at the same time as controls. This may indicate the need for low CO_2 levels (0.03%, as in the air, or 1%) for rapid colony formation. On the other hand, high levels of CO_2 (60%) decreased mycelial dry weight, showing that CO_2 may affect both colony formation and mycelial dry weight. Oxygen affected mycelial dry weight only at low concentrations (1 and 5%). The number of colonies formed was not markedly affected by any controlled atmosphere treatment. Only in pure CO_2 or N_2 were no colonies formed during the whole period the spores were kept under these extreme conditions, but the spores retained their viability and grew when again exposed to air.

The effect of atmospheric gases on spore germination and fungal growth has been studied by numerous investigators and comprehensive reviews have been published (Smith and Berry, 1976; Tabak and Cooke, 1968). It is generally accepted that most fungi require molecular oxygen for growth. Obligate anaerobiosis has not been reported (Smith and Berry, 1976) but facultative anaerobic growth may occur in *Mucor racemosus*, *M. rouxianus* (Wood-Baker, 1955) and *Rhizopus stolonifer* (Wells and Uota, 1970). Even with low levels of O_2 (0.5%) given without CO_2, fungi belonging to the genera *Alternaria*, *Aspergillus* and *Penicillium* could grow slightly (Yang and Lucas, 1970). Studies with high CO_2 concentrations have shown that 16% CO_2 (given with 21% O_2) causes spore germination of *R. stolonifer*, *Botrytis cinerea* and *Cladosporium herbarum* to be inhibited by 90%, while germination of *A. tenuis* and *Fusarium roseum* was significantly decreased only by concentrations > 32% (Wells and Uota, 1970).

Yang and Lucas (1970) divided the fungi they tested into two categories based on their ability to grow in atmospheres enriched with CO_2: (i) CO_2-sensitive, which comprise all the fungi which were inhibited in an atmosphere of 50% CO_2/50% O_2, and (ii) CO_2-tolerant, comprising those which grew under these conditions.

On the other hand, when the O_2 concentration was low enough (1 or 2%) to limit germination and mycelial growth, low concentrations of CO_2 (< 4%) could reverse the inhibition in most of the fungi tested. Similar results were obtained with *Phytophthora* spp. by Mitchell and Zentmyer (1971). It is generally accepted that CO_2 concentrations >95% are inhibitory to fungi (Tabak and Cooke, 1968). Recently, Paster *et al.* (1983) found that *A. ochraceus* was tolerant to quite high levels of CO_2. Only at >80% CO_2 was growth totally inhibited, although these concentrations were not lethal since, after transferring the colonies to air, normal growth resumed.

Summarizing the effects of gaseous environments on the growth of fungi, Tabak and Cooke (1968) stated that the effect of various levels of O_2 and CO_2 in either stimulating growth or being deleterious to fungi differs with the species. However, fungi are in general more sensitive to high CO_2 than to low O_2. Usually inhibition of fungal growth was due to the presence of CO_2 rather than the absence of O_2 and such was the case with *P. patulum*.

In our study, atmospheres containing 5 or 60% O_2 decreased patulin production but the addition of 1% CO_2 restored production to levels similar to those in 10 to 40% O_2 alone. This indicates that CO_2 also, in small amounts, is needed to allow production of patulin when extreme levels of O_2 are applied. Similarly, penicillin formation is enhanced by low levels of CO_2 (Lengyel and Nyiri, 1967). High levels of CO_2 ($\geq$20%), given in combination with 20% O_2, decreased the amount of patulin produced, while high levels of O_2 (60 and 70%) increased the time needed for maximal production (11 days instead of 7, as in the control). Beer and Woods (1977), working with *P. expansum*, concluded that contamination of apples by patulin might be decreased by low storage temperatures in controlled atmospheres containing 2–5% CO_2 and 3% O_2. These were the only combinations tested by them but our results suggest that other levels may also be useful.

Northolt *et al.* (1978), dealing with the effect of water activity and temperature on growth and patulin production by *Penicillium expansum*, *P. patulum*, and *Aspergillus clavatus*, found that the minimum water activity value for patulin production by *P. patulum* was 0.95 while the optimum temperatures for the toxin production by the fungus varied with the strain tested and were 8 or 31 °C. It therefore seems that temperature changes during storage and low water activities may also influence the compositions of controlled atmospheres required to inhibit patulin production.

The only report on penicillic acid was published by Lillehoj *et al.* (1972). They found that 40–60% CO_2, in combination with 20% O_2, diminished penicillic acid production by *P. martensii* by 21–44% when the temperature was decreased from 20 to 15 °C. No toxin was detected at either 5 or 10 °C. The effects of controlled atmosphere on ochratoxin production by *A. ochraceus* were reported recently by Paster *et al.* (1983). Ochratoxin production was completely inhibited by 30% CO_2 regardless of the O_2 levels given. Total inhibition of mycelial growth and ochratoxin production was obtained when the fungus was exposed for 14 days to 100% CO_2. However, after disconnection from the system, followed by 14 days of exposure to air, there was no significant difference in mycelial growth or ochratoxin production between the treated mycelia and controls.

There are several reports dealing with the effects of controlled atmospheres on aflatoxin production. Landers *et al.* (1967) reported that aflatoxin production in peanut kernels decreased with increasing concentrations of CO_2 from 0.03% to 100%. In general, decreasing the O_2 concentration resulted in smaller amounts of aflatoxin being produced. Sanders *et al.* (1968) found that visible growth and aflatoxin production by *A. flavus* were inhibited by 20% CO_2 at 17 °C and by 40% and 60% CO_2 at 25 °C. Like Landers *et al.* (1967), they found that at a constant temperature, an increase in CO_2 concentration caused a decrease in aflatoxin. Diener and Davis (1972) reported that aflatoxin production was reduced at ⩾20% CO_2 given in combination with 20% O_2, and almost completely inhibited when O_2 concentration was reduced to 5%. Wilson and Jay (1976) reported that aflatoxin production in peanuts was effectively controlled at <1% O_2. It therefore seems likely that mycotoxins could be controlled by enriching atmospheres with CO_2 or by decreasing O_2. However, controlled atmosphere, different substrates, relative humidities and temperatures may interact differently for each mycotoxin. Increasing CO_2 to 40% is now being used to control insects in stored products (Calderon and Navarro, 1979; Harein and Press, 1968). However our earlier studies, with those reported here, indicate that CO_2 concentrations required for insect control are much less than those needed for the control of storage fungi.

REFERENCES

AOAC (1980). *Official Methods of Analysis*, 13th edn. Arlington, VA, Sect. 26, pp. 111–116.

Bailey, S. W., and Banks, H. J. (1980). 'A review of recent studies of the effects of controlled atmospheres on stored product pests', in *Controlled Atmosphere Storage of Grains* (ed. J. Shejbal), pp. 101–118. Elsevier, Amsterdam.

Beer, S. V., and Woods, A. C. (1977). 'Effect of storage, temperature and atmosphere on patulin accumulation and lesion development in apples infected with *Penicillium expansum*', *Phytopathology*, **67**, 208 (abstr.).

Calderon, M., and Navarro, S. (1979). 'Increased toxicity of low oxygen atmospheres supplemented with carbon dioxide on *Tribolium castaneum* adults', *Entomologia exp. appl.*, **25**, 39–44.

Ciegler, A. (1977). Patulin, in *Mycotoxins in Human and Animal Health* (eds J. D. Rodricks, C. W. Hesseltine and M. A. Mehlman), pp. 609–624. Pathotox Publ. Co., College Park, MD.

Diener, U. L., and Davis, N. D. (1972). 'Atmospheric gases and aflatoxin production in peanuts', *Highlights Agric. Res.*, 19.

Epstein, E., Steinberg, M. P., Nelson, A. I., and Wei, L. S. (1979). 'Aflatoxin production as affected by environmental conditions', *J. Food Sci.*, **35**, 389–391.

Harein, P. K., and Press, A. F. (1968). 'Mortality of stored peanuts insects exposed to mixtures of atmospheric gases at various temperatures', *J. Stored Prod. Res.*, **4**, 77–82.

Jay, E. (1980). 'Low temperatures: effects on control of *Sitophilus oryzae* (L.) with modified atmospheres', in *Controlled Atmosphere Storage of Grains* (ed. J. Shejbal), pp. 65–71. Elsevier, Amsterdam.

Landers, K. E., Davis, N. D., and Diener, U. L. (1967). 'Influence of atmospheric gases on aflatoxin production by *Aspergillus flavus* in peanuts', *Phytopathology*, **57**, 1086–1090.

Lengyel, Z. L., and Nyiri, L. (1967). 'Automatic aeration and the action of carbon dioxide in fermentation processes', *Ind. Chim. Belge*, **32**, 798–800.

Lillehoj, E., Milburn, M. S., and Ciegler, A. (1972). 'Control of *Penicillium martensii* development and penicillic acid production by atmospheric gases and temperatures', *Appl. Microbiol.*, **24**, 198–201.

Mitchell, D. J., and Zentmyer, G. A. (1971). 'Effects of oxygen and carbon dioxide tension on growth of several species of *Phytophthora*', *Phytopathology*, **61**, 787–791.

Navarro, S., and Calderon, M. (1980). 'Integrated approach to the use of controlled atmospheres for insect control in grain storage', in *Controlled Atmosphere Storage of Grains* (ed. J. Shejbal), pp. 73–78. Elsevier, Amsterdam.

Navarro, S., and Donahaye, E. (1972). 'An apparatus for studying the effect of controlled low pressure and composition of atmospheric gases on insects', *J. Stored Prod. Res.*, **8**, 223–226.

Northolt, M. D., Van Egmond, A. P., and Paulsch, W. E. (1978). 'Patulin production by some fungal species in relation to water activity and temperature', *J. of Food Production*, **41**, 885–890.

Paster, N., Lisker, N., and Chet, I. (1983). 'Ochratoxin A production by *Aspergillus ochraceus* Wilhelm grown under controlled atmospheres', *Appl. Environ. Microbiol.*, **45**, 1136–1139.

Sanders, T. H., Davis, N. D., and Diener, U. L. (1968). 'Effect of carbon dioxide, temperature and relative humidity on production of aflatoxin in peanuts', *J. Am. Oil Chem. Soc.*, **45**, 683–685.

Smith, J. E., and Berry, D. R. (1976). *The Filamentous Fungi*, Vol. 2: *Biosynthesis and Metabolism*. Edward Arnold, London.

Tabak, H. H., and Cooke, W. B. (1968). 'The effects of gaseous environments on the growth and metabolism of fungi', *Bot. Rev.*, **34**, 126–252.

Wells, J. M., and Uota, M. (1970). 'Germination and growth of five fungi in low oxygen and high carbon dioxide atmospheres'. *Phytopathology*, **60**, 50–53.

Wilson, D. M., and Jay, E. (1976). 'Effect of controlled atmosphere storage on aflatoxin production in high moisture peanuts (groundnuts)', *J. Stored Prod. Res.*, **12**, 97–100.

Wood-Baker, A. (1955). 'Effects of oxygen-nitrogen mixtures on the spore germination of mucoraceous mould', *Trans. Br. mycol. Soc.*, **38**, 291–297.

Yang, H., and Lucas, G. B. (1970). 'Effects of N_2–O_2 and CO_2–O_2 tension on fungi isolated from damaged flue cured tobacco', *Appl. Microbiol.*, **19**, 271–277.

Trichothecenes and Other Mycotoxins
Edited by J. Lacey

CHAPTER 23

The effect of water activity and temperature on mycotoxin production by *Alternaria alternata* in culture and on wheat grain

N. Magan and J. Lacey

INTRODUCTION

The genus *Alternaria* is widely distributed in soil and on aerial plant surfaces, and many species are known to be pathogenic to plants. Species can often grow at low temperature and may be associated with extensive spoilage of fruit and vegetables during refrigerated transport and storage (Ellis, 1971). *Alternaria alternata* (Fr.) Keissler is ubiquitous and is abundant in the air spora, especially during ripening and harvesting of cereal crops (Darke *et al.*, 1976). Ripening ears of wheat are colonized by *A. alternata* soon after emergence (Magan, 1982) and it is reported to be the most common sub-epidermal fungus of wheat grain (Hyde and Galleymore, 1951). *A. alternata* alone or with other fungi, e.g. *Alternaria triticina*, can cause a conspicuous black or brown discoloration of wheat kernels called black point disease (Machacek and Greaney, 1938; Boyer, 1955; Bhowmik, 1969). This can result in decreased quality and yield of grain (Dickinson, 1981).

Because of the pathogenicity exhibited by many *Alternaria* spp. various workers have investigated the toxicity of crude preparations of cultures and infected substrates. Many isolates have been shown to be lethal to chicks, rats and mice (Doupnik and Sobers, 1968; Christensen *et al.*, 1968; Meronuck *et al.*, 1972). Extracts of *A. alternata* from cotton have been found to be highly toxic to brine shrimp, chick embryos and rats (Diener *et al.*, 1976).

The most important secondary metabolites of known mammalian toxicity produced by *A. alternata* are the dibenzo-α-pyrones altenuene (AE), alternariol (AOH), alternariol monomethyl ether (AME) and a derivative of tetramic acid, tenuazonic acid (Meronuck *et al.*, 1972; Pero *et al.*, 1973; Harvan and Pero, 1976). *A. alternata* has been demonstrated to produce some or all of these mycotoxins in various concentrations on tomato (Harwig *et al.*, 1979), sorghum (Sauer *et al.*, 1978) and pecans (Schroeder and Cole, 1977). Colonization of ears, black point disease and toxin production are all favoured by wet weather close to harvest.

Development of fungi is dependent on both abiotic and biotic variables (Sinha, 1973). Perhaps the two most important environmental factors are substrate water activity (a_w) and temperature. Other important factors are pH of the substrate and gas composition of the surrounding atmosphere (Troller and Christian, 1978; Magan and Lacey, 1984a,c). Knowledge of the influence of environmental factors on growth and mycotoxin formation can be an important aid in predicting mycotoxin contamination of grain, both in the field and in store. Further, the a_w and temperature limits for growth and mycotoxin production can sometimes be markedly different (Northolt and Bullerman, 1982).

This chapter reports the effect of a_w and temperature on the production of AE, AOH and AME by *A. alternata* on wheat extract agar and wheat grain.

MATERIALS AND METHODS

Culture methods

The isolate of *A. alternata* used in these studies was obtained from freshly harvested wheat grain directly plated on 2% malt agar in 1980 and chosen for its heavy sporulation. Cultures were stored on 2% malt agar at 5 °C and subcultured when required for experiments onto freshly prepared 2% malt agar plates incubated at 25 °C.

Wheat extract agar (pH 5.5) modified with glycerol to 0.98, 0.95 and 0.90 a_w, was used as the test medium (Magan and Lacey, 1984a). Petri dishes containing 13 ml of freshly prepared wheat extract agar were inoculated at the centre with a needle-point of spores obtained from the margin of 8–10-day-old colonies. Inoculated plates of the same a_w were sealed in sterile polyethylene bags, together with three or four uninoculated plates (Harris *et al.*, 1970). The water activity of representative samples of the media was checked with a dewpoint meter (Protimeter Ltd, Marlow, Bucks., U.K.).

Duplicate sets of each treatment were incubated at 5, 15, 25 and 30 °C. Every 5 days up to 30 days at 25 °C and 40 days at 15 °C; 2 Petri dishes were removed from each treatment, colony diameters were measured and then cultures extracted and analysed for AE, AOH and AME. The amount of each mycotoxin produced by a single colony of *A. alternata* was then calculated.

Wheat grain experiments

Flasks containing 150 g of winter wheat (cv. Maris Huntsman) were sterilized by autoclaving at 1.1 kg/cm (121 °C) for 25–30 min. After autoclaving, the initial water content of the grain (usually 8–10%) was determined using duplicate samples of approximately 5 g, oven-dried at 105 °C for 17–18 h (International Seed Testing Association, 1976). Sterile distilled water was then added to bring the water content to about 19, 22 and 26%, equivalent to 0.90, 0.95 and 0.98 a_w, respectively (Pixton, 1967; Muir, 1973). To allow equilibration of water content after adding water to the grain, the flasks were stored for 7–10 days at 5 °C and shaken vigorously each day. At the end of this period the moisture content was re-checked.

A single layer of sterile grain was placed on 7–10-day-old Petri dish cultures of *A. alternata*, the lids replaced and the plates shaken vigorously to coat the surfaces of the grain with spores and hyphae of the fungus. The inoculated grain was returned to the flask and thoroughly mixed with sterile grain remaining. The inoculated grain was then placed in four perforated sterile foil containers, each holding approximately 35–40 g grain. These were carefully placed on a wire gauze stand in a desiccator previously cleaned with alcohol and containing 250 ml of a glycerol/water mixture of the required a_w (Dallyn and Fox, 1980). The two halves of each desiccator were sealed with Vaseline. To allow air exchange, a small vent, plugged with cotton wool, was provided. Duplicate desiccators of each treatment were incubated for 3 weeks at 15, 25 and 30 °C.

Extraction

Samples of grain (40 g) or wheat agar cultures (25 g), were homogenized in 100 ml of methanol (AR) for 2–3 min. Thirty millilitres of the filtered extract was mixed with 60 ml of 20% ammonium sulphate in a 250 ml separating funnel. Distol grade hexane (2 × 30 ml) was added and the funnel shaken for 45 s before allowing the layers to separate. The upper hexane layer was discarded and the lower layer transferred to a 100 ml separating funnel. This was extracted with 4 × 5 ml of dichloromethane. The extracts were passed through a column of anhydrous sodium sulphate (granules, 30–40 mm × 8 mm) into a conical flask. The combined extracts were evaporated under a gentle stream of nitrogen. The residue was then dissolved in 2 ml methanol (60%) /0.2 M ammonium acetate pH 6 (40%) and diluted accordingly for analysis.

Analysis

The high-performance liquid chromatography (HPLC) system consisted of a column 250 × 4.6 mm packed with 5 μm Spherisorb ODS. The eluent consisted of 70% methanol and 30% ammonium acetate (0.2 M) pH 5.9 with a flow rate of 1 ml/min. A Gilson pump and manometric module (Model 302, 802 respectively) were used. The eluent was monitored using a Perkin-Elmer 2000 Fluorescence Spectrophotometer set with an excitation wavelength of 315 nm and an emission cut-off filter of 430 nm (for AE, AOH and AME) and an electrochemical detector with an excitation potential of 0.9 V (for AOH and AME).

Standards

Samples of AE, AOH and AME were kindly supplied by Dr P. M. Scott (Health and Welfare Canada, Ottawa, Ontario, Canada); Dr D. J. Harvan (National Institute of Health Research, Triangle Park, N.C., U.S.A.) and Dr R. M. Eppley (Food and Drug Administration, Washington D.C.,U.S.A.).

RESULTS

Cultural studies on wheat extract agar

The relationship between colony growth of *A. alternata* at different a_w levels and temperatures was compared with its ability to produce AE, AOH and AME. At 0.98 a_w and 15 °C colonies grew to 80 mm diameter after 20 days. AOH production increased rapidly between 10 and 20 days growth, reaching a maximum concentration of 51 μg/ml before decreasing by 50% during the next 5 days. Further decreases, to 25% of the maximum, occurred during the next 15 days (Figure 1a). AME concentration similarly increased to a maximum of 25 μg/ml after 25 days growth before decreasing again to only 10 μg/ml after 40 days. By contrast, AE production increased continuously once it was first detected after 10 days growth, to reach a maximum concentration of 57 μg/ml after 40 days.

At 0.95 a_w and 15 °C the isolate of *A. alternata* grew to a diameter of only 50 mm after 40 days. AOH was initially isolated after 10 days growth and then increased gradually, reaching a maximum concentration of 62 μg/ml after 40 days. AME and AE, both detected initially after 20 days growth, increased only slowly, reaching final concentrations of 10 and 5 μ/ml, respectively (Figure 1b). At 0.90 a_w growth and mycotoxin production were further decreased. Concentrations of all three metabolites were less than 5% of that at 0.95 a_w with AOH again produced in the highest concentration followed by AME and AE (Figure 1c).

Altering the temperature to 25 °C increased the growth of *A. alternata* and markedly increased the quantities of mycotoxins produced, compared to 15 °C. At 0.98 a_w colonies grew to 80 mm diameter in 10 days. Both AOH and AME were first detected after 5 days growth and reached maximum

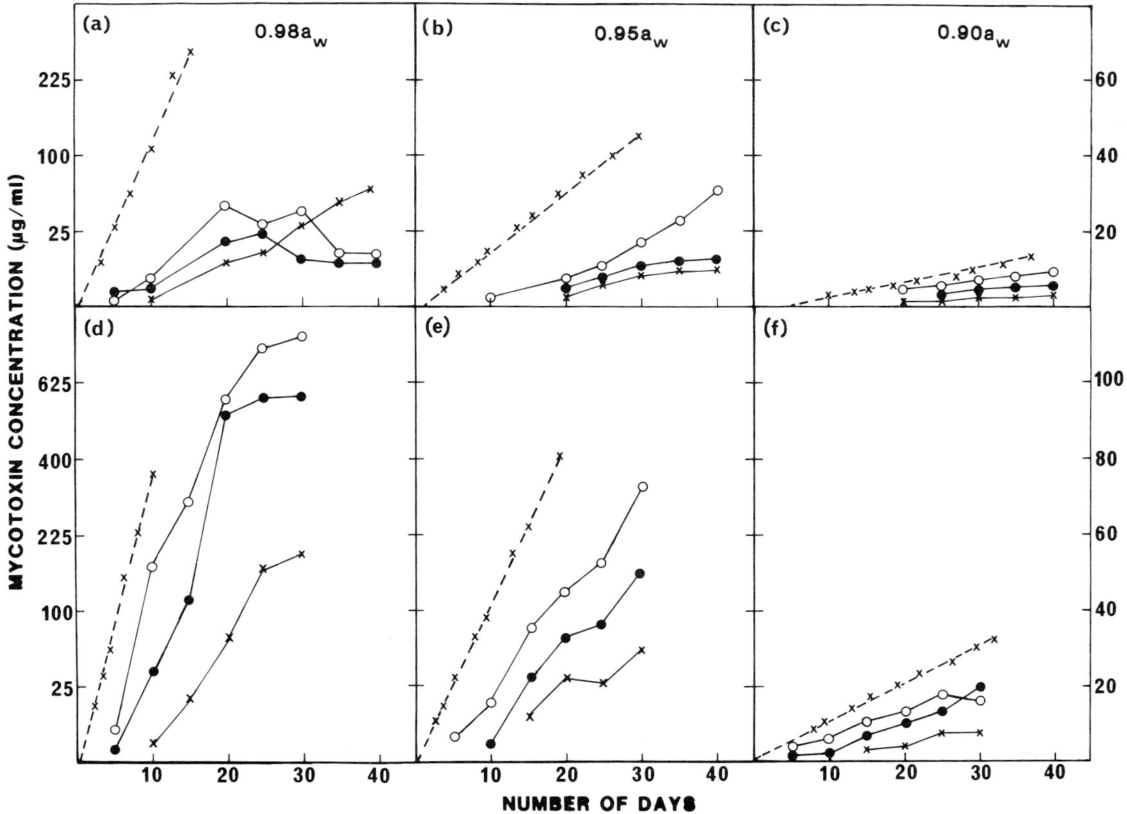

Figure 1 The relationship between growth of *A. alternata* on wheat extract agar and mycotoxin production at 15 (a,b,c) and 25 °C (d,e,f) and 0.98, 0.95 and 0.90 a_w (×---×, colony diameter; ●—●, alternariol; ○—○, alternariol monomethyl ether; ×—×, altenuene).

concentrations of 807 and 603 μg/ml, respectively, after 30 days (Figure 1d). Most mycotoxin was produced during the period 15–25 days after initial inoculation. No AE was detected until after inoculation, by which time the colony had grown to 80 mm diameter with 169 μg/ml present after 30 days.

At 0.95 a_w and 25 °C, colonies grew more slowly, reaching a diameter of > 80 mm after 20 days. At this a_w production of all three metabolites after 30 days was < 40% of that at 0.98 a_w (Figure 1e). AOH was still produced in the largest concentration, reaching 331 μg/ml after 30 days. AME, first detected in the growing colony after 10 days growth, increased to 158 μg/ml after 30 days. However, production of AE was only detected after 15 days growth when the colony had grown to 60 mm diameter but increased to a concentration of 55 μg/ml after 30 days.

At 0.90 a_w and 25 °C, growth of *A. alternata* was even slower, reaching only about 30 mm diameter after 30 days growth. At this a_w, maximum concentration of AOH was produced after 25 days growth (18 μg/ml) but concentration then decreased by about half in the next 5 days. By contrast, AME production increased continuously to a maximum of 22 μg/ml after 30 days. Very little AE was produced at this a_w and 25 °C (Figure 1f).

The effects of temperature on the production of AE, AOH and AME at different a_w levels after an incubation period of 30 days are summarized in Table 1. Maximum production of all three mycotoxins occurred at 25 °C and 0.98 a_w. Changing temperature and a_w from 25 °C and 0.98 a_w decreased mycotoxin production with the greatest decrease occurring when both were changed together. In general, AOH was produced in greatest concentration followed by AME and AE.

Experiments with wheat grain

Only small concentrations of AE were produced at 15 °C, e.g. less than 0.5 μg/g grain at 0.98 a_w. Production was much greater at 25 °C, with 24.3 μg/g at the same a_w. Further increasing the temperature to 30 °C led to a decline in production to only 1.0 μg/g grain. The effect of decreasing a_w was most notable at 25 °C. The amount of AE produced at 0.95 a_w was less than 5% of that at 0.98 a_w, while at 15 and 30 °C and the same a_w concentrations were about a quarter and one-tenth respectively. Only traces of AE were found at 0.90 a_w regardless of incubation temperature.

At the lowest temperature tested, 15 °C, and 0.98 a_w, 10.8 μg AOH/g grain was produced. Production was six times greater at 25 °C (Figure 2) but at 30 °C was again less than at 15 °C. AOH responded to changes in a_w similarly to AE with production declining by 90% between 0.98 and 0.95 a_w. Only a trace was detected at 0.90 a_w and 15 or 30 °C.

Production of AME at 0.98 a_w with increasing temperature from 15 to 25 °C changed from 0.6 μg to 194.8 μg/g grain, while at 30 °C there was again a

Table 1 Effect of temperature and water activity on production of mycotoxins by *A. alternata* on wheat extract agar (after 30 days incubation)

	Water activity								
	Mycotoxin[a] (μg/ml)			Mycotoxin[a] (μg/ml)			Mycotoxin[a] (μg/ml)		
Temperature (°C)	AE	AOH	AME	AE	AOH	AME	AE	AOH	AME
5	2.7	16.5	1.5	0.3	1.2	0.2	NG[b]	NG	NG
15	27.9	42.1	7.8	2.0	19.2	3.0	0.1	0.6	0.2
25	169.2	807.0	603.0	55.3	331.5	158.1	1.1	11.1	21.8
30	13.4	48.7	32.7	4.4	19.9	8.5	5.3	3.5	5.9

[a] AE, altenuene; AOH, alternariol; AME, alternariol monomethyl ether.
[b] NG, no growth.

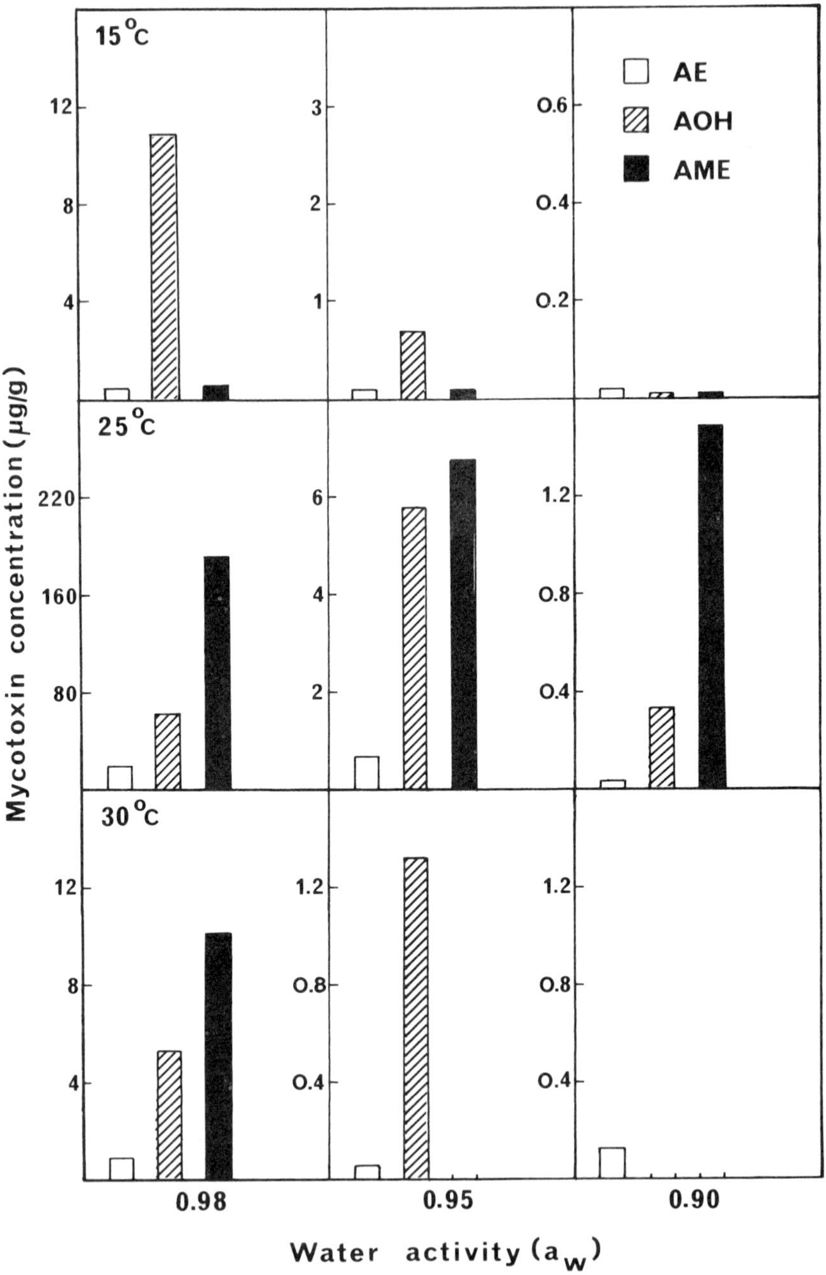

Figure 2 The influence of temperature and water activity on the production of mycotoxins by *A. alternata* on wheat grain (AE, altenuene; AOH, alternariol; AME, alternariol monomethyl ether).

marked decrease with production only 5% of that at 25 °C. Decreasing a_w to 0.95 resulted in only a trace of AME being produced at 15 .007C while production at 25 °C was only 10% of that at 0.98 a_w. At 30 °C no AME was detected at either 0.95 or 0.90 a_w. Like AE and AOH, only a trace of AME was found at 15 and 25 °C and 0.90 a_w.

As in experiments on wheat extract agar, the optimum conditions for production of all three metabolites on wheat grain was 25 °C and 0.98 a_w. However, by contrast with wheat extract agar, at 25 °C and 30 °C AME was produced in the largest concentration on wheat grain followed in decreasing order by AOH and AE, while at 15 °C there was most AOH, followed by AME and AE.

DISCUSSION

Temperature and a_w both affect the production of AE, AOH and AME by *A. alternata* on wheat extract agar and wheat grain. The optimum conditions for mycotoxin production on both substrates were 25 °C and 0.98 a_w. The concentrations of all three metabolites were decreased more markedly when temperature and a_w were altered together than when either factor was changed individually. On the wheat extract medium AOH was invariably produced in largest concentration followed by AME and AE, but on wheat grain at 25–30 °C, AME was present in greater concentration than AOH or AE.

AOH and AME are cytotoxic to both Hela and lymphoma cells with an ED_{50} of 6 and 8 μg/ml, respectively. Together, they can act synergistically against Hela cells, as an antibiotic against *Bacillus mycoides* and are possibly teratogenic to mice at 100 mg/kg. AE, on the other hand, is active only against Hela cells (Harvan and Pero, 1976).

The total amounts of AOH, AME and AE produced by a single colony of the *A. alternata* isolate tested in this study were high, especially at 25 °C and 0.98 a_w. Further, all three metabolites were produced over the temperature range 5–30 °C and the a_w range 0.98–0.90, although at marginal temperatures and 0.90 a_w little of any mycotoxin was produced. On wheat extract agar at 25 °C the minimum a_w for germination of *A. alternata* conidia is 0.85, and for growth is 0.88 (Magan and Lacey, 1984a). A comparison of the a_w values and temperatures over which linear growth and AE, AOH and AME production occurs on wheat extract agar is shown in Figure 3. Very small amounts of all three mycotoxins were produced at 0.90 a_w, the lowest tested, suggesting that slightly more water is necessary for mycotoxin production than growth, with optimum amounts being produced at > 0.95 a_w. Similar differences in a_w and temperature requirements for growth and production of mycotoxins by *Aspergillus* and *Penicillium* species have been reported by Northolt and Bullerman (1982).

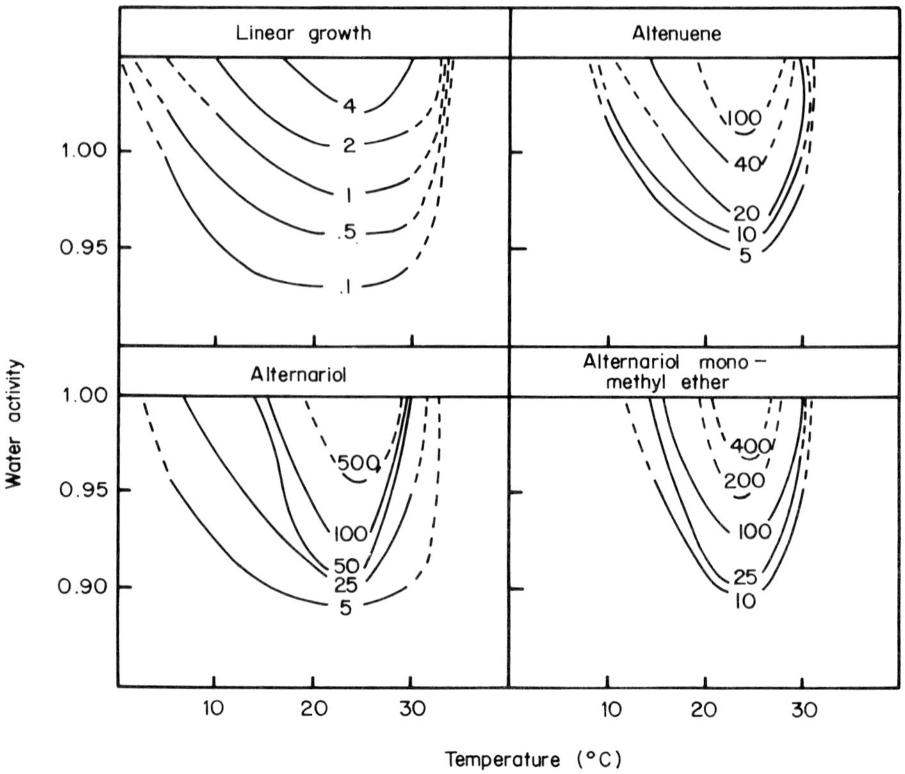

Figure 3 Comparison of water activity and temperature conditions favourable for linear growth of *A. alternata* (from Magan and Lacey, 1984a), and production of altenuene, alternariol and alternariol monomethyl ether. Numbers on the isopleths indicate growth rates in mm/day and mycotoxin concentration in μg/ml.

In culture, at all a_w values and temperatures, with the exception of 0.95 a_w and 15 °C, the increase in the diameter of the *A. alternata* colony was accompanied by a steady increase in production of AE, AOH and AME until the end of the experiment. At 0.95 a_w and 15 °C, maximum production of AOH and AME occurred after 20–25 days incubation, when the colony had grown to about 80 mm in diameter. However AE concentrations continued to increase until the end of the experiment (40 days). Aflatoxin B_1 production by *Aspergillus parasiticus* has been similarly observed to decrease when incubated in submerged culture (Northolt *et al.*, 1976). It was also noticeable that a lag time of 5–10 days often elapsed between inoculation and AME and AE production, especially at lower a_w values (0.95–0.90 a_w). This was less pronounced with AOH production. Other studies on solid substrates have shown that mycotoxin production by some *Aspergillus* and *Penicillium* species occurs soon after mycelial growth commences with lag times often less than 30 h (Northolt, 1979). However, longer lag times have been reported in liquid culture media (Sansing *et al.*, 1973).

Although experiments with wheat grain were limited to 21 days, the optimum conditions for production of AE, AOH and AME were the same as those on agar media i.e. 25 °C and 0.98 a_w. However AME was produced in larger concentrations than AOH at 25 and 30 °C. Further, no AME was detected in wheat grain at 30 °C and 0.95–0.90 a_w, indicating some differences from the production of this metabolite in agar. In the field, fluctuations of temperature, changes in relative humidity and rainfall all influence the colonization of developing grain by *A. alternata*. A minimum dew period of 3–6 h is necessary for black point infection and symptom development (Southwell *et al.*, 1980); a_w is particularly critical and in the field there is a period of 10–20 days after anthesis, when wheat grain is milky ripe (growth stage (GS) 70–79; Zadoks *et al.*, 1974) and the water content is about 70% (a_w = 1.00). During dough development (GS 80 onwards) the water content decreases rapidly to 20% (a_w = 0.92) or less as the grain ripens. While the water content is high, the a_w is close to 1.00 and ideal for germination and growth of *A. alternata*. Although Hyde and Galleymore (1951) found sub-epidermal mycelium when water content was 40% or less black point may occur at any stage from anthesis to GS 79, increasing with stage of grain development and inoculum density (Southwell *et al.*, 1980). The most susceptible period of the host plant therefore appears to be when a_w is close to 1.00, and provided high inoculum of *A. alternata* is present, both mycotoxin production and black point may increase. Our study provides information on the production of mycotoxins on wheat grain in pure culture only. No account was taken of the effect other fungi which colonize the ripening grain, such as *Cladosporium* spp., *Epicoccum nigrum* and *Fusarium* spp., may have on mycotoxin production. Interspecific interactions between these fungi and *A. alternata* can occur (Magan and Lacey, 1984b) and could decrease the production of AE, AOH and AME.

The only similar study with *Alternaria*, spp. has been on the effect of temperature and moisture content on the production of tenuazonic acid (TZA) by *Alternaria tenuissima* isolated from cottonseed (Young *et al.*, 1980). Maximum production of TZA on cottonseed occurred at 20 °C, and 37.5% water content (1.00 a_w) and minimum at a water content of 14.9% (0.85 a_w). Production of TZA was halved when a_w was decreased to 0.95. Young *et al.*, (1980) therefore suggested that for TZA production 20 °C and more than 0.90 a_w was required. Unfortunately, because of analytical problems, we were unable to study the water relations of TZA production on wheat grain and in culture. However, biosynthesis of different mycotoxins produced by *Alternaria* spp. may be favoured by different temperatures.

In our investigation an isolate of *A. alternata* which spored heavily in culture and showed little aerial hyphal growth was used in both cultural and wheat grain experiments with little evidence of any loss in their ability to produce AE, AOH and AME. However, there are frequent reports that *Alternaria* species can quickly lose their toxin-producing ability (Forgacs and Carll, 1962). Coombe *et al.* (1970) found that subculturing the original culture twice resulted in dehydroaltenusin production by *A. tenuis* dropping to almost zero. Häggblom (1981) found that high mycotoxin-producing strains of *A. alternata* had little aerial mycelium, and that appearing only after 7 days, while low mycotoxin-producing strains had much floccose mycelial growth even after 2 days.

Christensen and Kaufmann (1969, 1974) considered that grain harvested with a high percentage of *Alternaria* contamination was sound. However conducive environmental conditions and excessive rainfall, especially late in the growing season, may cause rapid development of *A. alternata* and increase not only black point of wheat grain but also the production of AE, AOH, AME and TZA, which could present a hazard if used for human consumption or as animal feedstuff.

ACKNOWLEDGEMENTS

We thank Dr D. E. Clegg and Mrs Y. Dunne for their assistance in developing the extraction and analysis procedures.

REFERENCES

Bhowmik, T. P. (1969). '*Alternaria* seed infection of wheat', *Plant Dis. Rep.*, **53**, 77–80.

Boyer, M. G. (1955). 'Effect of *Alternaria tenuis* on other common seed-borne fungi'. *Assoc. Off. Seed Anal. Proc.*, **45**, 53–54.

Christensen, C. H. and Kaufmann, H. H. (1969). *Grain Storage: the role of fungi in quality loss*. University of Minnesota Press, Minnesota, U.S.A.

Christensen, C. H. and Kaufmann, H. H. (1974). 'Microflora', in *Storage of Cereal Grains and their Products* (ed. C. H. Christensen), pp. 158–192. American Association of Cereal Chemists, Inc., St Paul.

Coombe, R. G., Jacobs, J. J., and Watson, T. R. (1970). 'Metabolites of some *Alternaria* species: the structure of altenusin and dehydroaltenusin', *Aust. J. Chem.*, **23**, 2343–2351.

Dallyn, H. and Fox, A. (1980). 'Spoilage of materials of reduced water activity by xerophilic fungi', in *Microbial Growth and Survival in Extremes of Environment* (eds G. H. Gould and J. E. L. Corry), pp. 129–139. Academic Press, London.

Darke, C. S., Knowelden, J., Lacey, J., and Ward, A. M. (1976). 'Respiratory diseases of workers harvesting grain', *Thorax*, **31**, 294–302.

Dickinson, C. H. (1981). 'Leaf surface micro-organisms as pathogen antagonists and as minor pathogens', in *Strategies for Control of Cereal Diseases* (eds J. F. Jenkyn and R. T. Plumb), pp. 109–121. Blackwell Scientific Publications, Oxford.

Diener, U. L., Wagener, R. E., Morgan-Jones, G., and Davis, N. D. (1976). 'Toxigenic fungi from cotton', *Phytopathology*, **66**, 514–516.

Doupnik, B., and Sobers, E. K. (1968). 'Mycotoxicoses: toxicity to chicks of *Alternaria longipes* isolated from tobacco', *Appl. Microbiol.*, **16**, 1596–1597.

Ellis, M. B. (1971). *Dematiaceous Hyphomycetes*. Commonwealth Mycological Institute, Kew.

Forgacs, J., and Carll, W. T. (1962). 'Mycotoxicoses', *Adv. Vet. Sci.*, **7**, 273.

Harris, R. F., Gardner, W. R., and Adebayo, A. A. (1970). 'Agar dish isopiestic equilibration method for constant water potential of solid substrates', *Appl. Microbiol.*, **19**, 536–537.

Harvan, D. I., and Pero, R. W. (1976). 'The structure and toxicity of *Alternaria* metabolites', *Adv. Chem. Service*, **149**, 344–355.

Harwig, J., Scott, P., Stoltz, D. R., and Blanchfield, B. J. (1979). 'Toxins of moulds from decaying tomato fruit', *Appl. Env. Microbiol.*, **38**, 267–274.

Hyde, M. B., and Galleymore, H. B. (1951). 'The sub-epidermal fungi of cereal grains. II. The nature, identity and origin of the mycelium in wheat', *Ann. Appl. Biol.*, **38**, 348–356.

Häggblom, P. (1981). 'Production of alternariol and alternariol monomethyl ether and morphology of *Alternaria alternata*, *Trans. Br. Mycol. Soc.*, **77**, 185–187.

International Seed Testing Assocation (1976). 'International rules for seed testing', *Seed Sci. Technol.*, **4**, 3–117.

Machacek, J. E., and Greaney, F. J. (1938). 'The "black point" or kernel smudge disease of cereals', *Can. J. Res. Sect. C.*, **16**, 84–114.

Magan, N. (1982). 'The microflora of wheat grain: ecology of the fungi and effect of fungicides', Ph.D. thesis, Reading University, U.K.

Magan, N. and Lacey, J. (1984a). 'The effect of temperature and pH on the water relations of field and storage fungi', *Trans. Br. Mycol. Soc.*, **82**, 71–81.

Magan, N., and Lacey, J. (1984b). 'The effect of water activity, temperature and substrate on interactions between field and storage fungi', *Trans. Brit. Mycol. Soc.*, **82**, 83–93.

Magan, N., and Lacey, J. (1984c). 'The effects of gas composition and water activity on growth of field and storage fungi and their interactions', *Trans. Brit. Mycol. Soc.*, **82**, 305–314.

Meronuck, R. A., Steele, J. A., Mirocha, C. J., and Christensen, C. H. (1972). 'Tenuazonic acid, a toxin produced by *Alternaria alternata*', *Appl. Microbiol.*, **23**, 613–617.

Muir, W. E. (1973). 'Temperature and moisture in grain storage', in *Grain Storage: Part of a System* (eds. R. N. Sinha and W. E. Muir), pp. 49–70. The Avi Publishing Co., Westport, CT, U.S.A.

Northolt, M.D. (1979). 'The effect of water activity and temperature on the production of some mycotoxins', Ph.D. thesis, Agricultural University of Wageningen, Holland.

Northolt, M. D. and Bullerman, L. B. (1982). 'Prevention of mold growth and toxin production through control of environmental conditions', *J. Food Protect.*, **45**, 519–526.

Northolt, M. D., Verhulsdonk, C. A. H., Soentoro, P. S. S., and Paulsch, W. E. (1976). 'Effect of water activity and temperature on aflatoxin production by *Aspergillus parasiticus*', *J. Milk Food Technol.*, **39**, 170–174.

Pero, R. W., Posner, M., Harvan, D., and Spalding, J. W. (1973). 'Toxicity of metabolites produced by *Alternaria*', *Env. Health Perspect.*, June, 87–94.

Pixton, S. W. (1967). 'Moisture content: its significance and measurement on stored products', *J. Stored Prod. Res.*, **3**, 35–47.

Sansing, G. A., Davis, N. D., and Diener, U. L. (1973). 'Effect of time and temperature on ochratoxin A production by *Aspergillus ochraceus*', *Can. J. Microbiol.*, **19**, 1259–1263.

Sauer, D. B., Seitz, L. M., Burroughs, R., Mohr, H. E., and West, J. L. (1978). 'Toxicity of *Alternaria* metabolites found in weathered sorghum grain at harvest', *J. Agric. Food Chem.*, **26**, 1380–1383.

Schroeder, H. W., and Cole, R. J. (1977). 'Natural occurrence of alternariols in discoloured pecans', *J. Agric. Food Chem.*, **25**, 204–206.

Sinha, R. N. (1973). 'Interrelations of physical, chemical and biological variables in the deterioration of stored grains', in *Grain Storage: Part of a System* (eds R. N. Sinha and W. E. Muir), pp. 15–48. The Avi Publishing Co., Westport, CT., U.S.A.

Southwell, R. J., Brown, J. F., and Wong, P. T. W. (1980). 'Effect of inoculum density, stage of plant growth and dew period on the incidence of black point caused by *Alternaria alternata* in durum wheat', *Ann. Appl. Biol.*, **96**, 29–35.

Troller, J. A., and Christian, J. H. B. (1978). 'Microbial growth', in *Water Activity and Food*, chap. 5, pp. 87–105. Academic Press, London.

Young, A. B., Davis, N. D. and Diener, U. L. (1980). 'The effect of temperature and moisture on tenuazonic acid production by *Alternaria tenuissima*', *Phytopathology*, **70**, 607–609.

Zadoks, J. C., Chang, T. T., and Konzak, C. F. (1974). 'A decimal code for the growth stages of cereals', *Weed Res.*, **14**, 415–421.

Trichothecenes and Other Mycotoxins
Edited by J. Lacey

CHAPTER 24

The influence on mycotoxin production of interactions between fungi and their environment

M. O. Moss and J. M. Frank

INTRODUCTION

In the laboratory study of mycotoxin production by pure cultures it is possible to control such factors as water activity, temperature and nutrient availability. However, on the growing crops, and during post-harvest storage, moulds are exposed to complex interactions with the host species, its physiological condition and growth stage, other micro-organisms, and biocides applied at different stages of crop development, or preservatives applied during storage (Hesseltine, 1983).

Although most, if not all, of the mycotoxins are secondary metabolites, it is still not clear how their production relates to the physiology and ecology of the producing organisms. Batt (1983) concluded that aflatoxin production by *Aspergillus parasiticus* and asexual reproduction were independent, although the two phenomena are often linked. Thus Torres *et al.* (1980) showed that 39 successive transfers of *Aspergillus flavus* on moist wheat resulted in both the loss of the ability to produce aflatoxin and a reduction in asexual sporulation, and Stutz and Krumperman (1976) have shown that they both require the same temperature regime, during studies on the effect of temperature cycling on the production of aflatoxin. However, sclerotial development appears independent of aflatoxin production (Bennett *et al.*, 1979) although sclerotia frequently contain tremorgenic indole metabolites, such as aflatrem and cyclopiazonic acid, and occasionally may contain aflatoxin (Wicklow and Cole, 1982). At the biochemical level aflatoxin production may be correlated with the ability of the fungus to produce nitrate from ammonia or amino nitrogen (nitrification) (White and Johnson, 1982).

The complexity of interactions between factors influencing aflatoxin production is shown by apparently conflicting reports of experimentally controlled physicochemical parameters. For instance, Bennett *et al.* (1978) reported that light seemed not to influence aflatoxin production by *A. flavus*, although continuous white light inhibited the formation of sclerotia, while Gunasekaran (1981) and Mashaly and El-Deeb (1983) found aflatoxin production was inhibited by light. Bennett *et al.* (1981) found that light may interact with temperature so that aflatoxin production may be inhibited by light at high and low, but not at intermediate, temperatures.

Strain variability in toxin production

It has long been known that secondary metabolite production is very strain-dependent (Hesseltine *et al.*, 1970). More specifically, Schroeder and Boller (1973) showed that such variability in aflatoxin production may be correlated with the source of the isolates examined. Thus 74–100% of isolates from peanut and cottonseed, but only 20–58% of isolates from rice, were shown to be producers. Comparable variability is known for other species producing other mycotoxins. Thus rubratoxin production by *Penicillium purpurogenum* (Moss and Hill, 1970) and mycophenolic acid by *Penicillium roquefortii* (Engel *et al.*, 1982) are further well-documented examples.

Changes in biosynthetic potential during laboratory culture of a single strain has been one of the problems dogging natural products chemists ever since they became interested in mould metabolites. Mayne *et al.* (1971) showed that as few as six weekly transfers of *Aspergillus parasiticus* resulted in a reproducible reduction in aflatoxin production, while non-aflatoxigenic variants are frequently isolated during laboratory culture (Bennett *et al.*, 1981).

The role of the host in mycotoxin formation

Fungi producing mycotoxins range from plant pathogens, such as *Claviceps purpurea* and *Rhizoctonia leguminicola*, to saprophytes, such as species of *Aspergillus* and *Penicillium*. The latter are traditionally associated with deterioration during storage, but *Aspergillus flavus* and related species may develop before harvest. That aflatoxin may be produced in the field has been known for several years (Lillehoj and Hesseltine, 1977) and recently Hesseltine *et al.*, (1981) showed that, of 238 field samples collected in North Carolina during 1977, 187 contained aflatoxin. Such stress conditions as drought (Jones *et al.*, 1981) or low nitrogen (Jones and Duncan, 1981) predispose maize either to infection by *Aspergillus flavus*, or to increased aflatoxin production following infection, or to both. Hill *et al.* (1983b) have shown that, with dry conditions, soil temperatures of 26–32 °C for groundnuts and air temperatures of 30–40 °C for maize predisposed these crops to infection and aflatoxin production.

The influence of plant genotype on aflatoxin formation in groundnut and maize has been reviewed by Mixon (1977) and Zuber (1977) respectively. In both instances it has been possible to obtain resistant varieties, although the nature of resistance may be indirect. Thus resistance to insect damage may decrease the risk of aflatoxin formation in the field.

Laboratory studies by Shannon *et al.* (1980) have demonstrated that varieties of maize vary in their ability to support zearalenone formation even when infection by *Fusarium graminearum* occurs. When a single strain of mould is able to produce more than one toxin the genetic composition of the plant may influence the production of each toxin differently. For instance, a strain of *Gibberella zeae* produced zearalenone in all 29 varieties of maize tested but deoxynivalenol in only seven (Hart *et al.*, 1982).

Interactions between fungal and host genomes are too complex to rationalize in biochemical terms at present but, as more descriptive and quantitative information becomes available, it should eventually be possible to do so.

An interesting aspect of mycotoxin occurrence in plants is the possibility of translocation. Zearalenone produced in the stalks of maize infected with *F. graminearum* may be translocated to the ears without the spread of the fungus (Sutton *et al.*, 1976). Also the macrocyclic trichothecene derivative, baccharin, which was first considered to be a metabolite of the Brazilian shrub *Baccharis megapotamica*, seems more likely to be a fungal metabolite produced by a soil fungus which is translocated, and perhaps altered by the plant (Jarvis *et al.*, 1981).

Interactions between micro-organisms

Inhibitory interactions are suggested by many experiments on mycotoxin production by one fungus in the presence of another. For instance, aflatoxin production by *A. parasiticus* was markedly inhibited by both *Aspergillus candidus* and *A. chevalieri*, although *A. parasiticus* grew well in the presence of both species (Boller and Schroeder, 1973, 1974). It was not possible to distinguish between several alternative hypotheses, which are not mutually exclusive, to account for these observations. The competing fungus may partially replace *A. parasiticus* in jointly infected kernels, it may metabolize aflatoxins as they are formed, it may affect the metabolism of *A. parasiticus*, or it may compete for a substrate required for aflatoxin production but not growth. Several studies have demonstrated the inhibitory effect of *A. niger*. Ashworth *et al.* (1965) showed that, if *A. niger* colonized groundnuts first, no aflatoxin is produced by subsequent infection by *A. flavus*. Wicklow *et al.* (1980) found that only *A. niger* and *Trichoderma viride* out of 13 species of fungi tested inhibited aflatoxin formation in maize. Such phenomena are not confined to laboratory studies. Hill *et al.* (1983a) demonstrated that, following drought stress, extensive fungal colonization of physically sound mature groundnut kernels may occur. A combination of quantitative mycological examination and aflatoxin analysis demonstrated that when the ratio of *A.*

flavus : *A. niger* exceeded 19 : 1 aflatoxin was present, but when it was less than 9 : 1 no aflatoxin could be detected. Bacteria also influence fungal growth and aflatoxin production, with *Brevibacterium linens* causing slight inhibition and *Acetobacter aceti* stimulation (Weckbach and Marth, 1977). A fascinating study by Shotwell *et al.* (1975) of a bin of maize accidentally contaminated with water, which developed extensive mould growth, showed that occurrence of mycotoxins can be highly localized. Even kernels that were part of a clump containing some highly contaminated kernels were free of aflatoxin. Of 140 kernels individually analysed, 16 were positive for aflatoxin (ranging from 230 to 3800 ppb) and 12 were positive for zearalenone (ranging from 9 to 1700 ppm) but none contained both toxins together.

The difficulties of studying mixed-culture systems may sometimes give rise to conflicting reports in the literature. Thus, Alderman *et al.* (1973) found that *Penicillium purpurogenum* inhibited aflatoxin production by *A. parasiticus*, while Moss and Badii (1982a) found a significant increase in aflatoxin production in dual cultures of these organisms (Table 1). Rubratoxin, a toxic metabolite of *P. purpurogenum*, enhanced aflatoxin production even at concentrations which inhibited growth slightly (Moss and Badii, 1982b).

Table 1 The influence of *P. purpurogenum* on aflatoxin production by *A. parasiticus* growing on maize (from Moss and Badii, 1982a)

Time (days)	Aflatoxins (μg/flask)[a]	
	Control	+*P. rubrum*
5	11 (13)	58 (27)
7	435 (251)	1672 (890)
9	2278 (555)	3922 (1193)
13	702 (708)	1330 (624)

[a] Means and standard deviations of five replicate flasks.

Another mould metabolite reported to enhance aflatoxin production is the unsaturated fatty acid amide cerulenin (Figure 1) produced by a *Cephalosporium* sp. (Fanelli *et al.*, 1982).

INFLUENCE OF AGROCHEMICALS ON THE PRODUCTION OF *FUSARIUM* TOXINS IN CEREALS

Since antimicrobial activity, as measured by a reduction in biomass, may not always be associated with decreased mycotoxin formation, we have started a

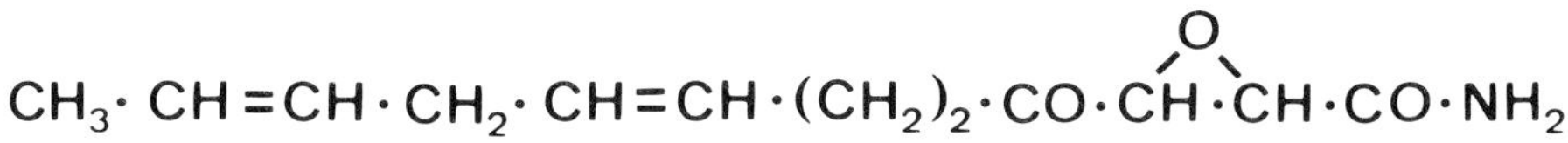

Figure 1 Cerulenin, a metabolite of *Cephalosporium*.

study of the effects of agricultural biocides on the growth and toxin production of fungi. Tridemorph, Zineb and Carbendazim (Figure 2) are fungicides used in the control of pathogens of cereals (Anon, 1983) and represent three distinct chemical types.

Materials and methods

Culture

Strain NRRL 3299 was used for these preliminary studies. This strain was originally referred to as *Fusarium tricinctum* or *F. poae*, but has subsequently been identified as *F. sporotrichioides* (Ichinoe and Kurata, 1983).

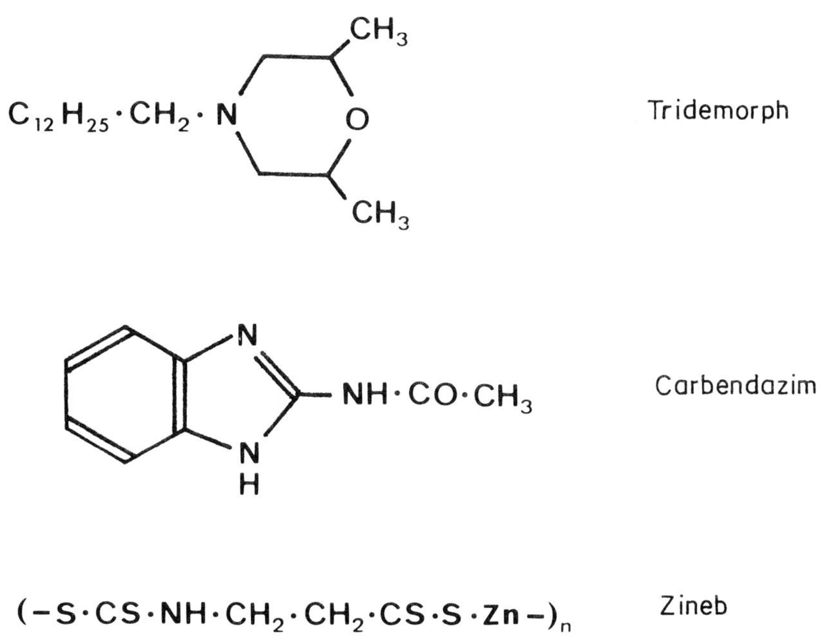

Figure 2 Fungicides used in the treatment of cereal crops.

Cultures were obtained from single spores using the method of Booth (1977) and maintained for routine use on potato sucrose agar.

Biocides

Tridemorph, Zineb and Carbendazim were kindly provided by Miss Jean Robb of the Edinburgh School of Agriculture. Tridemorph was in the form of a solution containing 75% active ingredient, Zineb and Carbendazim were wettable powders containing 65 and 50% active ingredients respectively. The biocides were not sterilized but were dissolved, or suspended, in sterile distilled water and finally diluted to the required concentration in sterile growth medium.

Growth medium

The medium was modified from that of Borrow *et al.* (1961) and contained (g l^{-1}): sucrose (62.2), glycine (3.5), NH_4NO_3 (3.5), KH_2PO_4 (2.0), $MgSO_4 . 7H_2O$ (0.38), micronutrient solution (2 ml). The micronutrient solution contained (g l^{-1}): $ZnSO_4 . 7H_2O$ (1.0), $FeSO_4 . 7H_2O$ (1.0), $CuSO_4 . 5H_2O$ (0.15), $MnSO_4 . 4H_2O$ (0.1), K_2MoO_4 (0.1). The medium was adjusted to pH 5.5, dispensed into 250 ml conical flasks (50 ml) and sterilized by autoclaving at 121 °C for 15 min. Flasks were inoculated with 1 ml of a spore suspension containing approximately 10^6 spores ml^{-1} and incubated at 25 °C on an orbital shaker rotating at 150 rpm.

Dry weight determination

The mycelium was filtered onto a tared filter paper, washed with distilled water and dried in a forced-air oven at 85–95 °C overnight. Before weighing they were allowed to equilibrate to room temperature for 15 min and placed in a microwave oven for 1.5 min.

Toxin analysis

The filtrates from three replicate flasks were bulked and treated with silica gel (100–200 mesh) and polyvinyl polypyrrolidone (Sigma) at the rate of 1 g of each per 100 ml filtrate. The mixture was shaken for 15 min, adjusted to pH 7.5 with saturated aqueous Na_2CO_3 and filtered through a glass fibre filter (GF/C). The filtrate (100 ml) was extracted twice with 15 ml chloroform: ethyl acetate (2 : 1 v/v) and the combined extracts passed through anhydrous Na_2SO_4, reduced to a small volume in a rotary evaporator under reduced pressure, and finally to dryness in a small sample tube under a stream of nitrogen. The dried sample was taken up in 1 ml of ethyl acetate and T-2 toxin estimated by thin-layer chromatography using chloroform : methanol (95 : 5 v/v) on aluminium-backed silica gel 60 plates (Merck 5553). Plates were sprayed with sulphuric acid (1 part 6N + 1 part MeOH) and heated at 110 °C until charring was just evident. Fluorescence under UV (360 nm) was compared with that of a range of standards chromatographed on the same plate, and estimations were only attempted when the intensity of an appropriate dilution of the sample fell within that of the range of standards (30–120 ng per spot).

Results

Our initial studies have shown that these fungicides have quite different effects on the production of T-2 toxin by *Fusarium sporotrichioides* (NRRL 3299) grown in shake flasks. Tridemorph increased toxin production significantly despite inhibiting biomass formation (Table 2; Figure 3). Carbendazim, the most active antifungal agent against *Fusarium*, decreased toxin production proportionally to the decrease in growth (Table 3), Zineb suppressed toxin formation, even at concentrations which had little effect on growth (Table 4). In practice these biocides are applied in mixtures at several stages during crop development. Further studies in the field will be essential, but growth of pure isolates in the laboratory under controlled conditions give an opportunity to demonstrate the possible interactions. Similar interactions have been observed in the enhancement of aflatoxin production in moist grain inadequately treated with propionic acid as a preservative (Al-Hilli and Smith, 1979; Pettersson *et al.*, 1978). Conversely, Bauer *et al.* (1983) cited propionic acid as an example of a preservative which gave reduced aflatoxin

Table 2 Influence of tridemorph on growth and T-2 toxin production of *Fusarium sporotrichioides* (NRRL 3299)

		Dry weight (mg/flask)			T-2 toxin (μg/flask)		
Time (h)	Conc. (ppm) Tridemorph	0	8	48	0	8	48
55		392	406	280	90	30	650
71		611	575	470	200	60	1000
79		596	677	604	130	70	925
95		827	802	761	175	80	1075
103		877	833	767	175	40	800
167		774	608	679	155	<10	600
263		624	499	435	100	60	tr

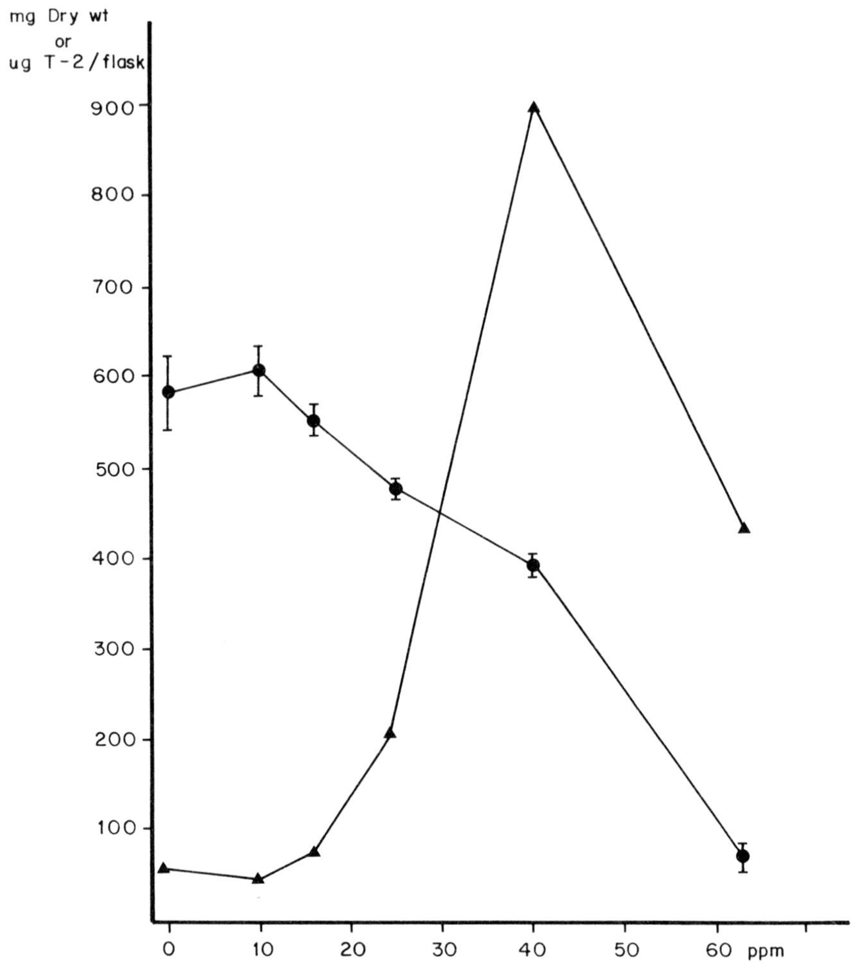

Figure 3 The influence of tridemorph on *Fusarium sporotrichioides* (NRRL 3299) after 77 h in shake flasks. ●—● Dry weight (mg/flask) mean and standard deviations of three replicates; ▲—▲ T-2 toxin (μg/flask) filtrates of three flasks bulked.

production, whereas others such as *p*-cresol caused enhanced production, suggesting other factors may also be involved.

An interesting example of a complex ecological interaction following the application of a herbicide is described by Penn and Lynch (1982). The application of *N*-phosphonomethyl glycine to destroy couch grass, prior to planting barley, successfully controlled the couch grass but acetic acid formed in the decaying rhizomes of this weed influenced the activity of soilborne

Table 3 Influence of Carbendazim on growth and T-2 toxin production of *Fusarium sporotrichioides* (NRRL 3299)

		Dry weight (mg/flask)			T-2 toxin (μg/flask)		
Time (h)	Conc. (ppm) Carbendazim	0	0.5	5.0	0	0.5	5.0
48		188	93.4	9	23	20	10
67		533	528	—	—	—	—
75		607	604	7	—	—	—
90		753	778	—	400	350	—
198		784	777	8	120	100	13
672		—	—	343	—	—	200

— = Not determined.

Table 4 Influence of Zineb on growth and T-2 toxin production of *Fusarium sporotrichioides* (NRRL 3299)

		Dry weight (mg/flask)			T-2 toxin (μg/flask)		
Time (h)	Conc. (ppm) Zineb	0	1	60	0	1	60
48		188	104	11	23	15	7.5
67		533	533	73	—	—	—
75		607	609	139	—	—	—
90		753	728	498	400	200	200
198		784	754	767	120	70	70

— = Not determined.

Fusarium culmorum, leading to a reduced growth of barley subsequently planted. Although there may be no involvement of mycotoxins in this particular case it demonstrates the multifactorial nature of disease.

ACKNOWLEDGEMENTS

We gratefully acknowledge financial support provided by the Ministry of Agriculture, Fisheries and Food.

REFERENCES

Alderman, G. G., Emeh, C. O., and Marth, E. H. (1973). 'Study of interactions between *A. parasiticus* with *P. italicum, Lactobacillus plantarum* and *P. rubrum*', *Z. Lebensm. Unters. Forsch.*, **153**, 305–311.

Al-Hilli, A. L., and Smith, J. E. (1979). 'Influence of propionic acid on growth and aflatoxin production by *Aspergillus flavus*, *FEMS Microbiol. Lett.*, **6**, 367–370.

Anon (1983). 'Use of fungicides and insecticides on cereals, 1983', ADAS, Ministry of Agriculture Fisheries and Food, Booklet 2257 (83).

Ashworth, L. J., Schroeder, H. W., and Langley, B. C. (1965). 'Aflatoxins: environmental factors governing occurrences in Spanish peanuts', *Science*, **148**, 1228–1229.

Batt, C. (1983). 'Relationship of asexual development to aflatoxin production in *Aspergillus parasiticus*', *J. Food Safety*, **5**, 31–40.

Bauer, J., Montgelas, A. V., and Gedek, B. (1983). 'Aflatoxin B, production in presence of preservatives and antimicrobial agents', *Proc. Int. Symp. Mycotox.* National Research Centre, Cairo, pp. 249–255.

Bennett, J. W., Dunn, J. J., and Goldsman, C. I. (1981). 'Influence of white light on production of aflatoxins and anthraquinones in *Aspergillus parasiticus*', *Appl. Environ. Microbiol.*, **41**, 488–491.

Bennett, J. W., Fernholz, F. A. and Lee, L. S. (1978). 'Effect of light on aflatoxins, anthraquinones, and sclerotia in *Aspergillus flavus* and *A. parasiticus*', *Mycologia*, **70**, 104–116.

Bennett, J. W., Horowitz, P. C., and Lee, L. S. (1979). 'Production of sclerotia by aflatoxigenic and non aflatoxigenic strains of *Aspergillus flavus* and *A. parasiticus*, *Mycologia*, **71**, 415–422.

Bennett, J. W., Silverstein, R. B., and Kruger, S. J. (1981). 'Isolation and characterization of two non aflatoxigenic classes of morphological variants of *Aspergillus parasiticus*', *J. Amer. Oil Chem. Soc.*, **58**, 952A–955A.

Boller, R. A., and Schroeder, H. W. (1973). 'Influence of *Aspergillus chevalieri* on production of aflatoxin in rice by *Aspergillus parasiticus*', *Phytopathology*, **63**, 1507–1510.

Boller, R. A., and Schroeder, H. W. (1974). 'Influence of *Aspergillus candidus* on production of aflatoxin in rice by *Aspergillus parasiticus*', *Phytopathology*, **64**, 121–123.

Booth, C. (1977). *Fusarium, Laboratory Guide*. Commonwealth Mycological Institute, Kew.

Borrow, E. G., Jefferys, E. G., Kessell, R. H. J., Lloyd, E. C., Lloyd, P. B., and Nixon, I. S. (1961). 'The metabolism of *Gibberella fujikuroi* in stirred culture', *Can. J. Microbiol.*, **7**, 227–276.

Engel, G., von Milczewski, K. E., Prokopiek, D., and Teuber, M. (1982). 'Strain-specific synthesis of mycophenolic acid by *Penicillium roqueforti* in blue-veined cheese', *Appl. Environ. Microbiol.*, **43**, 1034–1040.

Fanelli, C., Fabbri, A. A., Finotti, E., Bellincampi, D., Gualandi, G., and Passi, S. (1982). 'Induction of the biosynthesis of aflatoxins in *Aspergillus parasiticus* by cerulenin'. *Proc. V Int. IUPAC Symp. on Mycotoxins and Phycotoxins*, Vienna, pp. 166–169.

Gunasekaran, M. (1981). 'Optimum culture conditions for aflatoxin B_2 production by a human pathogenic strain of *Aspergillus flavus*', *Mycologia*, **73**, 697–704.

Hart, L. P., Braselton, W. E. Jr, and Stebbins, T. C. (1982). 'Production of zearalenone and deoxynivalenol in commercial sweet corn varieties', *Plant Dis.*, **66**, 1133–1135.

Hesseltine, C. W. (1983). 'Conditions leading to contamination by aflatoxins', *Proc. Int. Symp. Mycotox.* National Research Centre, Cairo, pp. 47–69.

Hesseltine, C. W., Rogers, R. F., and Shotwell, O. L. (1981). 'Aflatoxin and mould flora in North Carolina', *Mycologia*, **73**, 216–228.

Hesseltine, C. W., Shotwell, O. L., Smith, M., Ellis, J. J., Vandegraft, E., and Shannon, G. (1970). 'Production of various aflatoxins by strains of the *Aspergillus flavus* series', *Proc. First U.S. Jpn. Conf. Toxic. Microorg*, pp. 202–210.

Hill, R. A., Blankenship, P. D., Cole, R. J., and Sanders, T. H. (1983a). 'Effects of soil moisture and temperature on preharvest invasion of peanuts by the *Aspergillus flavus* group and subsequent aflatoxin development', *Appl. Environ. Microbiol.*, **45**, 628–633.

Hill, R. A., Wilson, D. M., Cole, R. J., Sanders, T. H., and Blankenship, P. D. (1983b). 'Ecology of the *Aspergillus flavus* group and aflatoxin formation in maize and groundnut', *Proc. Int. Myc. Symp.*, University of Sydney, 12–15 August 1983.

Ichinoe, M., and Kurata, H. (1983). 'Trichothecene-producing fungi', in *Trichothecenes, Chemical, Biological and Toxicological Aspects* (ed. Y. Ueno), pp. 73–82. Kodansha, Tokyo.

Jarvis, B. B., Midiwo, J. O., and Tuthill, D. (1981). 'Interaction between the antibiotic trichothecenes and the higher plant *Baccharis megapotamica*, *Science*, **214**, 460–462.

Jones, R. K., and Duncan, H. E. (1981). 'Effect of nitrogen fertilizer, planting date, and harvest date on aflatoxin production in corn inoculated with *Aspergillus flavus*', *Plant Dis.*, **65**, 741–744.

Jones, R. K., Duncan, H. E. and Hamilton, P. B. (1981). 'Planting date, harvest date and irrigation effects on infection and aflatoxin production by *Aspergillus flavus* in field corn', *Phytopathology*, **71**, 810–816.

Lillehoj, E. B., and Hesseltine, C. W. (1977). 'Aflatoxin control during growth and harvest of corn', in *Mycotoxins in Human and Animal Health* (eds J. V. Rodricks, C. W. Hesseltine and M. A. Mehlman), pp. 107–119. Pathotox, Illinois.

Mashaly, R. I., and El-Deeb, S. A. (1983). 'Effect of oxygen, illumination and sodium chloride on the aflatoxins produced by aspergillus', *Proc. Int. Symp. Mycotox.* National Research Centre, Cairo, pp. 477–483.

Mayne, R. Y., Bennett, J. W., and Tallant, J. (1971). 'Instability of an aflatoxin producing strain of *Aspergillus parasiticus*', *Mycologia*, **63**, 644–648.

Mixon, A. C. (1977). 'Influence of plant genetics on colonization by *Aspergillus flavus* and toxin production (peanuts)', in *Mycotoxins in Human and Animal Health* (eds J. V. Rodricks, C. W. Hesseltine and M. A. Mehlman), pp. 163–172. Pathotox, Illinois.

Moss, M. O., and Badii, F. (1982a). 'The influence of *Penicillium rubrum* on aflatoxin production by *Aspergillus parasiticus* on maize', *Proc. V Int. IUPAC Symp. on Mycotoxins and Phycotoxins*, Vienna, pp. 188–191.

Moss, M. O., and Badii, F. (1982b). 'Increased production of aflatoxins by *Aspergillus parasiticus* Speare in the presence of rubratoxin B', *Appl. Environ. Microbiol.*, **43**, 895–898.

Moss, M. O., and Hill, I. W. (1970). 'Strain variation in the production of rubratoxins by *Penicillium rubrum* Stoll', *Mycopath. Mycol. Appl.*, **40**, 81–88.

Penn, D. J., and Lynch, J. M. (1982). 'The effect of bacterial fermentation of couch grass rhizomes and *Fusarium culmorum* on the growth of barley seedlings', *Plant Path.*, **31**, 34–43.

Pettersson, H. B., Goransson, K. H., Tideman, K., and Johansson, T. (1978). 'Aflatoxin in a Swedish grain sample', *Nord. Vet. Med.*, **40**, 482–484.

Schroeder, H. W., and Boller, R. A. (1973). 'Aflatoxin production in species and strains of the *Aspergillus flavus* group isolated from field crops', *Appl. Microbiol.*, **25**, 885–887.

Shannon, G. M., Shotwell, O. L., Lyons, A. J., White, D. G., and Garcia Agurre, G (1980). 'Laboratory screening for zearalenone formation in corn hybrids and inbreds', *J. Assoc. Off. Anal. Chem.*, **63**, 1275–1277.

Shotwell, O. L., Goulden, M. L., Bothast, R. J., and Hesseltine, C. W. (1975). 'Mycotoxins in hot spots in grains. I: Aflatoxin and zearalenone occurrence in stored corn', *Cereal Chem.*, **52**, 687–697.

Stutz, H. K., and Krumperman, P. H. (1976). 'Effect of temperature cycling on the production of aflatoxin by *Aspergillus parasiticus*', *Appl. Environ. Microbiol.*, **32**, 327–332.

Sutton, J. C., Baliko, W., and Funnell, H. S. (1976). 'Evidence for translocation of zearalenone in corn plants colonized by *Fusarium graminearum*', *Can. J. Plant Sci.*, **56**, 7–12.

Torres, J., Guarro, J., Suarez, G., Sune, N., Calvo, M. A., and Ramirez, C. (1980). 'Morphological changes in strains of *Aspergillus flavus* (Link) and *Aspergillus parasiticus* (Speare) related with aflatoxin production', *Mycopathology*, **72**, 171–174.

Weckbach, L. S., and Marth, E. M. (1977). 'Aflatoxin production by *Aspergillus parasiticus* in a competitive environment', *Mycopathology*, **62**, 39–45.

White, J. P., and Johnson, G. T. (1982). 'Aflatoxin production correlated with nitrification in *Aspergillus flavus* group species', *Mycologia*, **74**, 718–723.

Wicklow, D. T. and Cole, R. J. (1982). 'Tremorgenic indole metabolites and aflatoxins in sclerotia of *Aspergillus flavus*: an evolutionary perspective', *Can. J. Bot.*, **60**, 525–528.

Wicklow, D. T., Hesseltine, C. W., Shotwell, O. L., and Adams, G. L. (1980). 'Interference competition and aflatoxin levels in corn', *Phytopathology*, **70**, 761–764.

Zuber, M. S. (1977). 'Influence of plant genetics on toxin production in corn', in *Mycotoxins in human and animal health* (eds J. V. Rodricks, C. W. Hesseltine and M. A. Mehlman), pp. 173–179. Pathotox, Illinois.

SYNTHESIS OF MYCOTOXINS

Trichothecenes and Other Mycotoxins
Edited by J. Lacey

CHAPTER 25

Mutants and mycotoxins: aflatoxins as a model system

J. W. Bennett

INTRODUCTION

Aflatoxins are the best characterized mycotoxins. Their production is limited to one genus of imperfect fungi *(Aspergillus)* and to only a few species within that genus. Depending on the taxonomist, aflatoxigenic aspergilli are classified into two species *(A. flavus* and *A. parasiticus*) or three species (*A. flavus*, *A. parasiticus* and *A. toxicarius*) (Murakami, 1971; Christensen, 1981). Acceptance of *A. toxicarius* as a separate species has been slow, and many contemporary workers view it as a variety of *A. parasiticus*. Even within these species, aflatoxin production varies between isolates. In a summary of thirteen published surveys of aflatoxin production by *A. flavus/parasiticus*, 1857 of 3343 isolates (56%) were aflatoxigenic (Bennett, 1982). In an earlier survey, Diener and Davis (1969) calculated that 60% of strains from agricultural commodities were aflatoxigenic. Also, loss or attenuation of aflatoxigenicity occurs among strains maintained in the laboratory (Kulik and Holaday, 1966; Mayne *et al.*, 1971). The variation in aflatoxin production by different strains is presumably genetic, but little is known about the hereditary control mechanism.

In contrast, the literature on other aspects of the aflatoxin problem is voluminous. Chemists have elucidated a family of aflatoxin compounds which share the common structural features of a bisfuran ring system fused to a substituted coumarin nucleus. The major aflatoxins are called B_1, B_2, G_1 and G_2 because of their respective blue and green fluorescence under long-wave ultra-violet light, and their relative chromatographic mobilities. Over a dozen related compounds, largely the products of biotransformations and chemical degradation, have also been characterized. These are extensively reviewed by Heathcote and Hibbert (1978). The structures of aflatoxins B_1, B_2, B_{2a}, G_1, G_2 and G_{2a} are presented in Figure 1.

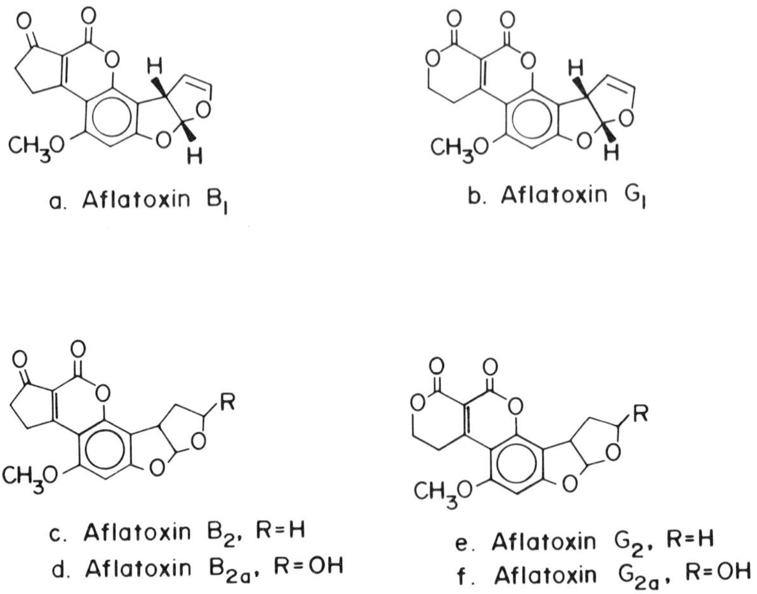

Figure 1 Structures of aflatoxins B_1, B_2, B_{2a}, G_1, G_2 and G_{2a}.

The aflatoxin family displays biological activities which, in general, are harmful for vertebrates. Most of the studies linking aflatoxin with carcinogenicity, mutagenicity, teratogenicity and toxicity utilize preparations that are largely or wholly aflatoxin B_1, usually the most abundant aflatoxin. Natural and experimentally induced toxicoses are documented for many animal species. Both acute and chronic effects vary with species, sex, nutritional status, age and mode of delivery; within a given species the young tend to be more susceptible than adults and males more susceptible than females (Newberne, 1973; Wyllie and Morehouse, 1977).

Aflatoxin B_1 is highly active as a mutagen and hepatocarcinogen (McCann *et al.*, 1975; Garner, 1980). The epidemiological data linking aflatoxin ingestion and primary liver cancer in certain human populations are compelling (Shank, 1977; Carlborg, 1979; Linsell, 1979). Metabolic activation is a prerequisite for carcinogenicity and an epoxide derivative is the hypothesized active intermediate (Kuczuk *et al.*, 1978). In general, the most pharmacologically active form of aflatoxin is aflatoxin B_1. Aflatoxin G_1 is somewhat less toxic, mutagenic and carcinogenic than aflatoxin B_1, while aflatoxins B_2 and G_2 are considerably less active even than aflatoxin G_1. The relationship between structure and biological activity has been reviewed by Chu (1977) and Stark (1980).

Aflatoxin B_1 has been used as a model compound for studying the chemical basis of carcinogenicity. The covalent bonding to DNA is believed to have a role in its carcinogenic effect (Croy *et al.*, 1978) and an aflatoxin–guanine adduct has been characterized (Essigman *et al.*, 1977). More recently, aflatoxin has been tested with a synthetic DNA polymer of alternating deoxyguanine and deoxycytosine residues. Normally the polymer can be converted from the right-handed B-DNA form to the left-handed Z-DNA form at high salt concentrations. Reactions with aflatoxins and subsequent adduct formation strongly inhibit the salt-induced conversion of this polymer from B-DNA to Z-DNA, even at relatively low binding levels (Nordheim *et al.*, 1983). Aflatoxin and certain other carcinogens may act at a regulatory region whereby the changes in the equilibrium between the B and Z forms of DNA interferes with normal levels of transcription (Nordheim *et al.*, 1983).

Further aspects of the mycological, agricultural, clinical and economic aspects of the aflatoxin problem are available in reviews by Goldblatt (1969), Detroy *et al.* (1971), Roberts (1974) and Heathcote and Hibbert (1978).

AFLATOXINS AND SECONDARY METABOLISM

'Natural product' is a general name for an organic substance produced by living organisms (Thomas, 1979). Natural products may be divided, somewhat arbitrarily, into two major groups: primary metabolites and secondary metabolites. Primary metabolites are widely distributed in nature, are present during most of the life cycle of the organism, have well-defined physiological roles, and are essential for growth. Examples of primary metabolites are amino acids, lipids, nucleic acids, intermediates of the citric acid cycle, and other such 'intermediary' metabolites familiar to students of biochemistry.

By contrast, secondary metabolites are of restricted taxonomic distribution, usually are produced at only one stage of the life cycle, have obscure physiological function, and are not essential to the growth of the producing organism. Secondary metabolites may have complex, even bizarre, chemical structures, but they are all synthesized from a few simple precursors such as acetate or isoprene. The best-known secondary metabolites are toxins and antibiotics; however, many compounds without detectable pharmacological activity are also known. The economic importance of secondary metabolites, their elusive function in nature, and the beguiling complexity of their chemical architecture has provoked many reviews and much speculation about their evolution and function (Bu'Lock, 1961, 1965, 1967; Bentley and Campbell, 1968; Weinberg, 1971; Demain, 1974; Bu'Lock *et al.*, 1982; Bennett and Ciegler, 1983).

All mycotoxins are secondary metabolites. By definition, mycotoxins display potent biological activity against vertebrates. But what does mycotoxin production do for the fungus? Why are some fungal strains highly toxigenic, while other strains in the same species produce little or no toxin? We do not know the answer to these questions.

Mycotoxins arise via a variety of different secondary metabolic pathways; aflatoxins are polyketides (Steyn, 1980). The polyketide pathway is the most common secondary metabolic route among fungi and most known polyketides are of fungal origin (Turner, 1971; Turner and Aldridge, 1983). Polyketide biosynthesis is analogous to fatty acid biosynthesis, but lacks the systematic dehydration and reduction steps. An enzyme-bound acetyl-CoA is condensed with a variable number of malonyl-CoA units, with concomitant decarboxylation. Detailed reviews of polyketide biosynthesis are available (Richards and Hendrickson, 1964; Turner, 1971; Packter, 1973; Mann, 1978; Turner and Aldridge, 1983).

Polyketides may be classified according to the number of component two-carbon units: C_6 compounds are triketides, C_8 compounds are tetraketides, etc. Aflatoxins B_1, B_2, G_1 and G_2 are all C_{17} compounds originally classified as nonaketides. Recent evidence indicates that they arise biosynthetically via C_{20} anthraquinone precursors and they have been re-classified as decaketides (Bennett and Christensen, 1983; Turner and Aldridge, 1983).

GENETICS OF AFLATOXIGENIC MOULDS

Aspergillus flavus and *A. parasiticus* are both anamorphic (imperfect) species for which no telemorph (perfect form) has ever been isolated. However, somatic recombination via the parasexual cycle has been demonstrated in both species and induced mutations affecting spore colour, auxotrophy, and mycelial pigmentation are isolated with ease (Papa, 1973, 1978; Gussack *et al.*, 1977; Bennett, 1979).

Both ultraviolet light and nitrosoguanidine treatment have been used to isolate mutants blocked in the aflatoxin pathway (Bennett and Goldblatt, 1973; Donkersloot *et al.*, 1972), some of which accumulate brightly coloured anthaquinone pigments that have been characterized as norsolorinic acid (Lee *et al.*, 1971), averufin (Donkersloot *et al.*, 1972); versicolorin A (Lee *et al.*, 1976), and averantin (Bennett *et al.*, 1980a). Because of their coloration, these mutants are easily screened in progeny from parasexual crosses (Bennett, 1979, 1982).

Another interesting class of mutants have been independently isolated in four laboratories. In these mutants aflatoxin B_1 production is low or absent, while high levels of aflatoxin B_2 are formed (van Walbeeck *et al.*, 1968; Schroeder and Carlton, 1973; Papa, 1977a; Gunasekaran, 1981).

Two additional classes of aflatoxinless mutants have been isolated after serial transfer of aflatoxigenic strains in defined media. Loss of aflatoxigenicity is associated with conditions in which the inocula for serial transfers contain hyphae rather than conidiospores (Bennett, 1981). These mutant classes have been named *fan* and *fluff* because the mutational lesion is pleiotropic, affecting colony morphology and sporulation as well as polyketide production (Bennett *et al.*, 1982; Bennett, 1982). Neither *fan* nor *fluff* segregates in a heterokaryon test. When an auxotrophic form of a *fan* or *fluff* mutant is crossed with an auxotroph not bearing the *fan* or *fluff* morphology, both anthraquinone and aflatoxin production may be observed in some of the progeny from the heterozygous diploid (Bennett *et al.*, 1982).

Formal genetic analysis is most advanced in *A. flavus* where eight linkage groups have been established (Papa, 1976, 1977b). Two independently isolated aflatoxinless mutants have been mapped to linkage groups II and VII (Papa, 1979) and one of the mutants that accumulates more aflatoxin B_2 than B_1 was mapped to linkage group VIII (Papa, 1977c). Unfortunately, these parasexual genetic analyses are laborious and many of the mutant markers are not recovered with expected random frequencies. Compounding the difficulty of genetic analysis is the inability to score haploid and diploid segregants by conidial diameters. In *Aspergillus* species with uninucleate conidia, such as *A. nidulans*, diploid spores have a larger diameter than haploid spores due to the doubling of DNA content (Roper, 1966). Wild-type haploid *A. flavus* and *A. parasiticus* have multinucleate conidia (Yuill, 1950). In diploids of these species, fewer nuclei are found than in haploid spores, thereby equalizing DNA content. Hence, in *A. flavus* and *A. parasiticus*, haploid, diploid and heterokaryotic spores have similar nuclear diameters (Leaich and Papa, 1975; Bennett *et al.*, 1980b).

In summary, neither *A. flavus* nor *A. parasiticus* is a particularly good species for conducting traditional genetic analyses. Nevertheless, several of the mutants isolated during elucidation of the parasexual cycle have been invaluable tools in dissecting the pathway of aflatoxin biosynthesis. This work is briefly described below, emphasizing the application of mutants in these studies.

AFLATOXIN BIOSYNTHESIS

Most of the biosynthetic studies on secondary metabolic pathways utilize isotopically labelled intermediates; other methods such as cell-free enzyme preparations, specific metabolic inhibitors, and blocked mutants are less commonly employed. The aflatoxin pathway is unusual in that all these approaches are available. As a result, the pathway of aflatoxin biosynthesis is among the best understood of mycotoxins. Only patulin and ergot biogenesis are known with a similar level of sophistication.

The availability of the blocked anthraquinone-accumulating mutants isolated during genetic studies was the key to discovering the unexpected decaketide origin of the aflatoxin family. See Maggon *et al.* (1977), Heathcote and Hibbert (1978), Steyn (1980), Bennett and Christensen (1983) and

Turner and Aldridge (1983) for comprehensive reviews of aflatoxin biosynthesis.

When ^{14}C-labelled anthraquinones are presented to wild-type *A. parasiticus* cultures in a replacement medium lacking nitrogen, label is recovered in aflatoxin B_1 for ^{14}C-labelled norsolorinic acid (Hsieh *et al.*, 1976), ^{14}C-labelled averufin (Lin and Hsieh, 1973), and ^{14}C-labelled versicolorin A (Lee *et al.*, 1976). When cultures of blocked mutants are presented with different anthraquinone intermediates the subsequent production or non-production of aflatoxin indicates whether the mutant block is before or after that particular compound. Appropriately designed 'feeding' studies of this sort can be used to establish a tentative sequence for the intermediates in the aflatoxin pathway. Currently, the generally accepted order is thought to be norsolorinic acid → averanthin → averufin → versicolorin A (Singh and Hsieh, 1977; Bennett *et al.*, 1980a).

In addition to these four anthraquinones identified from mutants, two other compounds are known intermediates in aflatoxin biosynthesis. Versicolorin hemiacetal acetate is accumulated when the wild-type or the versicolorin A-accumulating mutant are treated with the insecticide dichlorovos (Yao and Hsieh, 1974; Bennett *et al.*, 1976). Label from the xanthone sterigmatocystin, a natural metabolite of *A. versicolor*, is also incorporated into aflatoxin B_1 (Hiseh *et al.*, 1973). In the current view of the aflatoxin pathway, versicolorin hemiacetal acetate appears before and sterigmatocystin immediately after versicolorin A in the metabolic sequence.

The use of ^{13}C NMR (nuclear magnetic resonance) has revolutionized biosynthesis research. Once natural abundance spectra are available for a given metabolite, comparison with enriched, ^{13}C-labelled metabolites under controlled conditions allows direct determination of all the carbon atoms in that metabolite. ^{13}C NMR studies have been applied extensively to the aflatoxin pathway. The role of averufin as an intermediate has received particular attention in an attempt to resolve several competing hypotheses as to the rearrangement of the averufin side-chain in the construction of the aflatoxin in B bisfuran moiety (Townsend *et al.*, 1982).

The ^{13}C NMR literature has been reviewed by Steyn (1980) and Bennett and Christensen (1983). In general, it corroborates the earlier ^{14}C data and unambiguously establishes the biogenesis of aflatoxin B_1 via a series of the decaketide anthraquinone intermediates.

Biochemists' work with cell-free enzyme preparations has not been nearly as extensive as organic chemists' work with ^{13}C NMR. Wan and Hsieh (1980) have demonstrated the cell-free conversion of versiconal hemiacetal acetate, and Singh and Hsieh (1976) have reported a similar NADPH-dependent enzyme system that converts sterigmatocystin to aflatoxin B. Surprisingly, a cell-free extract from protoplasts of *A. flavus* did not convert versicolorin A to aflatoxin, although sterigmatocystin was accepted as a substrate in this system (Anderson and Dutton, 1980).

PERSPECTIVE AND PREDICTIONS

Auxotrophic mutants have always been one of the cornerstones of biochemistry in unravelling primary metabolic pathways. In the study of secondary metabolic pathways the emphasis has been with isotopic labels. The relative shortage of mutational probes is due to the different nature of the metabolic pathways. By definition, secondary metabolites are not required for growth in the producing organism. This means that in looking for mutations affecting secondary metabolic pathways, the entire population of mutagenized survivors in a mutant hunt must be screened. The conditional lethal protocols which are so readily available to the study of primary metabolism are adapted only with great difficulty to secondary metabolism.

Nevertheless, the aflatoxin model can inspire research for other mycotoxins. The fortuitous association of mycelial pigmentation and blocks in biosynthesis will not be found in most pathways, but do demonstrate the importance of having an easily screened phenotype. Although the aflatoxins remain the best-studied mycotoxins, the trichothecenes are gaining in interest and importance. The innate instability of many trichothecene-producing species, especially in the genus *Fusarium*, provides both an impediment and a challenge. Modern technologies allow by-passing of many traditional genetic approaches. 'Strain degeneration', so commonly encountered in *Fusarium* and among many mycotoxin-producing species, has many tantalizing similarities to the insertional inactivation associated with certain transposable elements studied in *Drosophila*, yeast, and maize (Shapiro, 1983). Increased collaboration between geneticists, biochemists, and molecular biologists will enrich mycotoxicology and provide new insights into secondary metabolism.

REFERENCES

Anderson, M. S., and Dutton, M. F. 'The use of cell free extracts from fungal protoplasts in the study of aflatoxin biosynthesis', *Appl. Environ. Microbiol.*, **40**, 706–709 (1980).

Bennett, J. W. (1979) 'Aflatoxins and anthraquinones from diploids of *Aspergillus parasiticus*', *J. Gen. Microbiol.*, **113**, 127–136.

Bennett, J. W. (1982). Genetics of mycotoxin production with emphasis on aflatoxins, in *Overproduction of Microbial Products* (eds V. Krumphanzl, B. Sikyta, and A. Vanek, pp. 549–561. Academic Press, London.

Bennett, J. W., and Christensen, S. B. (1983). 'New perspectives on aflatoxin biosynthesis', *Adv. Appl. Microbiol.*, **29**, 53–92.

Bennett, J. W., and Ciegler, A. (eds) (1983). *Secondary Metabolism and Differentiation in Fungi*. Marcel Dekker, New York.

Bennett, J. W., and Goldblatt, L. S. (1973). 'The isolation of mutants of *Aspergillus parasiticus* and *A. flavus* with altered aflatoxin producing ability', *Sabouraudia*, **11**, 235–241.

Bennett, J. W., Lee, L. S., and Cucullu, A. F. (1976). 'Effect of dichlorous on versicolorin A production in *Aspergillus parasiticus*', *Bot. Gaz.*, **137**, 318–324.

Bennett, J. W., Lee, L. S., Shoss, S. M., and Boudreaux, G. H. (1980a). 'Identification of averantin as an aflatoxin B_1 precursor: placement in the biosynthetic pathway', *Appl. Environ. Microbiol.*, **39**, 835–839.

Bennett, J. W., Vinnett, C. H., and Goynes, Jr., W. R. (1980b). 'Aspects of parasexual analysis in *Aspergillus parasiticus*', *Can. J. Microbiol.*, **26**, 706–713.

Bennett, J. W., Silverstein, R. B., and Kruger, S. J. (1981). 'Isolation and characterization of two nonaflatoxigenic classes of morphological variants of *Aspergillus parasiticus*', *Am. Oil Chem. Soc.*, **58**, 952A–955A.

Bennett, J. W., Wheeler, D. G., and Dunn, J. J. (1982). 'Genetic analysis of aflatoxin production in *Aspergillus parasiticus*', in *Advances in Biotechnology Fermentation Products*, pp. 409–415 (eds M. Moo Young, C. Vezina and K. Singh). Pergamon, Oxford.

Bentley, R., and Campbell, I. M. (1968). 'Secondary metabolism in fungi', in *Comprehensive Biochemistry*', Vol. 20, pp. 415–487. (eds M. Florkin and E. H. Stolz). Elsevier, Amsterdam.

Bu'Lock, J. D. (1961). 'Intermediary metabolism and antibiotic synthesis', *Ann. Rev. Appl. Microbiol.*, **3**, 293–342.

Bu'Lock, J.D. (1965). *The Biosynthesis of Natural Products*. McGraw-Hill, New York.

Bu'Lock, J. D. (1967). *Essays in Biosynthesis and Microbial Development*. Wiley, New York.

Bu'Lock, J. D., Nesbit, L. J., and Winstanley, D. J. (eds). (1982). *Bioactive Microbial Products*. Academic Press, New York.

Carlborg, F. W.(1979). 'Cancer, mathematical models and aflatoxin', *Food Cosmet. Toxicol.*, **17**, 159–166.

Christensen, M. (1981). 'A synoptic key and evaluation of species in the *Aspergillus flavus* group', *Mycologia*, **73**, 1056–1084.

Chu, F. S. (1977). 'Mode of action of mycotoxins and related compounds', *Adv. Appl. Microbiol.*, **22**, 83–143.

Croy, R. G., Essigman, J. M., Reinhold, V. N., and Wogan, G. N. (1978). 'Identification of the principal aflatoxin B_1-DNA adduct formed *in vivo* in rat liver', *Proc. Natl. Acad. Sci. (U.S.A.)*, **75**, 1745.

Demain, A. L. (1974). 'How do antibiotic-producing micro-organisms avoid suicide?', *Ann. N.Y. Acad. Sci.*, **235**, 601–612.

Detroy, R. W., Lillehoj, E. B., and Ciegler, A. (1971). 'Aflatoxin and related compounds', in *Microbial Toxins. Fungal Toxins*, pp. 3–178 (eds A. Ciegler, S. Kadis, and J. Ajl). Academic Press, New York.

Diener, J. L., and Davis, N. D. (1969). 'Aflatoxin formation by *Aspergillus flavus*', in *Aflatoxin. Scientific Background, Control and Implications*, pp., 13–54. (ed. L. A. Goldblatt). Academic Press, New York.

Donkersloot, J. A., Matales, R. I., and Yang, S. S. (1972). 'Isolation of averufin from a mutant of *Aspergillus parasiticus* impaired in aflatoxin production', *Biochem. Biophy. Res. Comm.*, **47**, 1051–1055.

Essigman, J. M., Croy, R. G., Nadzan, A. M., Busby, W. F., Reingold, V. N., Buchi, G., and Wogan, G. N. (1977). 'Structural identification of the major DNA adduct formed by aflatoxin B_1 in vitro', *Proc. Natl. Acad. Sci. U.S.A.*, **74**, 1870–1874.

Garner, R. C. (1980). 'Carcinogenesis by fungal products', *Br. Med. Bull.*, **36**, 47–52.

Goldblatt, L. A. (ed.) (1969). *Aflatoxin. Scientific Background, Control and Implications*. Academic Press, New York.

Gunasekaran, M. (1981). 'Optimum culture conditions for aflatoxin B_2 production by a human pathogenic strain of *Aspergillus flavus*', *Mycologia*, **73**, 697–704.

Gussack, G., Bennett, J. W., Cavalier, S., and Yatsu, L. (1977). 'Evidence for the parasexual cycle in a strain of *Aspergillus flavus* containing virus-like particles', *Mycopathologia*, **61**, 159–165.

Heathcote, J. G., and Hibbert, J. R. (1978). *Aflatoxins: Chemical and Biological Aspects*. Elsevier, Amsterdam.

Hsieh, D. P. H., Lin, M. T., and Yao, R. C. (1973). 'Conversion of sterigmatocystin to aflatoxin B_1 by *Aspergillus parasiticus Biochem*', *Biophys. Res. Comm.*, **52**, 992–997.

Hsieh, D. P. H., Lin, M. T., Yao, R. C., and Singh, R. (1976). 'Biosynthesis of aflatoxin. Conversion of norsolorinic acid and other hypothetical intermediates into aflatoxin B_1', *J. Agr. Food Chem.*, **24**, 1171–1174.

Kuczuk, M. H., Benson, P. M., Heath, H., and Hayes, A. W. (1978). 'Evaluation of mutagenic potential of mycotoxins using *Salmonella typhimurium* and *Saccharomyces cervisae*', *Mutation Res.*, **53**, 11–20.

Kulik, M. M., and Holaday, C. E. (1966). 'Aflatoxin: a metabolic product of several fungi', *Mycopath. Mycol. Appl.*, **30**, 137–140.

Leaich, L. L., and Papa, K. E. (19175). 'Identification of diploids of *Aspergillus flavus* by the nuclear condition of conidia', *Mycologia*, **67**, 674–678.

Lee, L. S., Bennett, J. W., Goldblatt, L. A., and Lundin, R. E. (1971). 'Norsolorinic acid from a mutant strain of *Aspergillus parasiticus*', *J. Am. Oil Chem. Soc.*, **48**, 93–94.

Lee, L. S., Bennett, J. W., Cucullu, A. F., and Ory, R. L. (1976). 'Biosynthesis of aflatoxin B_1: conversion of versicolorin A to aflatoxin B_1 by *Aspergillus parasiticus*', *J. Agr. Food Chem.*, **24**, 1167–1171.

Lin, M. T., and Hsieh, D. P. H. (1973). 'Averufin in the biosynthesis of aflatoxin B_1', *J. Am. Chem. Soc.*, **95**, 1668–1669.

Linsell, C. A. (1979). 'Decision on the control of a dietary carcinogen-aflatoxin', in *Carcinogenic Risks: Strategies for Intervention* (eds W. David, and A. C. Rosenfeld). WHO.

Maggon, K. K., Gupta, S. K., and Venkitasubramaniam, T. A. (1977). 'Biosynthesis of aflatoxins', *Bacterial Rev.*, **41**, 822–855.

Mann, J. (1978). *Secondary Metabolism*. Oxford University Press, London.

Mayne, R. Y., Bennett, J. W., and Tallant, J. (1971). 'Instability of an aflatoxin producing strain of *Aspergillus parasiticus*', *Mycologia*, **63**, 644–648.

McCann, J., Spingarn, N. E., Kobori, J., and Ames, B. N. (1975). 'Detection of carcinogens as mutagens: bacterial tester strains with R factor plasmids', *Proc. Natl. Acad. Sci. U.S.A.*, **72**, 979–983.

Murakami, H. (1971). 'Classification of the koji mold', *J. Gen. Appl. Microbiol.*, **17**, 281–309.

Newberne, P. M. (1973). Chronic aflatoxicosis. *J. Am. Vet. Med. Assoc.*, **163**, 1259–1302.

Nordheim, A., Hao, W. M., Wogan, G. N., and Rich, A. (1983). 'Salt-induced conversion of B-DNA to Z-DNA inhibited by aflatoxin B_1', *Science*, **219**, 1434–1436.

Packter, N. M. (1973). *Biosynthesis of Acetate-Derived Compounds*. Wiley, New York.

Papa, K. E. (1973). 'The parasexual cycle in *Aspergillus flavus*', *Mycologia*, **75**, 1201–1205.

Papa, K. E. (1976). 'Linkage groups in *Aspergillus flavus*', *Mycologia*, **68**, 159–165.

Papa, K. E. (1977a). 'Mutant of *Aspergillus flavus* producing more aflatoxin B_2 than B_1', *Appl. Environ. Microbiol.*, **33**, 206.

Papa, K. E. (1977b). 'Genetic analysis of a mutant of *Aspergillus flavus* producing more aflatoxin B_2 than B_1', *Mycologia*, **69**, 556–562.

Papa, K. E. (1977c). 'Genetics of aflatoxin production in *Aspergillus flavus*: linkage between a gene for a high B_2 : B_1, ratio and the histidine locus on linkage group

VIII', *Mycologia*, **69**, 1185–1190.

Papa, K. E. (1978). 'The parasexual cycle in *Aspergillus parasiticus*', *Mycologia*, **70**, 766–773.

Papa, K. E. (1979). 'Genetics of *Aspergillus flavus*: complementation and mapping of aflatoxin mutants', *Genet. Res.*, **34**, 1–9.

Papa, K. E. (1980). 'Dominant aflatoxin mutant of *Aspergillus flavus*', *J. Gen. Microbiol.*, **118**, 279–282.

Richards, J. H., and Hendrickson, J. B. (1964). *The Biosynthesis of Steroids, Terpenes, and Acetogenins*. Benjamin, New York.

Roberts, J. C. (1974). 'Aflatoxins and sterigmatocystins', *Fortsch. Chem. Org. Naturst.*, **31**, 119–151.

Roper, J. A. (1966). 'The parasexual cycle', in *The Fungi*, Vol. II: *The Fungal Organism*, pp. 586–617 (eds G. C. Ainsworth and A. S. Sussman), Academic Press, New York.

Schroeder, H. W., and Carlton, W. W. (1973). 'Accumulation of only aflatoxin B_2 by a strain of *Aspergillus flavus*', *Appl. Microbiol.*, **25**, 146–148.

Shank, R. C. (1977). 'Epidemiology of aflatoxin carcinogenesis', *Adv. Mod. Toxicol.*, **3**, 291–318.

Shapiro, J. (ed.). (1983). *Mobile Genetic Elements*. Academic Press, New York.

Singh, R., and Hsieh, D. P. H. (1976). 'Enzymatic conversion of sterigmatocystin into aflatoxin B_1 by cell free extracts of *Aspergillus parasiticus*', *Appl. Environ. Microbiol.*, **31**, 743–745.

Singh, R., and Hsieh, D. P. H. (1977). 'Aflatoxin biosynthetic pathway: elucidation by using blocked mutants of *Aspergillus parasiticus*', *Arch. Biochem. Biophys.*, **178**, 285–292.

Stark, A. A. (1980). 'Mutagenicity and carcinogenicity of mycotoxins: DNA binding as a possible mode of action', *Annu. Rev. Microbiol.*, **34**, 235–262.

Steyn, P. S. (ed.) (1980). *The Biosynthesis of Mycotoxins. A Study in Secondary Metabolism*. Academic Press, New York.

Thomas, R., (1979). 'Biosynthesis', in *Comprehensive Organic Chemistry. The Synthetic Reactions of Organic Compounds*, Vol. 5, *Biological Compounds*, (eds. D. Barton, W. E. Olls and E. Haslam). Pergamon Press, Oxford.

Townsend, C. A., Christensen, S. B., and Davis, S. G. (1982). 'Bisfuran formation in aflatoxin biosynthesis: the fate of the averufin side chain', *J. Am. Chem. Soc.*, **104**, 6152–6153.

Turner, W. B. (1971). *Fungal Metabolites*. Academic Press, New York.

Turner, W. B. and Aldridge, D. C. (1983). *Fungal Metabolites. II*, Academic Press, New York.

Van Walbeeck, W., Scott, P. M., and Thatcher, F. S. (1968). 'Mycotoxins from food-borne fungi', *Can. J. Microbiol.*, **14**, 131–137.

Wan, N. C., and Hsieh, D. P. H. (1980). 'Enzymatic formation of the bisfuran structure in aflatoxin biosynthesis', *Appl. Environ. Microbiol.*, **39**, 109–112.

Weinberg, E. D. (1971). 'Secondary metabolism: raison d'être, *Perspect. Biol. Med.*, **14**, 565–577.

Wyllie, T. D., and Morehouse, L. D. (eds) (1977). *Mycotoxic Fungi, Mycotoxins, Mycotoxicoses*. Marcel Dekker, New York.

Yao, R. C. and Hsieh, D. P. H. (1974). 'Steps of dichlorovos inhibition in the pathways of aflatoxin biosynthesis', *Appl. Microbiol.*, **28**, 52–57.

Yuill, E. (1950). 'The numbers of nucleii in conidia of *Aspergillus*', *Trans. Brit. Mycol. Soc.*, **33**, 324–333.

Trichothecenes and Other Mycotoxins
Edited by J. Lacey

CHAPTER 26

Biosynthetic studies on phomopsin A

Alan L. Payne, John A. Edgar, John L. Frahn and Claude C. J. Culvenor

INTRODUCTION

Phomopsin A, a chlorine-containing hepatotoxic metabolite of *Phomopsis leptostromiformis* (Kühn) Bubák ex Lind, is the cause of lupinosis, an important, often fatal, hepatic disease of sheep in Australia, South Africa and parts of Europe (Culvenor *et al.*, 1977). It occurs when animals ingest lupin (*Lupinus luteus, L. angustifolius* or *L. albus*) hay or stubbles on which *P. leptostromiformis* has grown as a saprophyte (van Warmelo *et al.*, 1970; van Warmelo and Marasas, 1972).

Phomopsin A was originally isolated from cultures of *P. leptostromiformis* grown on moist, autoclaved lupin seeds (Culvenor *et al.*, 1977), but later a liquid culture system for phomopsin A production was developed (Lanigan *et al.*, 1979) which greatly facilitated isolation of the compound and has enabled its biosynthesis to be studied.

Culvenor *et al.* (1983) have recently shown phomopsin A to be a cyclic hexapeptide in which all of the amino acids have been modified either by dehydrogenation or by various substitutions. Thus, the amino acids found in phomopsin A are 2,3-didehydroaspartic acid, 3-hydroxisoleucine, 2,3-didehydroisoleucine, 3,4-didehydroproline, 3,4-didehydrovaline and *N*-methyl [3-chloro-4,5-dihydroxy] phenylserine.

The incorporation of the ^{14}C labelled amino acids isoleucine, phenylalanine and proline and ^{36}Cl HCl into phomopsin A, and lack of incorporation of several other compounds, was reported prior to elucidation of the structure of phomopsin A (Payne, 1983). Due to incomplete knowledge of structure, no attempt was made to relate these incorporations to phomopsin A structure in detail. This chapter describes some more recent biosynthetic experiments of a similar type, and determination of the site of the radiolabel incorporations.

METHODS

The methods used for growth of the organism, incorporation of radiolabelled precursors, assay and isolation of phomopsin A and measurement of label incorporation are as described previously (Payne, 1983). To locate the site of label incorporation, phomopsin A was hydrolysed, with or without prior reduction of the dehydroamino acid units (Culvenor *et al.*, 1983), and the resultant fragments separated by thin-layer chromatography (TLC) on silica gel (Kieselgel 60) plates using propan-1-ol: water (70 : 30, v/v) as developing solvent, or by paper electrophoresis (PE) (Lanigan *et al.*, 1979).

RESULTS AND DISCUSSION

Precursors previously found to be readily incorporated were [^{36}Cl] HCl, L-[U-^{14}C] isoleucine, L-[U-^{14}C] phenylalanine and L-[U-^{14}C] proline. By use of similar methods, incorporation of L-[U-^{14}C] aspartic acid, L-[U-^{14}C] glutamic acid, [U-^{14}C] glycine, [2-^{14}C] malonic acid, and L-[U-^{14}] valine have been measured. All cultures produced 75–100 mg/l phomopsin A after addition of label, and efficiency and specificity of label incorporation were determined using both paper electrophoresis (PE) and thin-layer chromatography (TLC) for separation of phomopsin A. Only the L-[U-^{14}C] valine was found to be incorporated to any significant extent (1.6% of added label). The distributions of radiolabel after PE and TLC separations of culture fluid extracts are shown in Figure 1. The recovered label in this extract was virtually confined to the phomopsin A spot. In extracts of cultures containing labelled glutamic acid, little activity was detected on pherograms other than that associated with phomopsin A (Figure 2). The level of incorporation in this case, however, was comparatively small, and has precluded investigation of the site of label incorporation. In all other cases incorporation was low (0.05–0.25% of added label) and a number of labelled components were found in the extracts. In each case a peak of radioactivity was associated with the phomopsin A isolated, indicating a small amount of incorporation of the labelled compound. The results of all the incorporation experiments as assessed by both PE and TLC separations are given in Table 1.

For each of the four most significant incorporations (isoleucine, phenylalanine, proline and valine), the whole of the phomopsin A produced (approx. 2 mg) was separated by reverse-phase high-performance liquid chromatography (RPLC). The activity of the separated phomopsin A was measured by liquid scintillation counting of a portion of the total sample (results not shown). The results provided further confirmation of the incorporation of label into phomopsin A but were not quantitative due to conservative cutting of the eluted phomopsin A fraction to yield the purest product possible.

In the light of the recent structural information on phomopsin A the incorporation of [^{36}Cl] HCl, [^{14}C] isoleucine, [^{14}C] proline and [^{14}C] valine

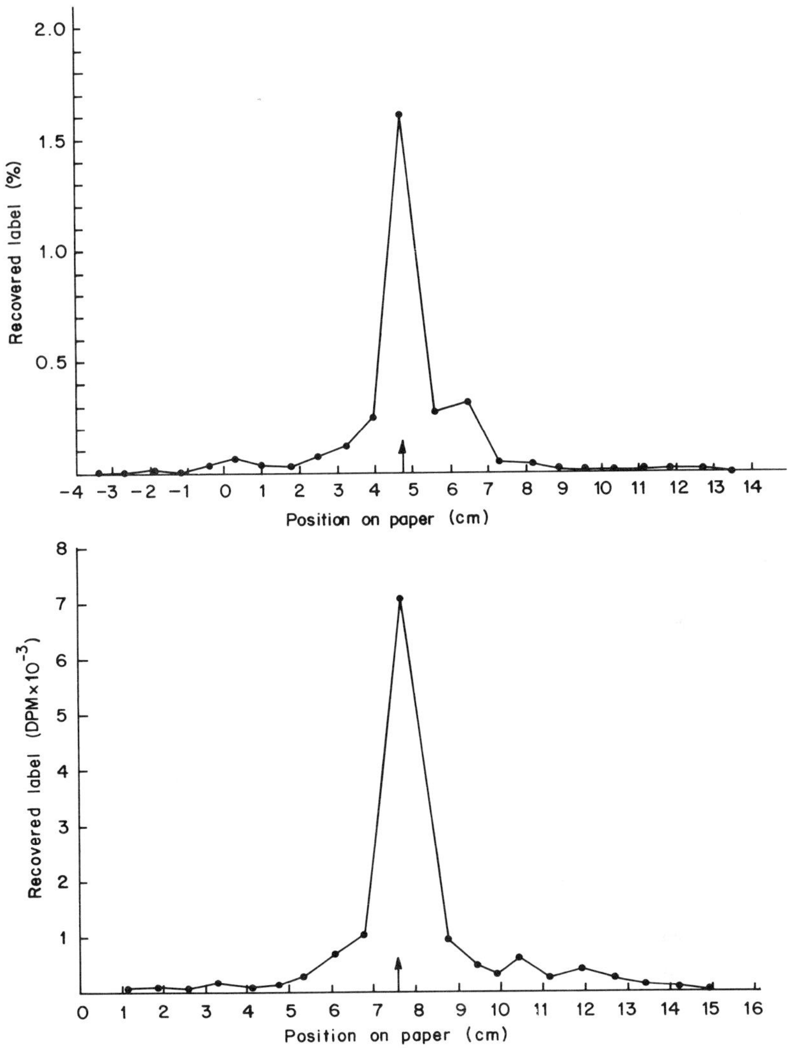

Figure 1 Distribution of ^{14}C label after paper electrophoretic and thin layer chromatographic separations of extracts of a culture labelled with [^{14}C] valine. Arrows indicate the position of phomopsin A.

are not at all surprising. The observed incorporation of [^{14}C] phenylalanine could also be reasonably expected despite its rather more extensive substitution in the phomopsin A molecule. The lack of significant incorporation of aspartic acid, however, was surprising since the presence of a 2,3-

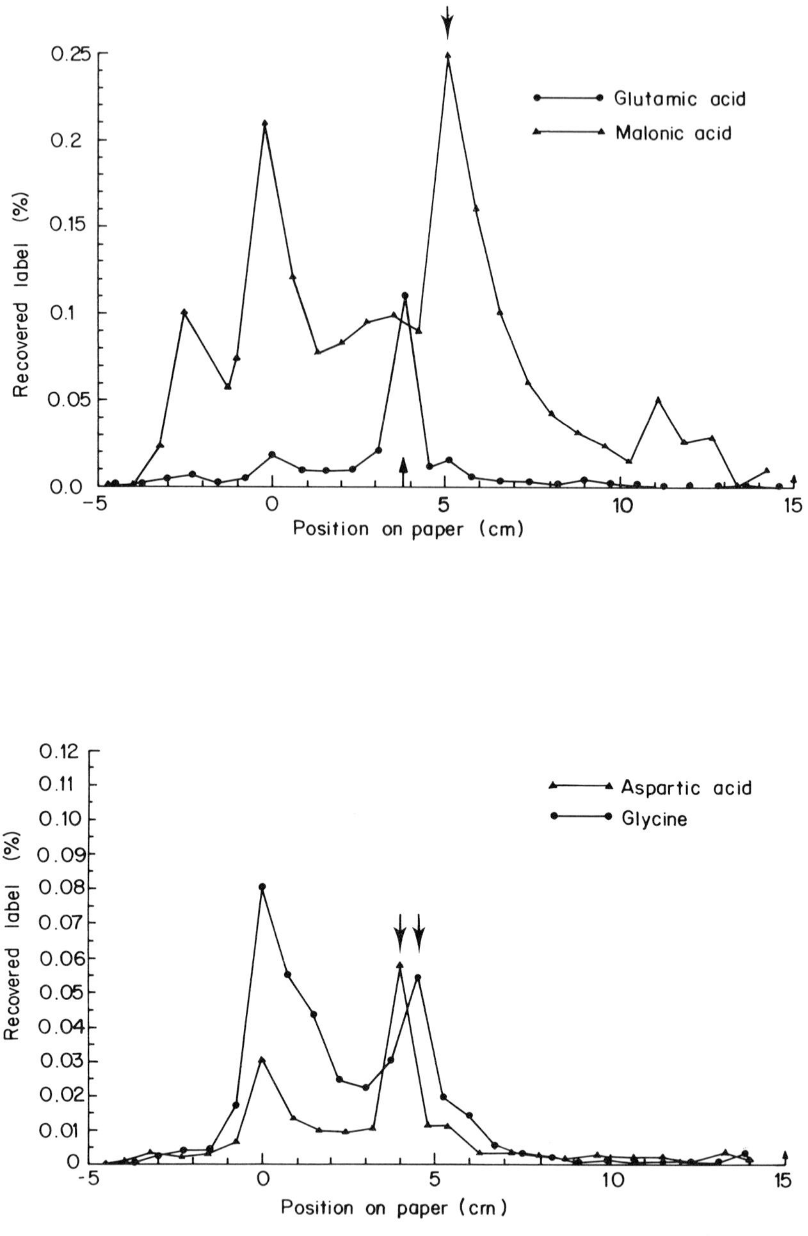

Figure 2 Radiolabel distribution after electrophoretic separation of extracts of cultures labelled with [^{14}C] aspartic acid, [^{14}C] glutamic acid, [^{14}C] glycine and [^{14}C] malonic acid. Arrows indicate the position of phomopsin A.

Table 1 Results of all radiolabelling experiments

Label added	Electrophoresis separations: Labelling efficiency[a] (%)	Labelling specificity[b] (%)	Specific activity (μCi/mmol)	Thin-layer chromatographic separations (DPM[c])
[^{14}C] acetic acid[d]	0.16	9	15	ND
[^{14}C] aspartic acid	0.06	31	6	ND
[^{14}C] glutamic acid	0.11	44	12	ND
[^{14}C] glycine	0.05	14	5	ND
[^{36}Cl] HCl[d]	0.22	36	17	ND
[^{14}C] isoleucine[d]	1.6	28	153	5.1×10^3
[^{14}C] malic acid[d]	0.14	12	13	ND
[^{14}C] malonic acid	0.25	14	24	ND
[^{14}C] phenylalanine[d]	1.5	29	116	6.8×10^3
[^{14}C] proline[d]	0.94	34	90	2.75×10^3
[^{14}C] propionic acid[d]	0.11	7	12	ND
[^{14}C] sucrose[d]	0.12	9	9	ND
[^{14}C] tryptophan[d]	0.19	3	15	ND
[^{14}C] valine	1.6	40	165	7.1×10^3

[a] Percentage of added label incorporated into phomopsin.
[b] Activity of phomopsin expressed as a percentage of total label recovered by XAD-2 extraction of culture filtrates.
[c] Activity of phomopsin spot (disintegrations per minute) for 1% of total samples.
[d] Payne, 1983
ND = Not done.

didehydroaspartic acid residue in phomopsin A has been demonstrated and is therefore the only major structural element in phomopsin A whose biosynthetic origin is unclear (Figure 3).

Earlier observations (Culvenor *et al.*, 1978) that mild hydrolysis of phomopsin A yielded oxaloacetic acid led to an attempt to label the appropriate portion of phomopsin A by addition of [^{14}C] malic acid to cultures (Payne, 1983). The relatively small incorporation achieved suggested that oxaloacetic acid was not present in phomopsin A as such, but was generated under mild hydrolysis conditions from some other chemical entity, now known to be the dehydroaspartic acid (Culvenor *et al.*, 1983), the origin of which is probably not oxaloacetic acid.

It is reasonable to assume that the amino acids are being incorporated directly into the appropriate part of the phomopsin A molecule. This has been demonstrated experimentally for the four most efficient incorporations

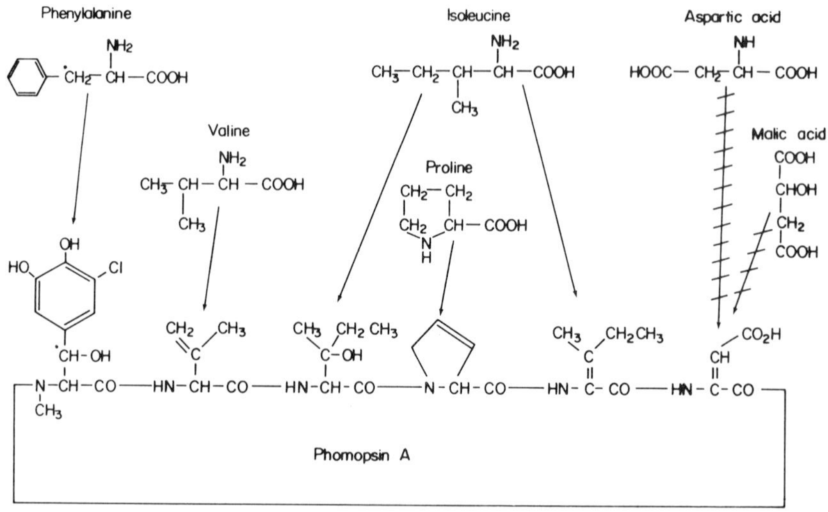

Figure 3 Structure of phomopsin A and relevant amino acids indicating the biosynthetic relationships that have been experimentally demonstrated and possible relationships which are not supported by biosynthetic data. The position of the ^{13}C label is indicated with a spot. Experimentally demonstrated →. Not supported by biosynthesis experiments ⇸.

(isoleucine, phenylalanine, proline and valine). Hydrolysis of labelled, RPLC purified phomopsin A with 6 N HCl after reduction yielded characteristic mixtures of the constituent amino acids. After TLC separation of the acids the relevant portions of the layer coating and similar-sized regions from the entire length of the plates were removed and extracted with water. The solvent was then evaporated and radioactivity measured by liquid scintillation counting (Payne, 1983). Distribution of radiolabel on electropherograms was determined as described earlier (Payne, 1983). In the cases of the isoleucine, proline and valine incorporations, the recovered activity was confined to the appropriate amino acid.

The substituted phenylalanine is not seen as a hydrolysis product of phomopsin A. 3-Chloro-4,5-dihydroxybenzaldehyde, sarcosine, methylamine and a substance believed, from the mass spectrum of its tetratrimethylsilyl derivative, to be 3-chloro-4,5-dihydroxyphenylpyruvic acid, are produced during hydrolytic degradation of this amino acid unit (Culvenor *et al.*, 1983). When ^{14}C phenylalanine-labelled phomopsin A was hydrolysed and products separated by PE, the activity was largely confined to a single component of low electrophoretic mobility. The identity of this fragment remains to be confirmed but is likely to be the 3-chloro-4,5-dihydroxyphenylpyruvic acid remnant of the *N*-methyl-3-[3-chloro-4,5-dihydroxy-phenyl] serine unit of phomopsin A. The phenylalanine origin of this part of the phomopsin A molecule was shown by labelling with *L*-[3-^{13}C] phenylalanine. The ^{13}C NMR spectrum of the phomopsin A produced showed a significant enhancement in the expected signal (Culvenor *et al.*, 1983) (Figure 3).

The levels of label incorporation from amino acids in these experiments was somewhat lower than may have been expected for a peptide product. When, however, the slow rate of phomopsin A accumulation (Lanigan *et al.*, 1979), the relative lack of incorporation of a range of other compounds, the possible metabolic fates of the amino acids and their various modifications in phomopsin A are considered, the incorporations achieved are quite significant.

The modifications to the structures of the amino acid subunits of phomopsin A (dehydrogenation and hydroxylation) are of interest and worthy of further study. For example, these experiments give no insight into whether the amino acids are modified prior to, or subsequent to, the assembly of the polypeptide chain, or its order or mode of assembly.

REFERENCES

Culvenor, C. C. J., Beck, A. B., Clarke, M., Cockrum, P. A., Edgar, J. A., Frahn, J. L., Jago, M. V., Lanigan, G. W., Payne, A. L., Peterson, J. E., Petterson, D. S., Smith, L. W., and White, R. R. (1977). 'Isolation of toxic metabolites of *Phomopsis leptostromiformis* responsible for lupinosis', *Aust. J. Biol. Sci.*, **30**, 269–277.

Culvenor, C. C. J., Smith, L. W., Frahn, J. L., and Cockrum, P. A. (1978). 'Lupinosis: chemical properties of phomopsin A, the main toxic metabolite of *Phomopsis leptostromiformis*', in *Effects of Poisonous Plants on Livestock* (eds R. F. Keeler, K. R. Van Kampen and L. F. James), pp. 565–573. Academic Press, New York.

Culvenor, C. C. J., Cockrum, P. A., Edgar, J. A., Frahn, J. L., Gorst-Allman, C. P., Jones, A. J., Marasas, W. F. O., Murray, K. E., Smith, L. W., Steyn, P. S., Vleggaar, R., and Wessels, P. L. (1983). 'Structure elucidation of phomopsin A, a novel cyclic hexapeptide mycotoxin produced by *Phomopsis leptostromiformis*', *J. Chem. Soc., Chem. Commun.*, **1983**, 1259–1262.

Lanigan, G. W., Payne, A. L., Smith, L. W., Wood, P.McR., and Petterson, D. S. (1979). 'Phomopsin A production by *Phomopsis leptostromiformis* in liquid media', *Appl.Environ. Microbiol.*, **37**, 289–292.

Payne, A. L. (1983). 'Biosynthesis of radiolabeled phomopsin by *Phomopsis leptostromiformis*', *Appl. Environ. Microbiol.*, **45**, 389–392.

van Warmelo, K. T., Marasas, W. F. O., Adelaar, T. F., Kellerman, T. S., van Rensburg, I. B. J., and Minne, J. A. (1970). 'Experimental evidence that lupinosis of sheep is a mycotoxicosis caused by the fungus, *Phomopsis leptostromiformis* (Kühn) Bubák', *J. S. Afr. Vet. Med. Assoc.*, **41**, 235–247.

van Warmelo, K. T., and Marasas, W. F. O. (1972). '*Phomopsis leptostromiformis:* the causal fungus of lupinosis, a mycotoxicosis, in sheep', *Mycologia*, **64**, 316–324.

CHEMISTRY AND BIOLOGICAL ACTIVITY OF MYCOTOXINS

Trichothecenes and Other Mycotoxins
Edited by J. Lacey

CHAPTER 27

Chemistry and biological activity of *Fusarium roseum* mycotoxins

Chester J. Mirocha, Robert J. Pawlosky, Zhu Tong-Xia and Yin-Won Lee

INTRODUCTION

The species concept in *Fusarium* is difficult to define and at best represents an attempt to set up a mechanical system of resolution between morphologically different isolates. Often the morphological characteristics used as species delineators can delineate variants of other 'species', i.e. they can occur in greater or lesser number in two or more isolates designated by some authors as different species. Therefore, in this chapter, the taxon *Fusarium roseum* 'Graminearum' has been used as defined by Snyder and Hanson and we consider this to be a reasonable method of describing and documenting the various toxic isolates we have found occurring on gramineae. Almost all of our isolates have come from maize, wheat and barley.

Wheat and barley infected with *F. roseum* was reported in the early Russian literature as causing a condition in man called 'Pijany chleb' or drunken bread. This mycotoxicosis was characterized by dizziness, vomiting, diarrhoea and headache. It seems probable that the cereals used for bread making were infected by *F. roseum* 'Graminearum', the common *Fusarium* scab-causing organism, which produces deoxynivalenol and, sometimes, nivalenol in laboratory culture together with zearalenone. It is not especially toxic compared to other isolates of *Fusarium* known to produce the potent T-2 toxin or diacetoxyscirpenol.

Epiphytotics of *Fusarium*-induced 'scab' have been reported in wheat in the U.S. and Canada in the years 1980 through 1982. Three samples of hard red winter wheat (1982 crop) obtained from Kansas analysed in our laboratory for deoxynivalenol (DON) contained, respectively, 9.1, 0.54 and 0.38 μg/g DON. When made into bread the finished product contained 6.1, 0.4 and 0.1 μg/g DON respectively. One and then two different slices of this bread was consumed by the senior author at two different times without any apparent ill effects. This was done in an attempt to confirm some of the signs and symptoms reported in the early literature associated with 'drunken bread'.

In this presentation we summarize the various toxins associated with the *F. roseum* group and the specific mycotoxicoses that they produce. Many of the toxins described as natural products of this group are not trichothecenes; examples: butenolide, fusaric acid, lycomarasmin. These will not be treated in this presentation except for a newly discovered metabolite (TDP-1) responsible for bone malformation in poultry. Those that will be described here are: zearalenone and the zearalenols; mono-, di- and tri-acetoxy-scirpenols; and deoxynivalenol, 3-acetoxy-deoxynivalenol, 15-acetoxy-deoxynivalenol and TDP-1.

It should be emphasized that when interpreting the signs of a mycotoxicosis one should be aware that 'causality is multiple'. More than one toxin may be present in the foodstuff suspected of causing the malady and may contribute to the signs or severity of signs noted. Thus, aflatoxin-containing maize may also contain cyclopiazonic acid and zearalenone-containing maize may also contain deoxynivalenol or another trichothecene.

ZEARALENONE AND HYPEROESTROGENISM

Production of zearalenone is not limited to the *F. roseum* group but it also occurs in species of other groups, e.g. *F. tricinctum*, *F. sporotrichioides* and, occasionally, in trace amounts in *F. moniliforme*. Isolates of *F. roseum* 'Culmorum', *F. roseum* 'Equiseti', *F. roseum* 'Gibbosum' and *F. roseum* 'Graminearum', cultivars given the status of separate species (respectively *F. culmorum*, *F. equiseti*, *F. gibbosum*, and *F. graminearum*) by other authors (Marasas *et al.*, chapter 1 in this volume) produce copious amounts of zearalenone in culture. It can thus be safely stated that the production of appreciable amounts of zearalenone both in laboratory culture and in the infection process of cereals under natural field conditions is most commonly associated with the *Fusarium roseum* group.

Hyperoestrogenism is a condition seen most often in swine consuming 'corn-on-the-cob' (maize) and maize stored in cribs open to the environment, i.e. wind and rain. It is common in the Midwest U.S.A. where maize is often stored in cribs on the farm. The most important effect on swine is economic, through infertility and lack of marketable offspring. In prepubertal gilts the vulva becomes swollen and oedematous and may result in prolapse; the uterus becomes enlarged, oedematous and tortuous, with atrophy of the ovaries; and the mammae become precociously enlarged. Zearalenone does not cause abortion but does contribute to infertility and decreased litter size (Chang *et al.*, 1979).

Zearalenone and its derivatives act by binding to specific oestrogen receptor protein of mammals and by competing for binding sites with oestradiol. This mechanism has been studied and reported on by Kiang *et al.* (1978).

The structure of zearalenone and its derivatives is shown in Figure 1. Activity of the molecule is most affected by modification of the 6′-ketone and the 1′–2′ double bond. Both partial reduction of the 6′-ketone to the hydroxyl and saturation of the 1′–2′ double bond enhance activity. The derivative resulting from these modifications as 'Zeranol' is used as an animal growth promotant and as 'Frideron' is used as a chemotherapeutant to alleviate postmenopause problems. Hydroxylation can also occur on the 5′ and 8′ positions; when it occurs in the 8′ position all biological activity is lost.

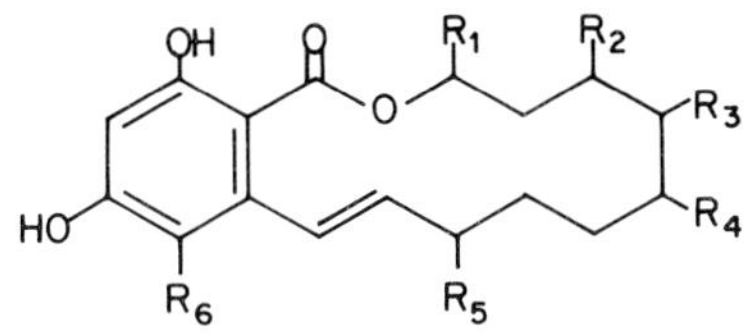

Compound	R_1	R_2	R_3	R_4	R_5	R_6
I	$-CH_3$(S),H	$=H_2$	$=H_2$	=O	$=H_2$	−H
III	$-CH_3$(S),H	−OH(R),H	$=H_2$	=O	$=H_2$	−H
IV	$-CH_3$(S),H	−OH(S),H	$=H_2$	=O	$=H_2$	−H
V	$-CH_3$,H	$=H_2$	$-H_2$	=O	−OH,H	−H
VI	$-CH_3$,H	$=H_2$	$=H_2$	=O	$=H_2$	−CHO
VII	$-CH_3$,H	−H	−H	=O	$=H_2$	−H
VIII	$-CH_3$,H	−OH,H	$-H_2$	−OH,H	$=H_2$	−H
IX	$-CH_3$,H	$=H_2$	$=H_2$	−OH,H	−OH,H	−H

I. zearalenone (F-2)
III. 8′(*R*)-hydroxyzearalenone (F-5-3)
IV. 8′(*S*)-hydroxyzearalenone (F-5-4)
V. 3′-hydroxyzearalenone (F-5-2)
VI. 5-formylzearalenone
VII. 7′-dehydrozearalenone
VIII. 6′,8′-dihydroxyzearalene
IX. 3′,6′-dihydroxyzearalene (Synthetic)

Figure 1 Naturally occurring zearalenone derivatives.

Hydroxylation of the 6′-ketone to the 6′-hydroxyl results in the formation of an asymmetric carbon atom and two isomers called alpha and beta zearalenols. The alpha isomer is more active than the beta. Zearalenol isomers (alpha and beta) can exist in either *trans* or *cis* configurations. The *cis* isomer of each derivative is always more active. Following this reasoning to its logical conclusion, I predicted that the most oestrogenically active derivative would be the *cis* isomer of alpha zearalenol and this has been confirmed by laboratory experimentation (unpublished data).

Trans-zearalenone is the predominant derivative found in naturally infected corn although alpha zearalenol can also be found infrequently in concentrations approximating 3% of the parent compound. Thus, if 10 μg of zearalenone/g is found, then one could expect to find 0.3 μg zearalenol/g. This metabolite is three times more active than its parent and hence its presence is significant. Some zearalenone-producing isolates of *Fusarium roseum* 'Graminearum' grown in laboratory culture on solid rice medium can produce a variety of derivatives. Thus one isolate produced *cis* and *trans* zearalenone together with *cis* and *trans* alpha and beta zearalenols. Such a varied production of zearalenones is unusual; nevertheless, it can be found. Likewise, most *Fusarium roseum* isolates produce zearalenone more copiously at 10–12 °C but in some isolates production is greatest at 20 °C on a rice solid medium. Thus the genetic potential of *Fusarium* is vast and imperfectly understood.

TRICHOTHECENES

The trichothecenes possess a tetracyclic-12,13-epoxytrichothec-9-ene skeleton as shown in Figure 2. Crotocin is the exception in that it has an extra oxirane in the 7–8 position of the A ring and hence is pentacyclic. Many naturally occurring derivatives of the trichothecenes have a hydroxyl or an ester at one or more of the positions C-3, C-4, C-7, C-8 and C-15. The approximately 60–70 derivatives that have been described can be divided into the following groups: (a) those having a functional group at C-8 other than a ketone, e.g. T-2 toxin; (b) those derivatives with a ketone at C-8, e.g. deoxynivalenol; (c) macrocyclic derivatives with a large ring between C-4 and C-15, e.g. verrucarin. The macrocyclic trichothecenes are distinguishable

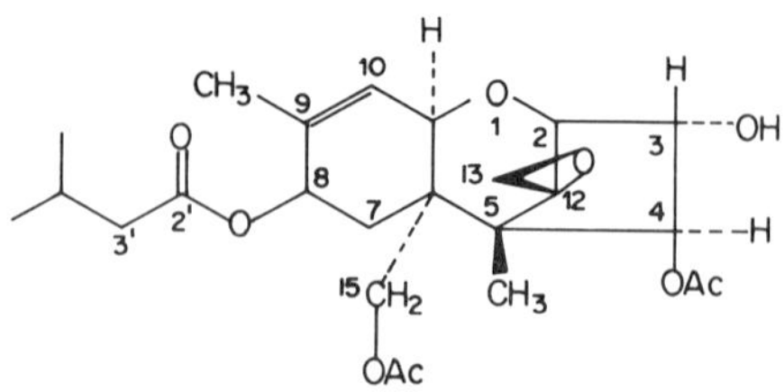

Figure 2 Basic structure of T-2 toxin (3-alpha-hydroxy-4,15-diacetoxy-8-alpha-isovaleroxy-12,13-epoxytrichothece-9ene). Most frequent substitution occurs at the 3,4, and 8 carbons.

from the non-macrocyclics on thin-layer chromatography because of their intense fluorescence when viewed by ultra violet illumination. The trichothecenes often associated with the *F. roseum* group are of the scirpene and nivalenol types. The scirpenes are best illustrated by scirpenetriol, diacetoxy and monoacetoxy scirpenols, and triacetoxyscirpene. The nivalenols are best exemplified by nivalenol, deoxynivalenol and fusarenone-X.

Scirpene trichothecenes

F. roseum 'Graminearum' and *F. roseum* 'Gibbosum' are good producers of diacetoxyscirpenol and other closely related derivatives. As an example, we will report our experience with *F. roseum* 'Gibbosum' when grown on a solid rice medium in the laboratory (Zhu and Mirocha, 1983).

When *F. roseum* 'Graminearum' is grown on a rice medium it produces several scirpene trichothecenes, the most important of which are monoacetoxyscirpenol (MAS), diacetoxyscirpenol (DAS), triacetoxyscirpene (TAS) and scirpentriol (SCT). Three isomers of MAS were found: 3-MAS, 4-MAS and 15-MAS. Of the three, 15-MAS is the most abundant and the one first described by Pathre *et al.* (1976). The resolution of the three isomers by thin-layer chromatography (TLC; silica-gel, Merck, ethyl acetate : toluene 3 : 1) and gas–liquid chromatography ((GLC; 30 m DB5 capillary, trifluoroacetate derivative) was obtained.

The positive chemical ionization (in methane) mass spectra of the TFA derivatives of the isomers are shown in Figure 3. The isomer 15-MAS is distinct from the others in its fragment ions and can easily be identified from among them. Fragment 457 (base peak) is very intense in 15-MAS and m/e^+ 343 is very weak. In contradistinction, isomers 3 and 4-MAS have a very intense 343 fragment (base peak) and a very weak 457. The isomers 3-MAS and 4-MAS are difficult to distinguish one from the other by mass spectra alone, although these are different, i.e. mostly in the relative intensities of their common fragments. As an example, fragment m/e^+ 517 (molecular ion) of 3-MAS is more intense than that of 4-MAS. This difference, plus the different retention time on column, serves to sort them out. Masses 403 and 343 are common to all the MAS isomers.

F. roseum 'Gibbosum' also produces three isomers of diacetoxyscirpenol. They, like the MAS isomers, can be easily resolved by TLC and GLC and also contain unique mass spectral fragment ions. The isomers of DAS are: 3,4-DAS, 3,15-DAS and 4,15-DAS. The isomer 3,15-DAS is unique in that its molecular ion (463) is its base peak and as a result it can be distinguished from the others very easily. The isomers 3,4-DAS and 4,15-DAS are similar to each other in fragment ions but can be resolved by chromatography. All three isomers contain m/e^+ 403 as a common fragment. The latter is diagnostic of DAS and its closely allied derivatives in positive chemical

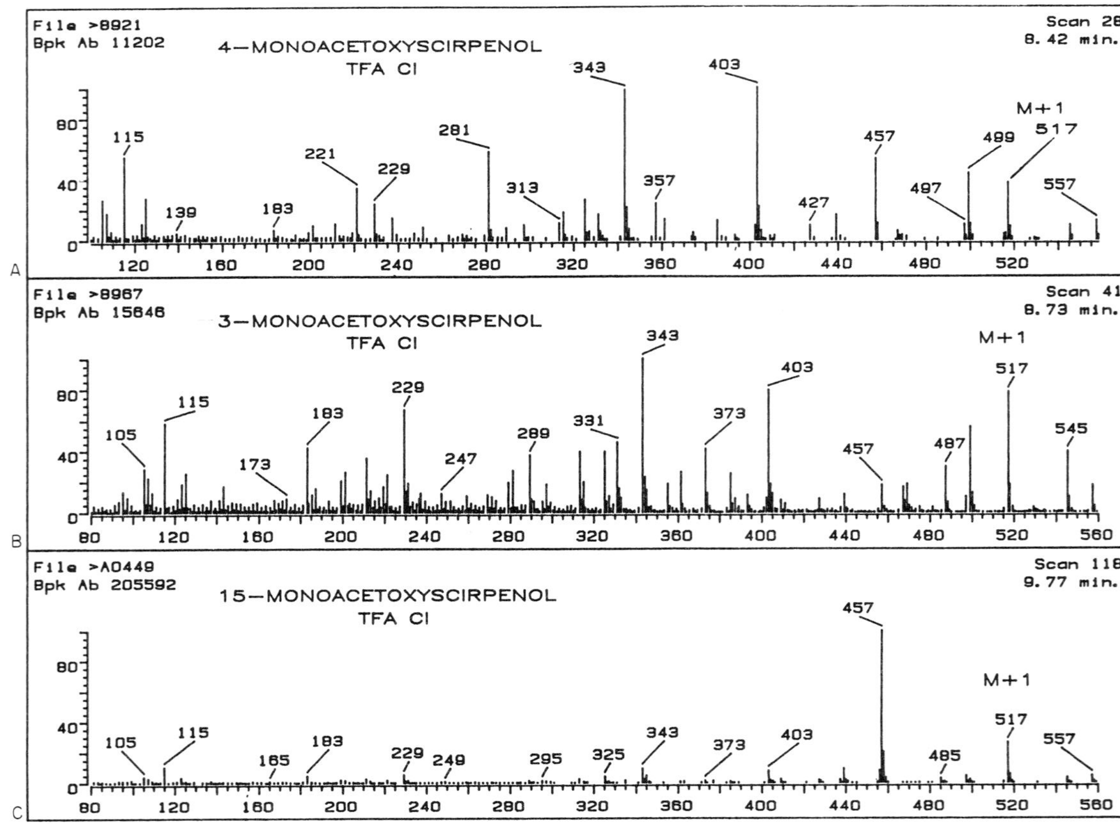

Figure 3 Positive chemical ionization (PCI) mass spectra of the trifluoroacetate (TFA) derivatives of monoacetoxyscirpenol. The spectra were obtained on a Hewlett Packard 5987 combination gas chromatograph–mass spectrometer (quadrupole) with methane gas at a pressure of 0.5 torr. **A**: PCI of 4-monoacetoxyscirpenol (base peak of 403); **B**: PCI of 3-monoacetoxyscirpenol (base peak of 343) and **C**: PCI of 15-monoacetoxyscirpenol (base peak of 457).

ionization in methane. The fragment 403 is also present in MAS but is not a major ion; conversely, mass 343 is present in all DAS isomers but is not very intense and is almost non-existent in 3,4-DAS.

Triacetoxyscirpene (TAS) is another major product in the scirpene group. It is an important metabolite because its production is appreciable and it contributes substantially to toxicity of the culture when tested on laboratory animals. It is usually overlooked in analyses of both field and culture substrates.

The mass spectrum of TAS in positive chemical ionization (methane) is excellent in terms of sensitivity and identification. The pseudo-molecular ion plus one is 409 and this mass is almost equal in intensity to the base peak which is 349. Other diagnostic masses of significant intensity are 391, 289 and 229.

The most important aspect of our work with the isomers of both MAS and DAS is the effect of substitution of acetate on various carbons (positional isomerization) and biological activity; the results will not be reported here. The yield of each isomer relative to the others is not great with usually one of the series predominating. Thus, of the MAS series, 15-MAS predominates followed by 4-MAS and 3-MAS in that order. Likewise, 4,15-DAS is the dominant species in culture followed by 3,4-DAS and 3,15-DAS.

Table 1 compares the activity of the scirpene trichothecenes when tested in 1-day-old chicks, in rats and in chick embryos. Of the derivatives, 15-MAS appears to be most active followed closely by 4,15-DAS and TAS. All three derivatives cause a severe skin reaction when painted on to the freshly shaven skin of rats. Although TAS appears to be the least toxic of the three, its skin vesicant action is equal to MAS and DAS. All three compounds will cause haemorrhaging in rats when administered by intubation or painted on to the skin. This property places the scirpenes in a category usually though of as more toxic than T-2 toxin and its derivatives.

Table 1 LD50 values of various scirpene trichothecenes

Compound	Chick[1/] mg/kg	Rat[2/] mg/kg	Egg Embryo[3/] ng/egg
Triacetoxyscirpene	8.0	2.3	232
4.15-diacetoxyscirpenol	5.9	1.6	123
15-monoacetoxyscirpenol	3.4	1.8	189

[1/] Broiler chicks weighing approximately 50 g were intubated into the crop with the toxin dissolved in 20% ethyl alcohol.
[2/] White female virgin rats (70 g) were intubated into the stomach with the toxicant dissolved in 20% ethyl alcohol.
[3/] Eggs used were those resulting from a cross of Rhode Island Red males. Toxicant was dissolved in dimethyl sulfoxide.

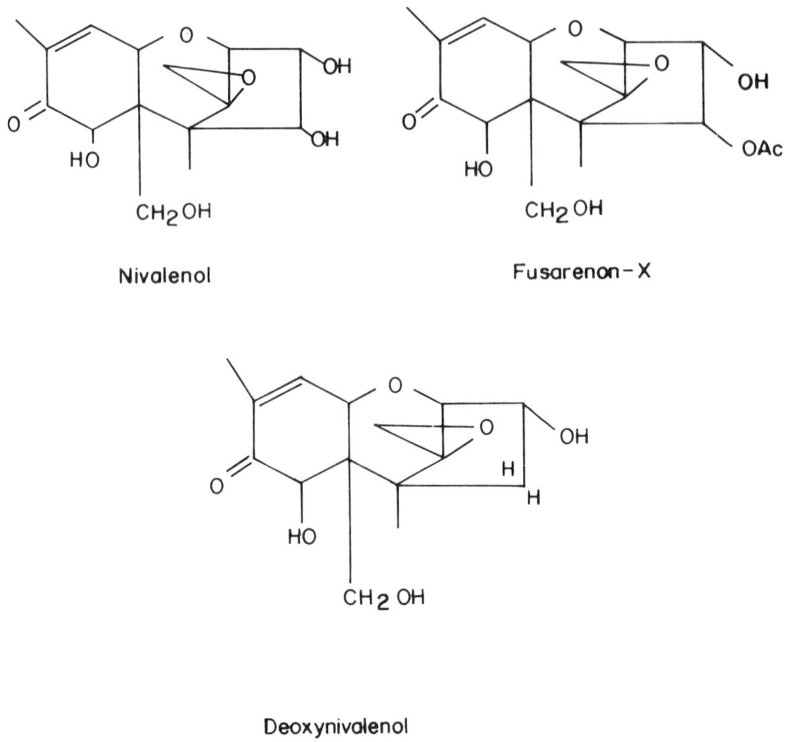

Figure 4 Structures of deoxynivalenol and related natural products: nivalenol and fusarenone-X. These compounds are best resolved as their trimethylsilyl ether derivatives of gas–liquid-chromatography.

Deoxynivalenol and derivatives

Deoxynivalenol is chemically related to nivalenol, and fusarenone-X; all three have a ketone substituted at C-8 and a hydroxyl at C-7. They also turn yellow on thin-layer chromatoplates when sprayed with a mixture of *p*-anisaldehyde and sulphuric acid in solution (Figure 4). Other derivatives are 7-dehydronivalenol, 7-dehydroxydeoxynivalenol and 3-monoacetoxydeoxynivalenol (3-acetoxy-DON) and 15-monoacetoxydeoxynivalenol (15-acetoxy-DON) (see Figure 5). All of the above have been found in laboratory culture and some, like the 3 and 15-acetoxy-DON isomers, have also been found in the field.

DON is important economically because it causes emesis and refusal of food by swine when ingested in their diet. It is far less toxic than T-2 or DAS trichothecenes. Its LD_{50} in mice (i.p.) is 70 mg/kg and in broiler chicks (p.o.) it is 140 mg/kg. By way of comparison, the LD_{50} of fusarenone-X in mice

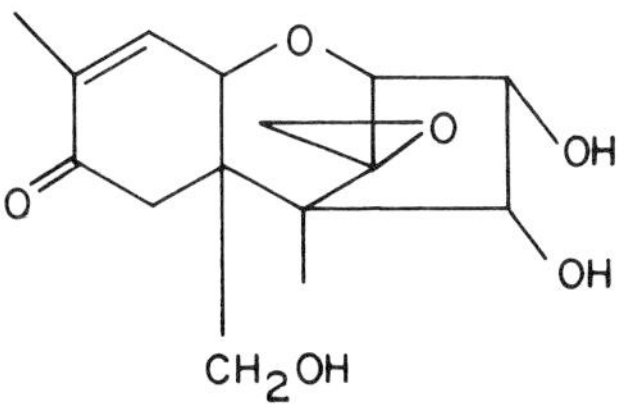

7-Dehydroxynivalenol

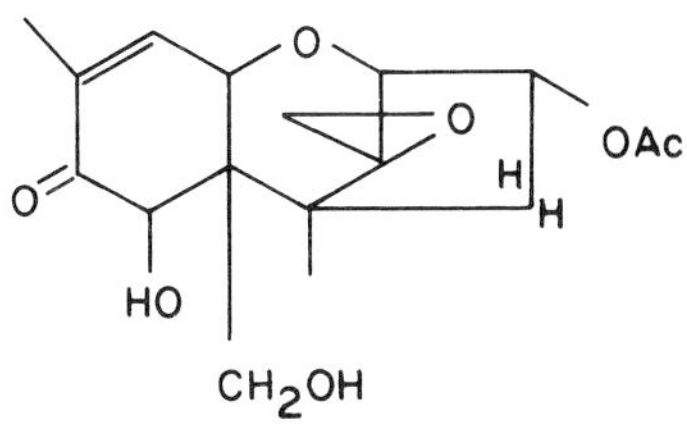

Monoacetoxydeoxynivalenol

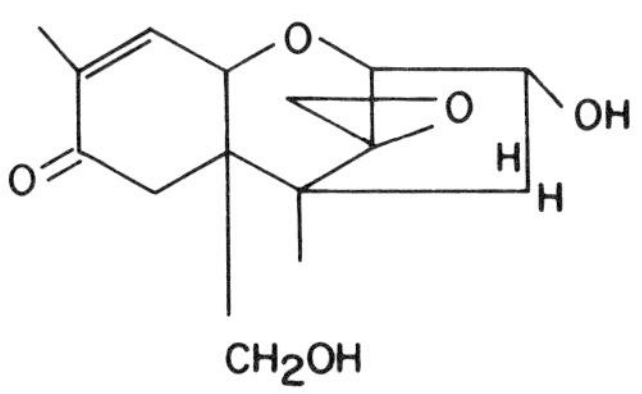

7-Dehydroxydeoxynivalenol

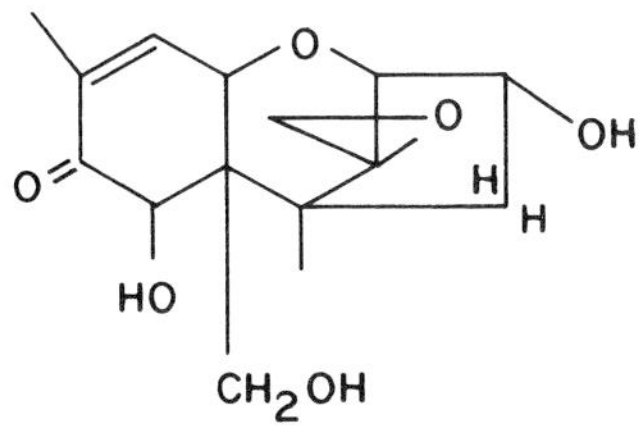

Deoxynivalenol

Figure 5 Deoxynivalenol and some of its derivatives found in laboratory culture of DON-producing *Fusarium* isolates.

(i.p.) is 5.6. When DON is painted onto the skin it does not cause skin lesions in the rat but it causes mild reddening and necrosis in the rabbit. Feed refusal by swine caused by DON in feedstuff is significant but DON alone does not account for all refusal of naturally contaminated corn. Other *Fusarium* metabolites are thought to contribute to the refusal activity. For example, the toxin herein referred to as TDP-1 causes rats, chicks and swine to refuse diets containing it.

DON is most often found in maize and wheat infected by *F. roseum* 'Graminearum'. Such maize is of high risk as an animal feed; it can contain zearalenone, DON and perhaps other undescribed trichothecenes. In 1982 infection of wheat ears with *Fusarium* scab or *Fusarium* blight caused a severe economic impact in Canada and the United States. The disease is prevalent in all major cereal-growing regions of the world but these outbreaks prompted tolerances and guidelines to be set by American and Canadian governments because of the vast acreage affected.

Analyses of *Fusarium*-blighted wheat in the U.S.A. and Canada usually detected DON and at time traces of zearalenone. Other trichothecenes were not found, either because they were not looked for or because efficient methods of detection have not been developed, or for both reasons. By contrast, Yoshizawa and Hosakawa (1983), have found both DON and nivalenol in *Fusarium*-infected barley and in commercial parched-barley flour. The presence of nivalenol in cereals and their products is important biologically because the latter is more toxic than DON. The LD_{50} of nivalenol in the adult male mouse (i.p.) is 6.9 mg/kg whereas the LD_{50} value for DON (i.p.) is 70 mg/kg.

Fusarium cultures that produce DON also produce, in varying amounts, 3-acetoxydeoxynivalenol (3-acetoxy-DON) and 15--acetoxydeoxynivalenol (15-acetoxy-DON). Although at the time of this writing the data are meagre, Drs R. Greenhalgh and D. Miller (Agriculture Canada, Ottawa, personal communication) have suggested that isolates of *Fusarium roseum* 'Graminearum' collected in the U.S.A. or Canada produce DON and 15-acetoxy-DON, whereas isolates collected in Japan produce DON and 3-acetoxy-DON. We have analyzed isolates collected in Norway and Italy and found that they, like the isolates from Japan, produce the 3-acetoxy-DON isomer. The number of isolates that have been analysed are few but the data suggest different genetic selection or origins of the *Fusarium* isolates in different regions. It should be pointed out that of the isolates examined, both isomers are formed by individual isolates but one of them will predominate and the other will be found in only trace (5–10 μg/kg) amounts. In addition to the above derivatives, Yoshizawa *et al.* (1983) described the occurrence of a novel DON derivative formed after metabolism by the rat. The metabolite is an oxidation product (oxirane ring is broken) and the proposed structure is 3,7,15,-trihydroxytrichothec-9,12-diene-8-one.

The chemical detection of DON and its derivatives (especially DON) has been pursued with earnest because of the economic importance of this group of trichothecenes and the need for analytical surveillance. Greatest attention has been given to rapid screening methods based on thin-layer chromatography although mass spectrometry has been most important in confirmation. The positive chemical ionization (in methane) mass spectrum of DON as the trimethylsilyl ether is shown in Figure 6A and is compared with the heptafluorobutyroimidazole derivative in Figure 6B. The mass spectrum in A shows an intense molecular ion at 513 which is approximately 60% of the base peak (m/e^+ 497). These two ions are very large and thus diagnostic for DON; when used for determination of DON by selected ion monitoring, 10 ng will give an electronic integration value of 20,913 for mass 497 and 11,056 for mass 513, indicating extreme sensitivity. By contrast the HFB derivative of DON shown in Figure 6B has a base peak at m/e^+ 215 and its molecular ion (m/e^+ 884) is only 4% of the base peak. Mass 884 is used for detection and quantitation of DON by many investigators because of its time-honoured use in electron capture detection. When the negative chemical ionization mass spectra of the TMS derivative of DON and its HFB derivative at a concentration of 100 ng is compared quantitatively by SIM, one finds a total area count of 12,000,000 for the former and 500,000 for the latter, a factor of 24 times greater sensitivity for TMS-DON "(Figure 7). TMS-DON is clearly a

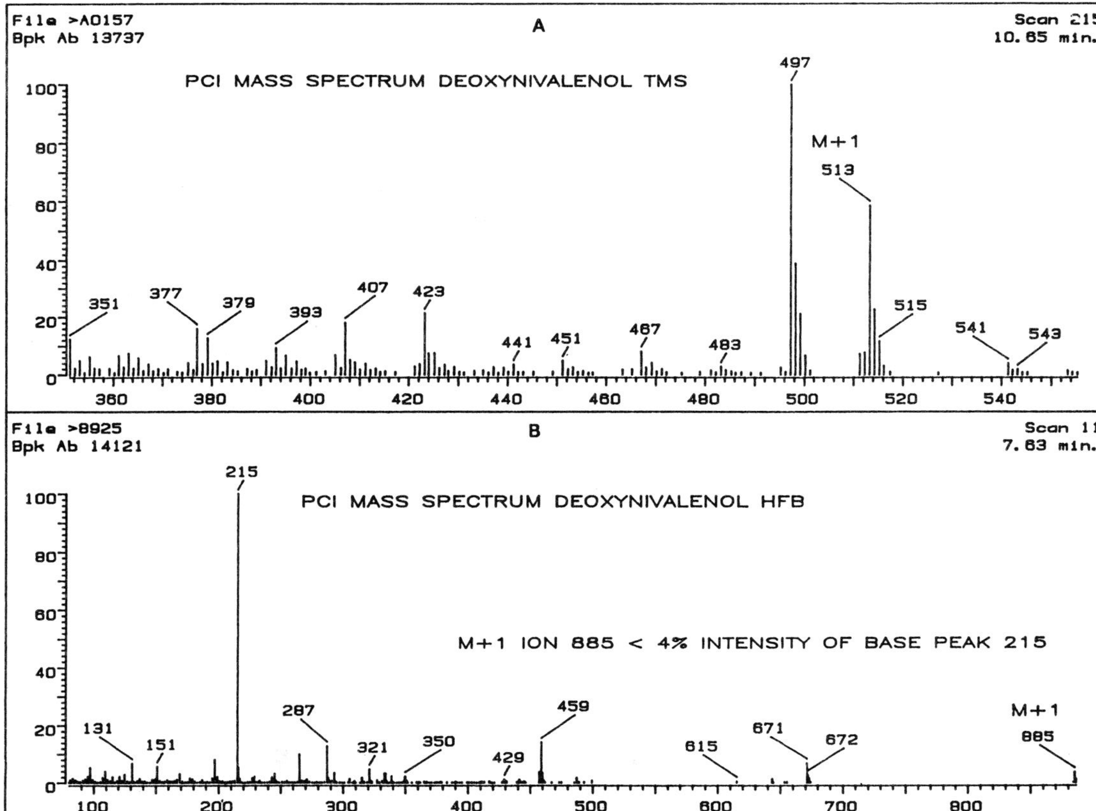

Figure 6 Positive chemical ionization mass spectrum of the trimethylsilyl ether derivative of deoxynivalenol and the heptafluorobutyroimidazole derivative of deoxynivalenol. The TMS ether derivative displays more abundant masses in the high mass region (497 and 513) and is more informative than the HFB-DON derivative.

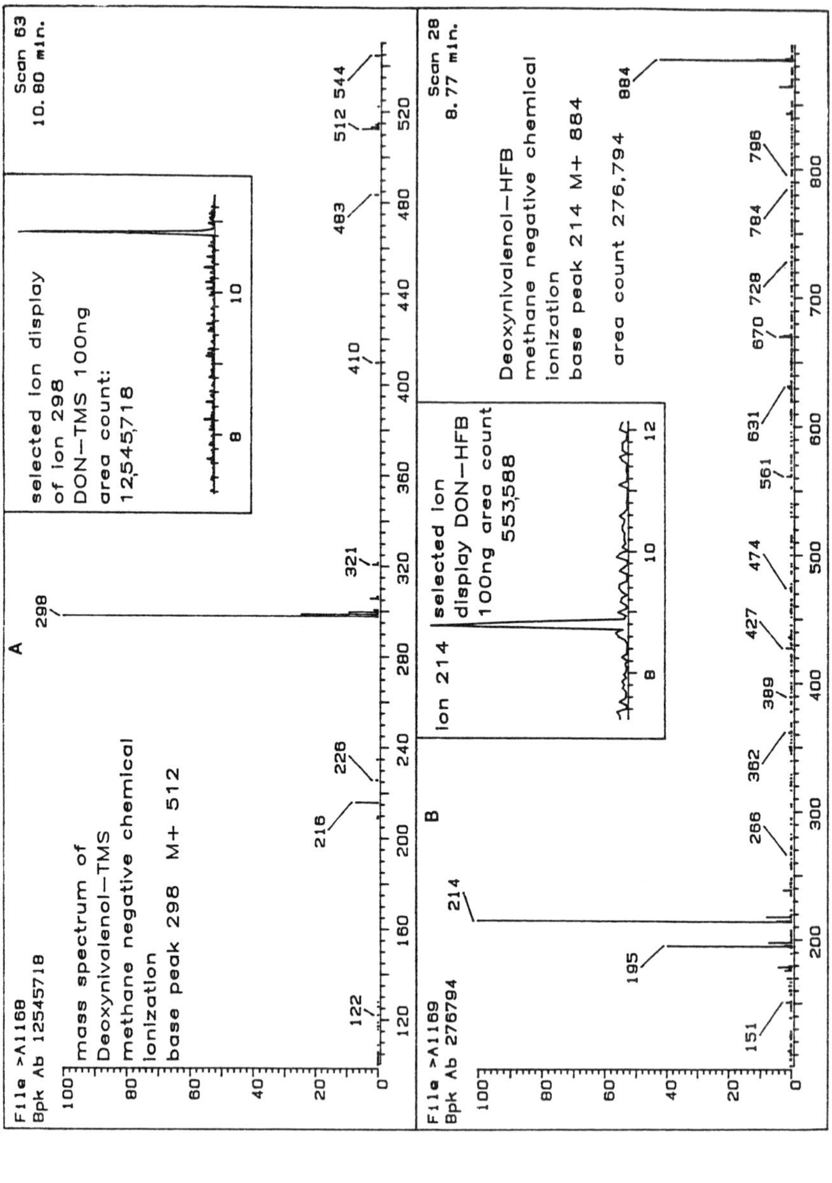

Figure 7 **A**: Mass spectrum of the trimethylsilyl ether derivative of deoxynivalenol in negative chemical ionization (NCI) using methane as the ionizing plasma gas. NCI is by far the most sensitive means of detection of TMS-DON. **B**: Mass spectrum of the heptafluorobutyroimidazole derivative of deoxynivalenol using negative chemical ionization with methane ionizing gas. The fragment 884 is the most intense ion but clearly less sensitive than the TMS-DON fragment of 298 shown above.

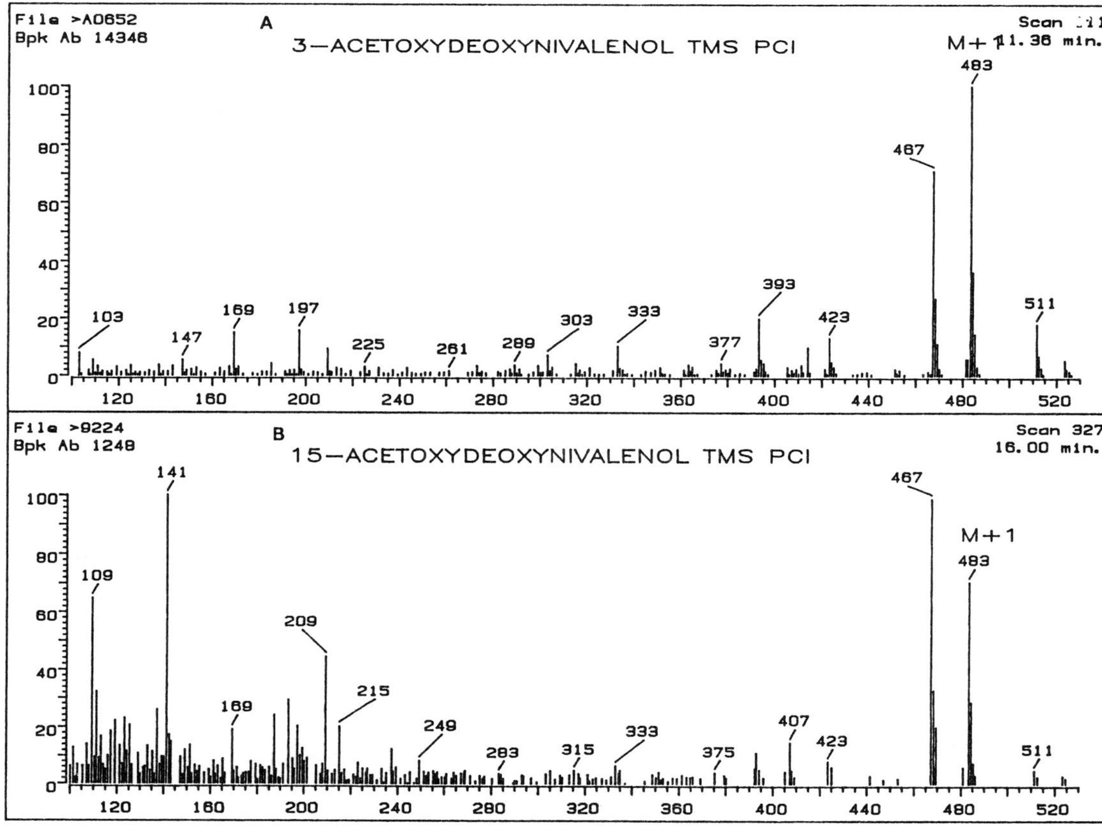

Figure 8 **A**: Mass spectrum of 3-acetoxydeoxynivalenol in positive chemical ionization using methane as the ionizing gas. **B**: Mass spectrum of 15-acetoxydeoxynivalenol in positive chemical ionization with methane plasma. The two positional isomers can be distinguished from each other by comparison of the high and low mass range fragments as well as the base peak (483 for the former and 467 for the latter).

superior derivative to use in mass spectrometry when using either electron impact, or negative and positive chemical ionization mass spectrometry. In addition to the greater sensitivity, it has a greater range of useful masses to use in identification; moreover the preparation of TMS ethers is easier and there are fewer steps in manipulation.

The positive chemical ionization (PCI) spectra of 3-acetoxy-DON and 15-acetoxy-DON isomers is shown in Figure 8A and 8B. The two isomers can be clearly distinguished from each other by their mass spectra. The PCI of 3-acetoxy-DON has a molecular ion (M + 1) of 483 and this fragment is also its base peak. Fragment 467 (M − 16) is also intense and useful. The PCI of 15-acetoxy-DON has a base peak of 467 and molecular ion plus 1 of 483. Masses 141, 209 and 407 are also useful in identification of the molecule. It is clear that the fragmentation pattern of these two isomers is entirely different and thus diagnostic.

MYCOTOXINS AFFECTING BONE FORMATION

The toxins produced by the *Fusarium roseum* group of fungi that are most visible in the mycotoxin literature are the zearalenones and the trichothecenes; they are all non-water-soluble with the exception of DON. Until recently little attention has been given to the water-soluble fractions. Some isolates of this group produce a water-soluble, ninhydrin-positive, polar and nitrogen-containing natural product that causes: (1) tibial dyschondroplasia in poultry, (2) reduction in hatchability, (3) haemorrhaging in rats and (4) refusal of feed in poultry, rats and swine. We have named this mycotoxin TDP-1 (Fusarochromone) and identified it as a chromone derivative (Pathre *et al.*, in press). Yin-Won Lee and Mirocha (1982) gave a preliminary report on its chemical properties.

We believe that the most important biological activity of TDP-1 is its effect on the development of the tibia in causing tibial dyschondroplasia. Walser *et al.* (1982) first described the development of a wide cartilaginous band extending distally from the proximal physis of the tibiotarsi of chicks receiving dietary levels of the TDP-1-producing *Fusarium* culture. A very wide zone of physeal hypertrophic cartilage was responsible for the lesion; there was also a reduction in chondroclasts. The lesion is characterized by the presence of a cone of cartilage which extends distally from the proximal tibiotarsal physis. Since that time we have reproduced this condition in chicks by using pure TDP-1, thus confirming our earlier observations.

It is of interest to note that Yang Jian Bo (1983), working in the Institute of Kaschin–Beck's Disease in Harbin, China, reported the correlation of Kaschin–Beck's disease in China with the presence of *Fusarium* in cereal grain. They have partially reproduced the disease in dogs and rats fed maize collected in areas of China where the disease is endemic, and also by feeding the animals cultures of *Fusarium oxysporum*. Kaschin–Beck disease is an endemic deforming osteoarthrosis. It is within the realm of possibility that TDP-1 may be involved in the disease as described in China.

The antihatchability activity of the TDP-1-producing *Fusarium* culture was tested by adding various proportions of the culture to the feed of laying hens and testing the viability of the fertilized eggs. On a 3% culture diet the hatchability of the eggs was reduced by 99% but returned to normal after the toxic feed was exchanged for a wholesome diet (Allen *et al.* 1982). Since that time the reduction in hatchability of the eggs was confirmed by use of purified TDP-1. The actual mode of action of the toxin is not understood although we know that the actual death of the embryo occurs during the first week of incubation of the eggs.

At present we are developing a method of analysis for TDP-1 and also experimenting with production methods in order to obtain greater quantities both for chemistry and testing in large animals.

REFERENCES

Allen, N. K., Jevne, R. L., Mirocha, C. J., and Yin-Won Lee. (1982). The effect of a *Fusarium roseum* culture and diacetoxyscirpenol on reproduction of white leghorn females. *Poultry Sci.* 61: 2172–2175.

Chang, G., Kurtz, H., and Mirocha, C. J. (1979). Effects of the mycotoxin zearalenone on swine reproduction. *Amer. J. Vet. Res.* 40: 1260–1267.

Kiang, D. T., Kennedy, B. J., Pathre, S. V., and Mirocha, C. J. (1978). Binding characteristics of zearalenone analogs to estrogen receptors. *Cancer Res.* **38**: 3611–3615.

Pathre, S. V., Lee, Y.-W., Mirocha, C. J. and Gleason, W. B. The structure of Fusarochromone: new mycotoxin from *Fusarium roseum* 'Graminearum' *Can. J. Chem.* (in press).

Pathre, S. V., Mirocha, C. J., Christensen, C. M., and Behrens, J. (1976). Monoacetoxyscirpenol. A new mycotoxin produced by *Fusarium roseum* 'Gibbosum'. *Agric. & Fd. Chem.* **24**: 97–103.

Walser, M. N., Allen, N. K., Mirocha, C. J., Hanlon, G. F., and Newman, J. A. (1982). Fusarium-induced osteochondrosis (Tibial Dyschondroplasia) in chickens. *Vet. Pathol.*, **191**, 544–550.

Yang Jian Bo. (1983). Etiology of Kashin–Beck's disease-mycotoxicosis caused by an unknown *Fusarium* toxin. *Abstr. Inter. Mycotoxin Symp.* held in Sydney NSW Australia, 12–15 Aug.

Yin-Won Lee, and Mirocha, C. J. (1982). A toxic metabolite of *Fusarium roseum* (Alaska 2-2) which reduced hatchability and causes tibial dyschondroplasia in poultry. (Abstract). *Phytopathology*, **72**: 2 p. 972.

Yoshizawa, T., and Hosokawa, H. (1983). Natural Co-occurrence of deoxynivalenol and nivalenol, trichothecene mycotoxins, in commercial foods. *J. Fd. Hyg. Soc. Jap.*, **24**: 413–415.

Yoshizawa, T., Takeda, H., and Ohi, T. (1983). Structure of a novel metabolite from deoxynivalenol, a trichothecene mycotoxin, in animals. *Agric. Biol. Chem.*, **47**, 2133–2135.

Zhu, Tong-xia and Mirocha, C. J. (1983). Scirpene trichothecenes produced by *Fusarium roseum* 'Gibbosum'. Abstract presented at 97th annual meeting of the Assoc. Off. Anal. Chem., Washington, D.C. Oct. 3–6, 1983.

Trichothecenes and Other Mycotoxins
Edited by J. Lacey

CHAPTER 28

Chemical and biological properties of trichothecenes from *Fusarium sporotrichioides*

Yoshio Ueno and Kenji Ishii

INCIDENCE OF *FUSARIUM SPOROTRICHIOIDES* TOXINS

Trichothecenes are a group of sesquiterpenoids produced by species of *Trichothecium*, *Trichoderma*, *Myrothecium*, *Cephalosporium*, *Fusarium*, *Stachybotrys*, *Verticimonosporium* and *Cylindrocarpon*. *Fusarium* is important because of the high incidence of contamination of foods and feed by this genus. Among *Fusarium* species, *F. graminearum* and *F. sporotrichioides* are frequently detected in foods and feed, and are very important as producers of trichothecenes.

Mycotoxicoses such as alimentary toxic aleukia (ATA) in Russia, mouldy corn toxicoses in the United States and red-mould disease in Japan were reported to be caused by fungi of the genus *Fusarium*. In relation to the mouldy corn toxicoses, Bamburg *et al.* (1968) isolated T-2 toxin from a culture of *F. tricinctum (F. sporotrichioides*) strain T-2. In Russia, scientists reported that the toxins causing ATA were steroidal compounds produced by *F. sporotrichioides* (Ueno *et al.*, 1972b). But Mirocha and Pathre (1973) analysed one of the toxins, 'poaefusarin', and detected trichothecenes such as T-2 toxin, neosolaniol and T-2 tetraol. We also observed that one isolate of *F. sporotrichioides* (NRRL 3510) from overwintered millet of the U.S.S.R. produced T-2 toxin, HT-2 toxin and neosolaniol (Ueno *et al.*, 1972b). *F. graminearum* is mainly responsible for red mould disease in Japan, and trichothecenes having a carbonyl group at position 8 of the skeleton (8-keto-trichothecenes, B-type) such as nivalenol, fusarenon-X and deoxynivalenol have been implicated. Nivalenol and fusarenon-X were isolated from *F. nivale* Fn-2B but the identification of this isolate has been questioned by some mycologists.

Another mycotoxicosis in Japan reportedly caused by *Fusarium* was the bean-hull poisoning of horses in Hokkaido. It is not certain whether this toxicosis was caused by trichothecenes because the intoxication has not been observed again. However, we isolated potent toxigenic *Fusarium* species from specimens of bean-hulls. Among the toxigenic *Fusarium* isolates, the best producer of trichothecene was a strain M-1-1. This strain was at first identified as *F. solani* but Dr Ichinoe (National Institute of Hygienic Sciences, Tokyo) reclassified it as *F. sporotrichioides* (Ichinoe, 1978). From a liquid culture of this strain we isolated T-2 toxin, neosolaniol, and two minor metabolites, NT-1 and NT-2 (Ueno *et al.*, 1972a; Ishii and Ueno, 1981).

Figure 1 shows the structures of trichothecenes reported from laboratories, including ours, to be produced by species of *F. sporotrichioides*, *F. poae* and *F. tricinctum*. To our knowledge these species do not produce 8-keto-trichothecenes (B-type, Figure 2). The main producer of the B-type trichothecenes seems to be *F. graminearum*.

Trichothecenes produced by *F. sporotrichioides* have been found occurring naturally in various substrates: T-2 toxin in maize, safflower, sorghum and mixed feed and diacetoxyscirpenol in maize, safflower and mixed feed, in many countries including the United States, Canada, India, West Germany, France, Scotland and Finland.

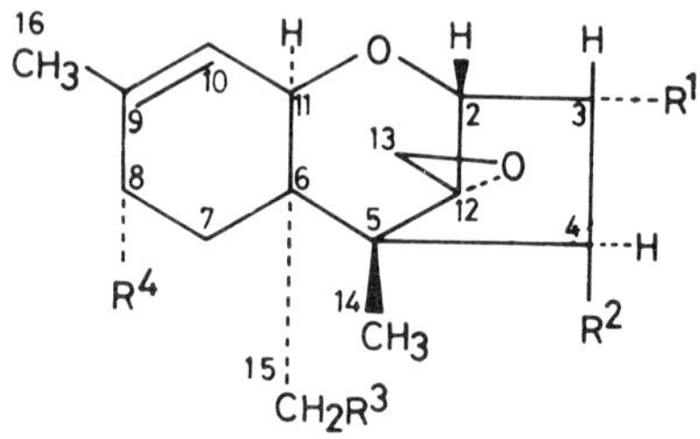

Compound	R^1	R^2	R^3	R^4
Acetyl T-2 toxin	$OCOCH_3$	$OCOCH_3$	$OCOCH_3$	$OCOCH_2CH(CH_3)_2$
T-2 toxin	OH	$OCOCH_3$	$OCOCH_3$	$OCOCH_2CH(CH_3)_2$
HT-2 toxin	OH	OH	$OCOCH_3$	$OCOCH_2CH(CH_3)_2$
8-Acetylneosolaniol	OH	$OCOCH_3$	$OCOCH_3$	$OCOCH_3$
Neosolaniol	OH	$OCOCH_3$	$OCOCH_3$	OH
NT-1	OH	$OCOCH_3$	OH	$OCOCH_3$
NT-2	OH	$OCOCH_3$	OH	OH
T-2 tetraol	OH	OH	OH	OH
Diacetoxyscirpenol	OH	$OCOCH_3$	$OCOCH_3$	H

Figure 1 Trichothecenes produced by *F. sporotrichioides*.

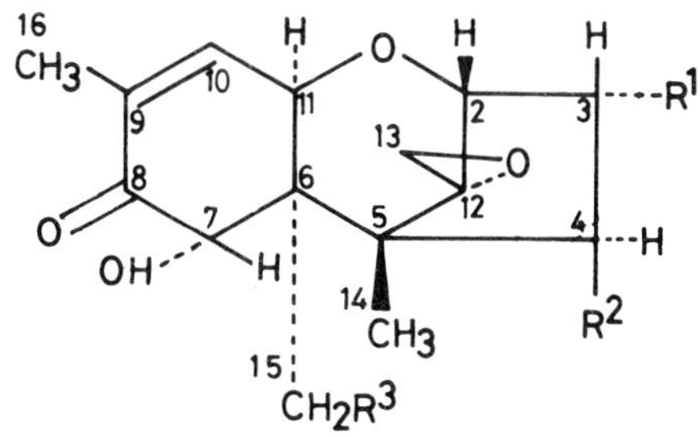

Compound	R^1	R^2	R^3
Nivalenol	OH	OH	OH
Fusarenon-X	OH	$OCOCH_3$	OH
Deoxynivalenol	OH	H	OH
3-Acetyldeoxynivalenol	$OCOCH_3$	H	OH

Figure 2 8-Keto-trichothecenes (B-type trichothecenes).

DECOMPOSITION OF TRICHOTHECENES

Chemically trichothecenes are fairly stable. The question of whether trichothecenes are decomposed in the environment has been studied by examining the microbial degradation of T-2 toxin. We isolated soil bacteria capable of assimilating the toxin and studied the pathway for transformation of the toxin. One such bacterium was *Curtobacterium* sp. strain 114-2. When this isolate was cultured in a medium containing T-2 toxin, HT-2 toxin and T-2 triol were formed as the amount of T-2 toxin decreased. HT-2 toxin was formed prior to T-2 triol. T-2 tetraol, which lacks all the acyl groups of T-2 toxin, was not detected in the medium. Thus the pathway for bacterial transformation of the toxin is believed to be through HT-2 toxin to T-2 triol but not to T-2 tetraol (Figure 3; Ueno *et al.*, 1983). We also conducted a degradation study using fungi isolated from air. In a medium containing T-2 toxin as a carbon source, one isolate of *Epicoccum* transformed it to neosolaniol and a strain of *Alternaria* sp. transformed it to the 15-acetate of T-2 tetraol (4-deacetylneosolaniol) through HT-2 toxin (Figure 4). In addition, one isolate of *Fusarium* sp. acetylated the toxin to acetyl T-2 toxin.

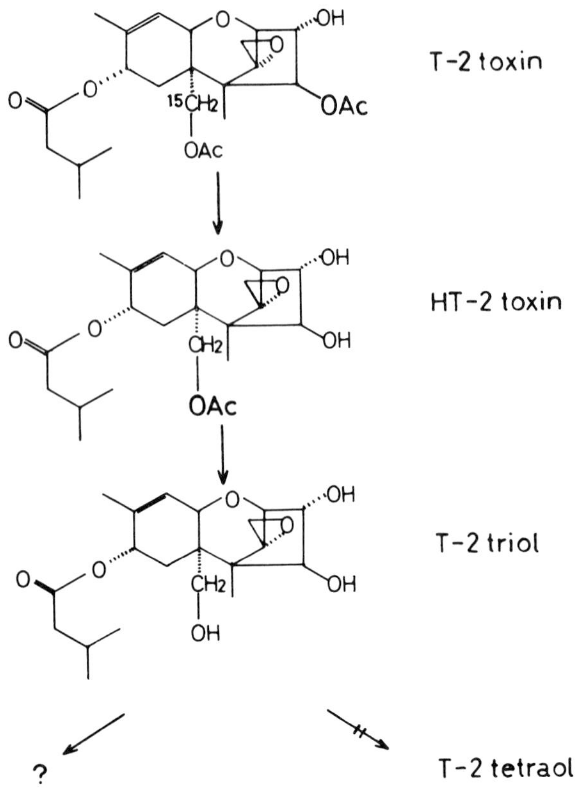

Figure 3 Bacterial transformation of T-2 toxin.

Table 1 LD_{50} values (mg/kg) of trichothecenes

	Mice				Rats			
Compounds	i.v.	i.p.	s.c.	p.o.	i.v.	i.p.	s.c.	p.o.
T-2 toxin		5.2		10.5				5.2
HT-2 toxin		9.2						
Diacetoxyscirpenol	12.0	23.0			1.3	0.75		7.3
8-Acetylneosolaniol						1.0		
Neosolaniol		14.0						
NT-1		28.5						
Nivalenol		4.1					0.9	
Fusarenon-X	3.4	3.4	4.2	4.5			0.5	4.4
Deoxynivalenol		70.0		46.0				
3-Acetyldeoxynivalenol		49.0		34.0				

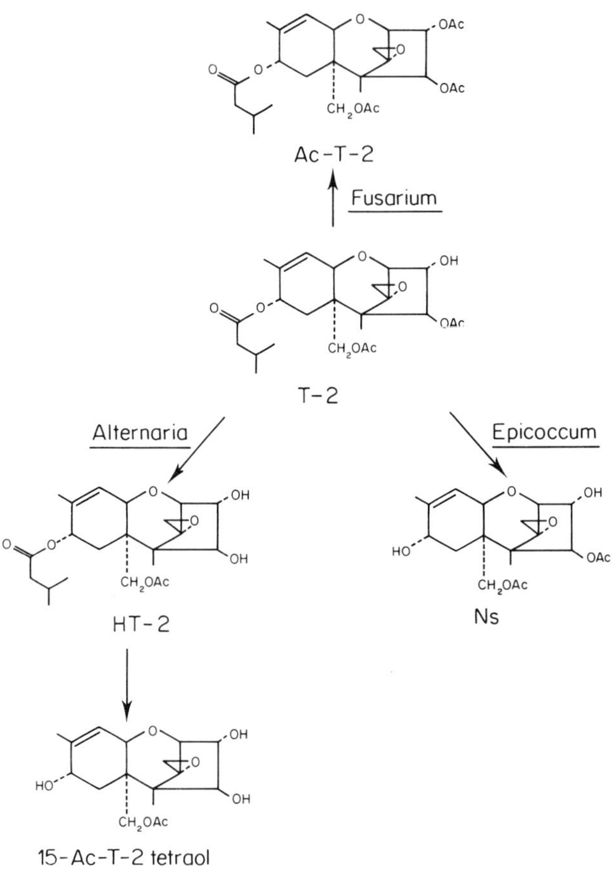

Figure 4 Transformation of T-2 toxin by fungi.

TOXICITY OF TRICHOTHECENES

The acute toxicitiy of trichothecenes has been studied in many laboratories using laboratory and farm animals such as mice, rats, guinea pigs, rabbits, cats, dogs, pigs, ducklings, chicks and trout. Half lethal doses of many trichothecenes have been determined with mice, making it possible to compare their toxicities. Table 1 shows the LD_{50} values in mice and rats of trichothecenes produced by *F. sporotrichioides* and four B-type toxins. The most toxic trichothecenes of *F. sporotrichioides* in mice is T-2 toxin (5.2 mg/kg, i.p.) and this is as toxic as nivalenol and fusarenon-X. Deoxynivalenol is now thought to be an important mycotoxin because many instances of contamination of foodstuff by this toxin have been reported. This toxin is

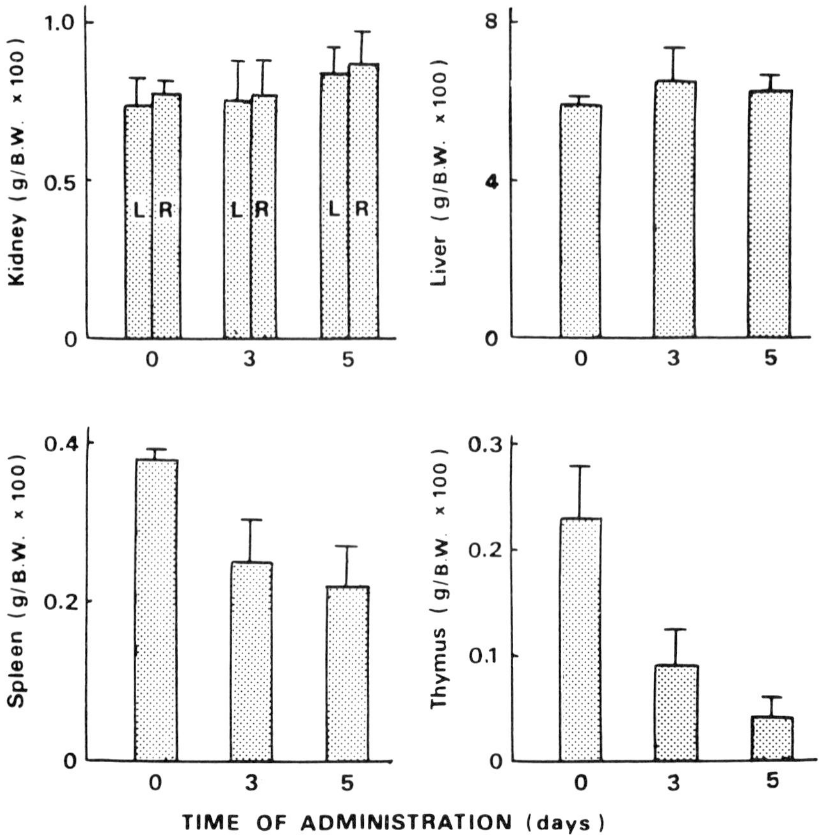

Figure 5 Effect of T-2 toxin on the tissue weight in mice. T-2 toxin was orally administered to 6-week-old male mice at a dose of 3 mg/kg/day for 3 and 5 days. Each value is an average and SD of four to eight mice.

much less toxic (70 mg/kg, i.p.) than T-2 toxin. In animals dosed with trichothecenes, cellular damage to the thymus, bone marrow, testis, ovary and intestine, has been found (Ueno *et al.*, 1972a).

Figures 5 and 6 show the subacute effects of T-2 toxin on the tissue weights and blood cell counts in mice (Sato *et al.*, 1978). The toxin was administered orally at a dose of 3 mg/kg for 3 and 5 days. The weights of spleen and thymus were markedly decreased while those of liver and kidneys were unchanged (Figure 5). The weight of thymus was decreased to 39 and 17% of that of control mice on days 3 and 5, respectively. Haematological analysis revealed that the numbers of white blood cells and platelets were decreased to 75 and 38% of controls, respectively, at day 3; and 22 and 21%, respectively, at day

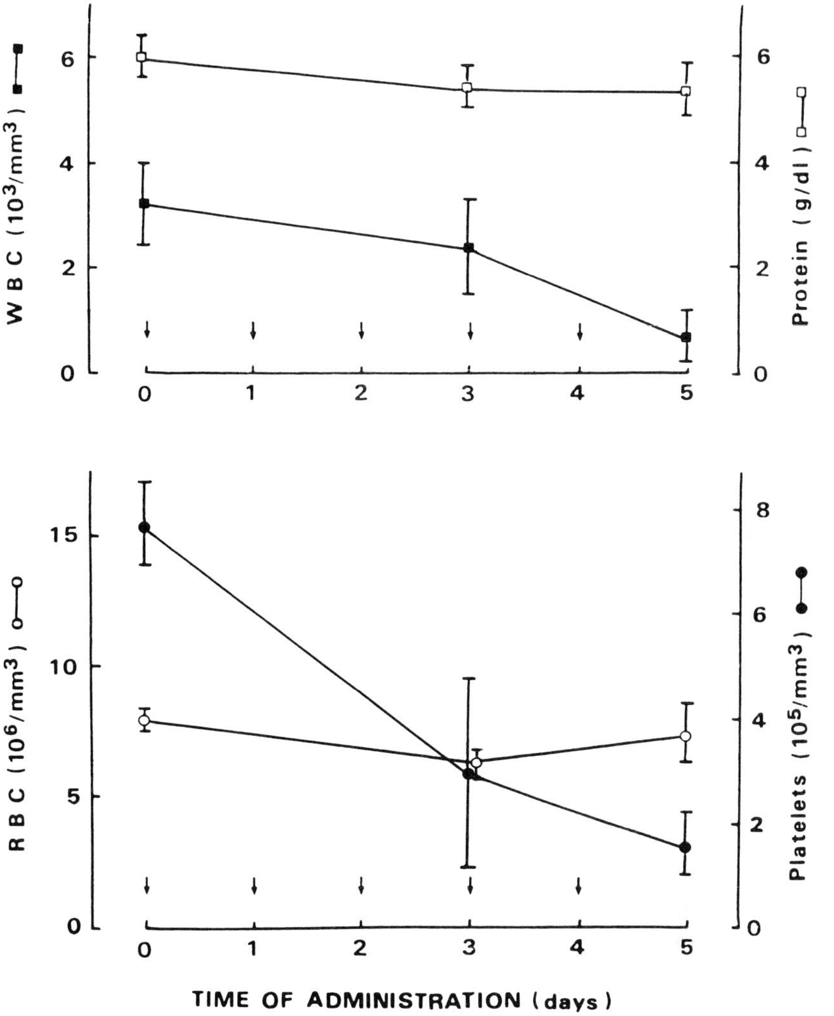

Figure 6 Effect of repeated administration of T-2 toxin on blood cell counts and plasma protein contents. The toxin was orally administered to male mice at a dose of 3 mg/kg/day for 3 and 5 days.

5, while numbers of red blood cells and plasma protein levels were unchanged (Figure 6).

Characteristic biological activities of trichothecenes are a skin necrotization and vomiting. Minimum effective doses of trichothecenes for skin reaction and for causing vomiting are shown in Table 2. T-2 toxin, HT-toxin and diacetoxyscirpenol induce a skin reaction on the back of a guinea pig at a

Table 2 Minimum effective doses for inducing skin necrotization and vomiting

Trichothecenes	Skin necrotization (μg)		Vomiting (mg/kg, s.c.)
	Guinea pigs	Mice	Ducklings
T-2 toxin	0.2	1	0.1
HT-2 toxin	0.2	1	0.1
Diacetoxyscirpenol	0.2	10	0.2
Neosolaniol	1	10	0.1
Nivalenol	10	100	1.0
Fusarenon-X	1	10	0.4
Deoxynivalenol		(75)*	13.5
3-Acetyldeoxynivalenol		(>1000)*	13.5

( )* Data obtained with rats.

dose of 0.2 μg. B-type trichothecenes are less active and 75 μg of deoxynivalenol is required for a reaction. Trichothecenes from *F. sporotrichioides* are more effective (0.1–0.2 mg/kg, s.c.) in causing vomiting in ducklings than are B-type trichothecenes. T-2 toxin is more than a hundred times as active as deoxynivalenol, although the latter toxin is called 'vomitoxin'.

Trichothecenes are cytotoxic to eukaryotic cells, including those of fungi, plants and animals (Ueno, 1983). The difference in activity between the

Table 3 Cytotoxicity of trichothecenes to cultured cells[a]

Trichothecenes	IC_{50} (μg/ml)		
	HeLa	HEK	HL
Acetyl T-2 toxin	1.0	0.8	0.03
T-2 toxin	0.01	0.02	<0.0003
HT-2 toxin	0.01	0.1	0.01
Neosolaniol	0.1	0.06	0.05
Diacetoxyscirpenol (DAS)	0.01	0.01	<0.001
Monoacetoxyscirpenol	0.1	0.1	0.3
Nivalenol	0.3	1.0	0.3
Fusarenon-X	0.1	1.0	0.3
Deoxynivalenol	1.0	3.0	0.5
Monoacetyldeoxynivalenol	10	10	10

[a] Data from Tanaka *et al.* (1977).

Table 4 Inhibition of protein synthesis in rabbit reticulocytes by trichothecenes and LD_{50} values in mice

Toxins	IC_{50} (μg/ml)	LD_{50} (mg/kg, i.p.)
T-2 toxin	0.01	5.2
HT-2 toxin	0.03	9.0
Neosolaniol	0.25	14.5
NT-1	0.25	28.5
NT-2	0.1	—[a]
T-2 triol	1.0	108.0
15-Ac-T-2 tetraol	2.0	—[a]
T-2 tetraol	2.5	—[a]

[a] Not examined.

trichothecenes from *F. sporotrichioides* and those of the B-type are also observed in cytotoxicity. Table 3 shows the IC_{50} values of trichothecenes in three lines of cultured cells reported by Tanaka *et al.* (1977).

Biochemically trichothecenes inhibit protein synthesis in eukaroytes (Ueno, 1983). We used a reticulocyte bioassay (Ueno *et al.*, 1969) to compare their activity in inhibiting protein synthesis. In Table 4, IC_{50} values in the bioassay and i.p. LD_{50} values in mice for the trichothecenes from *F. sporotrichioides* and for the microbial degradation products of T-2 toxin are shown. This table suggests that toxicity is related to the number of acyl groups, decreasing as the number of acyl groups decreases.

REFERENCES

Bamburg, J. R., Rigg, N. V., and Strong, F. M. (1968). 'The structures of toxins from two strains of *Fusarium tricinctum*', *Tetrahedron*, **24**, 3329–3336.

Ichinoe, M. (1978). 'Classification of mycotoxin-producing *Fusarium* species', *Proc. Jpn. Assoc. Mycotoxicol.*, **8**, 1–5.

Ishii, K., and Ueno, Y. (1981). 'Isolation and characterization of two new trichothecenes from *Fusarium sporotrichioides* strain M-1-1', *Appl. Env. Microbiol.*, **42**, 541–543.

Mirocha, C. J., and Pathre, S. V. (1973). 'Identification of toxic principle in a sample of poaefusarin', *Appl. Microbiol.*, **26**, 719–724.

Sato, N., Ito, T., Kumada, H., and Ueno, Y. (1978). 'Toxicological approaches to the metabolites of *Fusaria*. XIII. Hematological changes in mice by a single and repeated administrations of trichothecenes', *J. Toxicol. Sci.*, **3**, 335–356.

Tanaka, T., Matsuda, Y., Toyazakai, N., Ogawa, K., Matsui, Y., and Ueno, Y. (1977). 'Screening of trichothecene-producing *Fusarium* species from river sediments by mammarian cell culture techniques', *Proc. Jpn. Assoc. Mycotoxicol.*, **6**, 50–53.

Ueno, Y. (1983). 'General toxicology', in *Trichothecenes* (ed. Y. Ueno), pp. 135–146. Elsevier, Amsterdam.

Ueno, Y., Hosoya, M., and Ishikawa, Y. (1969). 'Inhibitory effects of mycotoxin on the protein synthesis in rabbit reticulocytes', *J. Biochem. (Tokyo)*, **66**, 419–422.

Ueno, Y., Ishii, K., Sakai, K., Kanaeda, S., Tsunoda, H., Tanaka, T., and Enomoto, M. (1972a). 'Toxicological approaches to the metabolites of *Fusaria*. IV. Microbial survey on "bean-hulls poisoning of horses" with the isolation of toxic trichothecenes, neosolaniol and T-2 toxin of *Fusarium solani* M-1-1', *Jpn. J. Exp. Med.*, **42**, 187–203.

Ueno, Y., Nakayama, K., Ishii, K., Tashiro, F., Minoda, Y., Omori, T., and Komagata, K. (1983). 'Metabolism of T-2 toxin in *Curtobacterium* sp. strain 144-2', *Appl. Env. Microbiol.*, **46**, 120–127.

Ueno, Y., Sato, N., Ishii, K., Sakai, K., and Enomoto, M. (1972b). 'Toxicological approaches to the metabolites of *Fusaria* V. Neosolaniol, T-2 toxin and butenolide, toxic metabolites of *Fusarium sporotrichioides* NRRL 3510 and *Fusarium poae* 3287', *Jpn. J. Exp. Med.*, **42**, 461–472.

Trichothecenes and Other Mycotoxins
Edited by J. Lacey

CHAPTER 29

The structure of Phomopsin A, a mycotoxin produced by *Phomopsis leptostromiformis* (Kühn) Bubák ex Lind

John A. Edgar, Claude C. J. Culvenor, John L. Frahn, Alan J. Jones, Charles P. Gorst-Allman, Walter F. O. Marasas, Pieter S. Steyn, Robert Vleggaar and Philippus L. Wessels

INTRODUCTION

Lupinosis is a disease of sheep and cattle grazing lupin stubbles (Gardiner, 1967). It is characterized by jaundice and yellow or ochre-colouring, very fatty livers which are atrophic and cirrhotic in the more chronic stages, and anorexia. Cells in the liver also show a colchicine-like arrest of mitosis (Gardiner and Petterson, 1972) and the liver becomes characteristically non-fatty and nodular in chronic lupinosis (Marasas, 1974). Death may occur within a few days of the appearance of the disease.

Investigation of outbreaks of lupinosis in Germany in the 1870s first suggested a fungal involvement in the aetiology of the disease (Zurn, 1879). It appeared in Western Australia around 1950 and in South Africa in 1954, and independent studies in both countries provided conclusive evidence that the fungus *Phomopsis leptostromiformis* (Kühn) Bubák ex Lind was responsible (Marasas, 1974). The principal mycotoxin, phomopsin A, was subsequently isolated in a crystalline state from *P. leptostromiformis* grown on lupin seed (Culvenor *et al.*, 1977) or liquid media (Lanigan *et al.*, 1979) and shown to reproduce the field disease when given to sheep (Culvenor *et al.*, 1977; Jago *et al.*, 1982). A collaborative study has now established a cyclic hexapeptide structure (**1**) for phomopsin A based on ^{1}H and ^{13}C nuclear magnetic resonance spectroscopy (NMR), mass spectrometry (MS) and hydrolysis (Culvenor *et al.*, 1983). This chapter will describe, in particular, the results of the hydrolytic studies which, within tight constraints set by MS and NMR data, led to the identification of the atypical amino acids present in the

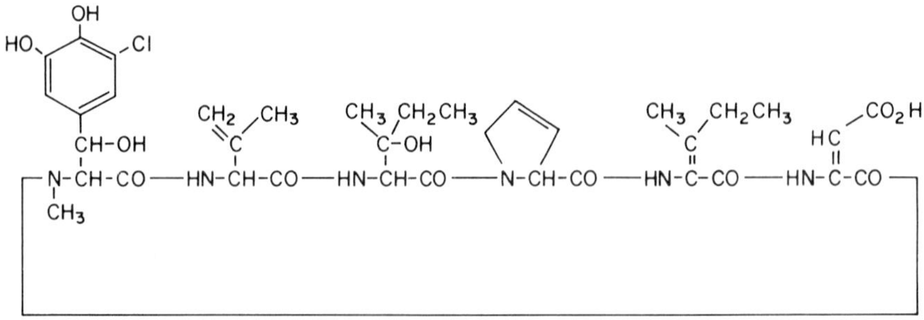

(1)

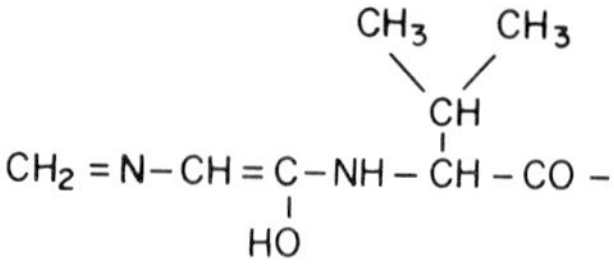

(2)

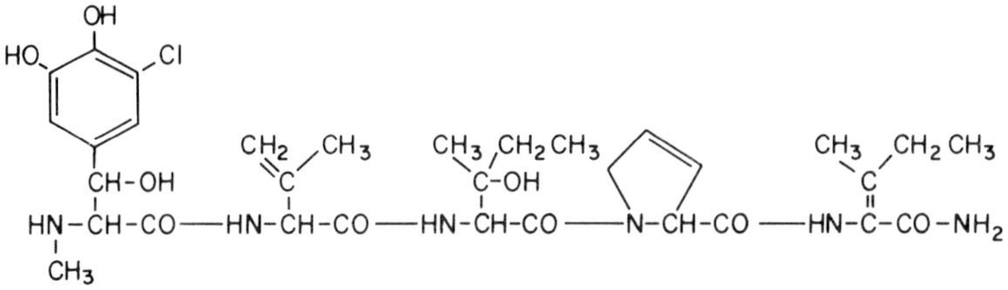

(3)

mycotoxin. A brief summary of the extensive NMR heteronuclear ^{13}C-[^{1}H] selective population inversion (SPI) results will also be presented.

Following its isolation (Culvenor *et al.*, 1977), attempts to determine the structure of phomopsin A by largely physical methods made only limited progress (Frahn *et al.*, 1983). Extensive ^{1}H and ^{13}C NMR data provided considerable knowledge of the functional groups and carbon, hydrogen groupings but little information on the molecular skeleton. Mass spectra, obtained under field desorption conditions, yielded ions which were observable for only an extremely short period (seconds) when critical conditions of desorption were established. This work gave an apparent molecular weight of 770/772 which, when combined with NMR and microanalytical data, suggested an elemental composition $C_{36}H_{43}N_{6}O_{11}Cl$ (Frahn *et al.*, 1983). Positive and negative ion fast atom bombardment MS, however, gave a molecular weight of 788/790 based on $[M+H]^+$, m/z 789/791; $[M-H]^-$, m/z 787/789;

M+Na]$^+$, m/z 811/813 and other confirmatory fragment ions. An exact mass measurement of the [M+H]$^+$, ^{35}Cl isotope, ion established an elemental composition of $C_{36}H_{45}N_6O_{12}Cl$ (found: 789.28232, requires 789.28618) for phomopsin A. Further support for the $C_{36}H_{45}N_6O_{12}Cl$ formulation came from ^{252}Cf plasma desorption mass spectrometry (we are grateful to Dr H. M. Fales for these results) which gave a series of Na and K adduct ions corresponding to loss of CO_2 from the molecular ion. The measured mass of the [M–CO_2+Na]$^+$ ion, m/z 767.28 (^{35}Cl isotope), was in good agreement with the calculated value, viz. m/z 767.278.

Early acid hydrolysis studies had resulted in the identification of oxalacetic acid (1 mole), a chlorodihydroxybenzaldehyde (1 mole), ammonia (2 moles) and methylamine (0.25 mole) but failed to produce substantive evidence of amino acids (Frahn *et al.*, 1983). An important advance came with the realization that labile 2,3-dehydroamino acids may be present in phomopsin A and recognition of small amounts of the amino acids glycine and valine and other, non-protein, amino acids in acid hydrolysates (6 M HCl, 110 °C 24 h). This led to a detailed capillary gas chromatography–mass spectrometry (GC–MS) analysis of the amino acids present in acid hydrolysates of phomopsin A and its reduction products.

The amino acids obtained on hydrolysis of phomopsin A (Figure 1, Table 1) were chromatographed on an OV101 SCOT column and a chiral phase (RSL-007) WCOT column as their *N*-trifluoroacetyl-*n*-butyl esters. Identification involved MS comparison and co-chromatography with authentic samples except in the case of the dehydrovalines and dehydroisoleucines which were identified initially from their mass spectra and subsequently by hydrogenation to the parent amino acids. The relative yield of the various amino acids (Figure 1, Table 1) indicated that reactions other than simple hydrolytic cleavage of peptide bonds were occurring. Only 3,4-dehydroproline was obtained in the expected yield, and therefore only it could be considered a unit of the mycotoxin molecule.

Sodium borohydride ($NaBH_4$) reduction of phomopsin A prior to acid hydrolysis resulted in the appearance of aspartic acid (from reduction of a 2,3-dehydroaspartic acid unit) and a mixture of isoleucine and alloisoleucine (from reduction of 2,3-dehydroisoleucine) as well as the amino acids obtained on hydrolysis of the unreduced mycotoxin. The aspartic acid and isoleucine/alloisoleucine were obtained in amounts equal to the yield of 3,4-dehydroproline. Thus 2,3-dehydroaspartic acid and 2,3-dehydroisoleucine were established along with 3,4-dehydroproline as units of the phomopsin A molecule—the former being the source of the oxalacetic acid seen in our earlier hydrolyses.

Catalytic hydrogenation (PtO_2/H_2) of phomopsin A followed by hydrolysis yields DL-valine, DL-isoleucine, L-proline and DL-aspartic acid in the ratio 1 : 1 : 1 : 1 (Figures 2a and 2b). The two dehydroisoleucines are still present

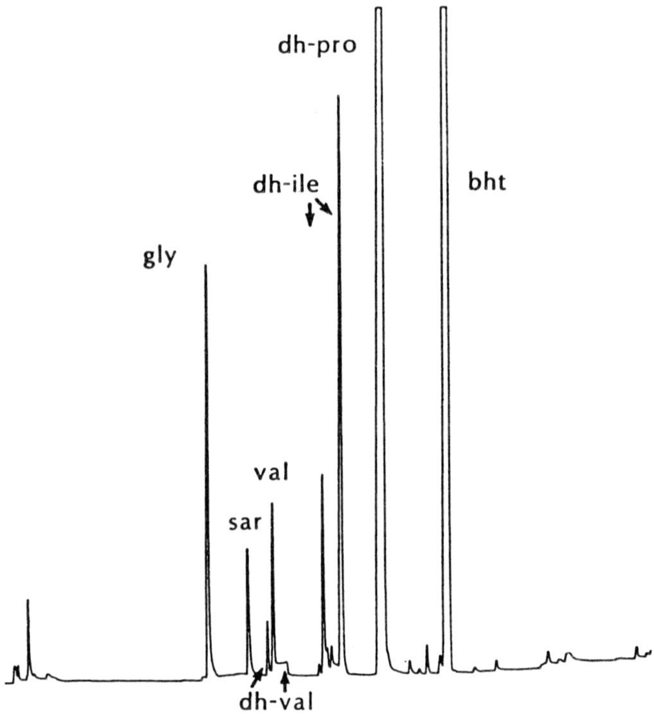

Figure 1 Capillary gas chromatogram (90 m, SCOT, OV101) of the *N*-trifluoroacetyl-*n*-butyl esters of the amino acids obtained on acid hydrolysis (6 M HCℓ, 24 h, 110 °C) of phomopsin A. The abbreviations are: gly, glycine; sar, sarcosine; val, valine; dh-val, dehydrovaline; dh-ile, dehydroisoleucine; dh-pro, 3,4-dehydroproline; bht, 2,5-ditertiarybutylhydroxy-toluene.

Table 1 Amino acids produced on hydrolysis of phomopsin A

Amino acid	Relative yield[a] (%)
Glycine	36
Sarcosine	11
DL-Valine	18
Dehydroisoleucine (a)	20
Dehydroisoleucine (b)	44
3,4-Dehydro-L-proline	100

[a] Relative to 3,4-dehydroproline.

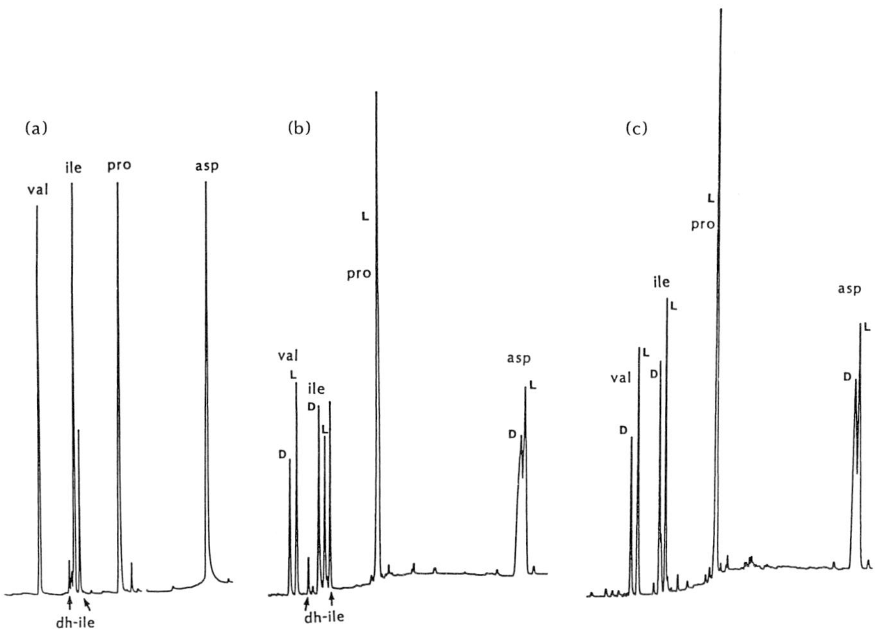

Figure 2 Capillary gas chromatograms of the *N*-trifluoroacetyl-*n*-butyl esters of (a) the amino acids obtained on hydrolysis of catalytically reduced (PtO_2/H_2) phomopsin A chromatographed on a 90 m SCOT OV101 column; (b) the same sample on a 25 m WCOT RSL-007 chiral phase column; and (c) after reduction (PtO_2/H_2) of the dehydroisoleucine components (25 m WCOT RSL-007 chiral phase column). The abbreviations are: val, valine; ile, isoleucine; pro, proline; asp, aspartic acid; dh-ile, dehydroisoleucine.

but the sarcosine and glycine seen previously are no longer detected in the hydrolysate. The presence of a 3,4-dehydrovaline subunit in phomopsin A, which is not reduced by sodium borohydride and undergoes racemization during catalytic hydrogenation, is in accord with these results. 3,4-dehydrovaline also contains the $C{=}CH_2$ grouping clearly indicated in ^1H NMR spectra. Unexplained, however, is the apparent instability of the 3,4-dehydrovaline unit during hydrolysis of phomopsin A and $NaBH_4$-reduced phomopsin A. The formation of DL-isoleucine and the virtual absence of allo-isoleucine (Figure 2) indicates that the E form of 2,3-dehydroisoleucine unit is present in phomopsin A. The two dehydroisoleucines, seen in the hydrolysates even after catalytic reduction (Figure 2), can be attributed to loss of water from a 3-hydroxyisoleucine subunit (or similar elimination) during hydrolysis. The L configuration of the α-carbon atom of the 3-hydroxyisoleucine was established by subsequent catalytic reduction of these dehydroisoleucines to L-isoleucine and a small amount of L-alloisoleucine (Figure 2c).

The phenylserine unit was deduced from two thus-far unexplained hydrolysis products, viz. the chlorodihydroxybenzaldehyde and the sarcosine. Aldol condensation of these molecules would yield the missing structural element (based on the $C_{36}H_{45}N_6O_{12}Cl$ formula) and their formation under hydrolytic conditions could therefore be attributed to an acid-catalysed retro-Aldol condensation. Support for this came from recognition of a chlorodihydroxyphenylpyruvic acid tetra-TMS (M^+, m/z 518, and appropriate fragmentation) among the products of conc. HCl hydrolysis. Such a product would arise by dehydration of the phenylserine unit to *N*-methyl-2,3-dehydro-3-(chlorodihydroxyphenyl)alanine and hydrolysis. Methylamine, previously recognized as a hydrolysis product (Frahn *et al.*, 1983), would also be produced. The substitution pattern of the aromatic ring in phomopsin A was established by selective population inversion (SPI) NMR experiments as 3-chloro-4,5-dihydroxy and MS and capillary GC comparison of an authentic sample of 3-chloro-4,5-dimethoxybenzaldehyde with the dimethyl ether of the product from phomopsin A showed them to be identical.

The origin of the glycine found in hydrolysates of phomopsin A (Figure 1) and $NaBH_4$-reduced phomopsin A, but absent in catalytically reduced phomopsin A (Figure 2a), warrants further investigation. Glycine could arise from the 3-hydroxyisoleucine unit via a retro-aldol mechanism, however this does not explain the complete absence of glycine in hydrolysates of catalytically reduced phomopsin A in which 3-hydroxyisoleucine is still present. An alternative pathway to glycine involves *N*-demethylation of the phenylserine-derived sarcosine. Sarcosine is not likely to undergo demethylation under the hydrolytic conditions involved, but demethylation may, in this case, be facilitated by electron movement from the phenylserine subunit to the adjacent dehydrovaline during hydrolysis. The resulting disproportionation could lead to formation, *inter alia*, of partial structure (**2**), hydrolysis of which would not only result in formation of glycine but also account for the low yield of the normally acid-stable 3,4-dehydrovaline and the presence of a substantial amount of DL-valine in acid hydrolysates of phomopsin A (Figure 1). Formation of glycine (and sarcosine) would be prevented after Pt-catalysed hydrogenation, during which both the dehydrovaline and the aromatic ring of the phenylserine are reduced and the electron movement proposed can no longer occur.

The amino acid units identified as described above, viz. 2,3-dehydroaspartic acid, (E)-2,3-dehydroisoleucine, 3,4-dehydro-L-proline, 3-hydroxy-L-isoleucine, 3,4-dehydrovaline and *N*-methyl-3-(3-chloro-4,5-dihydroxyphenyl)serine, fully account for the elemental composition $C_{36}H_{45}N_6O_{12}Cl$.

AROMATIC RING SUBSTITUTION AND AMINO ACID SEQUENCE

Of the extensive 500 MHz NMR studies made in Pretoria, only the briefest summary can be presented here. This work completed and confirmed the correlation of proton resonances with the signals of proton-bearing carbon atoms. Then a highly effective use was made of heteronuclear ^{13}C-[^{1}H], SPI experiments to establish the interconnection between groupings and thus the substitution pattern of the aromatic ring and the amino acid sequence (Culvenor *et al.*, 1983).

The SPI results defined the partial sequences 3,4-dehydroproline → 2,3-dehydroisoleucine → 2,3-dehydroaspartic acid and *N*-methyl-3-(3-chloro-4,5-dihydroxyphenyl) serine → 3,4-dehydrovaline → 3-hydroxyisoleucine. The linking of 2,3-dehydroaspartic acid to the substituted phenylserine follows from evidence related to a degradation product of phomopsin A, phomopsinamine (**3**). Thus mild acid hydrolysis (2 M HCl, room temperature, 24 h) of phomopsin A results in the selective loss of dehydroaspartic acid as oxalacetic acid and formation of the amine phomopsinamine (**3**) in which a two-bond coupling is observed between NH and N-CH_3 in the 500 MHz spectrum. The remaining link in the cyclic hexapeptide is by elimination, 3-hydroxyisoleucine → 3,4-dehydroproline.

STEREOCHEMISTRY AND CONFORMATION

The configurations of several chiral centres remain to be determined, as do the detail of strong intramolecular hydrogen bonding indicated in the NMR spectra. These factors may provide a rationale for certain experimental results which are not readily explained by the structure (**1**). These difficulties relate particularly to one of the phenolic OH groups. If phomopsin A is methylated with diazomethane, hydrolysed and the hydrolysate trimethylsilyated, a monomethyl monotrimethylsilyl derivative of 3-chloro-4,5-dihydroxybenzaldehyde (M^+, m/z 258) is formed. The inability to derivatize both phenolic OHs until after hydrolysis suggests that one of the phenolic OH groups in phomopsin A has reduced acid character, is extremely sterically hindered or is involved in bonding with a group elsewhere on the macrocyclic ring. Similarly, phomopsinamine ($[M-H_2O+Na]^+$ m/z 697/699, base peak m/z 171) is converted to a monomethyl derivative with diazomethane ($[M-H_2O+Na]^+$ m/z 711/713. base peak m/z 185) involving only one of the two phenolic OH groups. Phomopsin A (**1**) and phomopsinamine (**3**) also fail to complex with borate as would be expected if two ortho OH groups were present. Their eletrophoretic mobilities at pH 9.2 are unchanged whether the buffer is borate or carbonate. These details of structure may only be unequivocally established when suitable crystals become available for X-ray crystallography.

Note added in proof:

The structure of phomopsin A has recently been found to be linear (Edgar *et al.*, 1985). The amide bond between the phenylserine amino group and the dehydroaspartic acid, shown in (1), is not present.

REFERENCES

Culvenor, C. C. J., Beck, A. B., Clarke, M., Cockrum, P. A., Edgar, J. A., Frahn, J. L., Jago, M. V., Lanigan, G. W., Payne, A. L., Peterson, J. E., Petterson, D. S., Smith, L. W., and White, R. R. (1977). 'Isolation of toxic metabolites of *Phomopsis leptostromiformis* responsible for lupinosis', *Aust. J. Biol. Sci.*, **30**, 269–277.

Culvenor, C. C. J., Cockrum, P. A., Edgar, J. A., Frahn, J. L., Gorst-Allman, C. P., Jones, A. J., Marasas, W. F. O., Murray, K. E., Steyn, P. S., Vleggaar, R., Wessels, P. L. (1983). 'Structure elucidation of phomopsin A, a novel cyclic hexapeptide mycotoxin produced by *Phomopsis leptostromiformis*', *J. Chem. Soc., Chem. Commun.*, 1259–1262.

Edgar, J. A., Frahn, J. L., Cockrum, P. A. and Culvenor, C. C. J. (1985). 'Lupinosis. The chemistry and biochemistry of the phomopsins,' in *Mycotoxins and Phycotoxins* (eds. P. S. Steyn and R. Vleggaar) in press. Elsevier, Amsterdam.

Frahn, J. L., Jago, M. V., Culvenor, C. C. J., and Edgar, J. A. (1983). 'Chemical and biological properties of phomopsin, a toxic metabolite of *Phomopsis leptostromiformis*', *Toxicon*, Suppl. 3, 149–152.

Gardiner, M. R. (1967). 'Lupinosis', *Adv. Vet. Sci.*, **11**, 85–138.

Gardiner, M. R. and Petterson, D. S. (1972). 'Pathogenesis of mouse lupinosis induced by a fungus (*Cytospora* spp.) growing on dead lupins', *J. Comp. Path.*, **82**, 5–13.

Jago, M. V., Peterson, J. E., Payne, A. L., and Campbell, D. G. (1982). 'Response of sheep to varying doses of phomopsin', *Aust. J. Exp. Biol. Med. Sci.*, **60**, 239–251.

Lanigan, G. W., Payne, A. L., Smith, L. W., Wood, P.McR., and Petterson, D. S. (1979). 'Phomopsin A production by *Phomopsis leptostromiformis* in liquid media', *Appl. Environ. Microbiol.*, **37**, 289–292.

Marasas, W. F. O. (1974). '*Phomopsis leptostromiformis*', in *Mycotoxins* (ed. I. F. H. Purchase), pp. 111–127. Elsevier, Amsterdam.

Zurn, F. A. (1879). 'Massenerkrankung von Schafen durch den Genuss befallener Lupinen', *Vort. Thierärzte*, **7**, 251–277.

Trichothecenes and Other Mycotoxins
Edited by J. Lacey

CHAPTER 30

Recent advances in facial eczema (pithomycotoxicosis) research

B. L. Smith

INTRODUCTION

The disease facial eczema (pithomycotoxicosis) occurs in New Zealand, Australia and South Africa. It is caused by the fungus, *Pithomyces chartarum* (Berk. & Curt.) M. B. Ellis, which grows on pasture litter and produces the toxin sporidesmin. However, there has recently been considerable interest (Kellerman *et al.*, 1980; E. R. Aycardi, personal communication) in the possibility that *P. chartarum* may be associated with plants other than ryegrass to produce hepatogenous photosensitizations of ruminants which have a pathology differing slightly from that of facial eczema.

In New Zealand, facial eczema is one of the most important diseases of farmed livestock, especially ruminants. The disease varies in severity from year to year, causing major losses to the New Zealand livestock industry in usually 1 year in every 5. One such year was 1981, when the loss was estimated to be between $40 million and $100 million, or about 0.25% of New Zealand's GNP (Smith, 1983). The gravity of these losses is magnified by the fact that over 60% of New Zealand's export earnings are derived from farmed ruminants.

With this single mycotoxicosis causing such economic loss it is not surprising that a considerable research effort is mounted to combat facial eczema. Although the effort has been diluted somewhat by the recent economic recession and the current interest in ryegrass staggers there are still nine scientists working at least part-time on the disease in New Zealand. Excellent reviews of this field have recently been published (Mortimer *et al.*, 1978b,c; Mortimer and Ronaldson, 1983). The purpose of this chapter is to highlight certain areas where recent advances have been made.

GENETICS

It has been known for some time that the Merino is much more resistant to facial eczema than Romneys, Border Leicesters or their crosses (Smith *et al.*,

1980). This has since been confirmed using Merino crosses and both natural and experimental challenges (A. G. Campbell, personal communication; H. H. Meyer, personal communication).

Using Romney sheep, Campbell *et al.*, (1981) have exploited the considerable variation in susceptibility between individuals. They have shown by progeny testing that it is possible to select rams for their ability to reproduce resistant or susceptible lambs. Using the data from 1455 progeny it has been estimated that the heritability is 0.42 ± 0.09. A scheme for increasing the tolerance of Romney sheep by breeding from performance-tested rams is now being implemented in New Zealand.

IMMUNOLOGY

Jonas and co-workers have shown (Jonas and Ronaldson, 1974; Jonas and Erasmuson, 1977) that injection of complexes of sporidesmin or sporidesmin-like molecules with carriers (polylysine, bovine gamma globulin or various bacteria) can stimulate the formation of antibodies against sporidesmin in laboratory animals. In one experiment a hapten–bacteria complex was used successfully to immunize mice and significantly reduce a sporidesmin-induced bilirubinaemia (Jonas and Erasmuson, 1979).

In a recent Ruakura experiment antibodies were produced in sheep in response to a sporidesmin–bovine thyroglobulin complex but no protection against sporidesmin toxicity could be demonstrated (R. J. Fairclough, personal communication).

TOXICITY OF PASTURE

High-performance liquid chromatography has been used successfully to detect sporidesmin in washings from pasture samples (R. J. Fairclough, personal communication). Sporidesmin concentrations in sample washings correlated well with spore numbers during periods of rising and peak spore densities. Sporidesmin concentrations declined more quickly than spore numbers, confirming suspicions from field observations that old spores remaining on pastures after peak spore densities are less toxic than newly formed spores.

TOXICITY OF SPORIDESMIN

Recently Munday (1982) has thrown new light on the mechanism of toxicity of sporidesmin. The disulphide bridge of sporidesmin is essential for toxicity of the molecule, for example its cytotoxicity (Taylor, 1971). Munday (1982) has shown that autoxidation of the dithiol form of sporidesmin generates superoxide radical (O_2^-) and that superoxide may be further generated *in vivo* by the cyclic reduction/autoxidation of sporidesmin which is linked to glutathione and other thiols.

Munday (personal communication) has further identified a number of metals which will inhibit superoxide production from reduced sporidesmin and when sporidesmin is coupled with glutathione. However, no metals appeared more likely than zinc to be useful for the control of facial eczema in the field.

POTENTIATION AND THE KINETICS OF SPORIDESMIN DISPOSITION

Consecutive daily doses of extremely small amounts of sporidesmin (0.004 mg sporidesmin per kg per dose for 6 days of each week for 8 weeks; B. L. Smith, unpublished results) have been known for some time to be capable of causing considerable liver injury to sheep. It has now been shown (E. Manns, personal communication) that small doses of sporidesmin can potentiate the effect of later larger doses of sporidesmin. The effect lasts for up to about 3 weeks but its action has yet to be studied in detail under controlled conditions.

When small doses of sporidesmin are given to sheep on three consecutive days, peak sporidesmin concentrations in bile occur 1.5 to 5 hours after dosing, and decline to basal levels before 24 hours. Although not yet shown to be related to or causing the phenomenon of potentiation, it has been observed that on the third day animals susceptible to sporidesmin poisoning show a much greater maximum sporidesmin concentration in bile (Fairclough and Smith, 1983).

PREVENTING SPORIDESMIN TOXICITY

Several approaches to protecting laboratory animals and farmed ruminants have been made. Stimulation of the hepatic mixed-function oxidase system in sheep using hexachlorobenzene has been shown (Mortimer *et al.*, 1978a) to decrease liver injury caused by sporidesmin or by a natural field spore challenge. Likewise Munday *et al.* (1983) have shown that the antioxidant, butylated hydroxytoluene, has a liver injury sparing effect and that this, combined with the protective effect of zinc, produces almost complete protection. However these two substances and several like them have no practical application, mainly because they leave undesirable residues in meat.

The demonstration of the protective effect of zinc against sporidesmin poisoning in rats (Towers, 1977) followed by sheep (Smith *et al.*, 1977) and dairy cattle (Towers and Smith, 1978) has led to considerable research aimed at the control of the disease. The dose rates required to prevent the disease are high (20–30 times daily requirements) and concern has been expressed that the safety margins for some methods of administration and for some forms of zinc are insufficient (Smith *et al.*, 1979). However although some areas of concern have been identified (Smith, 1981) it is now felt, in the absence of better methods of control, that the advantages of using zinc for facial eczema control outweigh the risks and the alternative of clinical facial eczema. Current methods of zinc administration involve hand-dosing a zinc oxide slurry, spraying zinc oxide on pasture and administering zinc sulphate via drinking water to dairy cattle (Smith *et al.*, 1983).

It has been suggested that smaller amounts of zinc may be sufficient to prevent the potentiation caused by small amounts of sporidesmin. This has been tested and it has been found that the full dose of zinc is required to prevent potentiation (B. L. Smith, unpublished results).

Zinc has been shown to prevent the greater bile sporidesmin concentrations which occur on the third of three consecutive daily sporidesmin doses, but not the initial rise in bile sporidesmin after the first sporidesmin dose. There is indirect evidence that zinc may prevent sporidesmin toxicity by forming mercaptides with the two sulphur atoms of sporidesmin, thus preventing superoxide generation (R. Munday, personal communication). Munday has also suggested that zinc may act by removing 'free' copper from the superoxide-generating site where it is an effective catalyst, thus inhibiting the toxic action of sporidesmin. The prevention of zinc's prophylactic effect in sheep by the concurrent administration of copper has, in a pilot experiment (B. L. Smith and R. Munday, unpublished results), supported this hypothesis.

REFERENCES

Campbell, A. G., Meyer, H. H., Henderson, H. W., and Wesselink, C. (1981). 'Breeding for facial eczema resistance—a progress report'. *Proc. N.Z. Soc. Anim. Prod.*, **41**, 273–278.

Fairclough, R. J., and Smith, B. L. (1983). 'Sporidesmin in the bile of sheep resistant or susceptible to sporidesmin dosing', *Proc. N.Z. Soc. Anim. Prod*, **43**, 213–215.

Jonas, W. E., and Erasmusen, A. F. (1977). 'Immunological studies of sporidesmin: Production of antibodies to azo-linked derivatives of 2-amino-5-chloro-3,4-dimethoxy benzyl alcohol', *N.Z. Vet. J.*, **25**, 161–164.

Jonas, W. E., and Erasmusen, A. F. (1979). 'The effect of immunising mice with a derivative of 2-amino-5-chloro-3,4-dimethoxy benzyl alcohol coupled to some bacteria on sporidesmin-induced bilirubinaemia', *N.Z. Vet. J.*, **27**, 61–64.

Jonas, W. E., and Ronaldson, J. W. (1974). 'The production of rabbit antibodies to sporidesmin', *N.Z. Vet. J.*, **22**, 111–116.

Kellerman, T. S., Van der Westhuizen, G. C. A., Coetzer, J. A. W., Roux Cecilia, Marasas, W. F. O., Minne, J. .A., Bath, G. F., and Basson, P. A. (1980). 'Photosensitivity in South Africa. II. The experimental production of the ovine hepatogenous photosensitivity disease Geeldikkop (*Tribulosis ovis*) by the simultaneous ingestion of *Tribulus terrestris* plants and cultures of *Pithomyces chartarum* containing the mycotoxin sporidesmin', *Onderstepoort J. Vet. Res.*, **47**, 231–61.

Mortimer, P. H., Manns, E., and Coe, B. D. (1978a). 'Manipulation of liver metabolism in relation to ruminant toxicology', *Proc. N.Z. Soc. Anim. Prod.*, **38**, 59–64.

Mortimer, P. H., and Ronaldson, J. W. (1983). 'Fungal toxin induced photosensitization', in *Handbooks of Natural Toxins*, Vol. 1: *Plant and Fungal Toxins* (eds A. T. Tu and R. F. Keeler), Ch. 11, pp. 361–419. Marcel Dekker, New York.

Mortimer, P. H., di Menna, M. E., and White, E. P. (1978b). 'Pithomycotoxicosis "facial eczema" in cattle', in *Mycotoxic Fungi, Mycotoxins, Mycotoxicoses—an encyclopaedic handbook*, Vol. 2 (eds T. D. Wyllie and L. G. Morehouse), pp. 63–72. Marcel Dekker, New York.

Mortimer, P. H., White, E. P., and di Menna, M. E. (1978c). 'Pithomycotoxicosis "facial eczema" in sheep', in *Mycotoxic Fungi, Mycotoxins, Mycotoxicoses—an encyclopaedic handbook*, Vol. 2 (eds T. D. Wyllie and L. G. Morehouse), pp. 195–203. Marcel Dekker, New York.

Munday, R. (1982). 'Studies on the mechanism of toxicity of the mycotoxin, sporidesmin. I. Generation of superoxide radical by sporidesmin', *Chem-biol. Inter.*, **41**, 361–374.

Munday, R., Manns, E., and Mortimer, P. H. (1983). 'Effects of antioxidants on the toxicity of the facial eczema toxin, sporidesmin, in sheep', *Proc. N.Z. Soc. Anim. Prod.*, **43**, 209–212.

Smith, B. L. (1981). 'Special features of zinc toxicity', in *Zinc Toxicity and its Pathology in Relation to Facial Eczema Prophylaxis in Ruminants*, pp. 70–152. D Phil. thesis, University of Waikato, New Zealand.

Smith, B. L. (1983). 'The significance of mycotoxicoses in New Zealand', in *Proc. VIII Congress Internat. Soc. Hum. Anim. Mycol. (ISHAM)*, pp. 508–511.

Smith, B. L., Embling, P. P., and Pearce, M. G. (1983). 'Zinc sulphate in the drinking water of lactating dairy cows for facial eczema control'. *Proc. N.Z. Soc. Anim. Prod.*, **43**, 217–219.

Smith, B. L., Embling, P. P., Towers, N. R., Wright, D. F., and Payne, E. (1977). 'The protective effect of zinc sulphate in experimental sporidesmin poisoning of sheep', *N.Z. Vet. J.*, **25**, 124–127.

Smith, B. L., Reynolds, G. W., and Embling, P. P. (1979). 'Effect of method of oral administration of zinc sulphate on acute zinc toxicity in the sheep'. *N.Z. J. Exp. Agric.*, **7**, 107–110.

Smith, B. L., Stanbridge, T. A., and Embling, P. P. (1980). 'Sheep breed differences in pentobarbitone sleeping time and response to experimental sporidesmin intoxication', *N.Z. Vet. J.*, **28**, 35–36.

Taylor, A. (1971). 'The toxicology of sporidesmins and other epipolythiadioxopiperazines', in *Microbial Toxins*, Vol. VII (eds S. Kadis, A. Ciegler and S. J. Ajl), pp. 337–376. Academic Press, New York and London.

Towers, N. R. (1977). 'Effect of zinc on the toxicity of the mycotoxin, sporidesmin, to the rat', *Life Sci.*, **20**, 413–418.

Towers, N. R., and Smith, B. L. (1978). 'The protective effect of zinc sulphate in experimental sporidesmin intoxication of lactating dairy cows', *N.Z. Vet. J.*, **26**, 199–202.

Trichothecenes and Other Mycotoxins
Edited by J. Lacey

CHAPTER 31

Resistance to metabolic conversion of the epoxide group in trichothecenes

Angelika R. Bieger and Klaus P. Dose

INTRODUCTION

Fusaria and other trichothecene-producing fungi infect important foodstuffs, for example maize and other cereals. They have consequently become associated with human and animal intoxications throughout the world (Ueno, 1977). Important trichothecene-producing fungi are, e.g. *Fusarium* (Hsu *et al.*, 1972), *Verticimonosporium* (Eppley *et al.*, 1977), *Stachybotrys* (Eppley, *et al.*, 1977), *Trichothecium* (Godtfredsen and Vangedal, 1965) and *Cylindrocarpon* (Matsumoto *et al.*, 1977; Minato *et al.*, 1975).

Naturally occurring trichothecenes have a tetracyclic structure (Figure 1). Typical is the olefinic bond at C-9,10 and the epoxy group at C-12,13. These compounds may be characterized as 12,13-epoxytrichothecenes. Ueno (1977) has divided the known naturally occurring trichothecenes into four groups according to their chemical characters.

Variation of the substituents can substantially affect the toxicity of trichothecenes. After reductive removal of the epoxide group by chemical methods, however, the biological activity becomes completely lost (Patterson, 1973). Reports of the metabolic conversion of the epoxide group of trichothecenes are contradictory: Patterson (1973) has claimed that this group is converted by liver enzymes whereas Nakamura *et al.* (1977) emphasize its stability.

Such a resistance to metabolic conversion would be very remarkable because most naturally occuring epoxides are highly reactive. In mammals the conversion of aromatic and olefinic compounds often yields the intermediate formation of toxic epoxides. But these are usually quickly converted by enzymes such as glutathione-S-transferase, epoxide hydrolase and epoxide reductase.

Glutathione-S-transferases catalyse the addition of aliphatic epoxides and arene oxides to glutathione (Oesch *et al.*, 1980).

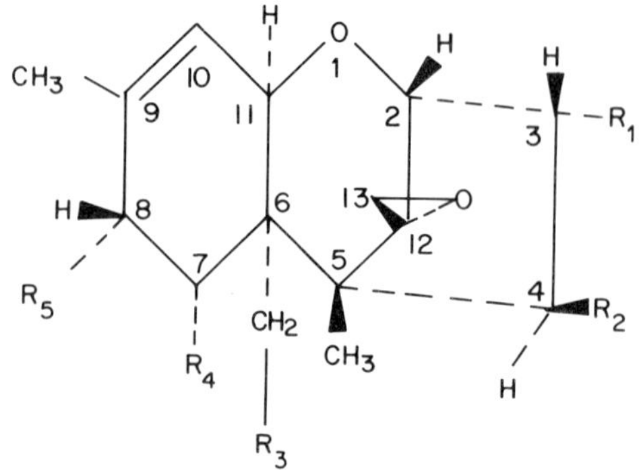

Trichothecene	R_1	R_2	R_3	R_4	R_5
T-2 toxin	OH	OAc	OAc	H	$OCOCH_2CH(CH_3)_2$
HT-2 toxin	OH	OH	OAc	H	$OCOCH_2CH(CH_3)_2$

Figure 1 T-2 and HT-2 toxin.

Epoxide hydrolases convert epoxides into 1,2-dihydrodiols (Nakamura *et al.*, 1977; Lu *et al.*, 1975).

The third group of enzymes, the vitamin K_1 epoxide reductases, play at least an essential role in metabolizing epoxides related to vitamin K_1 epoxide (Siegfried, 1978). In view of the significance of epoxide conversion for the understanding of molecular toxicology we have re-evaluated the stability of the trichothecene epoxide group towards enzymic conversion by liver enzymes.

EXPERIMENTAL PROCEDURES

The glutathione-S-transferase test

Materials

1-Chloro-2,4-dinitrobenzene (DNCB: 1 mM in 0.1 M potassium phosphate, pH 6.5 with 2% ethanol) was obtained from Sigma Chemical Co. GmbH, Munich, Germany. T-2 and HT-2 toxin were isolated from mouldy rice as described by Schmidt *et al.* (1981).

Test solutions

T-2 toxin: 1 mM in the sample; HT-2 toxin: 1 mM in the sample. Both toxins in crude or purified form; glutathione (GSH), 1 mM in 0.1 mm potassium phosphate pH 6.5 with 2% dimethylsulphoxide (DMSO). Various preparations of glutathione-S-transferase isolated from rat liver (Habig *et al.*, 1974) and kindly donated by Professor Dr F. Oesch and Dr T. Friedberg of the Pharmacological Institute, University, Mainz.

Procedure

The specific activity was 20 mM/min/mg. The protein concentration was measured by the method of Lowry *et al.* (1951). A typical test system with DNCB as substrate was set up as follows: 1 mM GSH, 1 mM DNCB, 0.1 mg enzyme in the sample; test volume: 2.2 ml 0.1 M potassium phosphate pH 6.5. Enzyme activity was determined by monitoring changes in absorbance (by a Perkin Elmer Spectrophotometer at 340 nm and 25 °C) caused by the conjugation of DNCB with GSH. A complete assay mixture without enzyme was used as a control (Habig *et al.*, 1974). At ΔE/min. = 0.025 the assay is linear. When trichothecenes were used as competitive substrates the concentrations in the test were as follows: 1 mM GSH, 1 mM DNCB, enzyme (0.20 mM/min/mg), 1 mM trichothecene in 0.1 mM potassium phosphate buffer with 2% DMSO.

The epoxide hydrolase test

The epoxide hydrolase activity was determined by applying a radiometric assay based on conversion of ^{3}H-styrene oxide to ^{3}H-styrene glycol as described by Oesch *et al.* (1971) at 37 °C. Only the inhibition of ^{3}H-styrene oxide conversion in the presence of T-2 toxin was investigated.

The vitamin K_1 epoxide reductase test

Vitamin K_1 epoxide reductase preparations have been isolated by Liptay (1982). The activity of the enzyme was measured as follows: 1 ml microsomal fraction, 2 mM dithiothreitol (buffer system: 0.025 M imidazole, 0.25 M sucrose, 0.08 M KCl; pH 7.2), 0,100 or 200 nmol T-2 toxin in 40 µl acetonitrile. Start of incubation: addition of 20 nmol ^{3}H − vitamin K_1 epoxide; incubation time: 4 min at 25 °C, stopping of incubation: addition of 5 ml isopropanol/*n*-hexane (4 : 6). The fraction with ^{3}H − vitamin K_1 epoxide was measured in a Zinsser scintillation counter (Isocap 3000).

Thin-layer chromatography (TLC)

T-2 toxin and HT-2 toxin were detected by TLC and subsequent fluorescence intensity measurement of T-2 toxin on the H_2SO_4-treated TLC plates as described before (Schmidt *et al.*, 1981).

RESULTS AND DISCUSSION

The data summarized in Table 1 indicate that the two trichothecenes do not compete with 1-chloro-2,4-dinitrobenzene for the active site of glutathione-S-transferases under the conditions of the test. In addition we have incubated extracts from mouldy rice containing T-2 toxin and HT-toxin with glutathione-S-transferases and 1 mM GSH for 1 h and for 24 h at 25 °C. The content of T-2 toxin (HT-2 toxin was not analysed) in the extracts was not changed during the incubation (see also Table 2).

Table 1 Conversion of 1-chloro-2,4-dinitrobenzene by glutathione-S-transferases in the absence and presence of trichothecenes (at 25 °C in 0.1 M potassium phosphate at pH 6.5; 1 mM GSH)

Substrate	$\triangle E$/min
1mM 1-chloro-2,4-dinitrobenzene	0.024
1mM 1-chloro-2,4-dinitrobenzene plus 1 mM T-2 toxin	0.0237
1 mM 1-chloro-2,4-dinitrobenzene plus 1 mM HT-2 toxin	0.024

Table 2 Quantitative TLC analysis of T-2 toxin on incubation in a GSH-S-transferase system, (same conditions as in Table 1)

Enzyme	Incubation (h)	T-2 toxin (ng/ml)
−	1	3500
+	1	3500
−	24	3500
+	24	3500

Tables 1 and 2 show that the epoxide group of T-2 toxin is not attacked by rat liver glutathione-S-transferases and cannot be conjugated with GSH under our experimental conditions. These results agree with those of Nakamura *et al.* (1977) in that a transfer of trichothecenes to GSH does not occur.

These findings are in disagreement with the previously published observations of Foster *et al.* (1975), who concluded that the glutathione-S-transferase test system may be a possible enzymatic assay for trichothecene mycotoxins in animal foodstuffs.

Our data also suggest that the two trichothecene epoxides are not bound to a catalytic or regulatory centre of glutathione-S-transferases, and that reduced glutathione does not react with T-2 or HT-2 toxin.

In a related series of experiments the interaction of trichothecenes with epoxide reductase activity from rat liver was tested. The data summarized in Table 3 clearly indicate that there is no interaction between T-2 toxin and the epoxide reductase in these preparations. In addition it could be shown that T-2 toxin exhibits no interaction with epoxide hydrolases from rat liver inasmuch as the conversion of ^{3}H-styrene oxide to ^{3}H-styrene glycol is not affected by 100 μM to 1 mM T-2 toxin. The toxin is therefore neither an inhibitor nor a substrate of rat liver epoxide hydrolases.

Table 3 Action of T-2 toxin on vitamin K_1 epoxide reductase activity (substrate: vitamin K_1 epoxide)

T-2 toxin nmol $\times$ ml^{-1}	Vitamin K_1 epoxide reductase activity (standard deviation of five experiments) nmol $\times$ min^{-1} $\times$ (g liver)$^{-1}$	
0	0.27	±0.04
100	0.27	±0.01
200	0.28	(only one experiment)

Thus, it must be concluded that the epoxide group in trichothecenes is extraordinarily resistant to metabolic degradation by rat liver enzymes.

ACKNOWLEDGEMENT

We are are grateful to Dr C. R. Wolf and Ms I. Liptay for having performed the epoxide hydrolase and the vitamin K_1 epoxide reductase tests at the Pharmacological Institute of the University of Mainz. We thank Professor Dr F. Oesch from the same institute for all his ideas and technical support.

REFERENCES

Eppley, R. M., Mazzola, E. P., Highet, R. J., and Bailey, W. J. (1977). 'Structure of satratoxin H, a metabolite of Stachybotrys atra', *J. Org. Chem.*, **42**, 240–243.

Foster, P. M. D., Slater, T. F., and Patterson, D. S. P. (1975). 'A possible enzymatic assay for trichothecene mycotoxins in animal feedstuffs, *Biochem. Soc. Trans.*, **3**, 875–878.

Godtfredsen, W. O., and Vangedal, S. (1965). 'Trichodermin, a new sesquiterpene antibiotic', *Acta Chem. Scand.*, **19**, 1088–1102.

Habig, W. H., Pabst, M. J., and Jakoby, W. B. (1974). 'Glutathione-S-transferases', *J. Biol. Chem.*, **249**, 7130–7139.

Hsu, I. C., Smalley, E. B., Strong, F. M., and Rubelin, W. E. (1972). 'Identification of T-2 toxin in mouldy corn associated with lethal toxicosis in dairy cattle', *Appl. Microbiol.*, **24**, 684–690.

Liptay, I. (1982). 'Rolle der DTT-abhängigen mikrosomalen Epoxid-Reduktase-Aktivität im Stoffwechsel fremdstoffabgeleiteter Epoxide', Diploma thesis, University of Mainz.

Lowry, O. H., Rosenbrough, N. J., Farr, A. L., and Randall, R. J. (1951). 'Protein measurement with the Folin phenol reagent', *J. Biol. Chem.*, **193**, 265–275.

Lu, A. Y. H., Ryan, D., Jerina, D. M., Daly, J. W., and Levin, W. (1975). 'Liver microsomal epoxide hydratase solubilisation, purification and characterization', *J. Biol. Chem.*, **250**, 8283–8288.

Matsumoto, M., Minato, H., Tori, K., and Ueyama, M. (1977). 'Structures of isororidin E, epoxyisororidin E, and epoxy-and diepoxyroridin H, new metabolites isolated from Cylindroncarpon species', *Tetrahedron Lett.*, **47**, 4093–4096.

Minato, H., Katayama, T., and Tori, K. (1975). 'Vertisporin, a new antibiotic from *Verticimonosporium diffractum*', *Tetrahedron Lett.*, **45**, 2579–2582.

Nakamura, Y., Ohata, M., and Ueno, Y. (1977). 'Reactivity of 12,13-epoxytrichothecenes with epoxide hydrolase, glutathione-S-transferase and glutathione', *Chem. Pharm. Bull.*, **25**, 3410–3414.

Oesch, F., Jerina, D. M., and Daly, J. (1971). 'A radiometric assay for hepatic epoxide hydrase activity with 7-[^{3}H]styrene oxide', *Biochim. Biophys. Acta*, **227**, 685–691.

Oesch, F., Tegtmeyer, F., Kohl, F.-V., Rüdiger, H., and Glatt, H. R. (1980). 'Interindividual comparison of epoxide hydratase and glutathione-S-transferase activities in cultured human fibroblasts', *Carcinogenesis*, **1**, 305–309.

Patterson, D. S. P. (1973). 'Mycotoxins: metabolism and liver injury', *Biochem. Soc. Trans.*, **1**, 917–922.

Schmidt, R., Bieger, A., Ziegenhagen, E., and Dose, K. (1981). 'Bestimmung von T-2 Toxin in pflanzlichen Nahrungsmitteln 1. T-2 in verschimmeltem Reis und Mais', *Fresenius Z. Anal. Chem.*, **308**, 133–136.

Siegfried, C. M. (1978). 'Solubilization of vitamin K epoxide reductase and vitamin K dependent carboxylase from rat liver microsomes', *Biochem. Biophys. Res. Commun.*, **83**, 1488–1495.

Ueno, Y. (1977). 'Trichothecenes: overview address', in *Mycotoxins in Human and Animal Health* (eds J. V. Rodricks, C. W. Hesseltine and M. A. Mehlmann), pp. 189–207. Pathotox Publications, Park Forest South, IL.

Trichothecenes and Other Mycotoxins
Edited by J. Lacey

CHAPTER 32

Some biochemical responses to PR toxin, a mycotoxin from *Penicillium roquefortii*

Ru-Dong Wei, Yan-Hwa Wu Lee and Yau-Huei Wei

INTRODUCTION

PR toxin is a mycotoxin produced from certain cultures of *Penicillium roquefortii* (Wei *et al.*, 1973; Wei and Liu, 1978, 1982) that have been used in food processing since AD 500 (Matheson, 1921). The chemical structure (Wei *et al.*, 1975) (Figure 1) and absolute configuration (Baert *et al.*, 1980) of the toxin are known. The toxin is lethal to rats and mice by either oral, i.p. or i.v. administration (Table 1). Recently, several biochemical parameters from the tissues of the PR toxin-treated animals were examined in our laboratories (Chen *et al.*, 1982). PR toxin produced acute toxic effects via an increase of capillary permeability, which may lead to severe systemic dehydration and a decrease in blood volume, and direct damage of the lungs, heart, liver and kidneys.

Soon after the structure of the toxin had been identified, Wei *et al.*, (1974) demonstrated that PR toxin has an unusually broad spectrum of inhibition, inhibiting DNA, RNA and protein biosynthesis in both animals and isolated

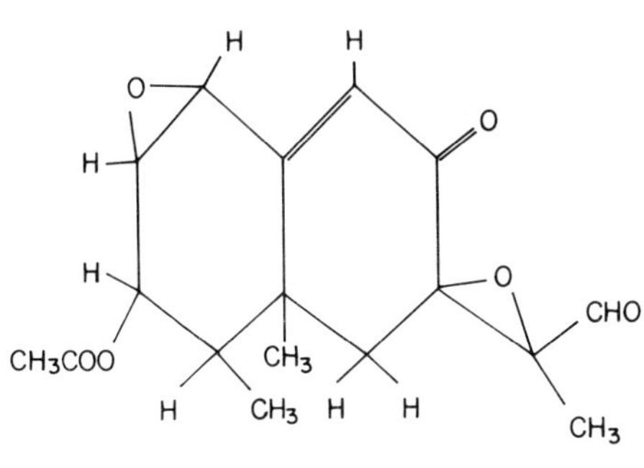

Figure 1 Structure of PR toxin.

Table 1 LD_{50} values for PR toxin in mice and rats

Test animal	Route	LD_{50} (mg/kg)	Reference
Rat, albino, male and both sexes	i.p.	11–14.5	Wei *et al.*, 1973; Polonelli *et al.*, 1978
Rat, Sprague-Dawley, male	i.v.	8.2	Chen *et al.*, 1982
Rat, albino, male	oral	115	Wei *et al.*, 1973
Mouse, ICR, male	i.p.	5.8	Wei *et al.*, 1976; Chen *et al.*, 1982
Mouse, C57, male and female	i.p.	1–2	Arnold *et al.*, 1978
Mouse, Swiss-Webster, male and female	i.p.	3.5–4.6	Arnold *et al.*, 1978
Mouse, C57, male and female	oral	72–100	Arnold *et al.*, 1978

cells. PR toxin at the concentration of 3.12×10^{-5} M inhibited the uptake of ^{14}C-thymidine, ^{3}H-uridine and ^{3}H-valine in the cultures of Ehrlich ascites tumour cells without any observed lag phase (Wei *et al.*, 1976; Wei and Chang, 1976). The molecular mechanisms of PR toxin inhibition on transcription and translation processes have been thoroughly investigated (Moulé *et al.*, 1976, 1977a, 1978). Recent studies indicate that PR toxin affects the genetic material by induction of DNA protein cross-links (Moulé *et al.*, 1980), formation of single-strand breaks, inhibition of replication and repair processes (Aujard *et al.*, 1979a,b). This chapter reports our recent findings on some biochemical responses of PR toxin.

MATERIAL AND METHODS

PR toxin

PR toxin was prepared as described previously (Wei *et al.*, 1973) and its purity was checked by high-performance liquid chromatography (Moreau *et al.*, 1979). The toxin was dissolved in DMSO (Sigma) and diluted in the same solvent to the desired concentrations before use.

Preparation of rat liver DNA polymerases

DNA polymerase β was purified from rat liver as described by Chang (1973). This procedure omitted gel filtration and partially purified the DNA polymerase β by chromatography on denatured DNA-agarose columns after separating DNA polymerase β from α and γ on DEAE-cellulose following a phosphocellulose column.

DNA polymerases α and γ were first purified by the same phosphocellulose and DEAE-cellulose columns as described above for the DNA polymerase β. Then, polymerase α was further resolved from DNA polymerase γ by hydroxyapatite chromatography as described previously (Mechali *et al.*, 1979). DNA polymerases α and γ eluted from this hydroxyapatite column were further purified through chromatography on denatured DNA-agarose column.

The specific activities of DNA polymerases α, β and γ with activated DNA, $(dA)_n \cdot (dT)_{12-18}$ and $(rA)_n \cdot (dT)_{12-18}$ as template primers were 60, 1824 and 4560 units per mg protein. One unit of enzyme activity is defined as 1 nmole of radioactive deoxyribonucleotide incorporated per hour. The purified DNA polymerases have similar properties to those reported for α, β and γ forms with regard to the pH optimum, template-primer specificity, sedimentation values and the sensitivity to *N*-ethylmaleimide and salts (Chang, 1973; Mechali *et al.*, 1979; Weissbach *et al.*, 1975; Smith and Gallo, 1972; Robert-Guroff *et al.*, 1977; Chang, 1976; Lewis *et al.*, 1974; Bolden *et al.*, 1977).

Assay of DNA polymerase α

The reaction mixture (50 μl) contained 50 mM Tris-HCl (pH 7.5), 10 mM $MgCl_2$, 0.5 mM dithiothreitol, 20 mM NaCl, 4 μg activated DNA, 0.1 mM concentrations of dATP, dGTP, dCTP and ^{3}H-dTTP (150 cpm/pmole), and the enzyme fraction. Incubation was carried out at 37 °C for 10–30 min.

Assay of DNA polymerase β

The reaction mixture (50 μl) contained 50 mM Tris-HCl (pH 8.0), 0.6 mM $MnCl_2$, 0.5 mM dithiothreitol, 1 μg of $(dA)_n \cdot (dT)_{12-18}$, 0.1 mM of ^{3}H-dTTP (150 cpm/pmole), 60 mM NaCl and the enzyme fraction. Incubation was carried out at 30 °C for 10–30 min. When activated DNA was used as template, the assay condition was similar to assay of DNA polymerase α except that 50 mM Tris-HCl (pH 8.8) and 0.1 N NaCl were contained in the assays.

Assay of DNA polymerase γ

The reaction mixture (50 μl) contained 50 mM Tris-HCl (pH 8.5), 20 mM KPO_4 (pH 8.5), 0.6 mM $MnCl_2$, 0.5 mM dithiothreitol, 1 μg of $(rA)_n \cdot (dT)_{12-18}$′ 0.1 N NaCl, 0.1 mM of (^{3}H)dTTP (150 cpm/pmole) and the enzyme fraction. Incubation was carried out at 30 °C for 10–30 min.

All DNA polymerase reactions were terminated by the addition of two drops of saturated sodium pyrophosphate and 2 ml of 10% cold trichloroacetic acid. The acid-insoluble radioactivity was measured as described previously (Lewis *et al.*, 1974).

Effects of PR toxin on DNA polymerases

To determine the effects of PR toxin on DNA polymerase activities, an appropriate amount of PR toxin was added to the assay mixture containing enzyme and assay salts at 4 °C before the enzymatic reaction started. All assays were performed in duplicate. For the control assays the same volume of DMSO was added instead of the PR toxin solution.

Preparations of mitochondria and the electron transfer complexes

Rat liver mitochondria were prepared according to the method of Vercesi *et al.* (1978), and usually exhibited a respiratory control ratio of 3.6–4.0 and an ADP : O ratio of 1.6 to 1.8 for succinate-supported respiration as determined by the method of Estabrook (1967). Succinate-cytochrome *c* reductase (SCR) and cytochrome *c* oxidase were prepared from mitochondria and assayed according to the methods of Yu *et al.* (1974, 1975) with modifications.

Interactions of PR toxin with mitochondria and the electron transfer complexes

The effect of PR toxin on the various functions of mitochondria was studied by incubation of the toxin with the mitochondria or the isolated enzyme complexes at room temperature for 3 min before activity assay. The perturbation of PR toxin on the optical spectra of cytochrome *b* and cytochrome c_1 in SCR was investigated with a Perkin-Elmer λ-5 double beam UV-visible spectrophotometer at 23 °C.

RESULTS AND DISCUSSION

Inhibition of DNA polymerase activities by PR toxin

To determine whether PR toxin exerted various degrees of inhibition on the activity of different DNA polymerases, experiments were performed with the purified rat liver DNA polymerases α, β and γ in the presence of various concentrations of PR toxin. Table 2 indicated that DNA polymerase α activity was strongly inhibited at PR toxin concentrations as low as 10^{-5} M, whereas DNA polymerase γ was inhibited less extensively. Conversely, DNA polymerase β was not inhibited at this concentration; but was slightly

Table 2 Effect of PR toxin on purified rat liver DNA polymerases α, β and γ

DNA polymerase	Template	Percentage activity remaining in the presence of indicated PR toxin concentrations (M)						
		Control	2×10^{-6}	5×10^{-6}	1×10^{-5}	2×10^{-5}	4×10^{-5}	1×10^{-4}
α	Activated DNA	100 (109.8)	70.3	47.3	25.8	13.9	5.3	4.6
β	Activated DNA	100 (87.0)	—	—	124.6	89.1	53.7	14.3
	$(rA)_n \cdot (dT)_{12\text{-}18}$	100 (103.2)	—	97.6	93.1	75.9	54.5	14.4
γ	$(rA)_n \cdot (dT)_{12\text{-}18}$	100 (283.0)	—	88.0	68.3	36.9	16.8	7.3

The assay conditions for different DNA polymerases under various template primers were described under Materials and Methods. In the assay system for polymerase β using $(rA)_n \cdot (dT)_{12\text{-}18}$ as template primer, the conditions were identical to that described for DNA polymerase γ except 20 mM KPO_4 (pH 8.5) was omitted. The values in parentheses refer to the pmoles of deoxyribonucleotide incorporated during the incubation in the control system.

stimulated. The PR toxin concentration required for 50% inhibition of the activities of DNA polymerase α, β and γ were estimated approximately at 5×10^{-6} M, 4×10^{-5} M and 1.5×10^{-5} M, respectively. It should be noted that the concentration of PR toxin required for 50% inhibition of the various DNA polymerases ranged between 5 and 40 μM, which is of the same order of PR toxin concentration as causes more than 50% inhibition of DNA synthesis in Ehrlich ascites tumour cells and rat liver cells in culture (Wei *et al.*, 1976; Aujard *et al.*, 1979a).

Mechanism of PR toxin inhibition on DNA polymerases

Three possible mechanisms may explain the inhibitory effect of PR toxin on DNA synthesis: (1) direct modification of the template primer itself, (2) direct alteration of the enzyme itself, (3) modification of both enzyme and template primer molecules. In order to test for the above three mechanisms we chose DNA polymerase β as a model enzyme and examined the concentration effects of enzyme and template primer on PR toxin inhibition. Table 3 showed that the inhibition by PR toxin was relieved progressively by the addition of increasing amounts of enzyme to the assay system. This result suggested that PR toxin inhibited the *in vitro* DNA synthesis by alteration of the DNA polymerase itself. Some interesting patterns, however, were revealed in Table 4. With over 16–64-fold variation in the template concentration, the degree of inhibition by PR toxin was altered or remained unchanged, depending on which templates used. Increasing $(rA)_n \cdot (dT)_{12-18}$ concentration from 0.25 μg to 16 μg per assay, reduced the extent of inhibition by PR toxin (4×10^{-5} M) dramatically from 66.4% to 16.6%. On the other hand, addition of increasing amounts of activated DNA or $(dA)_n \cdot (dT)_{12-18}$ did not relieve the inhibition of PR toxin. Thus our data

Table 3 Effect of progressively higher amounts of DNA polymerase β on the inhibition of replication by PR toxin

Amount of enzyme (μl)	Enzymatic activity (pmole/30 min)		Percentage inhibition
	Control	Toxin (4×10^{-5} M) treatment	
2	42.3	2.4	94.4
5	54.9	8.4	84.7
10	71.5	29.9	58.2

The enzymatic activity was determined with activated DNA as template primer.

Table 4 Effect of increasing template concentration on the inhibition of DNA polymerase β activity by PR toxin

Template	Percentage activity of the control at the indicated concentration of template-primer (μg)					
	0.25	0.50	1.0	4.0	8.0	16.0
Activated DNA	—	—	37.1 (46.4)	30.3 (91.2)	38.3 (134.8)	34.1 (185.2)
$(dA)_n \cdot (dT)_{12\text{-}18}$	50.0 (35.5)	48.1 (32.4)	54.8 (39.7)	46.5 (39.4)	51.8 (41.9)	58.3 (41.0)
$(rA)_n \cdot (dT)_{12\text{-}18}$	33.6 (39.3)	43.6 (65.0)	55.1 (80.4)	61.6 (88.2)	62.8 (99.8)	83.4 (117.3)

The assay conditions were similar to that described in Table 2. The final concentrations of PR toxin in the reaction mixtures were 4×10^{-5} M. The values in parentheses refer to the pmoles of deoxyribonucleotide incorporated in the control sample during the incubation of 20 min.

suggest that PR toxin acted preferentially on the synthetic RNA template $(rA)_n \cdot (dT)_{12-18}$ as compared to the DNA template $(dA)_n \cdot (dT)_{12-18}$ or activated DNA. This conclusion also agrees with what is known about the preference of action of PR toxin on DNA or RNA (Moulé *et al.*, 1980).

It is intuitively appealing to think about reasons explaining the higher sensitivity of DNA polymerase α activity to PR toxin inhibition as compared to the DNA polymerases β and γ. Interestingly, the order of their sensitivity to PR toxin inhibition is exactly in agreement with the sensitivity of DNA polymerase to sulphhydryl blocking reagents (Weissbach *et al.*, 1975; Smith and Gallo, 1972; Robert-Guroff *et al.*, 1977). Thus, PR toxins probably inactivate DNA polymerase by masking sulphhydryl groups of the active site as in the case of inhibition of RNA polymerase activity by PR toxin (Tashiro *et al.*, 1979). Verification of this interesting hypothesis, however, may require identification of the bound intermediate between PR toxin and DNA polymerase.

Mitochrondria membrane-destructive effect of PR toxin

In vivo studies have shown that PR toxin by either oral or i.p. administration causes breathing difficulties in rats and mice (Wei *et al.*, 1973, 1976) and decreases the respiration rate in cats (Chen *et al.*, 1982). We made a further study to check the biochemical effect of the toxin on the respiration and oxidative phosphorylation of the rat liver mitochondria. Table 5 shows that both the respiratory control and ADP : O ratio decreased concomitantly in a

Table 5 Effect of PR toxin on the respiration and oxidative phosphorylation of the rat liver mitochrondria

PR toxin concentration (M)	Respiratory control ratio	ADP : O ratio
0	4.0	1.6
6.25×10^{-7}	3.6	1.3
6.25×10^{-6}	3.2	1.1
3.12×10^{-5}	2.2	0.9

The assay system in a total volume of 2 ml contained 0.125 M sucrose, 50 mM KCl, 6 mM $MgCl_2$, 10 mM phosphate, 5 mM Hepes, 20 μM cytochrome *c*, pH 7.2. The mitochrondria (0.4–0.6 mg) were added to the assay mixture and incubated for 3 min at 23 °C in the reaction chamber of the oxygen monitor. The reaction was initiated by addition of 30 μl of 0.6 M succinate and the state 4 respiration was followed for 2 min, and then 20 μl of 20 mM ADP were introduced by a syringe to the reaction chamber to stimulate the respiration to state 3. The ratio between the respiration rates of the two states was calculated from the oxygen consumption curves, and the ADP : O ratio was determined from the same curve according to the method described by Estabrook (1967). For the inhibition experiments the mitochondria were incubated in the reaction chamber with 5 μl of PR toxin with different concentrations at 23 °C for 3 min before activity assay.

dose-dependent manner when PR toxin was added to the assay mixture. About 40% of both functions of the intact mitochondria were lost in the presence of 3.12×10^{-5} M PR toxin. These results indicate that PR toxin is a membrane-destructive compound, which impairs both the structural integrity and the ATP synthetase activity of the rat liver mitochondria.

Inhibitory site of PR toxin on the mitochondria respiratory chain

In order to understand the mechanism of action of PR toxin on mitochondrial respiration we employed the purified SCR and cytochrome *c* oxidase to locate the action site of the toxin. Figure 2 shows that PR toxin exhibited a dose-dependent inhibition on the SCR activity. The LD_{50} value was estimated to be about 2.1×10^{-5} M. The double reciprocal plot as shown in Figure 3 clearly indicates that PR toxin is a non-competitive inhibitor and the inhibitory constant as calculated from the plot was about 2.4×10^{-5} M. The toxin, however, did not show any inhibition of cytochrome oxidase activity. On the other hand, we found that the optical spectra of cytochrome a and a_3 of the cytochrome oxidase were not affected by the toxin but the spectral features of cytochromes b and c_1 were changed, significantly increased and the peak at 560 nm was slightly shifted to red, when PR toxin was present (Figure 4).

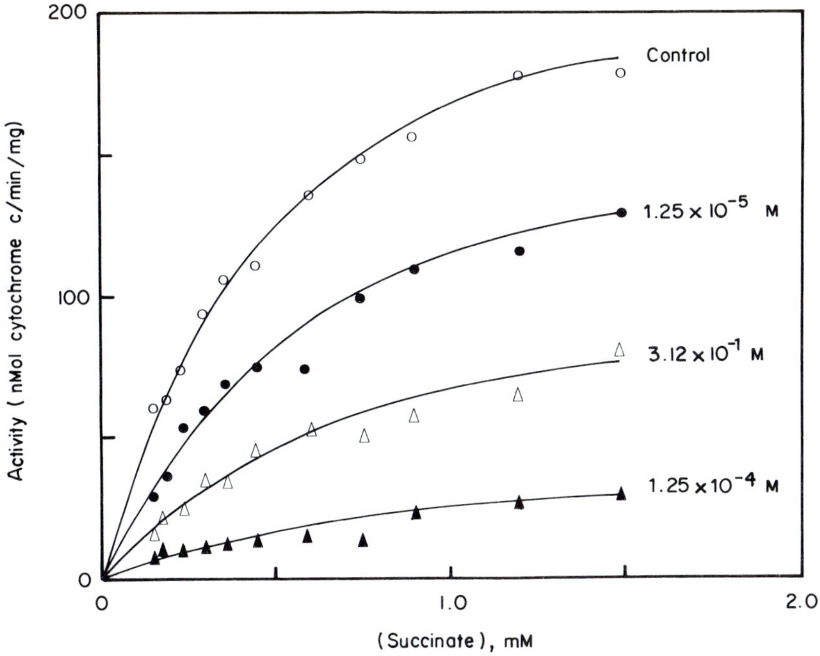

Figure 2 Inhibition of SCR activity by PR toxin. The assay mixture in a total volume of 1 ml contained 0.1 mM cytochrome *c*, 0.33 mM EDTA, 33 μM KCN, 0.1 M Na-K-phosphate buffer, pH 6.4, and various amounts of succinate. The reaction was initiated by addition of 20 μl (60 μg protein) enzyme preparation and the absorbance change at 550 nm was followed for 3–5 min. The PR toxin concentration was varied in the three experimental groups, 1.25×10^{-5} M, 3.12×10^{-5} M and 1.25×10^{-4} M. The controls were conducted with the addition of the same volume of DMSO.

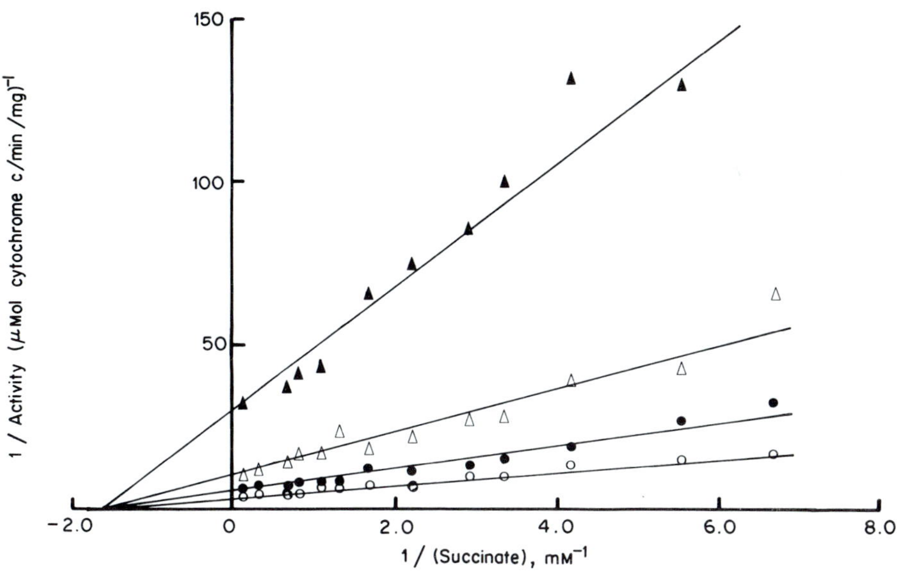

Figure 3 Double reciprocal plot of PR toxin inhibition on SCR activity. The toxin used: (○), control; (●), 1.25×10^{-5} M; (△), 3.12×10^{-5} M; (▲), 1.25×10^{-4} M. The pattern is typical of noncompetitive inhibition. The K_i was calculated to be 2.4×10^{-5} M. The K_m for succinate was 0.65 mM, which was not altered by PR toxin.

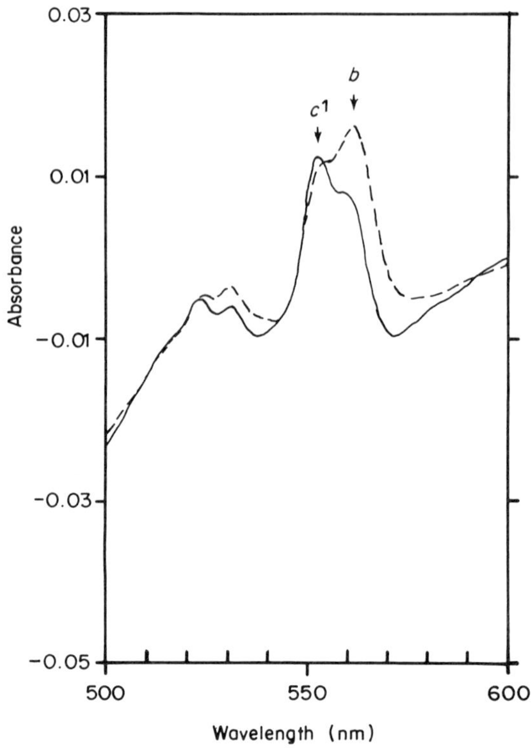

Figure 4 Effect of PR toxin on the different spectra of cytochrome *b* and cytochrome c_1 of succinate-reduced minus succinate-oxidized SCR. The freshly prepared SCR was dissolved in 0.25 M sucrose, 50 mM Na-K-phosphate, pH 7.4, and trace amounts of sodium cholate. The protein concentration was adjusted to 5.6 mg/ml by 50 mM Na-K-phosphate, pH 7.4. In 1 ml of the diluted SCR an excess amount of sodium succinate (about 0.1 mg) was added, and the optical spectrum (——) was then recorded from 600 to 500 nm. In the second experiment (– – –), 1.25×10^{-4} M PR toxin was added to 1 ml of the diluted SCR and an identical amount of sodium succinate was quickly added.

These findings suggested that PR toxin exerts its effects on the mitochondrial respiration and oxidative phosphorylation through actions on the membrane and the SCR complex of the mitochondrial respiratory chain.

ACKNOWLEDGMENTS

This work was partially supported by the National Science Council, Republic of China (NSC-70-0412-B010-34), and was performed while R. D. Wei and Y. H. Wu Lee had Research Chair Awards from the Tjing-Ling Medical Foundation.

REFERENCES

Arnold, D. L., Scott, P. M., McGuire, P. F., Harwig, J., and Nera, E. A. (1978). 'Acute toxicity studies on roquefortine and PR toxin, metabolites of *Penicillium roqueforti*, in the mouse', *Food Cosmet. Toxicol.*, **16**, 369–371.

Aujard, C., Chang, E. M., Icard, C., and Trincal, G. (1979a). 'Effects of PR toxin on liver cells in culture', *Toxicology*, **12**, 313–323.

Aujard, C., Moulé, Y., Moreau, S., and Darraq, N. (1979b). 'Persistence of DNA damage induced in cultured liver cells by PR toxin, a mycotoxin from *Penicillium roqueforti*', *Toxicol. Eur. Res.*, **2**, 273–278.

Baert, F., Foulon, M., Odou, G., and Moreau, S. (1980). 'The structure and absolute configuration of PR toxin', *Acta Cryst.*, **B36**, 402–406.

Bolden, A., Noy, G. P., and Weissbach, A. (1977). 'DNA polymerase of mitochondria is a γ-polymerase', *J. Biol. Chem.*, **252**, 3351–3356.

Chang, L. M. S. (1973). 'Low molecular weight deoxyribonucleic acid polymerase from calf thymus chromatin', *J. Biol. Chem.*, **248**, 3789–3795.

Chang, L. M. S. (1976). 'Phylogeny of DNA polymerase β', *Science*, **191**, 1183–1185.

Chen, F. C., Chen, C. F., and Wei, R. D. (1982). 'Acute toxicity of PR toxin, a mycotoxin from *Penicillium roqueforti*', *Toxicon*, 433–441.

Estabrook, R. W. (1967). 'Mitochondrial respiratory control and the polarographic measurement of ADP:O ratios', in *Methods in Enzymology* (eds R. W. Estabrook and M. E. Pullman), pp. 41–47. Academic Press, New York.

Lewis, B. J., Abrell, J. W., Smith, R. G., and Gallo, R. C. (1974). 'DNA polymerases in human lymphoblastoid cells infected with simian sarcoma virus', *Biochim. Biophys. Acta*, **349**, 148–160.

Matheson, K. J. (1921). 'Manufacture of cow's milk roquefort cheese', *U.S. Dept. Agr. Bull.*, No. 970, pp. 1–28.

Mechali, M., Abadiedebat, J., and Recondo, A. M. (1979). 'Eukaryotic DNA polymerase α', *J. Biol. Chem.*, **255**, 2114–2122.

Moreau, S., Masset, A., and Biguet, J. (1979). 'Resolution of *Penicillium roqueforti* toxin and eremofortins A, B, and C by high-performance liquid chromatography', *Appl. Environ. Microbiol.*, **37**, 1059–1062.

Moulé, Y., Jemmali, M., and Darracq, N. (1978). 'Inhibition of protein synthesis by PR toxin, a mycotoxin from *Penicillium roqueforti*', *FEBS Lett.*, **88**, 341–344.

Moulé, Y., Jemmali, M., and Rousseau, N. (1976). 'Mechanism of the inhibition of transcription by PR toxin, a mycotoxin from *Penicillium roqueforti*', *Chem.-Biol. Interactions*, **14**, 207–216.

Moulé, Y., Jemmali, M., Rousseau, N., and Darracq, N. (1977a). 'Action of monovalent cations on the biological properties of PR toxin, a mycotoxin from *Penicillium roqueforti*', *Chem.-Biol. Interactions*, **18**, 153–162.

Moulé, Y., Moreau, S., and Aujard, C. (1980). 'Induction of cross-links between DNA and protein by PR toxin', a mycotoxin from *Penicillium roqueforti*, *Mutation Res.*, **77**, 79–89.

Moulé, Y., Moreau, S., and Bousquet, J. F. (1977b). 'Relationships between the chemical structure and the biological properties of some eremophilane compounds related to PR toxin, *Chem.-Biol. Interactions*, **17**, 185–192.

Polonelli, L., Morace, G., Delle Monache, F., and Samson, R. A. (1978). 'Studies on the PR toxin of *Penicillium roqueforti*', *Mycopathologia*, **66**, 99–104.

Robert-Guroff, M., Schrecker, A. W., Brinkman, B. J., and Gallo, R. C. (1977). 'DNA polymerase γ of human lymphoblasts', *Biochemistry*, **16**, 2866–2873.

Smith, R. G., and Gallo, R. C. (1972). 'DNA dependent DNA polymerases I and II from normal human blood lymphocytes', *Proc. Natl. Acad. Sci. U.S.A.*, **69**, 2879–2884.

Tashiro, F., Hirai, K., and Ueno, Y. (1979). 'Inhibitory effects of carcinogenic mycotoxins on deoxyribonucleic acid-dependent ribonucleic acid polymerase and ribonuclease H', *Appl. Environ. Microbiol.*, **38**, 191–196.

Vercesi, A., Reynafarje, B., and Lehninger, A. L. (1978). 'Stoichiometry of H^+ ejection and Ca^{2+} uptake coupled to electron transport in rat heart mitochondria', *J. Biol. Chem.*, **253**, 6379–6385.

Wei, R. D., and Chang, Y. N. (1976). 'Studies on the inhibition of the growth of Ehrlich ascites tumors by PR toxin and related compounds', *Natl. Sci. Counc. Monthly, ROC*, **4**, 2464–2471.

Wei, R. D., and Liu, G. X. (1978). 'PR toxin production in different *Penicillium roqueforti* strains', *Appl. Environ. Microbiol.*, **35**, 797–799.

Wei, R. D., and Liu, G. X. (1982). 'Studies on the production and radioactive labelling of PR toxin by *Penicillium roqueforti*', *Natl. Sci. Counc. Monthly, ROC*, **10**, 437–442.

Wei, R. D., Schnoes, H. K., Hart, P. A., and Strong, F. M. (1975). 'The structure of PR toxin, a mycotoxin from *Penicillium roqueforti*, *Tetrahedron*, **31**, 109–114.

Wei, R. D., Schnoes, H. K., Smalley, E. B., Lee, S. S., Chang, Y. N., and Strong, F. M. (1974). 'Production, isolation, chemistry, and biological properties of *Penicillium roqueforti* toxin', *4th International Symposium on Animal, Plant, and Microbial Toxins (Tokyo)*.

Wei, R. D., Schnoes, H. K., Smalley, E. B., Lee, S. S., Chang, Y. N., and Strong, F. M. (1976). 'Production, isolation, chemistry, and biological properties of *Penicillium roqueforti* toxin', in *Animal, Plant, and Microbial Toxins, Vol. 2* (eds A. Ohsaka, K. Hayashi and Y. Sawai), pp. 137–144. Plenum, New York.

Wei, R. D., Still, P. E., Smalley, E. B., Schnoes, H. K., and Strong, F. M. (1973). 'Isolation and partial characterization of a mycotoxin from *Penicillium roqueforti*, *Appl. Microbiol.*, **25**, 111–114.

Weissbach, A., Baltimore, D., Bollum, F. J., Gallo, R., and Korn, D. (1975). 'Nomenclature of eukaryotic DNA polymerases', *Science*, **190**, 401–402.

Yu, C. A., Yu, L., and King, T. E. (1974). 'Soluble cytochrome b–c_1 complex and the reconstitution of succinate-cytochrome *c* reductase', *J. Biol. Chem.* **249**, 4905–4910.

Yu, C. A., Yu, L., and King, T. E. (1975). 'Studies on cytochrome oxidase—Interactions of the cytochrome oxidase protein with phospholipids and cytochrome *c*', *J. Biol. Chem.*, **250**, 1383–1392.

Trichothecenes and Other Mycotoxins
Edited by J. Lacey

CHAPTER 33

Uptake and subcellular distribution of aflatoxin B_1 by excised, cultured soybean roots

S. J. Walker, G. C. Llewellyn, E. B. Lillehoj and W. V. Dashek

INTRODUCTION

Aflatoxins (AFTs) are secondary metabolites of *Aspergillus flavus* and *A. parasiticus*, two fungi which can produce AFTs upon both autoclaved (Shotwell *et al.*, 1978) and field-grown (Stoloff, 1977) soybeans (*Glycine max*). The toxins are potent hepatocarcinogens, mutagens and teratogens (Butler, 1965; Carlborg, 1979; Carnaghan *et al.*, 1963; Shanks, 1977; Wogan, 1965; Wogan and Busby, 1980).

Although AFTS may affect germination, growth or elongation, chlorophyll synthesis, enzyme activities, amino acid uptake, protein or nucleic acid synthesis, and interactions with hormones or ultrastructure in plant systems (Dashek and Llewellyn, 1983), neither uptake by, nor intracellular distribution of, aflatoxin B_1 (AFB_1) within plants have been extensively investigated. This chapter reports uptake by, and distribution of AFB_1 within, excised, cultured soybean roots, which is of interest because aflatoxin-contaminated crops are occasionally ploughed (Mertz *et al.*, 1980).

MATERIALS AND METHODS

Germination conditions and incubation of excised roots

Glycine max, cv. 'Essex' (Southern States Cooperative, Inc., Richmond, VA) seeds (1979 and 1980 crops) were allowed to imbibe 3–4 days and the resulting roots were excised and incubated as described by Chao and Dashek (1973) except that the imbibition medium contained 50 μg/ml of either chloramphenicol or streptomycin. Roots were excised into an incubation medium consisting of 1% sucrose, 0.5 mM sodium citrate, 20 μg/ml streptomycin and 10 μg/ml 2,4-dichlorophenoxyacetic acid, pH 6.0 (Chao and Dashek, 1973).

Excised roots were blotted gently with Kimwipes and weighed before incubating on a New Brunswick gyratory shaker at 100–150 rpm in constant light for 18, 24 or 48 h at 24 ± 2 °C in 125–1000 ml Erlenmeyer flasks, adding 1 g wet root to 20 ml incubation medium. Thirty micrograms of AFB_1/ml or 65.63 μg AFTs (30 μg AFB_1/ml, 1.04 μg AFB_2/ml, 33.59 μg AFG_1/ml, 1.00 μg AFG_2/ml) were added to the incubation medium. The mixed AFTs were prepared using *Aspergillus parasiticus*, NRRL 2999, and quantified according to Llewellyn *et al.* (1973) but the AFB_1 was a gift from the USDA Southern Regional Research Center (New Orleans, LA). The AFB_1 yielded a single blue fluorescent spot upon thin-layer chromatography (see aflatoxin uptake).

Dry weight determinations

After incubation with or without either AFB_1 or AFTs, 1 g quantities of roots were harvested, dried after blotting at 40 ± 1 °C for 1.0, 2.0 and 2.5 h (to constant weight) in pre-weighed aluminium boats to obtain dry weights (balance sensitivity, 0.0001 g and data are expressed as percentages of the dry weight of roots at 0 h before AFTs were added).

Preparation of subcellular organelles

Following incubation, excised roots were homogenized on ice with a mortar and pestle into a medium consisting of 250 mM sucrose, 2 mM EDTA, 25 mM Tris-HCl, pH 7.2, at a ratio of 1 g fresh weight of roots to 4 ml homogenization medium. Homogenates were filtered with gentle suction through a single layer of Miracloth and centrifuged for 15 min at 1000 ×***g***. The supernatant from this was centrifuged for 30 min at 40,000 ×***g*** and the resulting supernatant at 80,000 ×***g*** for 30 min in an International Ultracentrifuge, Model-B 35, using an A-147 rotor at 4 °C or in a Sorvall RC2B which retained unbroken cells and cell wall debris.

Assays

ATPase

Centrifuged pellets were suspended in 2% sucrose, 1 mM $MgSO_4$ and 1 mM Tris, pH 7.2 to yield 1 μg protein/μl suspension medium. Resuspended organelles, Miracloth filtrates and 80,000 ×***g*** supernatants were dialysed (12–18 h at 4 °C) prior to assay for pH 6.0 and 9.0 K^+-stimulated ATPases (plasmalemma and mitochondrial marker enzymes, respectively) following Hodges and Leonard (1974). To ensure that the reaction was enzymatic, 10, 30, 50, 70 and 100 μg of organellar protein were used to assay ATPase activity. Miracloth filtrates and 80,000 ×***g*** supernatants were assayed similarly using 10, 30, 50, 70 and 100 μg of soluble protein.

IDPase

To quantify organellar inosine diphosphatase (IDPase) activity (dictyosome marker enzyme), isolated organelles were resuspended in 61 mM Tris-HCl, pH 7.5. Then, 3.70, 3.00, 1.50 and 0.75 μg organellar protein were utilized for determining enzyme activity following Dauwalder *et al.* (1969).

NADH-cytochrome c reductase

To assay NADH-cytochrome *c* reductase activity (endoplasmic reticulum marker enzyme), isolated organelles were resuspended in 0.05 M Tris-HCl, pH 8.0, 0.67 M sucrose and 0.001 M histidine to yield 1 mg/ml protein. Enzyme activity was assayed in samples containing 10, 30, and 60 μg of organellar protein following Hatefi and Rieske (1973).

Protein

Total protein was quantified colorimetrically utilizing Coomassie blue (Bradford, 1976) with bovine serum albumin (crystalline, Pfaltz and Bauer, Stamford, CT) for standard curve construction.

Analyses

Aflatoxin uptake

Incubation medium before and after incubation, Miracloth filtrates, organellar fractions and 80,000 ×***g*** supernatants of excised roots, incubated with chromatographically pure AFB_1, were partitioned once against chloroform within separatory funnels. The chloroform phase was removed, evaporated to dryness, reconstituted in 100 μl chloroform and spotted onto Adsorbosil+1 thin-layer plates. Then, 10 μg each of AFB_1, AFB_2, AFG_1, and AFG_2 (Applied Sciences Lab., Inc., State College, PA) were spotted onto the plates as reference standards. The plates were developed in unlined chambers containing 100 ml of chloroform–acetone–water (88.0 : 12.0 : 1.5) (Roebuck and Wogan, 1977). After development, plates were viewed with a long-wave UV-source, R_f values of fluorescent compounds recorded and the compounds scraped from the plates. The compounds were eluted from the Adsorbosil with methanol and quantified by measuring absorbance at 362 nm with a Gilford UV spectrophotometer. In addition, AFTs were quantified by fluorodensitometry directly upon the plates without elution. A Kontes fluorodensitometer (Vineland, New Jersey) in the long-wave UV mode was employed.

To confirm identification of AFB_1 in medium and tissue extracts by *R*f, some extracts were 'spiked' with authentic AFB_1 before thin-layer chromatography (TLC). Also TLC plates, with blue fluorescent spots with a similar

R_f to AFB_1, were sprayed with 2 g of 2,4-dinitrophenylhydrazine (Eastman Organic Chemicals, Rochester, NY) in 10 ml of 12 N HCl and 90 ml absolute ethanol (Crisan, 1968).

As a further test, homogenates, pellets and 80,000 $\times g$ supernatants from untreated roots were extracted with chloroform and the extracts chromatographed upon Adsorbosil+1 plates using authentic AFB_1 as a reference standard.

To study further the uptake and distribution of AFB_1 in cell fractions of excised roots, the latter were incubated for 18 h with 695,381 dpm [^{3}H]-AFB_1 (spec. act. 5 Ci/mmol, Moravek Biochemicals, City of Industry, CA). Following incubation, roots were homogenized and cell fractions prepared. Aliquots of roots homogenates and 80,000 $\times g$ supernatants and resuspended (1 ml, 1% $NaHCO_3$) 1000 and 40,000 $\times g$ pellets were added to 15 ml Redi-Solve and radioactivity counted twice in a Beckman LS-230 liquid scintillation counter for 20 min each time at a pre-set error of 0.2%. To study whether AFB_1 taken up by excised, incubated roots (see Results) would conjugate with protein, the 80,000 $\times g$ fraction from homogenates of certain roots incubated with [^{3}H]-AFB_1 were treated by the sodium dodecyl sulphate (SDS) procedure of Laemmli (1970). The SDS-released proteins were separated by gel permeation chromatography on a 2.2 × 12.2 cm Bio-Gel P-100 column. The columns, which were calibrated with Pharmacia Blue Dextran 2000, bovine serum albumin and L-proline, were eluted with 0.1 M, pH 6.5 phosphate buffer. Aliquots of column eluates were added to 10 ml Redi-Solve and the radioactivity quantified by liquid scintillation counting.

Statistical analysis

Following construction of means and standard deviations, significant differences were sought with a *t*-test where $N \leqslant 30$ (Mendenhall and Ott, 1972).

RESULTS

Aflatoxin-induced weight changes in cultured, excised roots

Roots incubated in medium containing AFTs consistently weighed less than roots incubated in medium without AFTs. The percentage difference in fresh weight between untreated and treated roots increased with time, although there did not appear to be an increase in wet weight difference between 18 and 24 h (Figure 1). However, roots incubated with 30 μg pure AFB_1/ml incubation medium often showed the same gain in fresh weight with time as untreated roots. To determine whether AFB_1 could affect root weight, the dry weights of toxin-treated and untreated–excised, cultured roots were compared.

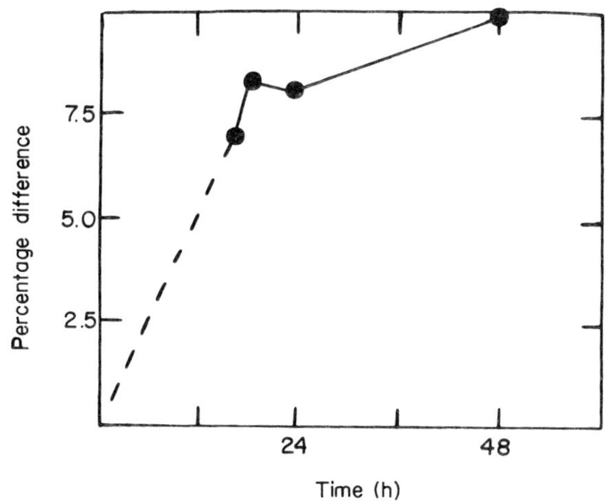

Figure 1 Percentage difference in fresh weight between mixed aflatoxins—treated and untreated roots.

When excised roots were incubated in a medium containing AFTs, their dry weights increased only up to 24 h and then decreased slightly (Figure 2B). In contrast, control roots continued to increase in weight up to 48 h. The dry weight of the untreated roots was significantly greater ($p = 0.05$) at each harvest time than that for treated roots.

Roots incubated in a medium containing AFB_1 alone increased their dry weight between 18 and 48 h (Figure 2A). The dry weight of control roots also increased with time. Except at 24 h, the dry weight of untreated roots was significantly greater ($p = 0.05$) than that for treated roots.

Uptake and distribution of aflatoxins

Approximately 24–66% of the AFB_1 added to the incubation medium was removed by excised roots incubated for 18 h. Of the unlabelled toxin taken up, 8.6, 10.3, 9.1 and 60.7% of that recovered within the Miracloth filtrate occurred, respectively, within the 1,000, 40,000 and 80,000 $\times \boldsymbol{g}$ pellets and 80,000 $\times \boldsymbol{g}$ supernatant (Table 1).

To enquire whether the unlabelled and labelled AFB_1 recovered within subcellular organelles were unchanged, the fractions were extracted with chloroform and then subjected to TLC. A diagram of a typical thin-layer chromatographic separation of AFTs present within the incubation media before and after incubation of excised roots and subcellular organelles is shown in Figure 3. Tentative identifications of spots using R_f values were

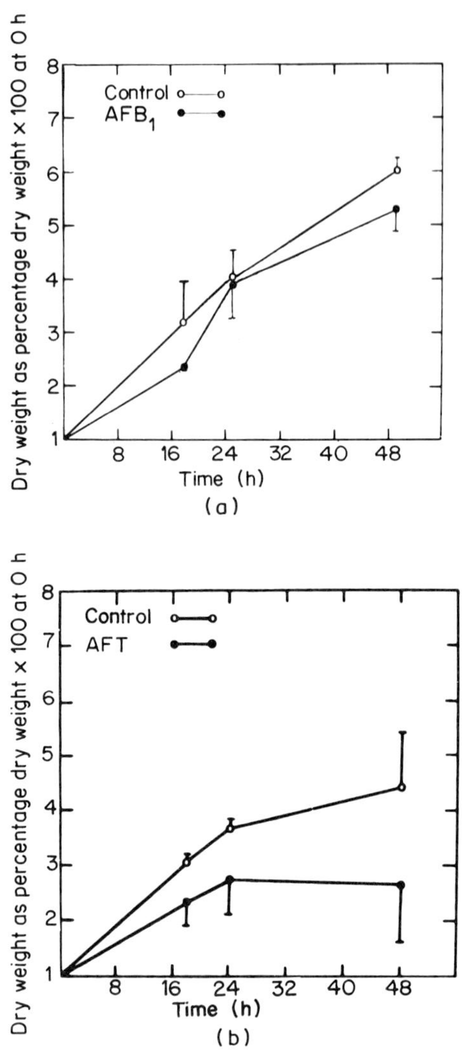

Figure 2 AFB_1 (a) and AFTs (b)—induced changes in dry weight of excised, incubated roots (N=3).

accomplished by spotting the plates with authentic AFB_1, AFB_2 AFG_1 and AFG_2. Homogenates and cell fractions from roots incubated in medium containing AFB_1 (sample 4), yielded a single blue-green fluorescent spot but the medium sometimes yielded four fluorescent spots (sample 5) after 18 h incubation of excised roots. The R_f values (Table 2) of these spots ranged

Table 1 Uptake and distribution of toxin within subcellular fractions of homogenates from excised roots in a medium containing AFB_1

Medium or subcellular fraction	Radioactive AFB_1 dpm[a]	Radioactive AFB_1 Percentage label recovered in fraction[b]	Non-radioactive AFB_1 μg AFB_1[c]	Non-radioactive AFB_1 Percentage recovered in fraction[d]
Medium:				
Prior to incubation	695,381 ± 9,307		1014.0 ± 137.3	
Following incubation	531,069 ± 16,892		353.5 ± 131.5	
Subcellular fraction:				
filtrate	45,307 ± 2,715[e]		88.0[e]	
1,000 × ***g*** pellet	2,682 ± 297	5.9	7.6 ± 4.9	8.6
40,000 × ***g*** pellet	4,300 ± 546	9.5	9.1 ± 3.5	10.3
80,000 × ***g*** pellet	1,097 ± 34	2.4	8.0 ± 5.5	9.1
80,000 × ***g*** supernatant	34,947 ± 1,844	77.1	53.4 ± 15.6	60.7

[a] Represent means of two experiments with three replicates in each experiment.
[b] Percentage of label in each fraction compared to that in filtrate.
[c] Represent means and standard deviations of three or more replicate experiments except for the filtrate which is an average of two.
[d] Percentage of AFB_1 in each fraction compared to that in filtrate.
[e] Presumably the remainder of the AFB_1 which was taken up from the medium was retained on the Miracloth in unbroken cells and/or cell walls.

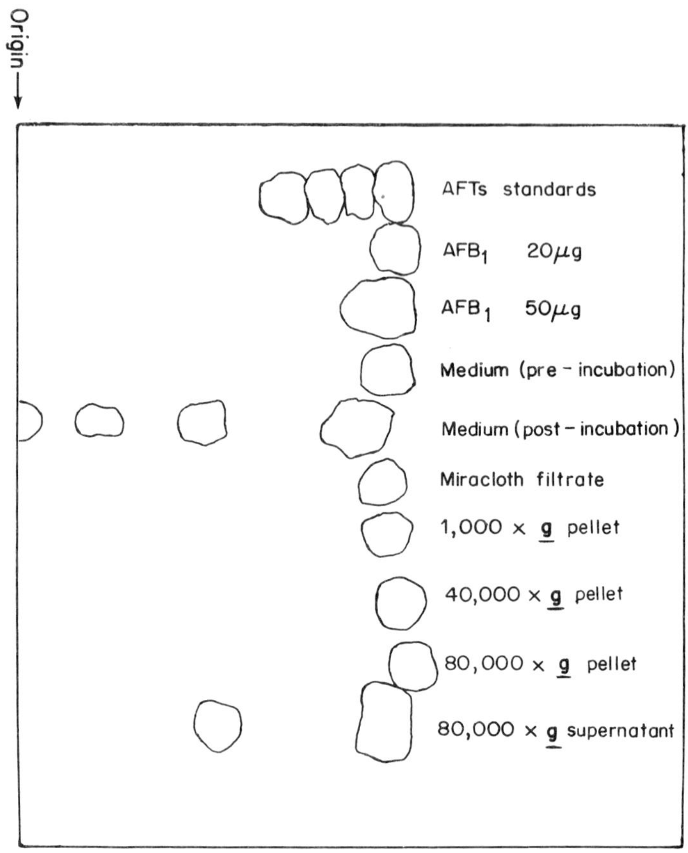

Figure 3 Diagrammatic representation of a thin-layer chromatographic separation of AFTs in growth medium and subcellular fractions.

from 0.02 to one nearly coincident with authentic AFB_1 (0.63). Each subcellular fraction contained either a blue or blue-green spot with an R_f comparable to AFB_1, except for the 80,000 $\times \boldsymbol{g}$ supernatant which appeared to contain AFB_1 and a fluorescent compound with an R_f less than that of the AFTs. Fluorescent spots were absent from the homogenates, pellets and 80,000 $\times \boldsymbol{g}$ supernatants of roots incubated without AFB_1. Verification that either the blue or blue-green fluorescent compounds of the medium and subcellular fractions were AFB_1 was accomplished by layering authentic AFB_1 over extracts of the medium and subcellular fractions on TLC plates. The compounds co-migrated with AFB_1. The occurrence of AFB_1 was also verified by a positive result (yellow-orange aflatoxin dinitrophenylhydrazone) with 2,4-dinitrophenylhydrazine (Crisan, 1968).

Table 2 R_f values of fluorescent compounds within the growth medium and cytoplasmic fractions of homogenates of roots incubated 18 h with AFB_1

Sample or standard	Fluorescent colour	R_f[a]
Medium		
Prior to root incubation	Blue-green	0.64 ± 0.050
Following root incubation	Blue	0.02 ± 0.002
	Blue	0.13 ± 0.003
	Green	0.31 ± 0.008
	Blue-green	0.63 ± 0.050
Root homogenate	Blue	0.63 ± 0.060
Miracloth filtrate	Blue	0.57 ± 0.080
1,000 × ***g*** pellet	Blue-green	0.62 ± 0.030
40,000 × ***g*** pellet	Blue	0.62 ± 0.000
80,000 × ***g*** pellet	Blue	0.63 ± 0.007
80,000 × ***g*** supernatant	Green	0.35 ± 0.020
	Blue	0.62 ± 0.020
AFB_1	Blue	0.64 ± 0.030
AFB_2	Blue	0.59 ± 0.020
AFG_1	Green	0.54 ± 0.020
AFG_2	Blue-green	0.48 ± 0.020

[a] Represent means and standard deviations for three replicate experiments.

Marker enzyme analyses of subcellular fractions prepared by differential centrifugation

The distributions of both the pH 6.0 and 9.0 K^+-stimulated ATPases within subcellular fractions are presented in Table 3. When expressed as the activity within the Miracloth filtrates, the percentage recoveries for pH 6.0 K^+-stimulated ATPase of untreated roots were 1.12 (1000 ×***g*** pellets), 8.63 (40,000 ×***g*** pellets), 2.87 (80,000 ×***g*** pellets) and 148.32 (80,000 × ***g*** supernatant). The corresponding values were 2.54, 5.61, 3.44 and 199.24 for the same sequence of fractions derived from untreated roots. For pH 9.0 K^+-stimulated ATPase, the percentage recoveries were 13.42 (1000 ×***g*** pellet) 15.89 (40,000 ×***g*** pellet), 1.05 (80,000 ×***g*** pellet) and 26.44 (80,000 ×***g*** supernatant) of that recovered within the Miracloth filtrate for treated roots. The percentage recoveries for untreated roots were 23.47 (1000 ×***g*** pellet), 27.62 (40,000 ×***g*** pellet), 1.23 (80,000 ×***g*** pellet) and 28.79 (80,000 ×***g*** supernatant). The pH 6.0 K^+-stimulated ATPase activity within the fractions was 160.9% (treated) and 210.8% (untreated) of that present in the Miracloth

Table 3 K^+-Stimulated ATPase activities of subcellular fractions obtained by differential centrifugation of homogenates of excised roots cultured in medium containing—or lacking—AFB_1

	Specific activity (μmol Pi h^{-1} mg $protein^{-1}$)		Total activity (μmol Pi released h^{-1})		Percentage recovery[b]	
	Plus AFB_1	Minus AFB_1	Plus AFB_1	Minus AFB_1	Plus AFB_1	Minus AFB_1
Filtrate						
pH 6.0	8.85 ± 0.32	6.40 ± 2.61	563.12 ± 70.39	449.12 ± 213.04	100.00	100.00
pH 9.0	3.97 ± 0.13	2.71 ± 0.34	251.70 ± 14.42	188.23 ± 37.09	100.00	100.00
1,000 × ***g*** pellet						
pH 6.0	1.21 ± 1.09	1.52 ± 1.06	6.80 ± 5.78	11.42 ± 8.56	1.21	2.54
pH 9.0	5.79 ± 0.47	6.05 ± 1.14	33.78 ± 0.10	44.18 ± 5.42	13.42	23.47
40,000 × ***g*** pellet						
pH 6.0	6.98 ± 0.66	2.96 ± 0.44	48.04 ± 8.97	25.20 ± 5.66	8.53	5.61
pH 9.0	5.89 ± 1.21	6.12 ± 0.77	40.00 ± 4.52	52.00 ± 10.41	15.89	27.62
80,000 × ***g*** pellet						
pH 6.0	13.19 ± 0.25	8.72 ± 3.63	16.17 ± 1.71	15.44 ± 5.30	2.87	3.44
pH 9.0	2.14 ± 0.40	1.30 ± 0.21	2.65 ± 0.72	2.32 ± 0.20	1.05	1.23
80,000 × ***g*** supernatant						
pH 6.0	22.81 ± 15.29	22.38 ± 9.30	835.20 ± 616.03	894.83 ± 430.67	148.32	199.24
pH 9.0	1.86 ± 0.27	1.37 ± 0.18	66.55 ± 15.48	54.20 ± 11.17	26.44	28.79

3–5 g fresh weight excised roots were incubated, harvested and organelles prepared; for assay of K^+-stimulated ATPase activity, organelles were suspended in 2% sucrose, 1 mM $MgSO_4$ and 1 mM Tris, pH 7.2 to yield 10 μg protein/10 μl suspension medium and then dialyzed (overnight at 4 °C) prior to assay for enzyme activity according to Leonard *et al.* (1973) and Hodges and Leonard (1974); to ensure that the reaction was enzymatic, 10, 30, 50, 70 and 100 μg of organellar protein were used for assaying ATPase activity; Miracloth filtrates and 80,000 × ***g*** supernatants were assayed in the same manner as the organelles using 10, 30, 50, 70 and 100 μg of soluble protein.

[a] Pure AFB_1 or AFT values containing 30 μg/ml AFB_1.

[b] Percentage recovery of total activity was calculated using the filtrate values as 100% of the total activity; data are means and standard deviations for $N = 5$.

filtrate. For the pH 9.0 K^+-stimulated ATPase the recoveries were 56.8% (treated) and 81.1% (untreated).

Only 60% of the ATP added was hydrolysed by the pH 6.0 ATPase present within the 80,000 $\times g$ pellet, indicating that substrate was in excess. This was substantiated by the linear increase in enzyme activity found when increasing amounts of enzyme were added to the reaction mixture.

Table 4 shows the occurrence of both NADH-cytochrome *c* reductase and IDPase within subcellular organelles. The 1000 $\times g$ pellet contained 55.0% (treated) and 55.8% (untreated) of the NADH-cytochrome *c* reductase activity recovered in the fractions which were assayed for this enzyme's activity. In contrast, reductase activity was absent from the 40,000 $\times g$ pellet while both the 80,000 $\times g$ pellet and its supernatant possessed equivalent enzyme activities. Statistically significant differences for mean enzyme activities between treated and untreated roots were not found for any cell fraction.

Whereas inosine diphosphatase was lacking in 1000 $\times g$ pellets derived from either treated or untreated roots, it did occur in both the 40,000 and 80,000 $\times g$ pellets. The 40,000 $\times g$ pellets contained 45.5% (treated) and 46.0% (untreated) of the IDPase activities recovered from the three fractions which were assayed for this enzyme. The 80,000 $\times g$ pellets had 54.5% (treated) and 54.0% (untreated) of the IDPase activity. The ATPase, NADH-cytochrome *c* reductase and IDPase activities increased linearly as the concentration of organellar protein was raised.

DISCUSSION

Aflatoxin-induced inhibition of elongation of excised, cultured roots

Inclusion of AFB_1 within the incubation medium inhibited the time-dependent increase of both fresh and dry weights at each harvest time. With dry weight, inhibition by AFB_1 was statistically significant ($p = 0.05$) except at 24 h. Young *et al.* (1978) also observed that the dry weight of AFB_1-treated roots was significantly different from that of untreated roots at each harvest time. However, it must still be established whether this reduction in dry weight occurs within the cytoplasm and/or cell wall.

Organellar composition of subcellular fractions

These, together with the rationale for assigning organelles to fractions, are presented in Table 5. An assignment of plastids to a fraction(s) was not possible since plastid marker enzyme analyses (Quail, 1979) were not performed. In addition, cellular vacuoles were most likely lysed by the homogenization procedures which were employed.

Table 4 Inosine diphosphatase and NADH-cytochrome *c* reductase activities of subcellular organelles isolated from excised, incubated soybean roots[a]

Fraction	IDPase^{-1} (mg Pi released h^{-1} mg protein^{-1})		NADH-cytochrome c reductase (A min^{-1} mg protein^{-1})	
	Plus aflatoxin	Minus aflatoxin	Plus aflatoxin	Minus aflatoxin
1,000 × ***g*** pellet	0.00 ± 0.00	0.00 ± 0.00	1.11 ± 0.36	1.03 ± 0.16
40,000 × ***g*** pellet	9.97 ± 3.16	9.65 ± 3.21	0.00 ± 0.00	0.00 ± 0.00
80,000 × ***g*** pellet	12.48 ± 3.88	11.71 ± 4.37	0.45 ± 0.07	0.41 ± 0.11
80,000 × ***g*** supernatant			0.45 ± 0.04	0.41 ± 0.19

[a] Data from Danley *et al.*, 1981.

Table 5 Rationale for tentative identification of organelles composing subcellular fractions obtained from homogenates of excised roots

Fraction	Presumed organelle content	Identifying markers	Reference
1,000 × ***g*** pellet	Nuclei with nuclear envelope–endoplasmic reticulum connections	High NADH-cytochrome *c* reductase activity[a] Lack of IDPase activity[b] Moderate RNA levels High DNA levels	Danley *et al.* (1981)

40,000 × ***g*** pellet	Dictyosomes—Mitochondria with nuclei and plasmalemma	Lack of NADH-cytochrome *c* reductase activity	Danley *et al*. (1981)
		pH 6.0 and 9.0 K^+-stimulated ATPase activity	Present investigation
		Moderate DNA and RNA levels	Danley *et al*. (1981)
		High IDPase activity	Danley *et al*. (1981)
80,000 × ***g*** pellet	Plasmalemma with mitochondria, dictyosomes and endoplasmic reticulum	pH 6.0 and 9.0 K^+-stimulated ATPase activities	Present investigation
		High NADH-cytochrome *c* reductase activity	Danley *et al*. (1981)
		High IDPase activity	Danley *et al*. (1981)
		Low RNA levels	Danley *et al*. (1981)
		Lack of DNA (Nagahashi *et al*., 1973)	Danley *et al*. (1981)
		Vesicles of dimensions of the plasmalemma (Berkowitz and Travis, 1979; Travis and Berkowitz, 1980)	Walker *et al*. (unpublished)
80,000 × ***g*** supernatant	Endoplasmic reticulum, ribosomes, cytosol	High NADH-cytochrome *c* reductase activity	Danley *et al*. (1981)
		High RNA levels	Danley *et al*. (1981)
		Very low DNA	Danley *et al*. (1981)

[a] Marker enzyme for endoplasmic reticulum (Hatefi and Rieske, 1973).
[b] Marker enzyme for dictyosomes (Dauwalder *et al*., 1969).
[c] Marker enzymes for the plasmalemma and mitochondria (Leonard *et al*., 1973).

Uptake and distribution of aflatoxins

Uptake of 24–66% of AFB_1 by excised roots is similar to that found by Young *et al.* (1978) for 18 h incubated roots, but its distribution within subcellular fractions of plants has not been reported before.

Table 1 indicates that only 13.3–24.5% of the AFB_1 which was depleted from the medium following incubation of the roots with either labelled or unlabelled toxin could be recovered within a Miracloth filtrate of root homogenates. Two possibilities, or even a combination of these, might account for the remaining 75.5–86.7%. These are that the bulk of the toxin was retained upon the Miracloth filtrate by being 'trapped' within unbroken cells and/or cell walls, or alternatively that the toxin was intracellularized but that the majority of it was metabolized. The first alternative could be tested by comparing the amount of radioactivity which was retained upon Miracloth to that which was recovered within its filtrate. The possibility that the AFB_1 was intracellularized and then subsequently metabolized could be explored by following the fate of label derived from radioactive AFB_1.

The recovery of AFB_1 in what appears to be a nuclear fraction (Table 1) is not surprising since Jacquet *et al.* (1971) reported that application of AFTs to meristematic cells of *Allium cepa* and *A. ascalonicum* can produce chromosomal abnormalities. In addition, AFB_1 can bind to DNA (Lutz *et al.*, 1980) and aflatoxin–nucleic acid adducts have been found within animal systems (cf. Danley *et al.*, 1981). To our knowledge such adducts have not been reported for plant systems and an investigation of whether the AFB_1 presumably within the nuclear fraction obtained from homogenates of excised roots contains macromolecular-bound AFB_1 would be of interest.

The recovery of AFB_1 within the 80,000 $\times g$ pellet suggests that the toxin may bind to the plasmalemma and/or another membraneous component within the pellet. Strobel and Hess (1974) demonstrated that plasmalemma fractions of sugar cane can bind helminthosporal, but whether this occurs with AFB_1 requires substantiation. A way of providing this substantiation might be to employ binding assays utilizing high specific activity [^{3}H]-AFB_1 similar to those carried out with labelled indole-3-acetic acid, a plant hormone (Batt and Venis, 1976). Gel permeation chromatography on P-100 of SDS-released membraneous proteins from 80,000 $\times g$ pellets of homogenates derived from excised roots cultured in medium containing [^{3}H]-AFB_1 yielded label in the monomer (free proline) region of the column rather than the void volume. This suggests that AFB_1 did not bind to a high molecular weight membrane protein. However, it is still possible that the AFB_1 was ionically bound and that the linkage was labelled with SDS. The possibility of AFB_1 affecting membranes is supported by Phillips (1982), who reported that several lactone-containing mycotoxins can exert their actions through either disruption or alteration of membrane transport systems by reaction at accessible membrance sulphhydryl receptors.

The bulk of the AFB_1 was recovered within the 80,000 $\times g$ supernatant. Since the supernatant was not further fractioned into microsomes/ribosomes, it is not known whether this AFB_1 was associated with them, the cytosol or both. An association with the microsomes would be significant since it is widely accepted that AFB_1 is metabolized by microsomes in animal systems (Hsieh *et al.*, 1977). The metabolites resulting from microsome action upon AFB_1 are AFB_{2a}, AFH, AFM_1, AFP_1, AFQ_1, and $AFB_{2,3}$ oxide. These metabolites were not apparent upon TLC plates but the former may have been below the detection limits (2 μg/kg) that we employed. A time-dependent investigation is required to establish whether microsomes of cells from excised roots can metabolize AFB_1. This is important since AFB_1–2,3 epoxide appears to be the active carcinogenic form of AFB_1 (Phillips, 1982; Swenson *et al.*, 1973), which might be a health risk to humans.

In summary, cultured excised soybean roots can take up exogenous AFB_1 and compartmentalize it. The majority of the AFB_1 is associated with ER, ribosomes and/or soluble cytoplasm. In addition, the roots do not appear to metabolize exogenous AFB_1 to any great extent. The association of AFB_1 with the cytosol could be explained by fragmentation of organelles during cell breakage and/or by diffusion of AFB_1 since it is a low molecular weight compound. While the latter seems unlikely since AFB_1 tends to form complexes with macromolecules (Hsieh *et al.*, 1977), the former explanation is quite possible. One way of testing this would be to incorporate glutaraldelhyde into the homogenization medium to preserve membrane integrity (Morré and Mollenhauer, 1964).

ACKNOWLEDGEMENTS

We are grateful to Ms Sondra Kramer, Ms Marcella Gower, Mrs Joyce Costain and Ms Mary Wesley for typing the manuscript. The technical assistance of Mr Raymond Francis is acknowledged. This work was aided by a donation for Cancer Research from the West Virginia Veterans of Foreign Wars in Huntington. We thank J. M. Danley for the results in Table 4.

REFERENCES

Batt, S., and Venis, M. A. (1976). 'Separation and localization of two classes of auxin binding sites in corn coleoptile membranes', *Planta*, **30**, 130–135.

Berkowitz, R. L., and Travis, R. L. (1979). 'An electron microscope comparison of plasma membrane vesicles from meristematic and mature soybean root tissues', *Pl. Physiol.*, **63**, 1191–1197.

Bradford, M. M. (1976). 'A rapid and sensitive method for the quantitation of microgram quantities of protein utilizing the principle of protein dye binding', *Analyt. Biochem.*, **72**, 248–254.

Butler, W. H. (1965). 'Liver injury and aflatoxins', in *Mycotoxins in Foodstuffs* (ed. G. N. Wogan), pp. 175–186. MIT Press, Cambridge, Mass.

Carlborg, F. W. (1979). 'Cancer, mathematical models and aflatoxin', *Food Cosmet. Toxicol.*, **17**, 159–166.

Carnaghan, R. B. A., Hartley, R. D., and O'Kelley, J. (1963). 'Toxicity and fluorescence properties of the aflatoxins', *Nature*, **200**, 1101.

Chao, H., and Dashek, W. V. (1973). 'Hydroxyproline metabolism during mungbean and soya-bean seedling growth', *Ann. Bot.*, **37**, 95–105.

Crisan, E. V. (1968). 'A 2,4-dinitrophenylhydrazine spray for the identification of aflatoxin B_1 on thin layer chromatoplates', *Contrib. Boyce Thompson Inst.*, **24**, 37.

Danley, J. M., Staggers, S., Walker, S., Varner, A., Llewellyn, G. C., and Dashek, W. V. (1981). 'Aflatoxin-induced alteration in the levels of membrane chemicals of subcellular organelles isolated from excised, incubated *Glycine max*, cv. "Essex" roots. I. Nonenriched organelles', *Mycopathologia*, **74**, 149–161.

Dashek, W. V. and Llewellyn, G. C. (1983). 'Mode of action of the hepatocarcinogens aflatoxins, in plant systems', *Mycopathologia*, **81**, 83–94.

Dauwalder, M., Whaley, W. G., and Kephart, J. E. (1969). 'Phosphatases and differentiation of the Golgi apparatus', *J. Cell Sci.*, **4**, 455–497.

Hatefi, Y., and Rieske, J. S. (1973). 'The preparation and properties of DPNH-cytochrome reductase (Complex I–III of the respiratory chain', *Methods in Enzymology* **10** (ed. R. W. Estabrook and M. E. Pullman). Academic Press, New York.

Hodges, T. K., and Leonard, R. T. (1974). 'Purification of plasma membrane-bound adenosine triphosphatase from plant roots', *Methods in Enzymology*, 31B. (eds S. Fleisher and L. Packer), p. 392. Academic Press, New York.

Hsieh, D. P. H., Wong, J. J., Wong, Z. A., Michas, C., and Ruebner, B. H. (1977). 'Hepatic transformation of aflatoxin and its carcinogenicity', *Origins of Human Cancer*, pp. 679–707. Cold Spring Harbor Laboratory.

Jacquet, J., Boutibonnes, P., and Saint, S. (1971). 'Biological effects of the aflatoxins of *Aspergillus parasiticus* ATCC 15/517 II', *Plants. Rev. Immunol.*, **35**, 159–189.

Laemmli, U. K. (1970). 'Cleavage of structural proteins during the assembly of the hood of bacteriophage T4', *Nature*, **227**, 680.

Leonard, R. T., Hansen, D., and Hodges, T. K. (1973). 'Membrane-bound adenosine triphosphatase activities of oat roots', *Pl. Physiol.*, **51**, 749–775.

Llewellyn, G. C., Carlton, W. W., Robbers, J. E., and Hansen, W. G. (1973). 'A rapid and simplified method for preparation of pure aflatoxin B_1', *Dev. Ind. Microbiol.*, **14**, 325–330.

Lutz, W. K., Joggi, W., Lutly, J., Sogelsdroff, P., and Schlatter, C. (1980). 'In vivo covalent binding of aflatoxin B_1 and aflatoxin M, to liver DNA of rat, mouse and pig', *Chem. Biol. Interactions*, **32**, 249–256.

Mendenhall, W., and Ott, L. (1972). *Understanding Statistics*. Duxbury Press, Belmont, CA.

Mertz, O., Lee, D., Zuber, M., and Lillehoj, E. (1980). 'Uptake and metabolism of aflatoxin by *Zea mays*', *J. Agr. Food Chem.*, **28**, 963–966.

Morré, D. J., and Mollenhauer, H. H. (1964). 'Isolation of the Golgi apparatus from plant cells', *J. Cell Biol.*, **23**, 295–305.

Nagahashi, G., Leonard, R. T., and Thomson, W. W. (1978). 'Purification of plasma membranes from roots of barley. Specificity of the phosphotungstic acid-chromic acid stain', *Pl. Physiol.*, **61**, 993–999.

Phillips, T. D. (1982). 'Toxicological activities of mycotoxins', *J. Am. Oil Chemists' Soc.*, **59**, 148.

Quail, P. H. (1979). 'Plant cell fractionation', *Ann. Rev. Plant Physiol.*, **30**, 425–484.

Roebuck, B. D., and Wogan, G. N. (1977). 'Species comparison of *in vitro* metabolism of aflatoxin B_1', *Cancer Res.*, **37**, 1649–1656.

Shanks, R. C. (1977). 'Metabolic activation of mycotoxins by animals and humans, An overview', *J. Toxicol. Environ. Health*, **2**, 1229–1244.

Shotwell, O. L., Vandegraft, E. E., and Hesseltine, C. W. (1978). 'Aflatoxin formation on sixteen soybean varieties', *J. Assoc. Off. Anal. Chem.*, **61**, 574–577.

Stoloff, L. (1977). 'Aflatoxins—an overview', in *Mycotoxins and Human Health* (ed. J. V. Rodricks, C. W. Hesseltine and M. A. Mehlman), pp. 7–28. Academic Press, New York.

Strobel, G. A., and Hess, W. M. (1974). 'Evidence for the presence of the toxin-binding protein on the plasma membrane of sugarcane cells', *Proc. Natl. Acad. Sci.*, U.S.A., **71**, 1413.

Swenson, D. H., Miller, J. A., and Miller, E. C. (1973). '2,3-dihydro 2,3-dihydroxyaflatoxin B_1: an acid hydrolysis product of an RNA-aflatoxin B_1 adduct formed by hamster and rat liver microsomes *in vitro*', *Biochem. Biophys. Res. Comm.*, **53**, 1260.

Travis, R. L., and Berkowitz, R. L. (1980). 'Characterization of soybean plasma membrane during development', *Pl. Physiol.*, **65**, 871–875.

Wogan, D. H. (1965). 'Mycotoxins in Foodstuffs', *Massachusetts Institute of Technology Press*, Cambridge, MA.

Wogan, G. N., and Busby, W. F. Jr. (1980). 'Naturally occurring carcinogens', in *Toxic Constituents of Plant Foodstuffs* (ed. I. E. Liener), pp. 329–369. Academic Press, New York.

Young, J. W., Dashek, W. V., and Llewellyn, G. C. (1978). 'Aflatoxin B_1 influence on excised soya-bean growth, ^{14}C-leucine uptake and incorporation', *Mycopathologia*, **66**, 91–97.

Trichothecenes and Other Mycotoxins
Edited by J. Lacey

CHAPTER 34

Inhibition of isoprenoid biosynthesis by fungal metabolites

Ming-Shi Shiao and Yun-Yun Hao

INTRODUCTION

The great diversity of structures and biological activities of secondary metabolites makes them good tools in the study of plant physiology. Some fungal metabolites have been proved to be particularly useful in this aspect (Cole and Cox, 1981). Recently mevinolin and compactin, two polyketides isolated from *Aspergillus terreus* (Alberts *et al.*, 1980) and *Penicillium citrinum* (Brown *et al.*, 1976; Endo *et al.*, 1976) were demonstrated to be specific inhibitors of mammalian 3-hydroxy-3-methylglutaryl coenzyme A reductase (HMG-CoA reductase; EC 1.1.1.34) (Figure 1). Their high specificity in cholesterol biosynthesis (Brown *et al.*, 1978) suggests that they may also be useful for the study of isoprenoid formation in plants and microorganisms. In fact, their inhibitory effects on tobacco pith callus (Hashizume *et al.*, 1983), root elongation of ethiolated radish seedlings (Bach and Lichtenthaler, 1982), phytosterol synthesis in suspension culture of sycamore cells (Ryder and Goad, 1980), and yeast (Bach and Lichtenthaler, 1983) have been reported recently.

The isoprenoid pathway is responsible for the formation of many kinds of plant hormones such as gibberellins (GAs), abscisic acid (ABA), and the side chain of cytokinins which regulate the dynamic and very complicated process of plant growth (Sembdner *et al.*, 1980) (Figure 2). Although the physiological effects of these plant hormones have been studied extensively their modes of action, particularly the way they regulate hormone formation, remain obscure. A possible approach is to block endogenous formation of these isoprenoid hormones simultaneously at predictable steps using specific inhibitors. With adequate precautions the biochemical action of individual isoprenoid hormone can be observed through exogenous supplementation. Inhibitors blocking isoprenoid pathway at early common steps, such as compactin and mevinolin on HMG-CoA reductase, are good probes for this particular purpose. As to HMG-CoA reductase, it is known that this key enzyme in

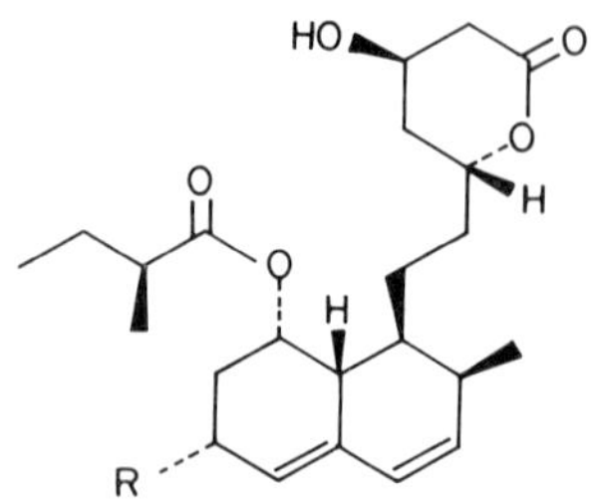

Figure 1 Chemical structure of mevinolin ($R=CH_3$) and compactin (R=H).

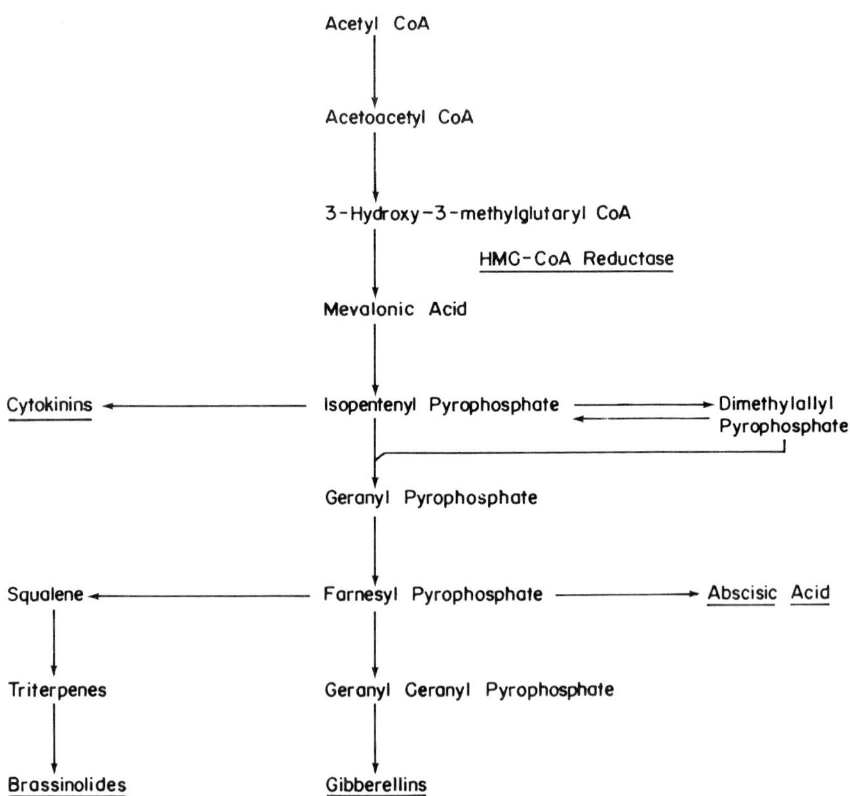

Figure 2 Biogenesis of isoprenoid plant hormones.

mammalian systems is highly regulated and quickly responds to changes in various physiological states (Brown and Goldstein, 1980; Schroepfer, 1981). In plant HMG-CoA reductase has been studied in sweet potato roots (Suzuki, 1975), maize seedlings (Berry *et al.*, 1972), peas (Brooker and Russell, 1975), and radish plants (Bach and Lichtenthaler, 1983). Phytochrome control of reductase activity in pea seedlings (Wong *et al.*, 1982) and enzyme distribution in *Nepeta cataria* leaf tissue (Arebalo and Mitchell, 1984) have also been reported recently. However, the role of HMG-CoA reductase on the formation of isoprenoid plant hormones is highly speculative and has not been well characterized. It is desirable to elucidate the potential of mevinolin as inhibitor of HMG-CoA reductase on the production of isoprenoid hormones in plants. In a preliminary study mevinolin was first utilized in our laboratory to study its inhibition of gibberellin formation in *Gibberella fujikuroi* (Shiao, 1983; Lewer and MacMillan, 1983; Dockerill and Hanson, 1978). It was also used on *Oryza sativa* cv. Tainon 67 to study its effect on seed germination and plant growth.

EXPERIMENTAL

Production of mevinolin by *Aspergillus terreus*

Aspergillus terreus (ATCC 20542), originally maintained on PDA slants, was cultured for mevinolin production (Alberts *et al.*, 1980). Isolation and purification were carried out by silica gel column chromatography, preparative TLC and recrystallization, following the procedure described in Shiao (1983).

Inhibition of gibberellin formation in *Gibberella fujikuroi* by mevinolin

Gibberella fujikuroi (ATCC 12616) was cultured to produce gibberellins in the medium described by Dockerill and Hanson (1978). Pure mevinolin in methanol (2 mg/ml) was added to filter paper (Whatman no. 1) and after evaporation of solvent aseptically the filter paper was added to 3-day-old liquid culture of *G. fujikuroi* (100 ml of broth in each of fifteen 300 ml culture flasks) to a final concentration of 5 μM. After another 6 days also at 27 ± 1.5 °C in an incubator shaker (140 rpm) the contents of the flasks were pooled and filtered to remove mycelium. The filtrate was acidified to pH 3.0 with 6 N HCl and extracted three times with equal volume of ethyl acetate. The combined extract was dried over sodium sulphate and the solvent evaporated by rotary evaporation. The brown preparation was passed through a silica gel column (3.1 × 44 cm) by eluting with a mixture of EtOAc : MeOH : HOAc (14 : 2 : 1, v/v). Fractions containing GA_3 were pooled and the volume of solvent reduced by evaporation. Acetic acid was greatly reduced by a repeated partition between ethyl acetate and water. GA_3 was further purified by preparative silica gel TLC (20 × 20 cm, 2 mm thickness; solvent, EtOAc : MeOH : HOAc = 14 : 2 : 1, R_f 0.67) and HPLC (RP-18, 25 cm, solvent : H_2O : CH_3CN = 1 : 1, v/v, flow rate 1 ml/min, refractometer detector, R_t 6.3 min). Identification of GA_3 with authentic sample was carried out both by TLC and mass spectroscopy (Jeol TMS D-100, 12 and 75 eV ionization potential).

Treatment of *Oryza sativa* with mevinolin

Sterilized seeds of *Oryza sativa* cv. Tainon 67 were incubated in distilled water with mevinolin and/or plant growth hormone (ABA and GA_3) in the darkness at 28 ± 1.5 °C for 5 days. Mevinolin and growth hormones were applied by filter paper method as described before. Seeds were transferred to Petri dishes containing distilled water for another 6 days in the same conditions. Inhibition of rice seedlings by mevinolin was measured by the retardation of radicle root and shoot growth. At least 50 seeds were used for each test and tests were repeated three times. In a separate experiment, 6-day-old seedlings of *O. sativa* were treated with 5 μM mevinolin for 5 days and radicle root and shoot growth was measured.

RESULTS AND DISCUSSION

Gibberellic acid (GA_3) formation in *Gibberella fujikuroi* is strongly inhibited by mevinolin. Addition of 5 μM mevinolin decreased GA_3 formation to 7% of that in untreated cultures although biomass was hardly affected (Shiao, 1983). Due to a large reduction of gibberellins in the culture only the most abundant species, GA_3, was isolated for characterization by HPLC and mass spectroscopy. Recently Lewer and MacMillan (1983) also reported that compactin at 10 μM can block the incorporation of [2-^{14}C]-acetate into GA_3 by 64.8% in *G. fujikuroi* GF-la. Their observation that the incorporation of [2-^{14}C]-mevalonate into GA_3 was also reduced by 38.5% implies that compactin also has an inhibitory effect beyond mevalonate formation. Kaurene synthase seems to be one of the regulatory sites of the isoprenoid pathway leading to the gibberellin formation (Bearder *et al.*, 1975; Hedden *et al.*, 1977). Our study suggests that mevinolin is a potent inhibitor and potential tool to elucidate the role of HMG-CoA reductase on the biosynthesis of gibberellins in *G. fujikuroi*.

Mevinolin (10^{-6}–10^{-4} M) was tested to assess the effect on seed germination of *Oryza sativa* cv. Tainon 67. With 10 μM mevinolin radicle length was inhibited by 81% although shoot growth was only decreased by 13%. With

100 μM mevinolin the growth of seedlings eventually ceased if it was maintained throughout the test period. As in *O. sativa* the root growth of radish seedlings is also strongly inhibited by mevinolin (Bach and Lichtenthaler, 1982). Addition of 2.5 μM GA_3 alone stimulated shoot growth by 13%, as expected, but simultaneous treatment with 2.5 μM GA_3 and 5 μM mevinolin resulted in inhibition of shoot and root growth to 67% and 32% respectively. Inhibition by mevinolin at low concentration, 5 μM and less, could be partially reversed only if mevinolin containing incubation solution was replaced with distilled water or 5 μM mevalonolactone was added subsequently. At greater concentrations, 10 μM and above, mevalonolactone alone was also inhibitory to the growth of *O. sativa*. Simultaneous treatment of 1 μM ABA and 5 μM mevinolin for 5 days in an 11-day test period decreased shoot length to 76% and root length to 31% of control values (mean values of shoot and root lengths in the control group were 4.6 cm and 7.4 cm respectively). This combined treatment also darkened the seeds while shoots were pale compared to control seedlings. Since the gibberellin content of *O. sativa* was low in treated seedlings quantitation by TLC and HPLC cannot be achieved for further mass spectrometric characterization. The inhibitory effect of mevinolin was even more pronounced if it was used to treat *O. sativa* later in the 11-day test period. With 5 μM mevinolin present in the last 5 days of test period lengths of radicles and shoots were 5% and 35% of control respectively. Treated seedlings were not pigmented while untreated seedlings showed pale yellow shoots. This suggested that the endogenous formation of isoprenoid plant hormones becomes insufficient to maintain the growth of seedlings when their requirement is greater later in the test period. Lack of pigmentation of seedlings indicated that the formation of pigment compounds, including carotenoids, is also inhibited by mevinolin.

Until now, variations of abscisic acid, gibberellin and cytokinin contents of rice plant during the entire growth cycle have not been reported (Osada *et al.*, 1973). However, the full extent of inhibition by mevinolin and its relation to endogenous concentration of these isoprenoid hormones cannot be assessed because of the extremely low concentrations in treated seedlings of *O. sativa* cv. Tainon 67. Further quantitative analyses of abscisic acid and gibberellins in conjunction with mevinolin treatment are currently in progress.

ACKNOWLEDGEMENTS

We thank Dr A. W. Alberts, Department of Biochemical Regulation, Merck Sharp & Dohme Research Laboratory, for a generous gift of reference sample of mevinolin. The technical assistance from Miss M-C Cheng is also highly appreciated. This work was supported by the National Science Council, Republic of China.

REFERENCES

Alberts, A. W., Chen, J., Kuron, G., Hunt, V., Huff, J., Hoffman, C., Rothrock, J., Lopez, M., Joshua, H., Harris, E., Patchett, A., Monaghan, R., Currie, S., Stapley, E., Albers-Schonberg, G., Hensens, O., Hirshfield, J., Hoogsteen, K., Liesch, J., and Springer, J. (1980). Mevinolin: a highly potent competitive inhibitor of hydroxymethylglutaryl coenzyme A reductase and a cholesterol-lowering agent, *Proc. Nat'l. Acad. Sci. USA*, **77**, 3957–3961.

Arebalo, R. E. and Mitchell, E. D. Jr. (1984). Cellular distribution of 3-hydroxy-3-methylglutaryl coenzyme A reductase and mevalonate kinase in leaves of *Nepeta cataria*, *Phytochemistry*, **23**, 13–18.

Bach, T. J., and Lichtenthaler, H. K. (1983). Mechanism of inhibition by mevinolin (MK 803) of microsome-bound radish and of partially purified yeast HMG-CoA reductase (EC 1.1.1.34), *Z. Naturforsch.*, C: Biosci., **38C**, 212–219.

Bach, T. J., and Lichtenthaler, H. K. (1982). Inhibition by mevinolin of mevalonate formation and plant root elongation. *Naturwissenschaften*, **69**, 242.

Bearder, J. R., MacMillan, J., and Phinney, B. O. (1975). Fungal Products. Part XIV, Metabolic pathway from ent-kaurenoic acid to the fungal gibberellins in mutant B1-41a of *Gibberella fujikuroi*, *J. Chem. Soc., Perkin Trans.*, **1**, 721–726.

Berry, D. L., Singh, B., and Salunkhe, D. K. (1972). Incorporation of β-hydroxy-β-methylglutaric acid into carotenoids and other ether soluble compounds in maize seedlings, *Plant Cell Physiol.*, **13**, 157–165.

Brooker, J. D., and Russell, D. W. (1975). Properties of microsomal 3-hydroxy-3-methylglutaryl coenzyme A reductase from *Pisum sativum* seedlings, *Arch. Biochem. Biophys.*, **167**, 723–729.

Brown, M. S., and Goldstein, J. L. (1980) Multivalent feedback regualtion of HMG-CoA reductase, a control mechanism coordination isoprenoid synthesis and cell growth, *J. Lipid Res.*, **21**, 505–517.

Brown, M. S., Faust, J. R., and Goldstein, J. L. (1978). Induction of 3-hydroxy-3-methylglutaryl coenzyme A reductase activity in human fibroblasts incubated with compactin (ML-236B), a competitive inhibitor of the reductase, *J. Biol. Chem.*, **253**, 1121–1128.

Brown, A. G., Smale, T. C., King, T. J., Hasenkamp, R., and Thompson, R. H. (1976). Crystal and molecular structure of compactin, a new antifungal metabolite from *Penicillium brevicompactum*, *J. Chem. Soc., Perkin Trans.*, **1**, 1165–1170.

Cole, R. J., and Cox, R. (eds) (1981). *Handbook of Toxic Fungal Metabolites*. Academic Press, New York.

Dockerill, B., and Hanson, J. R. (1978). The fate of C-20 in C-19 gibberellin biosynthesis, *Phytochemistry*, **17**, 701–704.

Endo, A., Kuroda, M., and Tsujita, Y. (1976). ML-236A, ML-236B, and ML-236C, new inhibitors of cholesterogenesis produced by *Penicillium citrinum*, *J. Antibiotics*, **29**, 1346–1348.

Hashizume, T., Matsubara, S., and Endo, A. (1983). Compactin (ML-236B) as a new growth inhibitor of plant callus, *Agric. Biol. Chem.*, **47**, 1401–1403.

Hedden, P., Phinney, B. O., MacMillan, J., and Sponsel, V. M. (1977). Metabolism of kaurenoids by *Gibberella fujikuroi* in the presence of the plant growth retardant, N,N,N-trimethyl-1-methyl-(2′,6′,6′-trimethylcyclohex-2′-en-1′-yl)prop-2-enylammonium iodide, *Phytochemistry*, **16**, 1913–1917.

Lewer, P., and MacMillan, J. (1983). Effect of compactin on the incorporation of mevalonolactone into gibberellic acid by *Gibberella fujikuroi*, *Phytochemistry*, **22**, 602–603.

Osada, A., Suge, H., Shibukawa, S., and Noguchi, I. (1973). Changes of endogenous gibberellin in rice plants as affected by growth stage and different growth conditions, *Proc. Crop Sci. Soc. Japan*, **42**, 41–45.

Ryder, N. S., and Goad, L. J. (1980). The effect of the 3-hydroxy-3-methylglutaryl CoA reductase inhibitor ML-236B on phytosterol synthesis in *Acer pseudoplantanus* tissue culture, *Biochim. Biophys. Acta*, **619**, 424–427.

Schroepfer, G. J. Jr. (1981). Sterol biosynthesis. *Ann. Rev. Biochem.*, **50**, 585–621.

Sembdner, G., Gross, D., Liebisch, H.-W., and Schneider, G. (1980). In *Encyclopedia of Plant Physiology, New Series*, Vol. 9 *Hormonal Regulation of Development*, **1**: *Molecular Aspects of Plant Hormones* (ed. J. MacMillan), pp. 281–444, Springer-Verlag, Berlin.

Shiao, M.-S. (1983). Inhibition of gibberellin biosynthesis in *Gibberella fujikuroi* and germination of *Oryza sativa* by mevinolin, *Bot. Bull. Acad. Sinica*, **24**, 135–143.

Suzuki, H. (1975). HMG-CoA reductase in higher plants. *Biosynthesis of Isoprenoids, Kagaku To Seibutsu*, **13**, 771–772.

Wong, R. J., McCormack, D. K., and Russell, D. W. (1982). Plastid 3-hydroxy-3-methylglutaryl coenzyme A reductase has distinctive kinetic regulatory features: properties of the enzyme and positive phytochrome control of activity in pea seedlings. *Arch. Biochem. Biophys.*, **216**, 631–638.

Trichothecenes and Other Mycotoxins
Edited by J. Lacey

CHAPTER 35

The use of a cytotoxicity test as a screening test for mycotoxins

Jean Robb and Mary Norval

INTRODUCTION

The ability of fungi to produce mycotoxins in forages and stored feedingstuffs is well known. As a result of this potential hazard, or because of specific disease problems or sub-optimal performance in poultry and animals, animal feedingstuffs are routinely screened for mycotoxins. Usually solvent extracts of feedingstuffs are subjected to a clean-up stage and then analysed by thin-layer, gas–liquid or high-pressure liquid chromatography. In many instances multimycotoxin methods (Gorst-Allman *et al.*, 1979) are used, making it possible to extract several mycotoxins of different chemical structures in the one procedure. These methods are often relatively insensitive, particularly for trichothecene toxins, and are only able to detect compounds of known chemical composition providing the necessary standards are available.

It is possible therefore that animals may eat feeds containing mycotoxins which routine laboratory methods cannot detect. Such a situation occurred in Scotland during the winter of 1980–81 when sub-optimal growth, poor feathering and behavioural abnormalities were seen in broilers (Robb *et al.*, 1982). Examination of poultry meal extracts by thin-layer chromatography (Patterson and Roberts, 1979) failed to demonstrate mycotoxins but bioassay with a tissue culture of HEpII cells enabled identification of the toxic component of the feed. Since then the sensitivities of the cytotoxicity and thin-layer chromatography methods have been compared using toxin standards and extracts from feedingstuffs. This chapter describes the results of these experiments.

METHODS

Thirty-three standard preparations of mycotoxins were used; they were aflatoxin G_1, cytochalasin B and D, scopoletin (Sigma), aflatoxin B_1, B_2, G_2

and M_1, citrinin, diacetoxyscirpenol, HT-2 toxin, luteoskyrin, 5-methoxysterigmatocystin, O-methyl sterigmatocystin, ochratoxin A, patulin, penicillic acid, PR-toxin, rubratoxin B, sterigmatocystin, T-2 tetraol, T-2 toxin and zearalenone (Makor Chemicals, Israel), roridin, T-2 triol, verrucarin A and verrucarol (Cambridge Research Biochemicals Ltd), 4-acetoxy-3,15-scirpendiol, 15-acetoxy-3,4-scirpendiol, scirpentriol, triacetoxyscirpene and moniliformin (all kindly donated by Dr P. S. Steyn) and deoxynivalenol (kindly donated by Dr M. Shepherd). Each was dissolved in acetone at 1 mg/ml and 10-fold serial dilutions made in acetone.

Animal feedingstuffs were extracted with acetonitrile, iso-octane and chloroform and then examined by thin-layer chromatography (Patterson and Roberts, 1979; Figure 1).

For TLC, plates (Merck Art 5554 DC-Alufolien Kieselgel 60F 254 and Art 5553 DC-Alufolien Kieselogel 60) were spotted with 5 μl samples of each concentration, run in toluene : ethyl acetate : formic acid 60 : 30: 10, and sprayed with 50% aqueous H_2SO_4 where necessary, before viewing in a UV cabinet (short-wave 254 nm, long-wave 366 nm). The lowest dilution at which spots could be detected visually was noted.

For biological assay, HEpII cells were used. The cells were maintained on coverslips in 1 ml Earles-based Eagles medium supplemented with 5% newborn calf serum, 200 i.u./ml penicillin and 20 μg/ml streptomycin, and subcultured by trypsinization when confluent. In the test, they were cultured on coverslips in 1 ml medium until semi-confluent. One microlitre (1 μl) of each dilution of the standard or 1 μl of feedingstuff extract was added and, after 48 hours incubation at 37 °C, the coverslips were removed, the cells fixed in ethanol, stained by Giemsa's and examined microscopically to assess cytotoxicity. The lowest dilution at which a cytocidal effect, i.e. marked inhibition of growth, cell-rounding and detachment from the surface, was apparent was noted. Observations were made in triplicate and cytotoxic levels were found to vary by a maximum of 1 dilution.

RESULTS

Table 1 shows that there was much variation in the level at which the standard mycotoxins were detected in the assay systems employed, from undetectable at 1000 ng to visible spots at 0.05 mg in TLC of citrinin and ochratoxin A respectively or cytotoxicity to HEpII cells at only 0.0001 ng of roridin A. While the TLC method detected most compounds, it was less sensitive than the cytotoxic assay for at least 13 of the 33 tested. This difference in sensitivity was most striking for the trichothecenes, e.g. T-2 toxin and verrucarin A.

The results in Table 2 show that, apart from the turkey feeds and the tapioca, no individual feedingstuff was consistently toxic to HEpII cells.

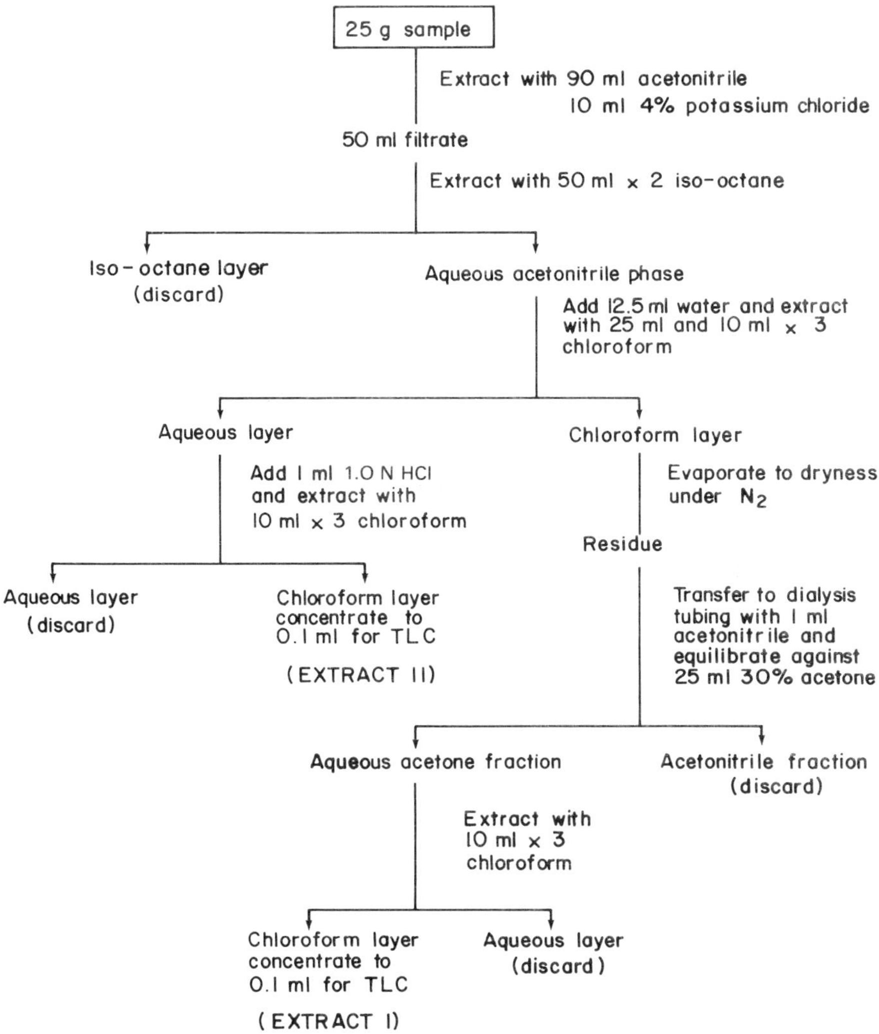

Figure 1 Extraction and clean-up procedure for the detection of mycotoxins in animal feeds.

However, few of these were tested, and these were directly related to a disease problem so that the results may not be representative. A supplement used in the poultry feeds was consistently toxic, but its effects in the final feed were diluted.

Table 1 Lowest levels at which mycotoxins were detected (ng) in various tests

Mycotoxin	Cytotoxicity HEpII	Thin-layer chromatography
4-acetoxy-3,15-scirpendiol	100	50
15-acetoxy-3,4-scirpendiol	100	500
Aflatoxin B_1	100	0.5
Aflatoxin B_2	100	0.5
Aflatoxin G_1	*	0.5
Aflatoxin G_2	1000	0.5
Aflatoxin M_1	10	0.5
Citrinin	10	0.05
Cytochalasin B	*	500
Cytochalasin D	10	50
Deoxynivalenol	100	500
Diacetoxyscirpenol	1	500
HT-2 toxin	10	500
Luteoskyrin	*	500
5-methoxysterigmatocystin	not done	5000
O-methyl sterigmatocystin	100	5000
Moniliformin	*	5000
Ochratoxin A	*	0.05
Patulin	100	500
Penicillic acid	1000	50
PR-toxin	1	50
Roridin A	0.0001	500
Rubratoxin B	*	*
Scirpentriol	100	50
Scopoletin	*	0.5
Sterigmatocystin	100	5000
T-2 tetraol	10	500
T-2 toxin	10	500
Triacetoxyscirpene	10	500
T-2 triol	100	500
Verrucarin A	0.1	5000
Verrucarol	1000	50
Zearalenone	100	50

* Not detected at 1000 ng.

DISCUSSION

Although TLC detected all the compounds used it was less sensitive than the cytotoxicity assay for 13 of the 33 mycotoxins tested. This was most striking for the trichothecene group. Thus this bioassay could provide a possible screening method for these mycotoxins. However, the test is not selective and there is always the possibility that some of the constituents of a mixed

Table 2 Cytotoxicity

Feed type	Cytotoxicity extract I		Cytotoxicity extract II	
	+ve	−ve	+ve	−ve
Barley	1	5	0	6
Wheat	7	14	1	20
Starter pellets	33	20	7	46
Maize	8	18	3	23
Starter crumbs	22	7	11	18
Broiler finisher	9	1	3	7
Turkey feed	2	0	2	0
Soya	6	5	3	8
Fat	1	2	1	2
Soffe	0	1	0	1
Supplement	3	0	1	2
Tapioca	1	0	0	1
Silage	6	18	0	22
Cattle cake	2	17	0	19
Grass	0	4	0	4
Sow mix	2	12	0	14
Potatoes	0	4	0	4
Draff	6	15	1	20
Oats	1	10	0	11
Straw	0	6	0	6

feedingstuff may be responsible for the toxicity. Untreated cereals or forages are less likely to contain interfering compounds. Apart from demonstrating sensitivity to the trichothecenes, the test will also indicate the presence of other mycotoxins, known and unknown, which are not included in the routine standards used for TLC or GLC analysis.

Lompe and Milezewski (1979) described the effect of 16 mycotoxins on three human and two porcine cell lines and found varying levels of sensitivity, suggesting that several cell lines of different origin might be used in screening. If extracts were passed through a clean-up cartridge such as Sep-Pak and aliquots tested against various cell lines, feedingstuffs which gave positive results could be re-extracted and analysed by the most sensitive chemical method available for the types of toxic compounds indicated by the cytotoxicity tests.

ACKNOWLEDGEMENTS

We wish to thank Dr T. Simpson for supplying some of the mycotoxins used in this work and for helpful advice. We are also grateful to Dr Brian Flannigan for helpful criticism.

REFERENCES

Gorst-Allman, C. P., and Steyn, P. S. (1979). Screening methods for the detection of thirteen common mycotoxins. *J. Chrom.*, **175**(2), 325–337.

Lompe, A., and Milezewski, K. E. (1979). Ein zellkulturtest für den nachweis von mykotoxinen. *Z. Lebensim Unters, Forsch.*, **169**, 249–254.

Patterson, D. S. P., and Roberts, B. A. (1979). Mycotoxins in animal feedstuffs: sensitive thin layer chromatographic detection of aflatoxin, ochratoxin A, sterigmatocystin, zearalenone and T-2 toxin. *J. Assoc. Off. Anal. Chem.*, **62**, 1265–1267.

Robb, J., Kirkpatrick, K., and Norval, M. (1982). Association of toxin-producing fungi with disease in broilers. *Vet. Rec.*, **111**, 389–390.

VETERINARY AND MEDICAL IMPLICATIONS OF MYCOTOXINS

Trichothecenes and Other Mycotoxins
Edited by J. Lacey

CHAPTER 36

Mycotoxins of veterinary importance in the United States

L. G. Morehouse

INTRODUCTION

During the past two decades mycotoxins have been implicated in a variety of disease syndromes in domestic food-producing animals and poultry in the United States. They may also affect productivity and performance through reduced growth rate and modified feed conversion, and there is now evidence that some toxic metabolites are immunosuppressive. These deleterious effects may stem from: (1) decreased palatability; (2) unavailability to animals of major nutrients used for the growth processes of moulds; and (3) accumulation of toxic metabolites in feedstuff.

At least 50 different genera of fungi in four major taxonomic groups are known to produce one or more toxic metabolites. However, the principal concern for animal health relates to toxins which are orally active or sufficiently poisonous by other routes to cause disease.

Clinical signs and post-mortem lesions are rarely pathognomonic in mycotoxicoses and differential diagnosis is necessary to eliminate infectious and toxic agents that may produce similar signs and lesions. History of exposure to mouldy feeds, along with actual identification of fungi, are helpful pointers but in themselves are not sufficient evidence to confirm the diagnosis. The presence of mycotoxins appropriate to the clinical signs and lesions is helpful for more informed diagnosis.

The laboratory diagnostician may be frustrated by a lack of appropriate samples of the original feed for testing. Consequently, failure to detect mycotoxins may be the result of sample rather than analytical deficiencies. Sampling is most effective when samples are obtained from a moving stream of grain at periodic intervals, but probe samples or spot samples can be satisfactory if at least ten locations are sampled throughout the suspect feedstuff to produce a meaningful composite sample (at least 5 kg). The sample should be taken randomly from the source if it is uniform in appearance and, additionally, where visible mould-damaged grain or forage is observed this material should be submitted in an air-tight container. Samples should be kept cool and dry after collection.

Considerable progress has been made over the past decade in the identification of mycotoxins in the United States; the establishment of full-service veterinary medical diagnostic laboratories (VMDLs) designed, equipped and staffed specifically to offer a complete diagnostic service, have been central to this area of clinical development. Nine of the new laboratories are in colleges of veterinary medicine, six in departments of veterinary science at universities, and three in State Department of Agriculture Centers. The majority of the laboratories are developing analytical capabilities for detecting mycotoxins, although some are restricting analytical efforts to those mycotoxins they feel represent the greatest health hazard to livestock and poultry populations in their specific locale. Twenty-two VMDLs in 20 states of the U.S.A. were contacted in early 1983 to determine the number of laboratories that were conducting analyses for the various mycotoxins and the analytical procedures being used.

The specific mycotoxins and the number of VMDLs that had developed analytical capability for them are as follows: aflatoxins – 22 (nine conducting analyses for B_2, G_1, and G_2 in addition to B_1); zearalenone – 16; ochratoxin – 14 (two for ochratoxin B); T-2 toxin – 12; deoxynivalenol (DON) – 9; diacetoxyscirpenol (DAS) – 8; ergot alkaloids – 4; citrinin – 3; sterigmatocystin – 3; slaframine – 1; penitrem A – 1. Our laboratory offers a multi-mycotoxin screen test for aflatoxin and ochratoxin in the *Aspergillus–Penicillium* group; T-2 toxin, diacetoxyscirpenol, vomitoxin, zearalenone and coumestrol in the *Fusarium* group. Sixteen of the laboratories utilized thin-layer chromatography (TLC) and nine included high-pressure liquid chromatography (HPLC) capability. Three laboratories had facilities for gas chromatography/mass spectrophotometry (GC/MS) as confirmatory procedures. Specimens for more detailed examinations frequently were referred to the National Veterinary Services Laboratories at Ames, Iowa, or to other major full-service VMDLs or research laboratories in the United States.

Aflatoxin was the most frequently encountered and significant mycotoxin at diagnostic laboratories in the midwestern, southern and southeastern United States. Deoxynivalenol (DON) and zearalenone were of greatest concern in the midwestern United States and ergot in the northwestern United States.

There seems little question that diagnostic laboratories remain hampered by lack of simple, rapid, inexpensive analytical procedures for mycotoxins. The detection of aflatoxins is facilitated by their fluorescence and absorbance qualities utilizing TLC or HPLC, but the fusarial toxins that may be involved in clinical problems (including zearalenone, T-2 and deoxynivalenol) do not have functional groups permitting their easy identification by spectrophotometric methods; extensive preparatory procedures are necessary prior to final analysis by TLC, GC and GC/MS.

A rapid screening method for aflatoxin B_1, zearalenone, ochratoxin A, T-2 toxin, diacetoxyscirpenol and vomitoxin in common grains has recently been described in this laboratory (Rottinghaus *et al.*, 1982). The mycotoxin extracts are decolorized through disposable alumina reverse-phase C18 and mycochar columns and identified by TLC. The sensitivity of the method is similar to other multi-mycotoxin screening methods, but less time-consuming. The procedure is used in five of the 22 full-service VMDLs and a further adoption of the technique within the diagnostic service should improve the diagnosis of some of the most frequently occurring mycotoxins in the United States.

Enzyme-linked immunosorbent assay (ELISA) tests, which are now under development for detection of various mycotoxins, may prove helpful in future diagnostic work. In this approach the mycotoxin is modified to produce a derivative with a functional group that can be conjugated to bovine serum albumin. Specific antibodies may then be raised by hyperimmunization of rabbits. Mycotoxins for which ELISA detection methodologies have been described include aflatoxin B_1 (Pestha *et al.*, 1980), aflatoxin M_1 (Pestha *et al.*, 1981a), ochratoxin A (Pestha *et al.*, 1981b), and trichothecenes in *Fusarium*-infected corn (Gendloff *et al.*, 1983).

Table 1 summarizes the pathological effects that have been attributed to mycotoxins, and illustrates that all body systems may be affected by individual, or combinations of, mycotoxin substances. A review of knowledge concerning the mycotoxicoses in domestic animals and poultry follows.

AFLATOXICOSIS

In the United States the presence of aflatoxin B_1 in corn has been emphasized, although this toxin may be produced on many other substrates. This emphasis has evolved because of the large acreages grown and the extensive use of maize in livestock feed; problems are encountered in both field and storage maize, particularly in the south-eastern United States. In the fall of 1977, for example, it was estimated that approximately 50% of the maize harvested in Florida contained more than 400 μg/kg aflatoxin B_1 (Edds, 1977). Table 2 summarizes the principal lesions of aflatoxicosis and indicates their likely etiology as reviewed by Patterson (1977).

Clinical manifestations may be due to the influence of the aflatoxin B_1 molecule on carbohydrate metabolism, mitochondrial function, lipid metabolism and protein biosynthesis and various mechanisms have been propounded to account for these effects (Hsieh, 1979).

Cattle

In cattle, a feature of aflatoxin poisoning is reduced growth rate and decreased milk production (Keyl, 1978; Garrett *et al.*, 1968). Young cattle are

Table 1 Pathological effects of mycotoxins on various body systems (Morehouse, 1981)

System	Pathological lesions or clinical signs	Toxins
Cardiovascular	Localized vasoconstriction	Ergotamine alkaloids
	Vascular fragility	Aflatoxin B_1
	Haematomas, haemorrhage into body cavities	Aflatoxin B_1, T-2 toxins, 4-Hydroxycoumarin (dicoumarol)
Digestive	Diarrhoea, atony, anorexia, tenesmus, liver necrosis, bile duct proliferation, fibrosis	Aflatoxin B_1
	Caustic effect on mucous membranes	T-2 and Stachybotrys toxins
	Intestinal haemorrhage	Aflatoxin B_1 and T-2 toxin
	Bile duct occlusion and resultant photosensitization	Sporidesmin
	Excessive salivation	Slaframine
	Vomition and feed refusal	Deoxynivalenol (DON)
Nervous	Tremors, inco-ordination, mania, ataxia, coma	Tremorgenic mycotoxins, ergotamine alkaloids, *Fusarium moniliforme* (leukoencephalomalacia)
Respiratory	Adenomatosis	4-ipomeanol
Urinary	Nephrosis, uremia	Ochratoxin A, Citrinin
Reproductive	Abortion, stillbirth	Aflatoxin B_1
	Infertility, prolonged oestrus	F-2, Zearalenone
	Small litters, weak pigs	T-2 toxin
Cutaneous	Inflammatory or necrotic change	T-2 toxin, butenolide
	Photosensitization	Sporidesmin

most susceptible; it has been shown experimentally that dietary levels as low as 200 μg/kg over a 3-month period decreased both weight gain and feed efficiency in 4-day-old calves (Allcroft, 1969). However, at 1 month of age the same insult for a 14-day period caused depression, inappetance and loss of weight gain after 14 days; 500 μg/kg per day proved fatal on day 14 (Pier *et al.*, 1976). Other studies have shown that 5-week-old calves with 2000 μg/kg aflatoxin in the diet had apparent loss of weight at 1 month of age and

Table 2 Common lesions associated with aflatoxicosis and their underlying cause (Patterson, 1977)

Sign or lesion	Underlying cause
Haemorrhages	Increased capillary fragility, increased prothrombin time
Hepatotoxic effects	
Fatty change	Lipid accumulation due to impairment of lipid transport from the liver
Liver necrosis	Degranulation of endoplasmic reticulum
Bruising of muscle	Induced membrane instability
Immunosuppressive effects	Humoral: suppression of protein synthesis Cellular: Generalized lymphoid depletion
Mutagenic effects:	
Hepatocarcinogenic action Teratogenic effects	B_1 or an active metabolite (epoxide) interacts with nucleic acid to induce mutation and interference with DNA template for protein synthesis

terminal signs developing at 4–6 months with evidence of abdominal pain and blood-stained diarrhoea (Allcroft and Lewis, 1963), whereas in older steers (6–8 months) aflatoxin dietary levels of 700–1000 μg/kg resulted in significant decreases in weight gain, food consumption and feed efficiency over the same period.

Lower levels of aflatoxin in calves may contribute to an ill-thrift syndrome; for example, oral doses of several aflatoxin metabolites (mixtures of B_1, B_2, G_1 and G_2) at levels of only 20–80 μg/kg daily for 6 weeks resulted in depression of nitrogen utilization and increase in urinary nitrogen excretion resulting from decreased feed intake and catabolic activity (Lynch *et al.*, 1973). The literature contains few reports of naturally occurring aflatoxicosis in adult cattle in the United States but the high levels of aflatoxin B_1 in field corn in the south-eastern states have been suggested to be responsible for unthrifty dairy cows, slow growth in replacement heifers, and marked drop in milk production (Edds, 1979). Experimentally, dairy cattle on dietary levels of 2300–4200 μg/kg aflatoxin from 3 to 4 weeks had a marked deterioration in health and marked reduction of feed consumption and milk production (Keyl, 1978).

The gross necropsy findings in bovine aflatoxicosis include generalized icterus, ascites, visceral oedema, congestion of intestines and kidneys, haemorrhages in the mesentery and fibrosis of the enlarged livers. Microscopically, there may be proliferation of bile duct epithelium, chronic obliterating endophlebitis of centrilobular and hepatic veins, a variation in size and shape of parenchymal cells, and a diffuse hepatic fibrosis. In a group of 6–8-month-old cross-bred Hereford steers given diets containing 100, 300, 700 and 1000 μg/kg of an aflatoxin B_1-fortified cottonseed meal ration over a 19–27-week period, gross changes in the liver were seen in only those animals given the 700 and 1000 μg/kg dosage levels. Four of 10 animals at the 700 μg/kg level had post-mortem evidence of hepatomegaly with fibrosis of the liver, while 6 of the 10 at the 1000 μg/kg level showed these changes. Microscopically, mild liver changes manifested primarily as 'irregularity of hepatocyte size' were discerned in 3 of 10 of the 300 μg/kg group, while livers from all animals in the 700 and 1000 μg/kg group showed some microscopic abnormalities ranging from irregularity of hepatocyte size to varying degrees of bile duct proliferation and fibrosis (Garrett *et al.*, 1968).

Swine

There are excellent descriptions of acute and chronic aflatoxicosis in swine from the United States and England (Cysewski *et al.*, 1968; Sisk *et al.*, 1968; Armbrecht *et al.*, 1978; Jones and Jones, 1969; Gagne *et al.*, 1968; Miller *et al.*, 1978). Swine have been reported to die within 48 hours with acute aflatoxicosis. Depression, muscular weakness, inco-ordination, anorexia, pale or yellow mucous membranes, haemorrhagic faeces and terminal convulsions are described as clinical features of acute aflatoxicosis in pigs.

In chronic aflatoxicosis, affected pigs show depression, anorexia and weight loss, and move with a stiff gait; the dramatic stunting effect of chronic aflatoxicosis in swine has been well documented (Sisk *et al.*, 1968). A number of research reports and recent experiences in observing the effects of various aflatoxin insults in swine suggest that the clinical effects are dose-related (Table 3).

Table 3 Effects of various levels of exposure of swine to aflatoxin*

Exposure level (μg/kg)	Effect
150–300	Reduced feed efficiency; decreased growth rate
200	Serum protein alteration
400–500	Organ change – gross pathology
800	Death loss – gradual
2000	Death loss – 1–3 days
450–1500	Breeding herds: abortions, stillbirths

* Adapted from *Aflatoxin in Corn and Other Commodities*, ESC-587, Extension Service, U.S. Department of Agriculture, Washington, D.C.

The liver is the principal target organ of aflatoxin B_1. The gross pathological changes are of a generalized 'fatty' tan liver with red foci. In the carcass generalized icterus may be observed along with gross haemorrhage on mucosal surfaces of intestinal tract (jejunum and ileum), heart and gluteal muscles. In some cases fresh blood may be found in the terminal ileum and large bowel. Microscopic examination of liver confirms fatty change with centrilobular congestion accompanied by necrosis of hepatocytes and haemorrhages; these changes may involve the centre of the lobules, resulting in numerous red foci and producing the 'nutmeg' appearance of the liver. In pigs chronically exposed to aflatoxin, there is an increase in fibrous connective tissue in the portal areas of the liver with proliferation of bile ductal epithelium and hypertrophy of the epithelial cells of the bile ducts.

Sheep

Experimental exposure of sheep to aflatoxin B_1 (4 mg/kg body weight) resulted in clinical signs of weakness, inappetance, diarrhoea, ruminal atony, haemorrhagic rectal discharge and fever in acute stages of the intoxication (Armbrecht, 1978). The lack of reports of field outbreaks of aflatoxicosis in adult sheep is probably because they are generally grazed on rangelands or pastures without exposure to aflatoxigenic fungi. However, the intoxication suffered by other species could occur if sheep are fed contaminated finishing or fattening rations in confinement.

Horses

Acute aflatoxicosis has been reported to occur in young Shetland ponies orally exposed to 1, 2 or 4 mg/kg body weight of aflatoxin B_1 (Edds, 1979). Naturally occurring episodes of acute aflatoxin poisoning in horses have been described (Angsubharkorn *et al.*, 1981). Histopathologically the lesions included encephalomalacia of cerebral hemispheres, fatty degeneration, bile duct hyperplasia and fibrosis of the liver, fatty infiltration of the kidney, haemorrhagic enteritis and myocardial degeneration.

Poultry

In their susceptibility to aflatoxin insult, poultry vary from the extremely sensitive duckling through turkey poults to chickens. The duckling manifests characteristic tissue responses (specifically bile duct hyperplasia) at very low dosage levels and this fact is the basis of a standard bioassay system for aflatoxin.

The turkey is reported to be ten times more sensitive to aflatoxin than the chicken (Morehouse, 1981). A detailed history of Turkey X disease in

England in the early 1960s later diagnosed as aflatoxicosis has been published, along with a general discussion of aflatoxicosis in turkeys and other poults (Austwick, 1978). Clinically the signs of aflatoxicosis in poultry include general lassitude, ruffling of feathers, diarrhoea, ataxia and opisthotonus with terminal convulsions in some acute cases. Upon gross examination there are haemorrhages in many tissues, generalized oedema, myocardial congestion, pericardial distension, and swollen kidneys with petechial haemorrhages. Gross changes in the liver vary from swelling and congestion to areas of haemorrhagic and necrotic foci to pale, shrunken appearance in chronic cases; microscopic examination reveals marked vacuolization of hepatic cells and diffuse necrosis. These changes may be superseded by a regeneration phase in which proliferation of the bile duct epithelium (characteristic of hepatic response in other species) is observed. Fibrosis and thickening of the hepatic capsule has been reported to be a characteristic response to aflatoxicosis in poultry which may be related to the sensitivity of the avian liver to toxic insult (Campbell, 1955–57).

The LD_{50} (mg/kg) varies widely even within strains of a given breed of chicken (e.g. 6.5–16.50 within six different strains of leghorns) (Edds, 1979); Barred Rocks and hybrid broilers are relatively resistant to aflatoxin, while New Hampshire chicks have been reported to be very sensitive (Vohara, extension release, University of California). Accounts of field outbreaks of aflatoxicosis in chickens in the U.S. have been published (Smith and Hamilton, 1970; Hamilton, 1971), and the effects of dietary aflatoxin on fertility, hatchability and progeny performance of layer hens has been well-defined experimentally (Austwick, 1978). Egg production was significantly decreased by the third week when diets of mature broiler breeder hens were supplemented with 5 and 10 μg/g aflatoxin B_1 (Howarth and Wyatt, 1976). Fertility did not seem affected at this level of dietary aflatoxin exposure.

OCHRATOXIN AND CITRININ TOXICOSIS

These mycotoxins, produced by several different species of the genera *Aspergillus* and *Penicillium*, are primarily nephrotoxic and to a lesser extent hepatotoxic. Ochratoxin A is the major ochratoxin metabolite of importance in animal disease.

Cattle

Decreased weight gains and reduction of feed intake may be observed; polyuria, with urine of low specific gravity, are salient clinical features. In gross examination the kidneys appear pale, and microscopically there is evidence of tubular degeneration and interstitial fibrosis. One-month-old

Jersey calves given 0.1 and 0.5 mg/kg body weight of ochratoxin A orally for 30 days had polyuria, while calves given 1–2 mg/kg body weight showed depression, reduced weight gain, low urine specific gravity and dehydration (Pier *et al.*, 1976). Holstein calves of the same age administered ochratoxin A by stomach tubes at levels of 11 and 25 mg/kg body weight died within 24 hours with epicardial haemorrhage as the only gross or microscopic abnormality (Ribelin *et al.*, 1978).

'Natural' outbreaks of ochratoxicosis in cattle would be expected to be transient or chronic, reflecting the generally low level of feed contamination. Experimentally, adult cattle fed more than 20 times the amount of ochratoxin likely to occur in nature exhibited transitory clinical signs of anorexia, diarrhoea and abrupt cessation of milk production (Ribelin, 1978a). However, it has been shown experimentally that naturally occurring levels of ochratoxin are sufficient to produce toxicosis in young preruminant calves (Pier *et al.*, 1976).

Swine

Naturally occurring ochratoxicosis in swine has been reported primarily in Denmark and Sweden, although unofficial reports indicate its occurrence in a number of countries including the United States. Classical ochratoxicosis in swine is a chronic renal disease with low mortality, although a more acute disease also characterized by polydipsia and polyuria has been reproduced with nephrotoxic fungal cultures of *Penicillium viridicatum* (Carlton and Tuite, 1970). This may be akin to a toxicosis described in newly weaned pigs with perirenal oedema and renal pathology identical to that of mycotoxic nephropathy (Larson *et al.*, 1962).

Clinically ochratoxicosis is characterized by ataxia, stiff arched back and abdominal distension in the paralumbar region. In gross examination the kidneys are enlarged and pale with an uneven surface; microscopically there is degeneration of tubular epithelium and formation of connecting tissue in the interstitium, and the glomeruli may be hyalinized with cysts in the cortex (Elling and Moller, 1973).

Renal lesions identical to those reported in field cases have been reproduced using ochratoxin A-contaminated feeds at levels of 200, 1000 and 4000 μg/kg (Krogh *et al.*, 1974). These levels corresponded to naturally occurring ochratoxin A concentration in animal feeds associated with porcine nephropathy (Krogh *et al.*, 1973), and were fed over a 4-month period. The lowest level resulted only in microscopic lesions, but the two higher levels resulted in the characteristic macroscopic renal lesions of the natural disease.

Ochratoxin A has been linked to foetal death in laboratory and domestic animals and teratogenic and toxic effects have been reported for this compound in rats (Still *et al.*, 1971; Brown *et al.*, 1976), but natural occurrences of such effects in domestic or food-producing animals have not been reported in the United States.

Poultry

Ochratoxin poisoning in poultry has been described experimentally in both acute and chronic manifestations (Peckham *et al.*, 1971). The acute toxicity of the ochratoxin compounds for poultry has been studied and is the subject of two reviews (Carlton and Krogh, 1979; Peckham, 1978). Single-dose oral LD_{50} values of ochratoxin A for 1-day-old chicks are reported to be in the range of 3.4 ± 0.19 mg/kg body weight (Prior *et al.*, 1976). Ochratoxin is more acutely toxic to chicks than either aflatoxin or T-2 toxin (Doerr *et al.*, 1974). The minimum dietary growth inhibitory dose for the young broiler chick is 2 mg/kg for ochratoxin, whereas 2.5 and 4.0 mg/kg are required for growth inhibition by aflatoxin and T-2 toxin respectively.

The clinical disease leads to emaciation and dehydration. Gross examination reveals liver and kidney haemorrhages with acute nephrosis manifested at the cellular level by necrosis and vacuolation of hepatocytes. Ochratoxicosis may also cause osteoporosis and increased tibial diameters related to poor mineralization (Huff *et al.*, 1977). Mature laying hens may be severely affected at exposure levels from 1 to 4 mg/kg ochratoxin A (Chowdhury *et al.*, 1971; Prior and Sisodia, 1978), but levels as low as 0.5 mg/kg can decrease egg production, and body weight has been reported to be affected by insults of 1 mg/kg and above (Prior and Sisodia, 1978). In recent years, ochratoxicosis has been documented in broilers, layers and turkeys with field outbreaks occurring in seven states in the U.S. and levels of ochratoxin ranging from 0.3 mg/kg to 16.0 mg/kg (Wyatt, 1979).

TRICHOTHECENE TOXICOSIS

The trichothecene toxins, produced by fungi of *Fusarium* and *Myrothecium* sp., are skin and mucous membrane irritants. Experimental animals develop diarrhoea, rectal haemorrhaging and necrotic lesions of the oral cavity following their administration. The principal trichothecene compounds of veterinary interest are T-2 toxin, diacetoxyscirpenol (DAS), and deoxynivalenol or vomitoxin (DON), the latter occurring recently with zearalenone in maize and wheat with great frequency in the midwestern United States.

Cattle

Reports of field outbreaks of *Fusarium* toxicosis in cattle have described inappetance, ataxia, scouring and marked decrease in milk yield in dairy cattle (Fisher *et al.*, 1967). Generalized mucosal, serosal and subcutaneous haemorrhages in adult feedlot cattle have been reported (Hibbs *et al.*, 1974) and uncontrollable haemorrhaging following castration and dehorning in calves on *Fusarium*-contaminated silage (Dahlgren and Williams, 1972). Russian workers have indicated haemorrhagic diathesis to be the most common characteristic sign of fusariotoxicosis in cattle, which is manifested as diffuse haemorrhage in subcutaneous tissue and under the serosa of parenchymatous organs (Kurmanov, 1978). T-2 toxin in commercially prepared cattle rations and in mouldy corn has been associated with haemorrhaging and poor weight gain in cattle (Mirocha *et al.*, 1974a; L. D. Kintner and G. D. Osweiler, personal communication, 1977), and has followed consumption of contaminated Brewer's grains (Petrie, 1977).

Experimentally, the parenteral administration of T-2 toxin led to haemorrhaging on mucosal surfaces in cattle, including ecchymotic haemorrhages in the epicardium and ruminal, abomasal and intestinal walls (Kosuri *et al*, 1970). The mechanism for the reported haemorrhagic syndromes in trichothecene toxicosis is not clearly understood but it is known that in the rat, clotting and prothrombin time are delayed and vascular permeability is increased with resultant widespread haemorrhages (Smalley *et al.*, 1970). It is also suggested that synthesis of vitamin K by rumen micro-organisms may be inhibited by the antibiotic activity of trichothecenes (Gustafsson *et al.*, 1962). The characteristic haemorrhagic syndrome reported to occur in cattle in the field with T-2 exposure, and by some experimental studies, has not been reported by other workers administering purified T-2 toxin in cattle (Pier *et al.*, 1976; Weaver *et al.*, 1977; Osweiler *et al.*, 1981), but responses that include inappetance, dehydration, abomasal ulcers and 'bloody faeces' have been observed experimentally following T-2 administration orally (0.16–0.64 mg/kg) in calves; the LD_{50} in young calves with non-functional rumens was established at 0.6 mg/kg body weight (Ribelin, 1978b).

Swine

Young swine showed no gross microscopic or biochemical changes when T-2 was incorporated into rations at 1, 2, 4 and 8 mg/kg and fed for 8 weeks; swine offered T-2 toxin in diets at 24 and 32 mg/kg refused to eat, and those offered diets containing 16 mg/kg T-2 ate only sparingly (Weaver *et al.*, 1978a). Pregnant sows on diets containing 12 mg/kg T-2 toxin for 220 days became repeat breeders and had small piglets and small litters (Weaver *et al.*, 1978b). T-2 toxin administered intravenously at levels of 0.21–0.41 mg/kg caused abortion in sows with no lesions produced (Weaver *et al.*, 1978c).

Diacetoxyscirpenol (DAS) caused lingual and buccal mucosal lesions when fed at concentrations of 2–10 mg/kg for 9 weeks with decreased weight gains and erosions in the oesophagus (Weaver *et al.*, 1977). The LD_{50} of D.A.S. (i.v.) for swine is 0.376 mg/kg body weight (Weaver *et al.*, 1978d).

Vomitoxin or DON is associated with emesis and feed refusal in swine (Vesonder *et al.*, 1976). The minimum emetic dose given orally to pigs has been reported to be 0.1–0.2 mg/kg and purified vomitoxin reduced feed consumption by 20% at the 3.6 mg/kg level and by 90% at the 40 mg/kg level (Forsyth *et al.*, 1977).

The high incidence of vomitoxin contamination of grain is a cause of concern in the midwestern U.S.; in 1981 and 1982 there were reports of contamination in the maize belt from Iowa to New York, and in 1982 portions of the winter wheat crop from Kansas, Nebraska, Missouri and Iowa were contaminated with vomitoxin. In our laboratory a survey of 43 grain samples from 30 Missouri counties found 39 (91%) to contain over 0.1 mg/kg, 32 (74%) over 1 mg/kg, and 8 (19%) over 10 mg/kg vomitoxin, with the highest level found being 61 mg/kg in the fields surveyed (G. E. Rottinghaus, personal communication, 1982). On one farm where feed refusal was being encountered in swine, more than 2000 bushels of wheat were found to be contaminated with levels as high as 25 mg/kg.

Feed refusal of naturally infected maize blended with sound maize was much greater than that of DON-spiked corn containing the same amount of mycotoxin, and it has been postulated that other metabolites present in the corn may also be active as feed refusal factors (Forsyth *et al.*, 1977).

The widespread contamination of the 1982 crops of winter wheat in the U.S. led the FDA to recommend that no more than 4 mg/kg vomitoxin should be allowed in wheat and wheat products to be used as animal feed ingredients, with an added recommendation that these ingredients should not exceed 10% of swine and pet diets nor 50% of the diets of ruminants and poultry.

Poultry

The trichothecene compounds T-2 and DAS both have pathological effects on poultry, but they have been found to be highly resistant to vomitoxin. A diet containing sufficient naturally contaminated corn to provide a 50 mg/kg vomitoxin concentration was fed to 6-day-old broiler cockerels for 6 days with no effect on growth or feed consumption; however, plaques were observed in the mouth along with gizzard erosion (Moran *et al.*, 1982). Similar diets fed to laying hens for 4 weeks had no recorded effect on production, food consumption, body weight, egg weight or shell strength (Moran *et al.*, 1983).

Lesions of T-2 and DAS (Chi and Mirocha, 1978) toxicosis in poultry include caseous, necrotic lesions of the oral cavity, specifically on the hard palate and underside of the tongue; there is fatty degeneration and haemorrhaging in internal organs and marked reduction in leukocyte and thrombocyte counts.

Experimental studies have shown the effects of this toxin to vary in severity with age of the birds and duration of exposure to toxic insult. For orally administered T-2 toxin, LD_{50} (mg/kg) of 5.25, 4.97, and 6.27 have been given for 1-day-old chicks, 8-day-old broiler chicks, and laying hens respectively (Chi *et al.*, 1977a). Oral lesions develop at very low exposures to T-2 toxin, i.e. 0.4 mg/kg fed to broiler chicks from 1 day to 9 weeks of age (Chi *et al.*, 1977b), and 0.6 mg/kg fed to laying hens for 3 weeks (Chi *et al.*, 1977c). Neural toxicity effects in chickens occurred at dosages of 4, 8 and 16 mg/kg, which also are growth-retarding concentrations (Wyatt *et al.*, 1973). Other effects of the toxin in laying hens include decreased feed consumption, egg production and egg shell thickness. Intensive haemorrhaging in organs and tissues in poultry has occurred in experiments in which *Fusarium* contaminated diets were fed alone or in combination with T-2 toxin. A review of emetic effects of *Fusarium* spp. and the T-2 toxin in poultry has been published (Joffe, 1978), and possible field outbreaks of T-2/DAS toxicity have been described (Wyatt *et al.*, 1972).

ZEARALENONE TOXICOSIS (F-2 TOXIN)

Zearalenone, or F-2, can be produced by at least four different *Fusarium* species and is best known for its hyperoestrogenic effects in swine, although it may also exert clinical effects in cattle and poultry. There has been increased concern regarding this mycotoxin throughout the midwest because of widespread *Fusarium* contamination of wheat grain. F-2 toxin was found at levels from 0.1 to 8 mg/kg in more than 750 samples analysed in this laboratory during 1982.

Cattle

Impairment of bovine reproductive function has been associated circumstantially with zearalenone toxicity, though experimental confirmation is limited (Morehouse, 1979). In one dairy 25 mg/kg of F-2 toxin was found in the meal feed of the animals; *Fusarium graminearum* and *Fusarium culmorum* were isolated from the feed, and there were clinical signs of oestrogenism (Korpinen, 1972). In another herd, mouldy hay had been the vehicle for zearalenone to dairy cattle which subsequently showed decreased fertility (Mirocha *et al.*, 1968). This laboratory recommends that no more than 5 mg/kg zearalenone be allowed in the diet of non-lactating cattle or virgin heifers (Osweiler, 1982).

Swine

A zearalenone-induced oestrogenic syndrome in swine has been recognized clinically for many years.

The clinical signs include tumefaction of the vulva, which becomes swollen and oedematous and may result in vaginal prolapse, especially in the prepubertal gilt. The mucosa of the entire reproductive tract and myometrium are all thickened due to oedema and a combination of cellular hypertrophy and hyperplasia of the elements composing the walls of these structures (Kurtz *et al.*, 1969). These changes are characterized histologically by interstitial oedema of the submucosa and hyperplasia of submucosal endometrial glands; ovaries of prepubertal gilts exposed to F-2 toxin are grossly hypoplastic with small follicles and no evidence of corpora lutea formation. Mammary glands may respond with moderate enlargement and squamous metaplasia of ductal elements in mammary tissue (Kurtz *et al.*, 1969).

Levels as low as 1–2 mg/kg zearalenone in the total diet may cause signs of oestrus in some young female swine and high levels are considered highly likely to induce oestrus signs in young pigs (Osweiler, 1982). Research at this university has shown that as little as 10 mg/kg zearalenone will cause anoestrus for up to 90 days if fed to gilts or sows of breeding age. Consequently, we recommend that no zearalenone-containing feed be used in rations for breeding swine (Osweiler, 1982).

A physiological explanation for the effects of F-2 on reproductive performance is not available and there is still conflict about the extent to which F-2 toxin may be foetopathic. The clinical confusion may result from the possible production of trichothecene toxins, depending on the species of *Fusarium* involved, leading to a broader 'fusariotoxicosis complex' with reproductive problems in swine. Potential for such combined effects was noted when 'mixed' feeds containing 7, 38 and 64 mg of zearalenone/kg of feed and 0.5, 2.5 and 4.5 mg of deoxynivalenol/kg respectively were fed to swine in early pregnancy; foetus numbers were decreased in treated animals at slaughter and clinical signs of hyperoestrogenism and partial feed refusal was noticed in some of the high-dosage group animals (Long *et al.*, 1982).

Decreased libido and some degree of testicular atrophy have been reported in young boars fed *Fusarium*-infected maize (Bristol and Djurikovic, 1971), but exposure to zearalenone at levels commonly found in contaminated rations did not affect the reproductive potential of mature boars (Ruhr *et al.*, 1978).

Poultry

The feeding of *Fusarium*-contaminated grain has been reported to impair spermatogenesis in young ganders, and to induce infertility in both ganders and turkey cocks (Joffe, 1978). Responses encountered in poultry exposed to F-2 toxin vary from 'no effects' on egg production or egg fertility in geese or laying hens when fed at 100 mg/kg level (Palyusik and Karlic-Kovacs, 1975;

Marks and Bacon, 1976); to reports of swelling of vents, eversion of the cloaca and enlargement of oviducts in 10–12-day-old turkey poults exposed to insults of 300–800 mg/kg (Sherwood and Peberdy, 1976; Mirocha *et al.*, 1974b). The same concentration in 10-day-old chicks caused cystic changes in the genital tracts (Mirocha *et al.*, 1974b). Reported manifestations of fusariotoxicosis in poultry include oestrogenic changes, reduced feed consumption and impaired egg production (Meronuck *et al.*, 1970; Speers *et al.*, 1973).

ERGOTISM

A variety of disease syndromes in domestic animals are attributed to alkaloids produced by the fungi *Claviceps purpurea* and *Claviceps paspali*, and conditions associated with these toxic substances have been described in both man and animals since biblical times.

Cattle

Ergotism is recognized in two distinct clinical forms in cattle: a gangrenous form with necrosis and sloughing of the extremities and a neurotoxic form characterized by staggers, ataxia, tremors and convulsions (Morehouse, 1979). The former is attributed to ergotamine and related to alkaloids produced by *Claviceps purpurea*, and the latter to ergotamine and lysergic acid derivatives produced by *Claviceps paspali*. Both fungi grow under warm, moist conditions with *C. purpurea* invading rye, oats, wheat and Kentucky bluegrass, and *C. paspali* parasitizing paspalum or Dallis grass. The fungus replaces the grass ovary with a dark brown to purple sclerotium (Buck *et al.*, 1973).

The clinical signs of gangrene in cattle associated with *C. purpurea* may be very similar to those described for fescue foot, and peripheral gangrene should not be used as a clinical basis for diagnosing ergotism unless there is supporting evidence that large amounts of ergot could have been ingested. Differential diagnosis of these conditions may be assisted by seasonal differences in their times of occurrence in that fescue foot occurs in cattle when ergot should not be prevalent. Also, forelimb involvement is more common in ergotism than in fescue foot.

The toxicity of ergot sclerotia and clinical responses elicited following their ingestion are related directly to the amount of alkaloid associated with them (Morehouse, 1981; Cunningham *et al.*, 1944). Clinical evidence has been reviewed suggesting that cold predisposes cattle to lameness on an ergotized diet, and under such conditions effects may be seen with lower levels of the mycotoxin (Mantle *et al.*, 1978).

Prevention of losses due to ergotism depends principally on restricting access to ergotized pastures, grasses or silage. As in fescue foot, early-stage lesions of lameness regress when toxic feeds are removed.

Sheep

The selective grazing habits of sheep, avoiding the coarser flowering heads of grasses, may provide some protection against toxins in ergotized pastures (Mantle *et al.*, 1978), perhaps accounting for the rarity of ovine ergotism. However, oral administration of ergotamine tartrate to sheep at a daily dose of 1.0 mg/kg resulted in death of animals in 10 days (Greatorex and Mantle, 1973) with autopsy evidence of tongue necrosis and haemorrhages on the lateral aspects of lower limbs.

Field outbreaks of paspalum staggers in sheep also are rare. Lysergic acid, the principal alkaloid produced by *C. paspali*, was shown to have an acute LD_{50} of about 4 mg/kg, but clinical signs did not coincide with all the classically observed symptoms of paspalum staggers (Mantle *et al.*, 1978). Recent demonstration that paspalum staggers is attributable to a group of tremorgenic mycotoxins present in sclerotia rather than the alkaloid itself may explain these differences (Mantle *et al.*, 1978).

Swine

Convulsive forms of acute ergotism in pigs appear to be more common than in cattle, and sudden death may be a sequel to such convulsions; additional signs of ergot toxicity in swine are attributable to inappetance and slower weight gain; this phenomenon has been observed in feeder pigs receiving more than 0.05% ergot in their diet and may in turn be related to unpalatability of the diet (Friend and MacIntyre, 1970; Harrold *et al.*, 1974). Agalactia in sows may result from ergot toxicity and rations fed to sows containing 0.5 and 1.0% ergot resulted in lack of udder development and milk secretion (Nordskog and Clark, 1945). A similar effect in mice was considered to result from the centrally mediated action of lysergic acid derivatives on pituitary hormonal activity (Mantle, 1968).

TREMORGENIC MYCOTOXINS IN CATTLE AND SHEEP

A tremorgenic mycotoxin was isolated from *Penicillium cyclopium* in 1968 (Wilson *et al.*, 1968), and since then at least nine similar mycotoxins produced by *Penicillium* and *Aspergillus* spp. have been described.

Penitrem A has produced neuropathic effects experimentally in calves, consisting initially of fine tremors, progressing to marked tremors, ataxia, paddling and lateral recumbency, and opisthotonus; animals recovered after administration of the toxin ceased (Cysewski *et al.*, 1975). A *Penicillium* species producing the Penitrem A was isolated from commercial feed suspected to be involved in the deaths of dairy cattle (Ciegler, 1969).

Tremorgenic mycotoxins from ergot sclerotia have been reported to cause tremors of the head in sheep, followed by general body tremors and a staggering gait; these symptoms were identical to those observed in the 'paspalum staggers' syndrome (Mantle *et al.*, 1978; Cysewski, 1973). Identical symptoms have been reported in the experimental reproduction of 'ryegrass staggers' with tremorgenic mycotoxins from *Penicillium cyclopium* (di Menna *et al.*, 1976) and field exposure to them (di Menna and Mantle, 1978).

SLAFRAMINE TOXICITY

Slaframine is capable of initiating an intense salivation response in cattle, sheep, goats and horses. The fungus *Rhizoctonia leguminicola* is responsible for black patch disease of red clover and for the production of the mycotoxin slaframine. The ingestion of this metabolite causes profuse salivation and lacrimation in all animal species (Crump, 1973) but no pathological effect is reported following short-term exposure.

FACIAL ECZEMA – SPORIDESMIN TOXICITY

Facial eczema is of major importance in cattle and sheep in Australia and New Zealand which have been exposed to the mycotoxin sporidesmin produced by *Pithomyces chartarum* (Mortimer *et al.*, 1978). Clinical signs include irritation, reddening and oedema of non-pigmented areas of skin, followed by serum exudation of swollen areas and eventual deep necrosis and sloughing of tissues.

Swine and poultry are susceptible to sporidesmin, but facial eczema is essentially a disease of ruminant animals grazing infected pastures. Sporidesmin causes fibrous occlusion of bile ducts leading to secondary accumulation in the circulation of phylloerythrin, a normal breakdown product of chlorophyll usually excreted in the bile (Clare, 1944). Phylloerythrin is photodynamic and leads to an increased sensitivity to sunlight with the typical clinical sequellae.

Isolates of *P. chartarum* in the United States apparently produce little or no sporidesmin, although epidemic photosensitization diseases similar to facial eczema have been reported (Tudor *et al.*, 1961). In Florida, epidemics of photosensitization occurred 3–8 weeks after frost had killed Bermuda grass (Kidder *et al.*, 1961). Moulds colonized the deteriorating grass and illness occurred when cattle ingested this material while eating the superimposed new growth of grass. Similar cases have been reported in cattle 'accidentally' ingesting mouldy crimson clover when grazing Bermuda and Dallis grass

(Gibbons, 1959). Epidemics of hepatogenous photosensitization of cattle have also been reported from Oklahoma state in cattle ingesting mould-damaged alfalfa which produces pathological changes similar to those described for facial eczema of sheep (Glenn *et al.*, 1964). An extensive review of bovine photosensitization on a global scale has recently been published (Marasas and Kellerman, 1978).

FESCUE FOOT

Fescue foot is a disease of cattle manifested by arching of the back, rough coat and hind limb, tail and ear-tip lesions progressing from necrosis to gangrene and sloughing. Numerous genera of fungi have been found on toxic fescue but *Fusarium tricinctum* is thought the likely toxigenic agent; the fungus appears to proliferate in the yearly accumulated growth of tall fescue in fall and early winter when temperatures of about 5 °C provide optimal conditions in this environment. A toxic butenolide compound has been identified from *F. tricinctum in vitro*, producing some but not all of the pathological lesions associated with the condition (Garner and Cornell, 1978). Fescue foot remains a serious problem in certain years in Missouri cattle, with the onset of clinical symptoms developing as soon as 1 week after exposure but depending on the level of toxicity in pastures.

The disease is widely distributed in the United States. Important potentiating factors include excessive nitrogen fertilizer, with low concurrent rainfall, grazing or feeding on late-cut tall forage in the coldest part of the year and feeding of fescue grasses with minimum supplementation (Garner and Cornell, 1978).

ADENOMATOSIS (ATYPICAL INTERSTITIAL PNEUMONIA)

Adenomatosis is a non-specific response of alveoli in the bovine lung to a variety of agents, one of which is mouldy feed. In affected lung, the cells lining the alveolar walls proliferated as much as ten-fold, thus preventing normal gas exchange. The disease is usually progressive and fatal. Sixty-nine cattle in a herd of 275 died as a result of atypical interstitial pneumonia after being fed mouldy sweet potatoes (Peckham *et al.*, 1972). The specific aetiology was attributed to a toxin, 4-ipomeanol, found in the mould-damaged sweet potatoes from which *Fusarium solani* was isolated. It was later postulated that a precursor to this toxic compound was produced by sweet potatoes in response to 'injuries', and it was suggested that the fungus was necessary to convert the precursor into lung toxic substances (Burks *et al.*, 1974); atypical interstitial pneumonia has now been experimentally reproduced with the toxin 4-ipomeanol (Doster *et al.*, 1978).

SWEET CLOVER DISEASE

Sweet clover disease of cattle and sheep occurs frequently in animals fed on mouldy sweet clover hay and has been recognized in Canada and the United States for more than 50 years; the compound 3′,3′-methylenebis (4-hydroxycoumarin) produced by growth of *Aspergillus* and *Penicillium* spp. on improperly cured sweet clover hay or silage was later defined as its causative agent (Campbell and Link, 1941). The disease is characterized by fatal haemorrhage which occurs subcutaneously and extensively in areas where normal activity produces mild trauma; the underlying cause of the uncontrolled haemorrhage is a failure of blood coagulation due to loss of prothrombin (Roderick, 1931).

STACHYBOTRYOTOXICOSES

Although the saphrophytic fungus *Stachybotrys atra* is found in the U.S., the major disease problems associated with it have been reported from Eastern Europe and the U.S.S.R. The fungus has a predilection for cellulose substrates such as straw and hay, and may produce trichothecene derivatives which are highly irritating to skin and mucous membranes producing clinical effects akin to those of T-2 toxin (Eppley, 1977); the acute and chronic effects of the toxin in cattle and swine have recently been reviewed (Hintikka, 1978).

LEUKOENCEPHALOMALACIA IN HORSES (LEM)

LEM is a fatal neurotoxic disease which has been associated with mouldy corn consumption by horses in the United States for more than a century; major losses were reported in the eastern states at the turn of the century (Buckley and MacCallum, 1901), and in the state of Illinois in the mid-1930s. The importance of this disease in the United States has been re-emphasized recently with the report of eight naturally occurring herd episodes of LEM involving 269 horses of which 38 were clinically affected and 34 died. The toxin has not been identified but the condition is associated with feed contaminated with *Fusarium* spp., principally *Fusarium moniliforme* (Badiali *et al.*, 1968; Buck *et al.*, 1979).

Affected horses develop blindness, ataxia and pharyngeal paralysis prior to collapse and coma. The principal lesion of LEM is one of liquefactive necrosis of the cerebral hemispheres associated with haemorrhage; there is necrosis in the thalamus, brainstem and cerebellum (Badiali *et al.*, 1968; Wilson and Amaronpot, 1971).

Associated lesions have been described including congested liver and kidneys, erosive lesions of the gastrointestinal mucosa, subendocardial and subepicardial haemorrhages, and haemorrhages in the urinary bladder (Beister *et al.*, 1940; Graham, 1935, 1936). Such observations are consistent with changes observed in other fusariotoxicoses of animals and poultry, although damage to blood vessels has been considered to be a significant primary factor (Wilson *et al.*, 1973).

EFFECTS OF MYCOTOXINS IN OTHER ANIMAL SPECIES

The primary discussion of this chapter has centred on mycotoxic diseases of farm animals and poultry encountered in the U.S., but an extensive review of effects of mycotoxins in laboratory animals has been published (Carlton and Szczech, 1978).

Mycotoxicoses are uncommon in small companion animals but cats are highly susceptible to aflatoxin. Epizootics of hepatitis in dogs in south-eastern states were probably attributable to aflatoxin in commercial dog food products (Siebold and Baily, 1952).

CONCLUSIONS

A number of mycotoxins are discussed in this chapter that have been implicated in various disease syndromes affecting food-producing animals and poultry in the United States. Decreased productivity and performance associated with a number of these are described.

Clinical signs and post-mortem lesions are rarely pathognomonic, necessitating elimination of other infectious and toxic agents for differential diagnoses. The importance of identification of the responsible toxin in establishing a firm diagnosis is emphasized.

More than 20 full-service veterinary medical diagnostic laboratories (VMDLs) have been built in the United States in the past decade. These laboratories offer rapid differential diagnosis when carcasses or specimens are submitted to them from farms throughout the country. Twenty-two of these laboratories in 20 states of the U.S.A. have developed some degree of analytical capability for mycotoxins.

Overall, aflatoxin is the mycotoxin most frequently reported in these laboratories, with deoxynivalenol (DON) and zearalenone of greatest concern in the midwestern and northern states and ergot in the north-western states. Ochratoxin, T-2 toxin, diacetoxyscirpenol (DAS), ergot alkaloids, citrinin, sterigmatocystin, slaframine and penitrem A were also reported.

Although many diagnostic laboratories remain hampered by lack of simple, rapid, inexpensive analytical procedures for mycotoxins, the development of some rapid screening methods referred to in this chapter, and the progress in ELISA detection methods, offers promise for improved recognition of mycotoxins and diagnosis of the mycotoxicoses.

Effects of the mycotoxins on the immune system and synergistic effects of multiple mycotoxin exposure or concurrent involvement with other microorganisms are areas of prime need for research.

REFERENCES

Allcroft, R. (1969). Aflatoxicosis in farm animals. In *Aflatoxin* (ed. L. A. Goldblatt). Academic Press, New York.

Allcroft, R., and Lewis, G. (1963). Ground nut toxicity in cattle—experimental poisoning of calves and a report on clinical effects of older cattle, *Vet. Rec.*, **75**, 493–495.

Angsubharkorn, S., Toomvises, P., Romruen, K., and Newberne, P. N. (1981). Aflatoxicosis in horses, *J. Am. Vet. Med. Assoc.*, **178**, 274–278.

Armbrecht, B. H. (1978). Aflatoxicosis in sheep. In *Mycotoxic Fungi, Mycotoxins, Mycotoxicoses – An Encyclopedic Handbook,* Vol. II, (eds T. D. Wyllie and L. G. Morehouse), pp. 189–195. Marcel Dekker, Inc., New York and Basel.

Armbrecht, B. H., Wiseman, H. G., Shalkop, W. P., and Galeta, J. N. (1978). Swine aflatoxicosis – an assessment of growth efficiency and other responses in growing pigs fed aflatoxin, *Environ. Physiol.*, **1**, 198.

Austwick, P. K. C. (1978). Aflatoxicosis in poultry. In *Mycotoxic Fungi, Mycotoxins, Mycotoxicoses – An Encyclopedic Handbook,* Vol. II, (eds T. D. Wyllie and L. G. Morehouse), pp. 279–301 Marcel Dekker, Inc., New York and Basel.

Badiali, L., Abou-Youssef, N. H., Radwan, A. I., Hamdy, F. M., and Hildebrandt, P. K. (1968). Moldy corn poisoning as a major cause of an encephalomalacia syndrome in Egyptian equidai, *Am. J. Vet. Res.*, **29**, 2029–2035.

Beister, H. E., Schwarte, L. H., and Reddy, C. H. (1940). Further studies on moldy corn poisoning (leucoencephalomalacia) in horses, *Vet. Med.*, **35**, 636–639.

Bristol, F. M., and Djurikovic, S. (1971). Hyperestrogenism in female swine as a result of feeding moldy corn, *Can. Vet. J.*, **12**, 132.

Brown, M. H., Syczeck, G. M. and Purmalis, B. P. (1976). Teratògenic and 'toxic' effects of ochratoxin A in rats, *Toxicol. Appl. Pharmacol.*, **37**, 331–333.

Buck, W. B., Haliburton, J. C., Thilsted, J. P., Lock, T. F., and Vesonder, R.F. (1979). Equine leucoencephalomalacia: comparative pathology of naturally occurring and experimental cases, *Proc. 22nd Ann. Mtg. Amer. Assoc. Vet. Lab. Diag.*, pp. 239–258.

Buck, W. B., Osweiler, G. D., and Van Gelder, G. A. (1973). Mycotoxicoses. In *Clinical and Veterinary Diagnostic Toxicology*, p. 162. Kendall Hunt Publishing Co., Dubuque, Iowa.

Buckley, S. S., and MacCallum, W. G. (1901). Acute hemorrhagic encephalitis prevalent among horses in Maryland, *Am. Vet. Rev.*, **25**, 99–101.

Burka, L. P., Kohnert, L., and Wilson, B. J. (1974). 4-hydroxymyoporone, a key intermediate in the synthesis of pulmonary toxins produced by *Fusarium solanii* infested sweet potatoes, *Tetrahedron Lett.*, **46**, 4017–4020.

Campbell, H. A., and Link, K. P. (1941). Studies on hemorrhagic sweet clover disease. IV. The isolation and crystallization of the hemorrhagic agent, *J. Biol. Chem.*, **138**, 21.

Campbell, J. G. (1955–57). An investigation of the hepatotoxic effects in the fowl of the ragwort (*Senecio Jacobaea* Linn.) with special reference to the induction of liver tumors with seneciophylline, *Proc. Roy. Soc. Edin.*, Sect. B, **66**, 111–130.

Carlton, W. W., and Krogh, P. (1979). Ochratoxins. In *Conference on Mycotoxins in Animal Feeds and Grains Related to Animal Health*, pp. 165–287. U.S. Dept. of Commerce, National Technical Information Service, sponsored by Bureau of Vet. Medicine Food and Drug Admin., Rockville, Maryland (June).

Carlton, W. W., and Szczech, G. M. (1978). Mycotoxicoses in laboratory animals. In *Mycotoxic Fungi, Mycotoxins, Mycotoxicoses – An Encyclopedic Handbook*, Vol. II, (eds T. D. Wyllie and L. G. Morehouse), pp. 333–462. Marcel Dekker, Inc., New York and Basel.

Carlton, W. W., and Tuite, J. (1970). Nephropathy and edema syndrome induced in miniature swine by corn cultures of *Penicillium viridicatum*, *Pathol. Vet.*, **7**, 68–80.

Chi, M. S., and Mirocha, C. J. (1978). Necrotic oral lesions in chickens fed diacetoxyscirpenol, T-2 toxin and crotocin, *Poultry Sci.*, **57**, 807–808.

Chi, M. S., Mirocha, C. J., Kurtz, H. J., Weaver, G. A., Bates, F. Y., Shimoda, W., and Burmeister, H. R. (1977a). Acute toxicity of T-2 toxin in broiler chicks and laying hens, *Poultry Sci.*, **56**, 103–116.

Chi, M. S., Mirocha, C. J., Kurtz, H. J., Weaver, G., Bates, F., and Shimoda, W. (1977b). Subacute toxicity of T-2 toxin in broiler chicks, *Poultry Sci.*, **56**, 306–313.

Chi, M. S., Mirocha, C. J., Kurtz, H. J., Weaver, G., Bates, F., and Shimoda, W. (1977c). Effects of T-2 toxin on reproductive performance and health of laying hens, *Poultry Sci.*, **56**, 628–637.

Chowdhury, H., Carlson, C. W. and Semeniuk, G. (1971). A study of ochratoxin toxicity in hens, *Poultry Sci.*, **50**, 1855–1859.

Ciegler, A. (1969). Tremorgenic toxin from *Penicillium palitans*, *Appl. Microbiol.*, **18**, 128–129.

Clare, N. P. (1944). Photosensitivity diseases in New Zealand. III. The photosensitizing agent in facial eczema. *N.Z. J. Sci. Technol.*, **25A**, 202–220.

Crump, N. H. (1973). Slaframine (slobber factor) toxicosis. *J. Am. Vet. Med. Assoc.*, **163**, 1300–1302.

Cunningham, I. J., McIntosh, I. G., and Swan, J. B. (1944). The non-toxicity of milk and meat from ergotized cattle, *N.Z. J. Sci. Technol.*, **26**, 125–136.

Cysewski, S. J. (1973). Paspalum staggers and tremorgen intoxication in animals, *J. Am. Vet. Med. Assoc.*, **163**, 1291–1292.

Cysewski, S. J., Baetz, A. L., and Pier, A. C. (1975). Penitrem A intoxication of calves. Blood chemical and pathological changes, *Am. J. Vet. Res.*, **36**, 53–58.

Cysewski, S. J., Pier, A. C., Engstrom, G. W., Richard, J. L., Dougherty, R. W., and Thurston, J. R. (1968). Clinical pathologic features of acute aflatoxicosis of swine, *Am. J. Vet. Res.*, **29**, 1577.

Dahlgren, R. R., and Williams, D. E. (1972). Clinical report – hemorrhagic syndrome in feedlot cattle, *Bov. Pract.*, **7**, 52–53.

Doerr, J. A., Huff, W. E., Tung, H. T., Wyatt, R. D., and Hamilton, P. B. (1974). A survey of T-2 toxin, ochratoxin and aflatoxin for their effects on the coagulation of blood in young broiler chickens, *Poultry Sci.*, **53**, 1728–1734.

Doster, A. R., Mitchell, F. E., Farrell, R. L., and Wilson, B. J. (1978). Effects of 4-ipomeanol, a product from mold damaged sweet potatoes on the bovine lung, *Vet. Pathol.*, **15**, 367–375.

Edds, G. T. (1977). Report of environmental residue committee. *81st Proc., U.S. Anim. Hlth. Assoc.*, p. 219.

Edds, G. T. (1979). Aflatoxins. In *Conference on Mycotoxins and Animal Feeds and Grains Related to Animal Health*, pp. 80–164. U.S. Dept. of Commerce, National Technical Information Service, TG-300, sponsored by FDA, Rockville, Maryland.

Elling, F., and Moller, T. (1973). Mycotoxic nephropathy in pigs. *Bull. WHO*, **49**, 411–418.

Eppley, R. M. (1977). Chemistry of stachybotryotoxicosis. In *Mycotoxins in Human and Animal Health*, pp. 285–293 (eds J. V. Rodericks, C. W. Hesseltine and M. A. Mehlman), Pathotox Publ., Inc., Park Forest South, Illinois.

Fisher, E. E., Kellock, A. W., and Wellington, N. A. M. (1967). Toxic strain of *Fusarium culmorum* from *Zea mays* associated with sickness in dairy cattle, *Nature*, **214**, 322.

Forsyth, D. M., Yoshizawa, T., Morooka, N., and Tuite, J. (1977). Emetic and refusal activity of deoxynivalenol to swine, *Appl. Environ. Microbiol.*, **34**, 547–552.

Friend, D. W., and MacIntyre, T. M. (1970). Effect of rye ergot on growth and N-retention and growing pigs, *Can. J. Comp. Med.*, **34**, 198–202.

Gagne, W. E., Dungworth, D. L., and Moulton, J. E. (1968). Pathological effects of aflatoxin in pigs, *Pathol. Vet.*, **5**, 370–384.

Garner, G. B., and Cornell, C. N. (1978). Fescue foot in cattle. In *Mycotoxic Fungi, Mycotoxins, Mycotoxicoses – An Encyclopedic Handbook*, Vol. II, (eds T. D. Wyllie and L. G. Morehouse), pp. 45–62. Marcel Dekker, Inc., New York and Basel.

Garrett, W. N., Heitman, H. Jr., and Booth, A. N. (1968). Aflatoxin toxicity in beef cattle, *Proc. Soc. Exp. Biol. Med.*, **127**, 188–190.

Gendloff, E. H., Pestha, J. J., and Hart, L. P. (1983). ELISA detection of trichothecenes in spiked and *Fusarium* infected corn extracts. Abstracts, American Phytopath. Soc.

Gibbons, W. J. (1959). Forage poisoning, *Mod. Vet. Pract.*, **40** (16), 43–47.

Glenn, B. J., Monlux, A. W., and Panciera, F. J. (1964). A hepatogenous photosensitivity disease of cattle. Experimental production and clinical aspects of the disease, *Path. Vet.*, **1**, 469–484.

Graham, R. (1936). Toxic encephalitis or non-virus encephalomyelitis of horses, *Vet. Med.*, **31**, 46–50.

Graham, R. P. (1935). Results of inoculating laboratory animals with equine brain-tissue suspensions in equine brain-tissue filters from spontaneous cases of so-called corn stalk disease, *J. Am. Vet. Med. Assoc.*, **86**, 778–780 (June, 1935).

Greatorex, J. C., and Mantle, P. G. (1973). Experimental ergotism in sheep, *Res. Vet. Sci.*, **15**, 337–346.

Gustafsson, B. E., Daft, F. S., McDaniel, E. G., Smith, J. C., and Fitzgerald, R. J. (1962). Effects of vitamin K-active compounds and intestinal microorganisms in vitamin K deficient germ free rats, *J. Nutr.*, **78**, 461.

Hamilton, P. B. (1971). A natural and extremely severe occurrence of aflatoxicosis in laying hens, *Poultry Sci.*, **50**, 1880–1882.

Harrold, R. L., Dinussion, W. E., Haugse, C. M., and Johnson, J. N. (1974). Ergot in growing finishing swine rations. North Dakota Agr. Exp. Stat. Farm Res. (Nov.–Dec.).

Hibbs, C. M., Osweiler, G. D., Buck, W. B., and Macafee, G. P. (1974). Bovine hemorrhagic syndrome related to T-2 mycotoxin, *Proc. 17th Ann. Mtg. of Amer. Assoc. of Vet. Lab. Diag.*, pp. 305–308.

Hintikka, E. L. (1978). Stachybotryotoxicosis in cattle, swine and poultry. In *Mycotoxic Fungi, Mycotoxins, Mycotoxicoses – An Encyclopedic Handbook*, Vol. II, (eds T. D. Wyllie and L. G. Morehouse), pp. 152, 268, 327. Marcel Dekker, Inc., New York and Basel.

Howarth, B. Jr, and Wyatt, R. D. (1976) Effect of dietary aflatoxin on fertility, hatchability and progeny performance in broiler and breeder hens, *Appl.*

Environ. Microbiol., **31**, 680.
Hsieh, D. P. H. (1979). Basic metabolic effects of mycotoxins. In *Interactions of Mycotoxins in Animal Production*. p. 43. National Research Council, National Academy of Sciences, Washington, D.C.
Huff, W. E., Doerr, J. A., and Hamilton, P. B. (1977). Decreased bone strength during ochratoxicosis and aflatoxicosis, *Poultry Sci.*, **45**, 1724.
Joffe, A. Z. (1978). *Fusarium* toxicosis in poultry. In *Mycotoxic Fungi, Mycotoxins, Mycotoxicoses – An Encyclopedic Handbook*, Vol. II, (eds T. D. Wyllie and L. G. Morehouse), pp. 309–321. Marcel Dekker, Inc., New York and Basel.
Jones, N. R., and Jones, B. D. (1969). Aflatoxins, *Voeding*, **30**, 330.
Keyl, A. C. (1978). Aflatoxicosis in cattle. In *Mycotoxic Fungi, Mycotoxins, Mycotoxicoses – An Encyclopedic Handbook*, Vol. II, (eds T. D. Wyllie and L. G. Morehouse), pp. 9–28. Marcel Dekker, Inc., New York and Basel.
Kidder, R. W., Beardsley, D. W., and Irwin, T. C. (1961). Photosensitization in cattle grazing frosted common Bermuda grass, *Ag. Exp. Sta. Bull.*, No. 630. Gainesville, Florida (February).
Korpinen, E. L. (1972). Natural occurrence of F-2 and F-2 producing *Fusarium* strains associated with field cases of bovine and swine infertility, *Proc. IUPAC-sponsored symposium on control of mycotoxins*, Göteborg, Sweden, p. 21.
Kosuri, N. R., Grove, M. D., Yates, S. G., Tallent, W. H., Ellis, J. J., Wolff, A. A., and Nichols, R. E. (1970). Response of cattle to mycotoxins of *Fusarium tricinctum* isolated from corn and fescue, *J. Am. Vet. Med. Assoc.*, **157** (7), 938–994.
Krogh, P., Axelsen, N. H., Elling, F., Gyrd-Hansen, N., Hall, B., Hyldgaard, J., Nensen, A. E., Larsen, A., Madsen, H. P., Mortensen, T., Moller, O. K., Petersen, U., Ravnskov, U., Rostgaard, M., and Aalund O. (1974). Experimental porcine nephropathy – changes of renal function and structure induced by ochratoxin A contaminated feed. *Acta Pathol. Microbiol. Scand.*, Sect. A, supplementum no. 246, pp. 1–21.
Krogh, P., Hall, B., and Pedersen, E. J. (1973). Occurrence of ochratoxin A and citrinin in cereals associated with mycotoxic porcine nephropathy. *Acta Pathol. Microbiol. Scand.*, Sect. B81, pp. 689–695.
Kurmanov, I. A. (1978). Pathological alterations accompanying fusariotoxicosis in cattle. In *Mycotoxic Fungi, Mycotoxins, Mycotoxicoses – An Encyclopedic Handbook*, Vol. II, (eds T. D. Wyllie and L. G. Morehouse), pp. 85–107. Marcel Dekker, Inc., New York and Basel.
Kurtz, H. J., Nairn, M. E., Nelson, G. H., Christensen, G. N., and Mirocha, C. J. (1969). Histologic changes in the genital tracts of swine fed estrogenic mycotoxins, *Am. J. Vet. Res.*, **30**, 551–556.
Larson, H. E., Aalund, O., and Nielsen, K. (1962). Perirenal edema in pigs, *Nord. Vet. Med.*, **14**, 338–355.
Long, G. G., Deikman, M., Tuite, J. L., Shannon, G. M., and Vesonder, R. F. (1982). Effects of *Fusarium roseum* corn culture containing zearalenone on early pregnancy in swine, *Am. J. Vet. Res.*, **43**, 1599–1603.
Lynch, G. P., Smith, D. F., Covey, F. C., and Gordon, C. H. (1973). Aflatoxin and nitrogen balance in the young calf, *J. Dairy Sci.*, **56**, 1154–1158.
Mantle, P. G. (1968). Inhibition of lactation in mice following feeding with ergot sclerotia (*Claviceps fusiformis* Loveless) from the bulrush millet (*Pennisetum typhoides* Staph. and Hubbard) in an alkaloid component, *Proc. Roy. Soc. B.*, **170**, 423–434.
Mantle, P. G., Mortimer, P. H., and White, E. P. (1978). Mycotoxic tremorgens of *Claviceps paspali* and *Penicillium cyclopium* – a comparative study of effects of

sheep and cattle in relation to natural staggers syndrome, *Res. Vet. Sci.*, **24**, 49–56.

Marasas, W. F. O., and Kellerman, T. S. (1978). Photosensitivity in cattle. In *Mycotoxic Fungi, Mycotoxins, Mycotoxicoses – An Encyclopedic Handbook*, Vol. II, (eds T. D. Wyllie and L. G. Morehouse), pp. 73–84. Marcel Dekker, Inc., New York and Basel.

Marks, H. L., and Bacon, C. D. (1976). Influence of *Fusarium* infected corn and F-2 on laying hens, *Poultry Sci.*, **55**, 1864–1870.

di Menna, M. E., and Mantle, P. G. (1978). The role of *Penicillia* in rye grass staggers, *Res. Vet. Sci.*, **24**, 347–351.

di Menna, M. E., Mantle, P. G., and Mortimer, P. H. (1976). Experimental production of a staggers syndrome in ruminants by tremorgenic *Penicillium* from soil, *N.Z. Vet. J.*, **24**, 45–46.

Meronuck, R. A., Garren, K. H., Christensen, C. M., Nelson, G. H., and Bates, F. (1970). Effects on poultry poults and chicks of rations containing corn invaded by *Penicillium* and *Fusarium* sp., *Am. J. Vet. Res.*, **31**, 551–555.

Miller, D. M., Stuart, B. P., Crowell, W. A., Coler, J. R. Jr., Goven, A. J., and Brown, J. (1978). Aflatoxicosis in swine – its effect on immunity and relationship to salmonellosis. *Proc. 21st Ann. Mtg. Amer. Assoc. Vet. Lab. Diagnosticians*, pp. 135–146.

Mirocha, C. J., Christensen, C. M., and Nelson, G. H. (1974b). F-2 (zearalenone) estrogenic mycotoxins from *Fusarium*. In *Microbial Toxins*, Vol. 7, (eds S. Kadis, A. Ceigler, and L. J. A. Ajl), pp. 107–138. Academic Press, New York.

Mirocha, C. J., Harrison, J., Nichols, A., and McClintock, N. (1968). Detection of a fungal estrogen (F-2) in hay associated with infertility in dairy cattle, *Appl. Microbiol.*,**16**, 797.

Mirocha, C. J., Schauerhamer, B., and Pathre, S. (1974a). Isolation, detection and quantitation of the zearalenone in maize and barley, *J. Assoc. Off. Anal. Chem.*, **57**, 1104–1110.

Moran, E. T., Hunter, R. B., Ferket, P. R., and Young, L. G. (1983). *Proc. of the Poultry Industry school, Ontario Agricultural College, University of Guelph*, pp. 13–14.

Moran, E. T., Hunter, B., Ferket, P., Young, L. G., and McGirr, L. G. (1982). High tolerance of broilers to vomitoxin from corn infected with *Fusarium graminearum*, *Poultry Sci.*, **61**, 1828.

Morehouse, L. G. (1979). Mycotoxicoses of the bovine. *Bov. Pract.* (November), pp. 175–180.

Morehouse, L. G. (1981). Mycotoxins – their toxicology and principal lesions, *Proc. 85th Ann. Mtg. U.S. Anim. Hlth. Assoc.*, St. Louis, Missouri, pp. 232–267.

Mortimer, P. H., di Menna, M. E., and White, E. P. (1978). Pithomycotoxicosis (facial eczema) in cattle. In *Mycotoxic Fungi, Mycotoxins, Mycotoxicoses – An Encyclopedic Handbook*, Vol. II, (eds T. D. Wyllie and L. G. Morehouse), pp. 63–72. Marcel Dekker, Inc., New York and Basel.

Nordskog, A. W., and Clark, R. T. (1945). Ergotism in pregnant sows, female rats and guinea pigs, *Amer. J. Vet. Res.*, **6**, 107–116.

Osweiler, G. D. (1982). Wheat scab toxins – a threat to Missouri livestock. *Veterinary Medical Review*, College of Veterinary Medicine and Cooperative Extension Service, Vol. 3, No. 4, pp. 2–3.

Osweiler, G. D., Hook, B. S., Mann, D. D., Buening, G. M., and Rottinghaus, G. E. (1981). Effects of T-2 toxin in cattle. *Proc. 85th Ann. Mtg. U.S. Anim. Hlth. Assoc.*, pp. 214–231.

Palyusik, M., and Karlic-Kovacs, E. (1975). Effect on laying geese of feeds containing the fusariotoxins T-2 and F-2. *Acta Vet. Acad. Sci. Hung. tomuls.*, **25**, 363–368.

Patterson, D. S. P. (1977). Aflatoxin and related compounds: biochemistry and physiology. In *Mycotoxic Fungi, Mycotoxins and Mycotoxicoses – An Encyclopedic Handbook*, Vol. 1 (eds T. D. Wyllie and L. G. Morehouse), pp. 175–181. Marcel Dekker, Inc., New York and Basel.

Peckham, J. C. (1978). Ochratoxicosis in poultry. In *Mycotoxic Fungi, Mycotoxins, Mycotoxicoses – An Encyclopedic Handbook*, Vol. II, *Mycotoxicoses of Domestic and Laboratory Animals, Poultry and Aquatic Invertebrates and Vertebrates* (eds T. D. Wyllie and L. G. Morehouse) pp. 301–307. Marcel Dekker, Inc., New York and Basel.

Peckham, J. C., Doupnik, B., and Jones, O. H. Jr. (1971). Acute toxicity of ochratoxins A and B in chicks, *Appl. Microbiol.*, **21**(3), 492–494.

Peckham, J. C., Mitchell, F. E., Jones, O. H. Jr., and Doupnik, B. Jr. (1972). Atypical interstitial pneumonia in cattle fed moldy sweet potatoes. *J. Am. Vet. Med. Assoc.*, **160**, 169–172.

Pestha, J. J., Gaur, P. K., and Chu, L. S. (1980). Quantitation of aflatoxin B_1 and aflatoxin B_1 antibody by an enzyme-linked immunosorbent microassay, *Appl. Environ. Microbiol.*, **40**, 1027–1031.

Pestha, J. J., Li, Y. K., Hurder, W. O.,, and Chu, L. S. (1981a). Comparison of a radio-immunoassay and an enzyme linked immunosorbent assay for the analysis of aflatoxin M_1 in milk, *J. Assoc. Off. Anal. Chem.*, **64**, 294–301.

Pestha, J. J., Steineit, B. W., and Chu, L. S. (1981b). Enzyme-linked immunosorbent assay for detection of ochratoxin A, *Appl. Environ. Microbiol.*, **41**(6). 1472–1474.

Petrie, L. (1977). The identification of T-2 toxin and its association with a haemorrhagic syndrome in cattle, *Vet. Res.*, **101**, 326.

Pier, A. C., Cysewski, S. J., Richard, J. L., Baety, A. L., and Mitchell, L. (1976). Experimental mycotoxicosis in calves with aflatoxin, ochratoxin, rubratoxin and T-2 toxin, *80th Ann. Mtg. U.S. Anim. Hlth. Assoc.*, pp. 130–148.

Prior, M. G., and Sisodia, C. S. (1978). Ochratoxicosis in white leghorn hens, *Poultry Sci.*, **57**, 619–623.

Prior, M. G., Sisodia, C. S., and O'Neil, J. B. (1976). Acute oral ochratoxicosis in day-old white leghorns, turkeys and Japanese quail. *Poult. Sci.*, **55**, 786–790.

Ribelin, W. R. (1978a). Ochratoxicosis in cattle. In *Mycotoxic Fungi, Mycotoxins, Mycotoxicoses – An Encyclopedic Handbook,* Vol. II, (eds T. D. Wyllie and L. G. Morehouse), p. 30. Marcel Dekker, Inc., New York and Basel.

Ribelin, W. R. (1978b). Trichothecene toxicosis in cattle. In *Mycotoxic Fungi, Mycotoxins, Mycotoxicoses – An Encyclopedic Handbook*, Vol. II, (eds T. D. Wyllie and L. G. Morehouse), pp. 36–43. Marcel Dekker, Inc., New York and Basel.

Ribelin, W. R., Fukushima, K., and Still, P. E. (1978). The toxicity of ochratoxin to ruminants, *Can. J. Comp. Med.*, **42**, 172–176.

Roderick, L. M. (1931). Coagulation of blood in sweet clover disease in cattle, *Am. J. Physiol.*, **96**, 413.

Rottinghaus, G. E., Olesen, B., and Osweiler, G. D. (1982). Rapid screening method for aflatoxin B_1, zearalenone, ochratoxin A, T-2 toxin, diacetoxyscirpenol and vomitoxin. *Proc. 25th Ann. Meeting Amer. Assoc. of Vet. Lab. Diagnosticians*, pp. 477–484.

Ruhr, L. P., Osweiler, G. D., and Foley, C. W. (1978). The effect of zearalenone on fertility in the boar, *Proc. 21st Ann. Mtg. of Amer. Assoc. of Vet. Lab. Diag.*, pp. 127–133.

Sherwood, R., and Peberdy, J. L. (1976). Effects of zearalenone in the developing male chicken, *Br. Poult. Sci.*, **14**, 127–131.

Siebold, H. R., and Bailey, W. S. (1952). An epizootic of hepatitis in the dog, *J. Am. Vet. Med. Assoc.*, **121**, 201–206.

Sisk, D. B., Carlton, W. W., and Curtin, T. M. (1968). Experimental aflatoxicosis in young swine. *Am. J. Vet. Res.*, **29**, 1591.

Smalley, E. B., Marasas, W. F. O., Strong, F. M., Bamburg, J. R., Nichols, R. E., and Kosuri, N. R. (1970). Mycotoxicosis associated with moldy corn, *Proc. 1st U.S.–Japan Conf. Toxic Microorganisms*, U.S. Dept. of Interior, unnumbered publication, pp. 163–174.

Smith, J. W., and Hamilton, P. B. (1970). Aflatoxicosis in the broiler chicken, *Poultry Sci.*, **49**, 207–215.

Speers, G. M., Mirocha, C. J., and Christensen, C. M. (1973). Effect of feeding *F. tricinctum* isolate 'oxyrose' invaded corn and the purified T-2 mycotoxin on S.C.W.L. hens, *Poultry Sci.*, **52**, 2088.

Still, P. E., Macklin, A. W., Ribelin, W. E., and Smalley, E. B. (1971). Relativity of ochratoxin A to foetal death in laboratory and domestic animals, *Nature*, **234**, 563–564.

Tookey, H. L., Yates, J. J., Ellis, J. J., Grove, M. D., and Nichols, R. E. (1972). Toxic effects of a butenolide mycotoxin and *Fusarium tricinctum* cultures in cattle, *J. Am. Vet. Med. Assoc.*, **160**, 15–22.

Tudor, R. W., Beardsley, D. W., and Erwin, T. C. (1961). Photosensitization in cattle grazing frosted common Bermuda grass, *Univ. Fla. Agric. Exp. Sta. Bull.*, **630**, 1–121.

Vesonder, R. J., Ciegler, A., Jensen, A. H., Rohwedder, W. K., and Weesleder, D. (1976). Co-identity of refusal and emetic principle from *Fusarium* infected corn. *Appl. Environ. Microbiol.*, **31**, 280–285.

Vohara, P. Aflatoxicosis in poultry. California Agriculture, Univ. of California, 385 University Hall Berkely, California 94720 (extension release).

Weaver, G. A., Kurtz, H. J., and Mirocha, C. J. (1977). The effect of *Fusarium* toxins on food producing animals, *Proc. 81st Ann. Mtg. U.S. Anim. Hlth. Assoc.*, pp. 215–218 (October).

Weaver, G. A., Kurtz, H. J., Bates, F. Y., Chi, N. S., Mirocha, C. J., Behrens, J. C., and Robison, T. S. (1978a). Acute and chronic toxicity of T-2 mycotoxin in swine, *Vet. Rec.*, **103**, 531–535.

Weaver, G. A., Kurtz, H. J., Mirocha, C. J., Bates, F. Y., Behrens, J. C., and Robison, T. S. (1978b). Effect of T-2 toxin on porcine reproduction. *Can. Vet. J.*, **19**, 310–314.

Weaver, G. A., Kurtz, H. J., Mirocha, C. J., Bates, F. Y., Behrens, J. C., Robison, T. S., and Gipp, W. F. (1978c). Mycotoxin induced abortions in swine, *Can. Vet. J.*, **19**, 72–74.

Weaver, G. A., Kurtz, H. J., Mirocha, C. J., Bates, F. Y., Behrens, J. C. (1978d). Acute toxicity of the mycotoxin diacetoxyscirpenol in swine, *Can. Vet. J.*, **19**, 267–271.

Wilson, B. J., and Amaronpot, R. R. (1971). Causative fungus agent of leucoencephalomalacia in equine animals, *Vet. Rec.*, **88**, 484–486.

Wilson, B. J., Amaronpot, R. R., and Hildebrandt, P. K. (1973). Equine leucoencephalomalacia, *J. Am. Vet. Med. Assoc.*, **163**, 1293–1295.

Wilson, B. J., Wilson, C. H., and Hayes, A. W. (1968). Tremorgenic toxin from *Penicillium cyclopium* grown on feed materials, *Nature*, **220**, 461–464.

Wyatt, R. D. (1979). Biological effects of mycotoxins (other than aflatoxin) on poultry. In *Interactions of Mycotoxins in Animal Production*, pp. 87–95. National Academy of Sciences, Washington, DC.

Wyatt, R. D., Colwell, W. M., Hamilton, P. B., and Burmeister, H. R. (1973). Neural disturbances in chickens caused by dietary T-2 toxin, *Appl. Microbiol.*, **26**, 757–761.

Wyatt, R. D., Harris, J. R., Hamilton, P. B., and Burmeister, H. R. (1972). Possible outbreak of fusariotoxicosis in avians, *Avian Dis.*, **16**, 1123–1130.

Trichothecenes and Other Mycotoxins
Edited by J. Lacey

CHAPTER 37

Pathophysiology of acute T-2 intoxication in the cynomolgus monkey and rat models

David L. Bunner, Robert W. Wannemacher, Harold A. Neufeld, Craig R. Hassler, Gerald W. Parker, Thomas M. Cosgriff, and Richard E. Dinterman

INTRODUCTION

Trichothecene toxins have been implicated in human disease since the 1930s (Joffe, 1971) when alimentary toxic aleukia was reported in Russia. The specific causal toxin remained unknown until advances in chemical analysis suggested that T-2 was the principal toxin involved (Bamburg and Strong, 1971; Joffe, 1978). This first description of human morbidity and mortality (Joffe, 1971) for the most part came from populations that repeatedly ingested spoiled grains over a period of weeks to months, although some of the subjects may have only been exposed for days. These ingestion patterns resulted in severe bone marrow suppression, anaemia, bleeding, and immune suppression with infectious complications leading to death after several weeks. However, some deaths occurred within days.

Early reports from natural exposure of animals including horses (Ueno *et al.*, 1972) and cattle (Hsu *et al.*, 1972) suggested a role for T-2 in animal disease also. Impairment of protein synthesis by trichothecenes was described as early as 1968 (Ueno *et al.*, 1968) as a significant aspect of T-2 toxicity. Poultry studies (Wyatt *et al.*, 1973) showed a reduction in weight gain and decreased spleen and bursa of Fabricius weight. T-2's potent emetic quality has also been well described (Ellison and Kotsonis, 1973). Foetal malformations and maternal mortality were documented in mice (Stanford *et al.*, 1975). More detailed pathophysiological effects of trichothecene toxins on blood pressure, intestinal peristalsis and emesis were documented by Matsuoka *et al.* in 1979. *In vivo* biochemical alterations have also been shown (Pearson, 1978). More recently altered mitochondrial function was suggested by Pace (1983). Limited lethality data in rats and even in monkeys has been reported (Rukmini *et al.*, 1980).

However, studies intended to clarify the basic pathophysiology of acute high-dose exposure and the mechanism of death are generally lacking. The primary goal of our efforts and this chapter are to describe the clinical and physiological signs of acute T-2 intoxication, and to better understand the mechanisms of morbidity and mortality.

MATERIALS AND METHODS

T-2 toxin was obtained from Mycolab (St Louis) and was shown by thin-layer chromatography to have less than 0.1% contamination with other mycotoxins. It was dissolved in a 95% ethanol solution prior to parenteral administration.

An LD_{50} was calculated from a total of eight untethered cynomolgus (*Macaca fascicularis*) monkeys given dosages ranging from 0.25 to 6 mg/kg intramuscularly (i.m.) using probit analyses (Finney, 1952).

Fisher-Dunning 344 rats (Microbiologic Associates, Walkersville, MD) were shaved, treated with a depilatory and 24 hours later were exposed to T-2 toxin in dimethyl sulphoxide (DMSO) over a 4 cm^2 area. Shielding was not used. Four monkeys similarly prepared were treated with T-2 toxin over an area of 16 cm^2.

Biochemical, haematological and coagulation data were generated by studying nine cynomolgus monkeys before and after i.m. injection of 0.65 mg/kg of T-2 toxin. Three of these nine monkeys died within 24 hours. The maximum percentage change was calculated using their own basal values as control. Three control monkeys injected with ethanol also showed no significant changes over the test period but their data were not used in the calculations. These primates were studied after surgical implantation of a temperature probe and a right atrial catheter. A leather jacket was used to cover the surgical sites and a cable and swivel protected the temperature probe and catheter (Bunner *et al.*, 1982). Periodic blood samples were collected and temperatures taken.

Skin sensitivity was tested in Fisher-Dunning 344 rats and cynomolgus monkeys using purified T-2 in ethanol. Two microlitre samples were applied to the skin of anaesthetized animals and allowed to dry. They were observed and lesions measured at daily intervals for 3 days.

Blood pressure and peripheral resistance measurements were done in 12 unanaesthetized but restrained cynomolgus monkeys given doses of T-2 toxin ranging from 0.33 to 2.0 mg/kg i.m. Blood pressure was measured by using an external cuff method previously reported (Hassler *et al.*, 1979). Peripheral resistance was calculated from echocardiographic data not reported here.

Fisher-Dunning 344 rats had arterial and venous catheters placed, and were jacketed and tethered. The heart rate, blood pressure and cardiac index were measured, the latter using a thermodilution method (Muller and Mannesmann, 1981). They were given 2 mg/kg of T-2 toxin i.m. at time zero.

Lethality studies in rats were done in Fisher-Dunning 344 rats using the i.m. route. Blood chemistries were determined using a multichannel analyser (Cobas-bio, Roche, Nutley, NJ) with methodologies established for the analyser by that company. Platelet counts were done using a Coulter Model ZBI particle counter (Hialeah, Fla.). Coagulation tests were performed using standard assays on an $MLA^{R}750$ coagulometer (Medical Laboratory Automation, Inc., Mount Vernon, NY) (Triplett and Harms, 1981).

RESULTS AND DISCUSSION

Lethality studies

Eight cynomolgus monkeys were given intramuscular (i.m.) T-2 toxin in ethanol in doses ranging from 0.25 to 6 mg/kg. The calculated LD_{50} was 0.8 mg/kg and the minimum lethal dose was 0.31 mg/kg. This is in the same range as the rat (LD_{50} 0.5 mg/kg), indicating that for acute intoxication and assessment of mortality the rat is a comparable model. The calculated minimum lethal dose (95 percentile) was 0.31 or 39% of the LD_{50}, and suggests that the biological variation in susceptibility among the cynomolgus monkeys is relatively small. Skin absorption studies in the cynomolgus monkeys have not been completed but four primates had up to 8 mg/kg in DMSO applied to an area of skin 16 cm^2 that was shielded to avoid oral ingestion in order to assess toxicity. None of the animals died, although minor changes in serum chemistry did occur. Thus, a dose 10-fold higher than the i.m. route was non-lethal, suggesting either very little skin permeability, skin metabolism or both. In the rat using the dermal route the LD_{50} with DMSO as the solvent was 1.5 mg/kg. However, these initial studies in the rat model were done without a shield and oral ingestion may well have been playing a role. Further work is needed to clarify dermal absorption in the rat and monkey models.

Skin sensitivity, appetite, body temperature, gastrointestinal effects

Skin testing of cynomolgus monkeys showed that 200 ng/spot was required to cause erythema. This is a significantly greater dose than required for rats (50 ng/spot). Primates refused food, were listless and had diarrhoea and emesis within 1–2 hours of doses as low as 0.25 mg T-2/kg. The rats similarly refused food when given a comparable dose. The rapid onset of emesis and/or food refusal suggests a central origin since histopathological gastrointestinal tissue changes had not been noted at this time. The minimum effective dose (MED) for diarrhoea was 0.79 mg/kg in the primate. Although the MED could not be calculated for hypothermia, there appeared to be a good correlation between the dose of T-2 and the severity of hypothermia. Animals that died were the most profoundly hypothermic. Shock, poor tissue perfusion and decreased oxygen consumption could be the cause of the hypothermia but a change in central temperature regulation cannot be excluded. Further studies in this area are ongoing. Regaining appetite and energy took the primates nearly 7 days but the return to normal body temperatures in surviving animals took only 1–2 days. These latter data are comparable to the rat model.

Haematology

The expected increase in white blood count during the first day and ultimate lymphopenia from the third to the seventh day were seen in both models. There were striking increases in prothrombin time (Figure 1) and partial thromboplastin time, as well as a decrease in platelet count (Figure 2) and the

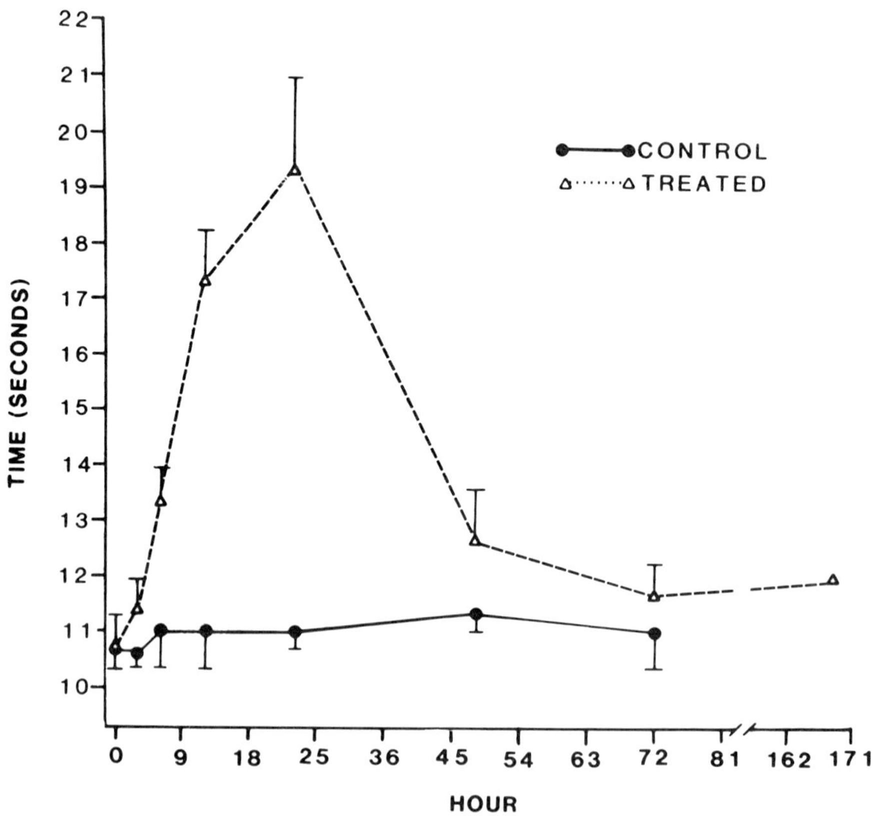

Figures 1 and 2 Prothrombin time and platelet count in cynomolgus monkeys (N=6) after intramuscular dose of 0.65 mg/kg T-2 toxin compared to vehicle only injected controls (N = 2).

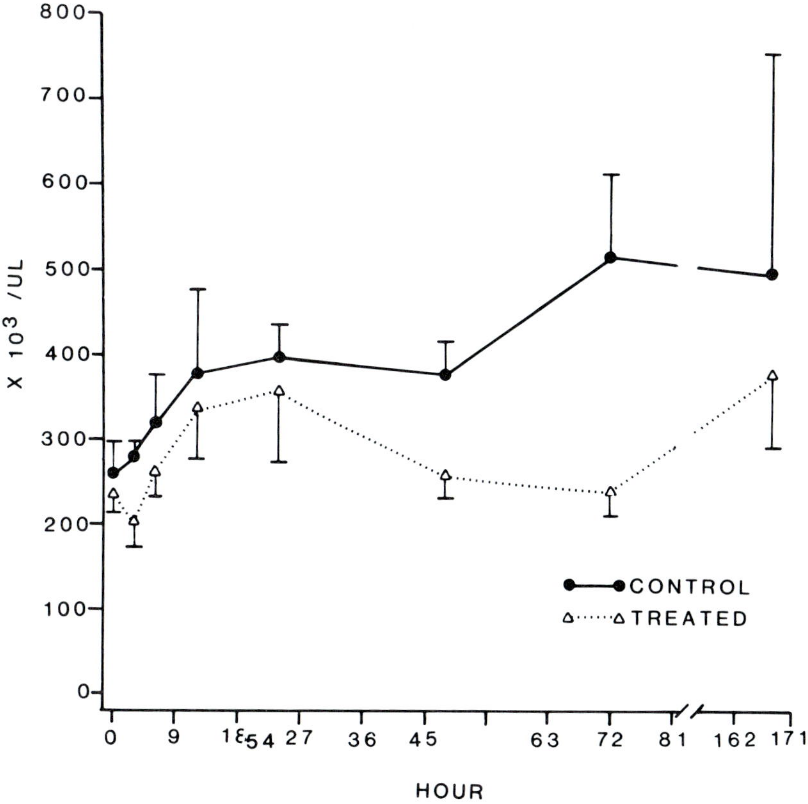

development of abnormal platelet function. All parameters had returned to normal within 3–7 days post-exposure. No overt bleeding occurred except for very occasional traces of blood in the stool or about the nose. No animals bled to death. However, the coagulation abnormalities could readily allow bleeding to persist if a lesion such as an ulcer or wound had occured. Coagulation data are not available in the rat for comparison.

Biochemical changes

Multi-channel chemistry studies were done and are listed with maximum percentage changes and times of occurrence in Table 1. There was a transient decrease in total protein and albumin, which might be explained by the known inhibition of protein synthesis although data on dilutional effects on the intravascular compartment and rates of protein degradation are not

Table 1 Percentage change (basal compared to post-toxin exposure) in clinical chemistry of cynomolgus monkeys after 0.65 mg/kg T-2 toxin intramuscularly ($N = 9$) with maximum percentage change, time of peak/change and significance level

Chemistry	Percentage change	Time in hours of maximum change	Level of significance
Sodium	−3	24	NSD
Potassium	+29	24	<0.05
Chloride	No change		NSD
Iron	+123	24	<0.05
Copper	−25		<0.05
Zinc	−47	48	<0.05
Calcium	+10	12	<0.05
	−12.3	48	<0.05
Phosphorus	+126	24	<0.05
Albumin	−23	48	<0.05
Total protein	−10.5	48	<0.05
Cholesterol	−37	24	<0.05
Triglyceride	−43	12	<0.05
	+194	48	<0.05
Uric acid	+857	24	<0.05
Glucose	+25	6	NSD
Creatinine	+229	24	<0.05
Blood urea nitrogen	+236	24	<0.05
Alkaline phosphatase	+48	12	<0.05
Serum glutamic pyruvic transaminase	+110	24	<0.05
5′ Nucleotidase	+44	12	<0.05
Gamma glutamyl transferase	+28	12	<0.05
Amylase	+1008	24	<0.05
Lactic dehydrogenase	+263	24	<0.05
Serum glutamic oxaloacetic transaminase	+583	24	<0.05
Creatine phosphokinase	+146.4	24	<0.05

known at this time. Phosphorus clearly increased and calcium probably decreased slightly. The slight decrease in calcium may have been in response to the increased phosphorus and decreased albumin levels. Cell injury might explain the increase in phosphorus and lesser increase in potassium that occurred. Tissue destruction can also explain the elevated blood urea nitrogen and creatinine, although impairment of renal blood flow and function may have also played a role but were not measured. Enzyme elevations included alkaline phosphatase, serum glutamic oxaloacetic transaminase, serum glutamic pyruvic transaminase, creatine phosphokinase, lactic dehydrogenase, 5′ nucleotidase, and amylase, and reflect widespread tissue injury. Although the specific organ sources of enzyme elevations are not absolutely certain, it is probable that they include the liver, skeletal and heart muscle, pancreas and intestine. Glucose levels were generally elevated when collected in sodium fluoride tubes but were significantly lower when collected as serum when there was a very high white blood count. This latter artifact was almost surely due to accelerated metabolism of glucose by the white blood cells after collection of the blood sample. Terminally, some primates became genuinely hypoglycaemic. This may have been caused by profound shock and terminal liver failure. Serum triglyceride elevations suggested either increased lipolysis or decreased lipid utilization. The serum iron was strikingly elevated, raising the possibility that the intoxicated animals might be more susceptible to infection because of an elevated serum iron in addition to the other known causes of immune impairment from trichothecene intoxication.

Cardiovascular effects in the cynomolgus monkey

Blood pressure and peripheral vascular resistance decreased. However, heart rate was decreased only at doses above the LD_{50} level. No consistent change in respiratory rate was seen. A decrease in cardiac contractility as calculated from echocardiographic data probably explains a portion of the hypotension and poor tissue perfusion. Figure 3 is an example of the results of one of the monkeys.

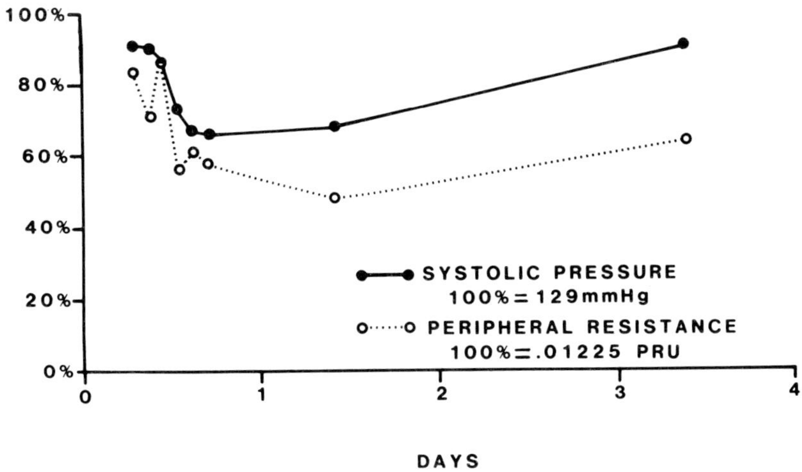

Figure 3 Systolic blood pressure and calculated peripheral resistance in single cynomolgus monkey after 0.65 mg/kg T-2 toxin intramuscularly.

Cardiovascular effects in the rat

At an LD_{50} dose of T-2 toxin, rats consistently showed a late decrease in blood pressure immediately before death and an increased heart rate followed by a decline in heart rate. A progressive decrease in cardiac output (Figure 4) was seen in those that died, and in those that lived a recovery toward normal cardiac output after 24–72 hours was seen. In contrast to the primates, the rats' peripheral vascular resistance increased. This will require further study to clarify the discrepancy. In addition to bradycardia, prolongation of the PR interval, QRS interval and the uncorrected QT interval occurred (Figure 5). Some of the rats also developed AV-dissociation. These findings suggest impaired cardiac conductivity as well as contractility.

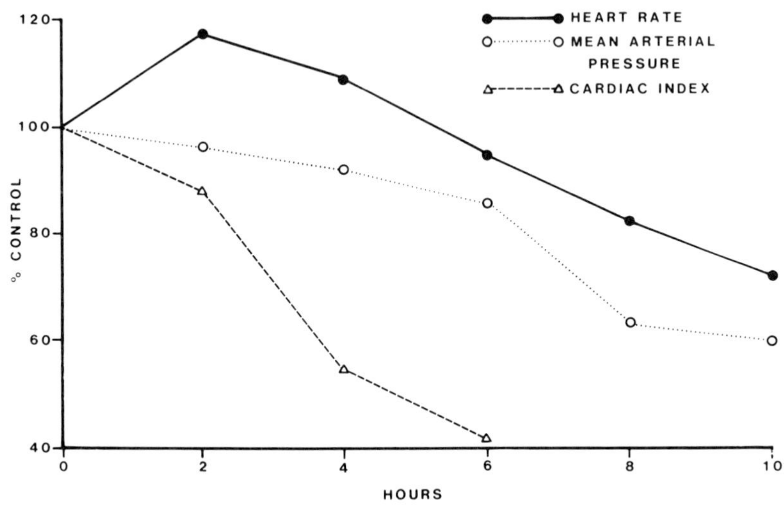

Figure 4 Mean heart rate, arterial pressure and cardiac index of rats after 2 mg/kg T-2 toxin intramuscularly ($N = 12$).

CONCLUSION

The reported pathophysiological studies of acute T-2 intoxication reported are most compatible with widespread tissue and organ injury including haematological, hepatic, renal, pancreatic, muscular and cardiac effects. The mechanism of death during acute intoxication seems to be cardiovascular with a decreased cardiac contractility, decreased cardiac output, and ultimately shock and death. Although changes in cardiac conductivity do occur, they appear to play only an ancillary role in the acute deaths. Secondary cellular

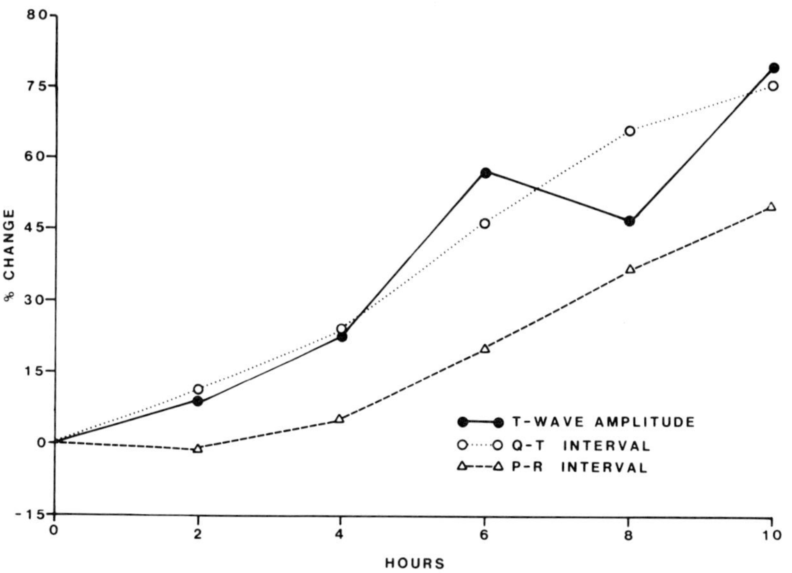

Figure 5 Mean T wave amplitude, Q–T interval and P–R interval in rats after 2 mg/kg T-2 toxin intramuscularly (N = 12).

effects (in addition to direct cellular toxin effects) from poor tissue perfusion, shock, tissue hypoxia and acidosis may play an important role.

In the areas of study in which data are available for both the cynomolgus monkey and the rat there is excellent agreement as to the qualitative and quantitative pathophysiological responses. Specifically, the haematological, biochemical, cardiovascular and clinical effects (emesis/food refusal, diarrhoea, lethargy, body temperature) are produced at very similar dosages. The lower cost, ready availability and preference for the rat model favours its use at this time. Proper laboratory animal use gives priority to conservation of primate use as well. Use of non-human primates has been, and will be, restricted to limited studies to verify the results in non-primate species in order to determine potential effects of toxins in man. Prior publications (Murphy *et al.*, 1978; Bukowski *et al.*, 1982) on the use of anguidine in human cancer trials can also be of help in these estimates.

ACKNOWLEDGEMENTS

In conducting the research described in this report the investigators adhered to the *Guide for the Care and Use of Laboratory Animals* as promulgated by the Committee on Care and Use of Laboratory Animals of the Institute of

Laboratory Animal Resources, National Research Council. The facilities are fully accredited by the American Association for Accreditation of Laboratory Animal Care. The views of the authors do not purport to reflect the positions of the Department of the Army or the Department of Defense.

REFERENCES

Bamburg, J. R., and Strong, F. M. (1971). 12, 13-Epoxy-Trichothecenes. In *Microbial Toxins* (eds S. Kadis, A. Ciegler and S. J. Ajl), Vol. VII, pp. 207–292. Academic Press Inc., New York.

Bukowski, R., Vaughn, G., and Bottomley, R. (1981). Phase II study of anguidine in gastroentestinal malignancies: a southwest oncology group meeting, *Cancer Treat. Rep.*, **66**, 381–383.

Bunner, D. L., McNamee, G. A., Jr., Dinterman, R. E., and Wannamacher, R. W., Jr. (1982). Lack of enhanced nocturnal growth hormone release in tethered cynomolgus monkeys, *Am. J. Physiol.*, **243**, R213–R217.

Ellison, R. A., and Kotsonis, F. N. (1973). T-2 toxin as an emetic factor in moldy corn, *Appl. Microbiol.*, **26** (4), 540–543.

Finney, D. J. (1952). *Probit Analysis*. Cambridge University Press, Cambridge.

Hassler, C. R., Lutz, G. A., Linebaugh, R., and Cummings, K. D. (1979). Identification and evaluation of noninvasive blood pressure measuring techniques, *Toxicol. Appl. Pharmacol.*, **47**, 193–201.

Hsu, I., Smalley, E. B., Strong, F. M., and Ribelin, W. E. (1972). Identification of T-2 toxin in moldy corn associated with a lethal toxicosis in dairy cattle, *Appl. Microbiol.*, **24**., 684–690.

Joffe, A. Z. (1971). Alimentary toxic aleukia. In *Microbial Toxins* (eds S. Kadis, A. Ciegler, and S. J. Ajl), Vol. VII, pp. 139–189. Academic Press Inc., New York.

Joffe, A. Z. (1978). *Fusarium poae* and *F. sporotrichoides* as principal causal agents of alimentary toxic aleukia. In *Mycotoxic Fungi, Mycotoxins and Mycotoxicoses: an Encyclopedic Handbook* (eds T. D. Wyllie and L. G. Morehouse), Vol. 3, pp. 21–86. Marcel Dekker Inc., New York.

Matsuoka, Y., Kubota, K., and Ueno, Y. (1979). General pharmacological studies of Fusarenon-X, a trichothecene mycotoxin from *Fusarium* species, *Toxicol. Appl. Pharmacol.*, **50**, 87–94.

Muller, B., and Mannesmann, G. (1981). Measurement of cardiac output by the thermodilution method in rats. II. Simultaneous measurement of cardiac output and blood pressure in conscious rats, *J. Pharmacol. Meth.*, **5**, 29–34.

Murphy, W. K., Burgess, M. A., Valdivieso, M., and Livingston, R. B. (1978). Phase I clinical evaluation of anguidine, *Cancer Treat. Rep.*, **62**, 1497–1502.

Pace, J. G. (1983). Effect of T-2 mycotoxin on rat liver mitochondria electron transport system, *Toxicon*, **21**(5), 675–680.

Pearson, A. W. (1978). Biochemical changes produced by *Fusarium* T-2 toxin in the chicken, *Res. Vet. Sci.*, **24**, 92–97.

Rukmini, C., Prasad, J. S., and Rao, K. (1980). Effects of feeding T-2 toxin to rats and monkeys, *Fd. Cosmet. Toxicol.*, **18**, 267–269.

Stanford, G. K., Hood, R. D., and Hayes, A. W. (1975). Effect of prenatal administration of T-2 toxin to mice, *Res. Commun. Chem. Path. and Pharm.*, **10**(4), 743–746.

Triplett, D. A., and Harms, C. S. (1981). *Procedures for the Coagulation Laboratory*. American Society of Clinical Pathology, Chicago.

Ueno, Y., Hosoya, M., Morita, Y., Ueno, I., and Tatsuno, T.(1968). Inhibition of the protein synthesis in rabbit reticulocytes by nivalenol, a toxic principle isolated from *Fusarium nivale*-growing rice, *J. Biochem.*, (*Tokyo*), **64**, 478–485.

Ueno, Y., Ishii, K., Sakai, K., Kanaeda, S., Tsunoda, H., Tanaka, T., and Enomoto, M. (1972). Toxicological approaches to the metabolites of *Fusaria*. IV. Microbial survey on 'Bean-hulls poisoning of horses' with the isolation of toxic trichothecenes Neosolaniol and T-2 toxin of *Fusarium solani* M-1-1, *Jap. J. Exp. Med.*, **42**(3), 187–203.

Wyatt, R. D., Hamilton, P. B., and Burmfisher, H. R. (1973). The effects of T-2 toxin in broiler chickens, *Poultry Sci.*, **52**, 1853–1859.

Trichothecenes and Other Mycotoxins
Edited by J. Lacey

CHAPTER 38

Dermal toxicity of T-2 toxin in guinea pigs, rats, and cynomolgus monkeys

Robert W. Wannemacher, Jr., David L. Bunner, Judith G. Pace, Harold A. Neufeld, Lucas H. Brennecke, and Richard E. Dinterman

INTRODUCTION

Trichothecenes are a group of chemically related, highly toxic metabolites of *Fusarium* species (Bamburg and Strong, 1971; Ueno, 1980). Of these, T-2 toxin is one of the more toxic. In addition to its parenteral toxicity, small amounts of T-2 mycotoxin when applied to the skin have strong inflammatory and irritative characteristics, which have been utilized to develop a sensitive bioassay for its detection (Wei *et al.*, 1972; Chung *et al.*, 1974; Eppley *et al.*, 1974; Hayes and Schiefer, 1979). Preliminary results indicated that when large amounts of T-2 toxins are applied to skin, death results within 48 hours without any noticeable skin reaction (Bamburg and Strong, 1971).

There is circumstantial evidence that trichothecene toxins have been used in the yellow rain attacks in east Asia and Afghanistan (Marshall,1982; Rosen and Rosen, 1982). Since skin necrosis and absorption of trichothecenes may have accounted for some of the severe toxicity observed in these attacks, various experimental models were investigated to evaluate the pathophysiology associated with topical applications of the fungal toxins.

The difficulty of interpreting data from experiments in which toxin is applied to the skin is well illustrated in the literature. For instance, another fungal toxin, aflatoxin, applied topically to the skin of rabbits, has been reported to cause severe lesions in the epidermis and liver. However, rabbits wearing collars, and unconscious rats, did not develop liver necrosis because ingestion of toxin from the hind foot pads during grooming was prevented (Joffe and Ungar, 1969; Ungar and Joffe, 1969; Purchase and Steyn, 1973). By contrast, even without precautions to prevent ingestion, no skin or liver changes were observed when aflatoxin was applied topically to the backs of gerbils for 35 days (Llewellyn, 1978).

MATERIAL AND METHODS

Determination of minimum effective dose (MED)

Twenty-four hours before use in an experimental study, male Fisher 344 rats (50–75 g); Hartley outbred guinea pigs (350–400 g), New Zealand white rabbits (2.5–2.9 kg), or cynomolgus monkeys (5–6 kg) had their backs shaved with an electric clipper and remaining hair was removed with a chemical depilatory Surgex hair-removing cream. Two microlitre samples of varying concentrations of T-2 mycotoxin in methanol were applied to the shaved backs of each species. The reactions were scored at 24, 48 and 72 h by the method of Draize *et al.* (1944), with both erythema and oedema being scored on a 0–5 scale of increasing intensity. A linear regression was determined between the reaction score and dose of mycotoxin for each species. The MED was defined as that dose of toxin that produced a score of 1.

LD_{50} and mean time to death determinations

Rats, guinea pigs, and cynomolgus monkeys were injected intramuscularly (i.m.) with 0.1 ml of varying doses of T-2 mycotoxin in ethanol : glycerol : water (2 : 3 : 5). Each animal was observed in order to determine time of death. Selected animals from each species who succumbed to the T-2 mycotoxin were necropsied and tissues saved for histopathological examination.

For topical studies in rats the back was shaved and hair removed with a depilatory 24 h before applying toxin. Varying concentrations of T-2 mycotoxin in either methanol or dimethylsulphoxide (DMSO) was applied as 0.1 ml sample in a 4 × 4 cm area on the back of the rat. They were observed for the next 7 days and time of death recorded for each rat. Selected rats that died were necropsied and tissues saved for histopathology.

Hair was shaved from the backs of guinea pigs and varying concentrations of T-2 mycotoxin were applied at 0.1 ml sample in either methanol or DMSO. Guinea pigs were observed for clinical signs and time of death. In another group of guinea pigs a barrier was applied to the skin at time of application of the toxin. The barrier consisted of a plastic sponge with a 4 × 4 cm hole sandwiched between a vinyl screen and a brass screen, held together with elastic tape. This barrier was placed over the area where the T-2 toxin had been applied and was held in place with the elastic tape. Guinea pigs with the barrier attached were observed for 14 days after application of the T-2 mycotoxin and time of death was recorded for individual animals.

A right atrial catheter was implanted in four cynomolgus monkeys, after immobilization with Ketamine (Ayerst Labs. Incorp., New York, NY). While still anaesthetized, monkeys were fitted with a leather jacket and the catheter was passed through a stainless-steel conduit. The flexible conduit was attached to a swivel located on the upper portion of the metabolic cage. A temperature probe was also implanted in the muscles of the back and passed, along with the catheter, to the swivel. The atrial catheter was attached to a second small swivel, which was located on the top of a big one. A plastic carrier was attached to the lower swivel, in which a temperature probe was maintained. The atrial catheter was kept patent by a slow infusion of 100 ml per day of a solution that contained 4 units per ml of heparin in half-normal saline. Six days after surgery, monkeys were again sedated, jackets opened, chest hair shaved, and either 1, 2, 4, or 8 mg/kg of T-2 mycotoxin in DMSO was applied to the chest. After the DMSO dried, the area was covered with a plastic screen attached to a ¼-inch plastic sponge. This barrier was held in place with elastic tape and the monkey was re-jacketed. Blood samples were obtained at 0, 3, 12, 24, 72 h and 7 days after exposure to the toxin. Various clinical parameters were monitored including body temperature, vomiting, bleeding, diarrhoea, excitability, listlessness, ability to stand, prone in cage, changes in breathing rate, coma, or death. In addition, daily food intake was recorded for each monkey. Seven days after applying the toxin, the monkeys were sedated, and their jacket and catheter were removed. Lesions on the chest were photographed and the monkeys observed for another 7 days before returning to the colony.

Metabolism of labelled T-2 mycotoxin

A group of 12 guinea pigs were injected i.m. with a solution that contained 1.0 mg of T-2 mycotoxin and 100 μCi of [^{3}H]T-2 mycotoxin per kg of body weight, in a mixture of ethanol : glycerol : water (2 : 3 : 5). Individual guinea pigs were placed in metabolic cages and daily urine and faecal collections were made for the next 28 days. Daily urine volumes were recorded and a sample was analyzed for radioactive content. Another sample of urine was analysed by thin-layer chromatography to determine radioactivity associated with the T-2 mycotoxin and metabolites (Pace *et al.*, 1983).

Another group of 24 guinea pigs had their backs shaved and 3.8 mg of T-2 toxin and 200 microcuries of [^{3}H]T-2 mycotoxin per kg by weight in methanol applied to the skin. Twelve of these guinea pigs had a screen barrier applied over the area of the skin on which the toxin had been painted.

Ability to remove T-2 mycotoxin from skin of the rat

Male, Fisher-344 rats weighing 75–100 g had their backs shaved and hair removed with a chemical depilatory. Twenty-four hours after removing the hair, 50 μl of a solution containing 5 mg of T-2 mycotoxin per ml of either methanol, ethanol : glycerol : water (2 : 3 : 5) or DMSO was applied to a 1 × 1 inch area of skin. Five rats were each treated 0, 5, 60, or 360 min after exposure to the toxin. Treatment consisted of wiping of the back with a soap-saturated 2 × 2 surgical sponge, followed by rinsing for 1 minute with stream of water. Another group of five rats which did not have their backs washed after exposure to toxin served as controls. The skin of each rat was observed on day 1, 2, and 3 after exposure and scored according to procedure of Draize *et al.*, (1944). These scores were utilized to calculate micrograms of toxin from a previously determined standard curve. Concentration of toxin in the treated rats was compared to the control non-treated group in order to determine percentage removal of toxin from the skin.

RESULTS AND DISCUSSION

The minimal effective dose (MED) of T-2 toxin to produce a skin response in different species is presented in Table 1. Of the species tested, rabbits were the most sensitive to dermal application of T-2 mycotoxin. Almost 2½ times as much T-2 mycotoxin was required to produce a similar response in rats and guinea pigs, as found previously by Hayes and Schiefer (1979). In contrast, the skin of a cynomolgus monkey was 10 times less sensitive to the T-2 mycotoxin than that of a rabbit. A MED of 200 ng in the monkey caused only very slight erythema, while a similar dose in the rabbit, rat, or guinea pig caused a severe erythema and oedema which was associated with definite skin necrosis. Variation in dermal response to the T-2 mycotoxin in these species may be related to its ability to penetrate the stratum corneum layer of skin. Regardless of the species, a very strong inflammatory and cytotoxic response was elicited when the toxin was applied to skin at higher doses.

The comparative systemic toxicity and lethality of T-2 mycotoxin of different species was assessed by the i.m. route or topical application (Table 2). By i.m. injection the lethal dose_{50} (LD_{50}) was between 0.5 and 1 mg/kg of

Table 1 Minimal effective dose (MED) of T-2 mycotoxin to produce a skin response in different species

Species	*N*	MED (ng/2 μl) (mean ± SE)	Significance
Rabbit	8	20.5 ± 8.8	$p < 0.01$, rat $p < 0.05$, guinea pig and monkey
Rat	10	51.3 ± 3.6	$p < 0.01$, rabbit $p < 0.05$, monkey
Guinea pig	10	53.2 ± 3.3	$p < 0.05$, rabbit
Monkey	4	209 ± 112	$p < 0.05$, rat and rabbit

MED = Dose to produce score 1.

Table 2 Comparative LD_{50} and mean time to death of T-2 mycotoxin by topical or i.m. routes in different species

			Topical				Topical + Barrier			
	i.m. Injection		Methanol		DMSO		Methanol		DMSO	
Species	LD_{50} (mg/kg)	MTD (h)	LD_{50} (mg/kg)	MTD (h)	LD_{50} (mg/kg)	MTD (h)	LD_{50} (mg/kg)	MTD (h)	LD_{50} (mg/kg)	MTD (h)
Rat	0.47	15.1	12.48	132	1.50	19.6				
Guinea pig	1.03	14.1	4.20	190	4.56	180	79.8	153 (4 of 6 died)	78.8	100 (4 of 6 died)
Cynomologous monkey	0.79	18.6							>8.0	

body weight for rat, guinea pig, and cynomolgus monkey, with a mean time of death (MTD) of 12–24 h. In the rat a similar LD_{50} was observed when the T-2 mycotoxin was applied to the skin in DMSO. In contrast, topical application of this mycotoxin in methanol on rats resulted in an LD_{50} of 12.5 mg/kg and a mean time of death of 5.5 days. When the toxin in DMSO was applied to the skin of rats, systemic lesions of necrosis of lymphoid tissues and intestinal epithelium were observed, with only mild dermal necrosis, but when T-2 toxin in methanol was applied to the skin the major lesion was skin necrosis with few detectable systemic effects. In contrast to the rat, LD_{50} values following topical application of T-2 mycotoxin to the skin of a guinea pig in either methanol or DMSO were similar, but 4 times that observed by the i.m. route. Mean time of death of guinea pigs after topical exposure to T-2 mycotoxin in either solvent was 7–8 days. With both solvents, toxin exposure resulted in severe skin necrosis and marked inflammation, characterized by oedema, redness and purulent discharge from the eyes, nose, and mouth, which started to develop about day 3 after applying the toxin to the skin. Histological examination revealed that in some cases the inflammatory discharge completely blocked the nasal passage. As the mucous membrane irritation developed the guinea pigs stopped eating or drinking water. No other gross histological lesions were observed by microscopic examination. Thus, it appears that the guinea pigs, during normal grooming procedures, were able to remove some of the toxin from their backs contaminating their eyes, noses, and mouths. The resulting conjunctivitis and rhinitis ultimately significantly contributed to death through dehydration, anorexia and/or suffocation. When a screen barrier was applied to the area of the back receiving the toxin, no guinea pigs died at a dose of 33 mg/kg of T-2 mycotoxin ($8 \times LD_{50}$ without barrier), while 4 out of 6 died in each group with approximately 80 mg/kg of toxin painted on the skin in either methanol or DMSO. Guinea pigs given toxin in DMSO died in approximately 4.2 days, while those that had it applied in methanol died in 6.3 days. With the barrier applied none of the guinea pigs developed the mucous membrane irritation that was observed without the barrier. Thus without the use of a barrier, guinea pigs were able to remove some of the T-2 mycotoxin from the back which contributed to their deaths. Further, from studies with the barrier, it can be concluded that this toxin is only very slowly absorbed through the skin of guinea pig.

When up to 8 mg/kg of T-2 toxin in DMSO was applied to the skin of cynomolgus monkey and the area covered with a screen and leather jacket, no deaths were observed. The monkeys developed a mild illness, indicated by slight elevations in serum CPK and mild reduction in food intake, together with severe skin lesions at the site of application of the toxin. Lesions were still evident 14 days after exposure but had almost completely resolved 28 days after. Since 10 times the i.m. LD_{50} dose for T-2 mycotoxin did not cause death of the monkey, it appears that less than 10% of the toxin applied to the skin was absorbed daily.

When labelled T-2 mycotoxin was injected i.m. in guinea pigs, radioactive compounds were rapidly excreted in the urine (Figure 1) but very little was T-2 mycotoxin. Most of the radioactivity was associated with more polar or conjugated trichothecenes. Half of the i.m.-injected radioactivity was excreted in urine over 28 days, even less if a screen barrier was placed over the site of application (Figure 1; Table 3). These data again support the conclusion that T-2 toxin is only slowly absorbed through the skin of guinea pigs.

Six hours after applying T-2 mycotoxin to the skin of the rat in either methanol or ethanol : glycerol : water solvent, the majority of toxin could be removed from the skin by washing with soap and water (Table 4). When applied in DMSO, washing with soap and water immediately or after 5 minutes removed most toxin, preventing death, but all rats died when washing was delayed by 60 min, suggesting that more than 50% of the dose had been absorbed. These data again indicate that T-2 mycotoxin was slowly absorbed through the skin when applied in methanol, but DMSO significantly increased its rate of absorption in the rat. Since the rats were sedated for 60

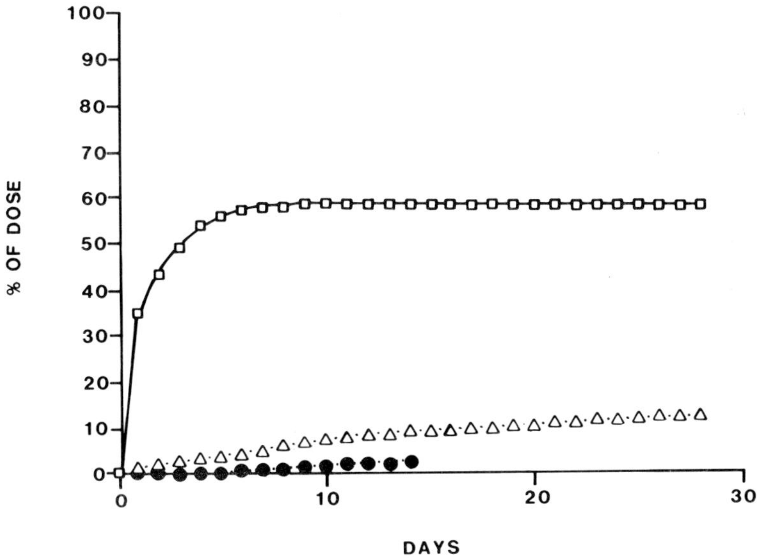

Figure 1 Cumulative radioactivity in guinea pig urine following a single exposure to labelled T-2 mycotoxin; □ = i.m. injection; △ = topical application; ● = topical application plus a screen barrier. The data are calculated as a percentage of radioactivity injected or applied to the skin.

Table 3 Cumulative rate of urinary excretion of radioactivity from a dose of [3-H]T-2 mycotoxin in guinea pigs

Route of administration	Percentage of dose applied	
	Day 7	Day 14
	Mean ± SE (N)	Mean ± SE (N)
Intramuscular	50.2 ± 2.9 (6)	52.3 ± 2.8 (6)
Topical	4.6 ± 0.52 (9)[a]	10.21 ± 0.65 (3)[c]
Topical + barrier	1.47 ± 0.42 (6)[b]	4.81 ± 0.81 (6)[d]

[a] $p < 0.001$, i.m. or topical + barrier.
[b] $p < 0.001$, i.m. or topical.
[c] $p < 0.001$, i.m.; $p < 0.01$, topical + barrier.
[d] $p < 0.001$, i.m.; $p < 0.01$, topical.

Table 4 Ability of soap and water to detoxify T-2 mycotoxin from skin of rats at various times after application in different solvents

Time after application (min)	Solvent		
	Methanol (percentage detoxified)	Ethanol : glycerol : water (percentage detoxified)	DMSO (percentage detoxified)
0	100 ± 0 [a]	97.1 ± 1.3	98.2 ± 2.4
5	93.9 ± 2.7	93.9 ± 2.6	75.8 ± 6.5
60	95.6 ± 1.8	92.0 ± 5.5	—[b]
360	58.2 ± 22.3	70.0 ± 15.1	—[b]

[a] Mean ± standard error of six rats.
[b] Lethal (all rats dead in 24 hours). Less than 50% of dose detoxified.

min after application of the toxin in DMSO, the death of the rats who had their skins washed after 60 minutes could not be explained by oral ingestion of toxin.

CONCLUSION

As observed by others (Bamburg and Strong, 1971) T-2 mycotoxin applied to the skin of various species can cause a spectrum of injury from erythema through necrosis. Of the species tested the rabbit was most sensitive to the toxin and the cynomolgus monkey least. Skin absorption of this toxin appeared to be slow compared to i.m. injection in the three species studied, except possibly when DMSO is used as a solvent on wet skin. As observed with aflatoxin (Purchase and Steyn, 1973; Llewellyn, 1978), a screen barrier was required to prevent oral ingestion and facial contamination with the T-2 toxin applied to skin elsewhere. With a screen barrier, very large doses of toxin were required in the guinea pig and cynomolgus monkey to cause ill effects or death. DMSO appears to increase percutaneous absorption of T-2 in rats but much less so in the guinea pig or monkey. The relationship to percutaneous absorption of T-2 mycotoxin through human skin will require additional studies which utilize an *in vitro* model (Franz, 1975, 1978; Bronaugh *et al.*, 1982) or human skin implants in the nude mouse. Soap-and-water wash appears to be an effective treatment for removal of toxin from the skin in most solvents. Additional data are required to determine what will be the maximum effective time if the toxin is applied in DMSO.

ACKNOWLEDGEMENTS

In conducting the research described in this report, the investigators adhered to the *Guide for the Care and Use of Laboratory Animals*, as promulgated by the Committee on Care and Use of Laboratory Animals of the Institute of Laboratory Animal Resources, National Research Council. The facilities are fully accredited by the American Association for Accreditation of Laboratory Animal Care.

REFERENCES

Bamburg, J. R., and Strong, F. M. (1971). 12,13-Epoxytrichothecenes. In *Microbial Toxins*, Vol. VII, *Algal and Fungal Toxins*, (eds S. Kadis, A. Ciegler, and S. Ajl), pp. 207–292. Academic Press, New York.

Bronaugh, R. L., Steward, R. T., Congdon, E. R., and Giles, A. L. (1982). Methods for *in vitro* percutaneous absorption studies. 1. Comparison with *in vivo* results, *Toxicol. Appl. Pharmacol.*, **62**, 474–480.

Chung C. W., Trucksess, M. W., Giles, A. L., and Freidman, L. (1974). Rabbit skin test for estimation of T-2 toxin and other skin-irritating toxins in contaminated corn. *J. Assoc. Off. Anal. Chem.*, **57**, 1121–1127.

Draize, J. H., Woodard, G., and Calvery, H. O. (1944). Methods for the study of irritation and toxicity of substances applied topically to skin and mucous membranes, *J. Pharmacol. Exp. Ther.*, **82**, 377–390.

Eppley, R. M., Stoloff, L., Trucksess, M. W., and Chung, C. W. (1974). Survey of corn for *Fusarium* toxins, *J. Assoc. Off. Anal. Chem.*, **57**, 632–635.

Franz, T. J. (1975). Percutaneous absorption. On the relevance of *in vitro* data, *J. Invest. Dermatol.*, **64**, 190–195.

Franz, T. J. (1978). The finite dose technique as a valid *in vitro* model for the study of percutaneous absorption in man. *Curr. Probl. Dermatol.*, **7**, 58–59.

Hayes, M. A., and Schiefer, H. B. (1979). Quantitative and morphological aspects of cutaneous irritation by trichothecene mycotoxin, *Food Cosmet. Toxicol.*, **17**, 611–621.

Joffe, A. Z., and Ungar, H. (1969). Cutaneous lesions produced by topical application of aflatoxin to rabbit skin, *J. Invest. Dermatol.*, **52**, 504–507.

Llewellyn, G. C. (1978). Toxic response in gerbils treated topically with aflatoxin B_1 and dimethylforamide, *Bull. Environm. Contam. Toxicol.*, **19**, 747–754.

Marshall, E. (1982). A cloudburst of yellow rain reports, *Sciences*, **218**, 1202–1203.

Pace, J. G., Watts, M. R., and Matson, C. F. (1983). Fate and distribution of T-2 toxin in guinea pigs, *Fed. Proc.*, **42**, 626.

Purchase, I. F. H. and Steyn, M. (1973). Absence of percutaneous absorption of aflatoxin, *Toxicol. Appl. Pharmacol.*, **24**, 162–164.

Rosen, R. T., and Rosen, J. D. (1982). Presence of four *Fusarium* mycotoxins and synthetic material in 'yellow rain.' Evidence for the use of chemical weapons in Laos, *Biom. Mass Spectrom.*, **8**, 443–450.

Ueno, Y. (1980). Trichothecene mycotoxins: mycology, chemistry, and toxicology, *Adv. Nutr. Res.*, **3**, 301–353.

Ungar, H., and Joffe, A. Z. (1969). Acute liver lesions resulting from percutaneous absorption of aflatoxins, *Pathol. Microbiol.*, **33**, 65–76.

Wei, R. D., Smalley, E. B., and Strong, F. M. (1972). Improved skin test for detection of T-2 toxin, *Appl. Microbiol.*, **23**, 1029–1030.

Trichothecenes and Other Mycotoxins
Edited by J. Lacey

CHAPTER 39

Dynamics of experimental trichothecene mycotoxicosis

H.B. Schiefer

INTRODUCTION

The trichothecenes, fungal metabolites of a number of fungi, especially *Fusarium* spp., are potent protein and DNA inhibitors (Ueno, 1980), and are known to cause radiomimetic lesions (Saito *et al.*, 1969), which characterize such diseases as 'alimentary toxic aleukia' of man (Joffe, 1978), 'mouldy corn toxicosis' in cattle (Hsu *et al.*, 1972), 'bean hulls poisoning' of horses (Ueno, 1980) and 'fusariotoxicosis' in chickens (Greenway and Puls, 1976). The scope of biological activities, metabolism, pharmacokinetics, mechanisms of inhibition of protein synthesis and other biochemical effects has been reviewed (Busby and Wogan, 1981) and there remains little doubt as to the fact that most diseases caused by the trichothecenes share a common feature: the immune response of all species is severely suppressed. Because of the depletion of the lymphoid and haemopoietic tissues, which characterize these toxicoses, affected individuals receiving sub-lethal doses are highly susceptible to infectious agents and die of complications caused by secondary bacterial, viral or parasitic invaders (Friend, 1982). If subjected to massive doses, death may occur in a matter of hours or days.

This summary of the dynamics of experimental trichothecene mycotoxicosis will be presented in a fashion akin to a complete toxicity study, i.e. the following aspects will be considered: acute toxicity, dermal toxicity, ocular toxicity, subacute and chronic toxicity (including immunotoxicity), effects on reproduction and fertility, teratology, mutagenicity and carcinogenicity. Most of the information presented here is derived from experiments with mice, using T-2 toxin as the toxic principle.

ACUTE TOXICITY

The LD_{50} for mice is given as ranging from 0.15 to 6.1 mg/kg body weight (Busby and Wogan, 1981), depending on route of administration. We chose an intragastric dose of 2.5 mg/kg of T-2 toxin in propylene glycol and mice were examined at intervals of 1, 2, 3, 4, 5, 6, 9, 12, 18, 24, 48 and 96 hours. T-2 toxin caused necrosis in all lymphoid tissues, haematopoietic tissues and in germinal regions of intestinal crypts. Mitotic activity in intestinal crypts ceased within 2 hours, after which numerous round bodies in the cytoplasm of intestinal crypt epithelial cells appeared by 3–4 hours. Many such bodies were evident by 12 hours, and often they formed into plugs of necrotic debris at the base of the crypts. Degenerative activity was evident by 9 hours and increased until crypt epithelium became hyperplastic between 24 and 48 hours, but villous atrophy was observed in some animals allowed to live for 9 days. Destruction of lymphoid cells in the lamina propria of the intestine occurred within 1 hour. Karyorrhexis in lymphoblasts of follicles in the spleen and lymph nodes occurred within 1–2 hours, after which complete necrosis was observed. This was followed by thymic atrophy, but by 96 hours regenerative activity was observed in the atrophic regions of the thymus. Mice given T-2 toxin developed a transient but marked lymphocytosis within 3 hours. Necrosis in haematopoietic tissue of the spleen and bone marrow was visible by 6 hours. Regeneration of megakaryocytes and other undifferentiated haematopoietic cells occurred in spleen and bone marrow by 48 hours.

A dose–response study (Hayes, 1979) utilized T-2 rates of 0.125, 0.25, 0.5, 1.0, 2.0, or 4.0 mg/kg. Lesions increased proportionally with dose above 0.5 mg/kg. Topical application of T-2 toxin in DMSO caused the same type of lesions (personal observation, unpublished).

DERMAL TOXICITY

Trichothecenes are highly irritant to the skin and to mucous membranes, and this property has been the basis for the development of a skin bioassay (see review, Hayes and Schiefer, 1979). The acute inflammatory reaction is characterized by hyperaemia, oedema and neutrophil exudation at lower concentrations and necrosis of the epidermis at higher concentrations. By comparing the intensity of reaction relative to reactions to a graded series of standard solutions of T-2 toxin applied to the same animal, one can measure concentrations of T-2 toxin in the range of 5–60 μg/kg accurately (Hayes and Schiefer, 1979). In dietary studies with T-2 toxin, perioral dermatitis and dermatitis of paws of mice is a common observation (Hayes, 1979; Friend, 1982).

OCULAR TOXICITY

We have not conducted any experiments of this nature, but there are reliable verbal reports of accidental contamination of the eye with trichothecenes, describing corneal opacity which follows rapidly after exposure.

SUBACUTE AND CHRONIC TOXICITY

Dietary T-2 toxin, at levels of 10 or 20 mg/kg, substantially reduces growth and food consumption. During the first 3 weeks of exposure to T-2 toxin, lymphoid tissues, bone marrow and splenic red pulp become hypoplastic, resulting in anaemia, lymphopenia and eosinopenia. Subsequently, during continued exposure to T-2 toxin, haematopoietic cells regenerate in bone marrow and splenic red pulp and become hyperplastic by 6 weeks. Granulopoiesis and thrombopoiesis resume in advance of erythropoiesis. All lymphoid tissues remain atrophic (Hayes *et al.*, 1980).

The interval before regeneration of haematopoietic cells will be prolonged by diets with reduced protein content (Hayes and Schiefer, 1980).

The effects of trichothecenes on the lymphoid tissues have been extensively studied in *in vitro* systems and after i.v. or i.p. administration of the toxins (see review in Friend, 1982), but rarely after dietary application of the toxins. Friend *et al.* (1983a) fed 5, 10 or 20 mg/kg of T-2 toxin to mice and measured the effect of the toxin on the immune system by counting total spleen cell numbers and the *in vitro* proliferative response of spleen cells to the T-cell mitogen, Concanavalin-A (Con-A) and to the B-cell mitogen, lipopolysaccharide (LPS). Consumption of 20 mg/kg of T-2 toxin depressed the spleen cell count longer than 3 weeks, whereas the other levels of T-2 toxin induced a temporary depression only. Consumption of 20 mg/kg of T-2 toxin by mice for 1–4 weeks depressed T-cell responses to Con-A, but the B-cell responses were depressed in mice fed 10 and 20 mg/kg of T-2 toxin as well as in mice fed a control diet at a restricted rate (Friend *et al.*, 1983a). T-2 toxin suppressed the severity of inflammatory response to acute Herpes simplex infection in mice, and it was determined that 20 mg/kg of T-2 toxin was more immunosuppressive than cyclophosphamide (Friend, 1983c).

Studies with sheep have shown that lambs treated with 0.6 mg/kg body weight daily for 21 days developed leukopenia on day 7, lymphopenia on days 17 and 14 and had depressed mitogenic responses to LPS on day 7, and prolonged prothrombin times.

At necropsy, lymphoid atrophy of mesenteric lymph nodes and spleens was most marked in lambs treated with 0.6 mg/kg body weight of T-2 toxin (Friend *et al.*, 1983b).

Ducks receiving 20 or 30 mg/kg of T-2 toxin for 2 or 3 weeks also show generalized atrophy of all lymphoid tissues (Hayes and Wobeser, 1983). Occasionally these ducks exhibited signs of bilateral wing paresis, which may suggest CNS involvement.

In our subacute and chronic studies we have encountered the occasional death due to haemorrhagic enteritis, and increase of liver weights, with no morphological changes observed, were regularly found. All other organs, particularly heart, lungs and kidneys, were not remarkable.

EFFECTS ON REPRODUCTION AND FERTILITY

Although there have been reports that the testicular cells may be damaged by T-2 toxin (Smalley and Strong, 1974), there are no reports on the reproductive ability of animals fed T-2 toxin over prolonged periods of time. Injection of trichothecenes (Ito *et al.*, 1980) or intragastric application can cause dose-related foetal death and abortion (Hood *et al.*, 1978; Stanford *et al.*, 1975; Hayes, 1981; Rousseaux, 1983).

TERATOLOGY

The trichothecenes are teratogenic (Hayes, 1981). Tail and limb abnormalities are produced following 0.5 mg/kg i.p. at the tenth day of gestation (Hood *et al.*, 1978), or 3.0 mg/kg after intragastric application (Rousseaux, 1983).

MUTAGENICITY

For the most part the trichothecenes do not demonstrate mutagenic activity in bacterial test systems (Hayes, 1981; Busby and Wogan, 1981), but there may be effects on yeasts (Schappert and Khachatourians, 1983).

CARCINOGENICITY

Little information exists on the possible carcinogenicity of the trichothecenes. Rats did not develop tumours after ingesting a diet containing 10 mg/kg of T-2 toxin over an 8-month period (Marasas *et al.*, 1969); however, one has to consider the fact that rats are more resistant to T-2 toxin than mice (Hayes and Schiefer, 1982). Schoental *et al.* (1979) have reported an increased incidence of tumours of the digestive tract and brain. Further studies are needed.

METABOLISM AND PHARMACOKINETICS

The tissue distribution and metabolism has not been extensively studied, although recent activities associated with the question of whether 'yellow rain' is a trichothecene mycotoxicosis will produce a plethora of reports on this topic.

Studies that are available indicate that the major excretory pathway is the urine and the faeces, and that the trichothecenes are fairly rapidly excreted (Chi *et al.*, 1978; Matsumoto *et al.*, 1978; Nakano *et al.*, 1979; Yoshizawa *et al.*, 1981; Robison *et al.*, 1979). However, when most studies ceased, after about 4 days, approximately 20–40% of T-2 toxin was unaccounted for. Thus, it is possible that residual amounts of T-2 or its metabolites could continue to exert effects longer than 4 days (Mirocha *et al.*, 1982).

'YELLOW RAIN' AND TRICHOTHECENES

Since the U.S. made the allegation that trichothecenes were used as chemical warfare agents, there has been an intensive discussion about whether trichothecenes might have been used. Besides the findings of unusual concentrations of trichothecenes in samples from alleged attacks (Rosen and Rosen, 1982; Mirocha *et al.*, 1982), the discussion had to make do with interpretations of second-hand reports, epidemiological considerations in interpretations of natural background levels and natural occurrence or absence of trichothecenes in South-east Asia (Schiefer, 1982, 1983; Humphreys and Dow, 1982).

It appears that neither mycotoxicoses nor other diseases occur in South-east Asia which might be able to cause the symptoms and effects on all sorts of life (human, animal and plant life) that are reported to occur. The symptoms described by victims are more consistent with stachybotryotoxicosis, a disease caused by macrocyclic trichothecenes (Mirocha, 1980) rather than with other, normocyclic trichothecenes, due to dietary exposure. However, the topical or inhalation routes of application have not been studied very well, and preliminary, unpublished studies give the impression that similar symptoms and lesions might be induced when suitable carriers, like DMSO, are used. Much more work is needed before final conclusions can be drawn.

CONCLUSION

The trichothecenes are, indeed, a group of fungal metabolites that deserve considerable attention and study, although the dangers due to occasional ingestion of quantities of trichothecenes may have been overstated. The task is complicated by the presence of potentiating factors, either other mycotoxins or low protein levels in the diet, and other xenobiotics.

To integrate our knowledge of occurrence with knowledge of mechanisms by which trichothecenes act, and are metabolized and modified, remains a formidable task before one can fully define the hazards of trichothecenes to human or animal populations or specify tolerance levels in the diet.

REFERENCES

Busby, W. F., and Wogan, G. N. (1981). Trichothecenes. In *Mycotoxins and N-Nitroso Compounds: Environmental Risks*, Vol. II, (ed. R. C. Shank), Chapter 3, pp. 29–45. CRC Press, Boca Raton, Fla.

Chi, M. S., Robison, T. S., Mirocha, C. J., Swanson, S. P., and Shimoda, W. (1978). Excretion and tissue distribution of radioactivity from tritium-labelled T-2 toxin in chicks, *Toxicol. Appl. Pharmacol.*, **45**, 391–402.

Friend, S. C. E.(1982). The effects of trichothecene mycotoxins on the immune system. Ph.D. thesis, University of Saskatchewan.

Friend, S. C. E., Babiuk, L. A., and Schiefer, H. B. (1983a). The effects of dietary T-2 toxin on the immunological function and Herpes simplex reactivation in Swiss mice, *Toxicol. Appl. Pharmacol.*, **69**, 234–244.

Friend, S. C. E., Hancock, D. S., Schiefer, H. B., and Babiuk, L. A. (1983b). Experimental T-2 toxicosis in sheep, *Can. J. Comp. Med.*, **47**, 291–297.

Friend, S. C. E., Schiefer, H. B., and Babiuk, L. A. (1983c). The effects of dietary T-2 toxin on acute Herpes simplex virus Type 1 infection in mice, *Vet. Pathol.*, **20**, 737–760.

Greenway, J. A., and Puls, R. (1976). Fusariotoxicosis from barley in British Columbia. I. Natural occurrence and diagnosis, *Can. J. Comp. Med.*, **40**, 12–15.

Hayes, A. W. (1981). *Mycotoxin Teratogenicity and Mutagenicity*. CRC Press, Boca Raton, Fla.

Hayes, M. A. (1979). Morphological and toxicological studies on experimental T-2 mycotoxicosis. Ph.D. thesis, University of Saskatchewan.

Hayes, M. A. and Schiefer, H. B.(1979). Quantitative and morphological aspects of cutaneous irritation by trichothecene mycotoxins, *Food Cosmet. Toxicol.*, **17**, 611–621.

Hayes, M. A. and Schiefer, H. B. (1980). Subacute toxicity of dietary T-2 toxin in mice: Influence of protein nutrition, *Can. J. Comp. Med.*, **44**, 219–228.

Hayes, M. A., and Schiefer, H. B. (1982). Comparative toxicity of dietary T-2 toxin in rats and mice, *J. Appl. Toxicol.*, **2**, 207–212.

Hayes, M. A., and Wobeser, G. A. (1983). Subacute toxic effects of dietary T-2 toxin in young Mallard ducks, *Can. J. Comp. Med.*, **47**, 180–187.

Hayes, M. A., Bellamy, J. E. C., and Schiefer, H. B. (1980). Subacute toxicity of dietary T-2 toxin in mice: Morphological and hematological effects, *Can. J. Comp. Med.*, **44**, 203–218.

Hood, R. D., Kuczuk, M. H., and Szczech, G. M. (1978). Effects in mice of simultaneous prenatal exposure to ochratoxin A and T-2 toxin, *Teratology*, **17**, 25.

Hsu, I. C., Smalley, E. B., Strong, F. M., and Ribelin, W. E. (1972). Identification of T-2 toxin in moldy corn associated with a lethal toxicosis in dairy cattle, *Appl. Microbiol.*, **24**, 684–690.

Humphreys, G. R., and Dow, J. (1982). An epidemiological investigation of alleged CW/BW incidents in Southeast Asia. Directorate of Preventive Medicine, Surgeon General Branch, National Defence Headquarters, Ottawa, 1982.

Ito, Y., Ohtsubo, K., and Saito, M. (1980). Effects of fusarenone-X, a trichothecene produced by *Fusarium nivale*, on pregnant mice and their foetuses, *Jap. J. Exp. Med.*, **50**, 167–172.

Joffe, A. Z. (1978). *Fusarium poae* and *F. sporotrichioides* as principal causal agents of alimentary toxic aleukia. In *Mycotoxic Fungi, Mycotoxins, Mycotoxicoses. An Encyclopedic Handbook*, Vol. 3 (eds T. D. Wyllie and L. G. Morehouse), pp. 21–86. Marcel Dekker, New York and Basel.

Marasas, W. F. O., Bamburg, J. R., Smalley, E. G., Strong, F. M., Ragland, W. L., and Degurse, P. E. (1969). Toxic effects on trout, rats and mice of T-2 toxin produced by the fungus *Fusarium tricinctum* (Cd). Snyd. et Hans., *Toxicol. Appl. Pharmacol.*, **15**, 471–482.

Matsumoto, H., Ito, T., and Ueno, Y. (1978). Toxicological approaches to the metabolites of *Fusaria*. XII. Fate and distribution of T-2 toxin in mice, *Jap. J. Exp. Med.*, **48**, 393–399.

Mirocha, C. J. (1980). Rubratoxin, sterigmatocystin and stachybotrys mycotoxins. In *Conference on Mycotoxins in Animal Feeds and Grains Related to Animal Health*

(June 13, 1980) (ed. W. Shimoda), pp. 152–232. Bureau of Vet. Med. FDA, Washington.

Mirocha, C. J., Watson, S., and Hayes, W. A. (1982). Occurrence of trichothecenes in samples from Southeast Asia associated with 'yellow rain'. *Proc. V. Internat. IUPAC Symposium on Mycotoxins and Phycotoxins*. Vienna, Austria, 1–3 September , pp. 130–133.

Nakano, N., Nagahara, A., Shimizu, T., Aibara, K., Fujimoto, Y., Morooka, N., and Tatsuno, T. (1979). The tissue distribution and the pattern of excretion of (^{14}C)-13-labelled 12,13-epoxytrichothec-9-ene in mice and rats, *Jap. J. Med. Sci. Biol.*, **32**, 269–279.

Robison, T. S., Mirocha, C. J., Kurtz, H. J., Behrens, J. C., Chi, M. S., Weaver, G. A., and Nystrom, S. D. (1979). Transmission of T-2 toxin into bovine and porcine milk, *J. Dairy Sci.*, **62**, 637–641.

Rosen, R. T., and Rosen, J. D. (1982). Presence of four *Fusarium* mycotoxins and synthetic material in 'yellow rain'. Evidence for the use of chemical weapons in Laos. *Biomed. Mass Spectrom.*, **9**, 443–450.

Rousseaux, C. (1983). Department of Veterinary Pathology, Saskatoon. M.Sc. Research in Progress, unpublished.

Saito, M., Enomoto, M., and Tatsuno, T. (1969). Radiomimetic biological properties of the new scirpine metabolites of *Fusarium nivale*, *Gann*, **60**, 599–603.

Schappert, K. T., and Khachatourians, G. G. (1983). Effects of fusariotoxin T-2 on *Saccharomyces cerevisiae* and *Saccharomyces carlsbergensis*. *Appl. Env. Microbiol.*, **45**, 862–867.

Schiefer, H. B. (1982). Study of the possible use of chemical warfare agents in Southeast Asia. A Report to the Department of External Affairs, Canada. University of Saskatchewan, 1982; UN Document A/37/308.

Schiefer, H. B. (1983). The possible use of chemical warfare agents in Southeast Asia, *Conflict Quarterly*, **3**, 32–40.

Smalley, E. B., and Strong, F. M. (1974). Toxic trichothecenes. In *Mycotoxins*, (ed. I. F. H. Purchase), Chapter 18. Elsevier, New York.

Schoental, R., Joffe, A. Z., and Yagen, B. (1979). Cardiovascular lesions and various tumors found in rats given T-2 toxin, a trichothecene metabolite of *Fusarium*, *Cancer Res.*, **39**, 2179–2189.

Stanford, G. K., Hood, R. D., and Hayes, A. W. (1975). Effect of prenatal administration of T-2 toxin to mice, *Res. Comm. in Chem. Path. Pharmacol.*, **10** (4), 743–746.

Ueno, Y. (1980). Trichothecene mycotoxins: mycology, chemistry, and toxicology. In *Advances in Nutritional Research*, Vol. 3, (ed. H. H. Draper), Chapter 10, pp. 301–353. Plenum Publ. Corp., New York.

Yoshizawa, T., Mirocha, C. J., Behrens, J. C., and Swanson, S. P. (1981). Metabolic fate of T-2 toxin in a lactating cow, *Food Cosmet. Toxicol.*, **19**, 31–39.

Trichothecenes and Other Mycotoxins
Edited by J. Lacey

CHAPTER 40

A summary of the effects of dietary T-2 toxin on acute herpes simplex virus type 1 infection in mice*

S. C. E. Friend, H. B. Schiefer, and L. A. Babiuk

INTRODUCTION

T-2 toxin, a trichothecene mycotoxin produced by *Fusarium* species (Ueno, 1977a,b), is potently cytotoxic to lymphoid and haemopoietic tissues of many species when administered parenterally (Rusch and Stähelin, 1965; Saito *et al.*, 1969; Sato *et al*, 1975; Weaver *et al.*, 1978; Rosenstein *et al.*, 1979), by gavage (Lutsky *et al.*, 1978; Lutsky and Mor, 1981), or in the diet (Hayes *et al.*, 1980).

Herpes viruses, as exemplified by herpes simplex virus type 1 (HSV-1), are important infectious agents causing disease in animals and man (Knotts *et al.*, 1973; Charlton *et al.*, 1977; Andrewes *et al.*, 1978). For many years mice have been used in the study of active and latent HSV-1 infections (Stevens and Cook, 1971; Knotts *et al.*, 1973; Cook and Stevens, 1973; Renis *et al.*, 1976). As many workers investigating the cytotoxic effects of T-2 toxin have used mice (Ueno *et al.*, 1970, 1972; Ohtsubo and Saito, 1977; Sato *et al.*, 1978; Hayes *et al.*, 1980, Lafarge-Frayssinet *et al.*, 1981), we chose a mouse model of HSV-1 to study the effects of dietary T-2 toxin on a viral disease. We suspected that the depletion of lymphoid and haemopoietic tissues caused by dietary T-2 toxin (Hayes *et al.*, 1980) might impair some or many of the immunological defence mechanisms which protect mice against HSV-1 infection (Rager-Zisman and Allison, 1976, 1980; Nash *et al.*, 1980a,b, 1981), resulting in a disseminated viral infection and death. Cyclophosphamide (Cy), a known immunosuppressive drug, increases the susceptibility of mice to HSV-1 infection (Rajcani *et al.*, 1974; Worthington *et al.*, 1980). Since the action of Cy on the murine immune system is well documented (Stockman *et*

* The details of this research work have been published in *Veterinary Pathology*, **20**, 737 (1983).

al., 1973; Turk and Poulter, 1972) we used this drug as a comparison for possible immunosuppressive effects of dietary T-2 toxin on HSV-1 infection in young adult mice.

MATERIALS AND METHODS

The experimental design is outlined in Table 1.

Three-week-old male white Swiss mice were fed control diet or diet supplemented with 10 or 20 mg T-2 toxin/kg for 2 or 3 weeks. These mice were then inoculated with herpes simplex virus type 1 (HSV-1) (9.6×10^6 plaque-forming units (PFU)) intraperitoneally (i.p.). This dosage was the geometric mean of the titre of HSV-1 which killed 10/10 (3.03×10^7 PFU) and 0/10 (3.03×10^6 PFU) 5-week-old male white Swiss mice. To compare the effects of T-2 toxin against a known immunosuppressive drug, Cy was injected i.p. at 150 mg/kg, 24 h after treatment with HSV-1, into 18 mice fed the control diet. Five mice in each group were to be killed and necropsied on the fourth, sixth and eighth days after HSV-1 challenge, and necropsies were performed on all mice which died prematurely. Half of the brain and liver were frozen at −70 °C for virus isolation. Tissues were fixed for microscopic examination and lesions were scored on a scale of 0 to 4. The grading of the

Table 1 Experimental design for the T-2 toxin/HSV-1 interaction study

Group	Number of mice	T-2 toxin Dose[a] (mg/kg)	T-2 toxin Time[b] (weeks)	HSV-1[c]	Cyclophosphamide[d]	Mortality rate[e]
A	16	20	2	+	−	12/16
B	20	20	3	+	−	15/20
C	20	10	2	+	−	3/20
D	14	10	3	+	−	3/14
E	15	0	2	+	−	0/15
F	19	0	3	+	−	0/19
G	15	0	3	+	+	1/15
H	5	0	2	MEM	−	0
I	4	20	3	MEM	−	0
J	3	10	3	MEM	−	0
K	3	0	3	MEM	+	0

[a] T-2 toxin in diet.
[b] Period of feeding time (weeks), before HSV-1 inoculation.
[c] 9.6×10^6 PFU of HSV-1 or Eagles minimal essential media (MEM) was injected i.p. into mice.
[d] 150 mg/kg inoculated i.p., 24 h after HSV-1 inoculation.
[e] Number of mice dying before the eighth day following HSV-1 challenge on day 1.

lesions in the central nervous system (CNS) was based on the degree of the inflammatory response, whereas the grading of hepatic and adrenal lesions was based on the severity and extent of necrosis. Splenic lesions were scored according to the degree of depletion of red and white pulp, while thymic lesions were graded according to the degree of depletion of cortical thymocytes.

RESULTS

The mortality rate for mice in the study is included in Table 1.

Five-week-old white Swiss mice which consumed a diet containing 20 mg/kg of T-2 toxin for 2 or 3 weeks were highly susceptible to HSV-1 infection and 27 of 36 (75%) died as a result of severe and extensive hepatic and adrenal necrosis. These mice had marked thymic and splenic atrophy and bone marrow depletion. Although HSV-1 was isolated from their livers and brains, there was little or no inflammatory response found in the adrenals, livers, spinal cords, brains, or ganglia. The necrotizing encephalomyelitis, which was observed in mice fed control diet and inoculated with HSV-1, was absent.

High levels of dietary T-2 toxin appeared to be more immunosuppressive than Cy because only one of 15 mice died after treatment with HSV-1 and Cy. Mice treated with Cy had changes in brain, spinal cord, spleens, thymus, and bone marrow which were similar to those fed 20 mg/kg T-2 toxin and infected with HSV-1; however, liver lesions were much less severe. HSV-1 was reisolated from the brain and liver of these mice also. HSV-1-infected mice on a diet containing 10 mg/kg of T-2 toxin had lesions, the severity of which was intermediate between those of HSV-1-infected mice fed either a diet with 20 mg/kg of T-2 toxin or a toxin-free control diet. Only six of 34 mice infected with HSV-1 and fed a diet containing 10 mg/kg of T-2 toxin died. Hepatic and adrenal necrosis was less extensive. The infrequent isolation of virus from liver and brain was consistent with the lack of intranuclear inclusion bodies and a more marked inflammatory response. Bone marrow and splenic red pulp were depressed only moderately by 10 mg T-2 toxin/kg diet. This may have enhanced the necrotizing encephalomyelitis observed in mice killed on days 6 and 8 after HSV-1 infection.

Hepatic lesions were mild and those of the adrenals moderate in HSV-1-infected mice fed toxin-free control diet. None of the mice died. The rare isolation of HSV-1 from the liver and brain, and the finding of a moderate to severe necrotizing encephalomyelitis in these mice, was consistent with an essentially functional immune system.

CONCLUSION

Immunosuppression of mice by dietary T-2 toxin or by Cy appeared to protect against the destruction of the spinal cord and brain by an HSV-1-induced inflammatory response characterized by a necrotizing meningo-encephalomyelitis. Mice infected with HSV-1 and fed 20 mg/kg of T-2 toxin in the diet and mice inoculated with HSV-1 and then Cy had few or no brain lesions, yet virus was isolated from brains as early as day 3. Where Cy and T-2 toxin abrogated the inflammatory response there were no lesions. Perhaps the inability of these mice to mount an immune response, due to the marked depletion of thymus, splenic white and red pulp and bone marrow, prevented the destruction of the CNS. Although T-2 toxin fed at levels of 20 mg/kg for 14 or 21 days, or the injection of Cy at 150 mg/kg reduced the necrotizing phase of the disease, it is evident that high levels of the toxin, even more than Cy, increased the susceptibility of HSV-1, allowed the virus to proliferate and ultimately resulted in the death of mice normally resistant to HSV-1.

REFERENCES

Andrewes, C., Pereira, H. G., and Wildy, P. (1978). Herpetoviridae. In *Viruses of Vertebrates*, 4th edn, pp. 312–355. Baillière Tindal, Toronto.

Charlton, K. M., Dulac, G. C., Thomas, F. C., and Mitchell, H. K. (1977). Necrotizing encephalitis in skunks caused by herpes simplex virus, *Can. J. Comp. Med.*, **41**, 460–465.

Cook, M. L., and Stevens, J. G. (1973). Pathogenesis of herpetic neuritis and ganglionitis in mice: evidence for intra-axonal transport of infection, *Infect. Immun.*, **7**, 272–288.

Hayes, M. A., Bellamy, J. E. C., and Schiefer, H. B. (1980). Subacute toxicity of dietary T-2 toxin in mice: morphological and hematological effects, *Can. J. Comp. Med.*, **44**, 203–218.

Knotts, F. B., Cook, M. L., and Stevens, J. G. (1973). Latent herpes simplex virus in the central nervous system of rabbits and mice, *J. Exp. Med.*, **138**, 740–744.

Lafarge-Frayssinet, C., Decloitre, F., Mousset, S., Martin, M., and Frayssinet, C. (1981). Induction of DNA single-strand breaks on T_2 toxin, a trichothecene metabolite of *Fusarium*. Effect on lymphoid organs and liver, *Mutat. Res.*, **88**, 115–123.

Lutsky, I., and Mor, N. (1981). Experimental alimentary toxic aleukia in cats, *Lab. Anim. Sci.*, **31**, 43–47.

Lutsky, I., Mor, N., Yagen, B., and Joffe, A. Z. (1978). The role of T-2 toxin in experimental alimentary toxic aleukia: a toxicity study in cats, *Toxicol. Appl. Pharmacol.*, **43**, 111–124.

Nash, A. A., Field, H. J., and Quartey-Papafio, R. (1980a). Cell-mediated immunity in herpes simplex virus-infected mice: induction, characterization and antiviral effects of delayed type hypersensitivity, *J. Gen. Virol.*, **48**, 351–357.

Nash, A. A., Phelan, J., and Wildy, P. (1981). Cell-mediated immunity in herpes simplex virus-infected mice: H-2 mapping of the delayed-type hypersensitivity response and the antiviral T cell response, *J. Immunol.*, **126**, 1260–1262.

Nash, A. A., Quartey-Papafio, R., and Wildy, P. (1980b). Cell-mediated immunity in herpes simplex virus-infected mice: functional analysis of lymph node cells during periods of acute and latent infection, with reference to cytotoxic and memory cells, *J. Gen. Virol.*, **49**, 319–322.

Ohtsubo, K., and Saito, M. (1977). Chronic effects of trichothecene toxins. In *Mycotoxins in Human and Animal Health* (eds J. V. Rodricks, C. W. Hesseltine and M. A. Mehlman), pp. 255–262. Pathotox Publ., Park Forest South, Illinois.

Rager-Zisman, B., and Allison, A. C. (1976). Mechanism of immunologic resistance to herpes simplex virus 1 (HSV-1) infection, *J. Immunol.*, **116**, 35–40.

Rager-Zisman, B., and Allison, A. C. (1980). The role of natural killer (NK) cells in resistance to herpes simplex virus 1 (HSV-1) infection, *Parasite Immunol.*, **2**, 71–84.

Rajcani, J., Gajdosova, E., and Mayer, V. (1974). Pathogenesis of herpesvirus hominis infection in immunosuppressed mice, *Acta Virol.*, **18**, 135–142.

Renis, H. E., Eidsen, E. E., Mathews, J., and Gray, J. E. (1976). Pathogenesis of herpes simplex virus types 1 and 2 in mice after various routes of inoculation, *Infect. Immun.*, **14**, 571–578.

Rosenstein, Y., Lafarge-Frayssinet, C., Lespinats, G., Loisillier, F., Lafont, P., and Frayssinet, C. (1979). Immunosuppressive activity of *Fusarium* toxins. Effects on antibody synthesis and skin grafts of crude extracts, T-2 toxin and diacetoxyscirpenol, *Immunology*, **36**, 111–117.

Rusch Von, M. E., and Stähelin, H. (1965) Über einige biologische Wirkungen des Cytostaticum Verrucarin A, *Arzneimittelforsch*, **15**, 893–897.

Saito, M., Enomoto, M., and Tatsuno, T. (1969). Radiomimetic biological properties of the new scirpene metabolites of *Fusarium nivale*, *Gann*, **60**, 599–603.

Sato, N., Ito, T., Kumada, H., Ueno, Y., Asano, K., Saito, M., Ohtsubo, K., Ueno, I., and Hatanaka, Y. (1978). Toxicological approaches to the metabolites of *Fusaria*. XIII. Hematological changes in mice by a single and repeated administration of trichothecenes, *J. Toxicol. Sci.*, **3**, 335–356.

Sato, N., Ueno, Y., and Enomoto, M. (1975). Toxicological approaches to the toxic metabolites of *Fusaria*. VIII. Acute and subacute toxicities of T-2 toxin in cats, *Jpn. J. Pharmacol.*, **25**, 263–270.

Stevens, J. F., and Cook, M. L. (1971). Latent herpes simplex virus in spinal ganglia of mice, *Science*, **173**, 843–845.

Stockman, G. D., Heim, L. R., South, M. A., and Trentin, J. J. (1973). Differential effects of cyclophosphamide in B and T cell compartments of adult mice, *J. Immunol.*, **110**, 277–282.

Turk, J. L., and Poulter, L. W. (1972). Selective depletion of lymphoid tissue by cyclophosphamide, *Clin. Exp. Immunol.*, **10**, 285–296.

Ueno, Y. (1977a). Mode of action of trichothecenes, *Pure Appl. Chem.*, **49**, 1737–1745.

Ueno, Y. (1977b). Trichothecenes: overview address. In *Mycotoxins in Human and Animal Health* (eds J. V. Rodricks, C. W. Hesseltine and M. A. Mehlman), pp. 189–207. Pathotox Publ., Park Forest South, Illinois.

Ueno, Y., Ishikawa, Y., Amakai, K., Nakajima, M., Saito, M., Enomoto, M., and Ohtsubo, K. (1970). Comparative study on skin-necrotizing effect of scirpene metabolites of *Fusaria*, *Jpn. J. Exp. Med.*, **40**, 33–38.

Ueno, Y., Sato, N., Ishii, K., Sakai, K., and Enomoto, M. (1972). Toxicological approaches to the metabolites of *Fusaria*. V. Neosolaniol, T-2 toxin and butenolide, toxic metabolites of *Fusarium sporotrichioides* NRRL 3510 and *Fusarium poae* 3287, *Jpn. J. Exp. Med.*, **42**, 461–472.

Weaver, G. A., Kurtz, H. J., Bates, F. Y., Chi, M. S., Mirocha, C. J., Behrens, J. C., and Robison, T. S. (1978). Acute and chronic toxicity of T-2 mycotoxin in swine, *Vet. Rec.*, **103**, 531–535.

Worthington, M., Conliffe, M. A., and Baron, S. (1980). Mechanism of recovery from systemic herpes simplex virus infection. 1. Comparative effectiveness of antibody and reconstitution of immune spleen cells on immunosuppressed mice, *J. Infect. Dis.*, **142**, 163–174.

Trichothecenes and Other Mycotoxins
Edited by J. Lacey

CHAPTER 41

Aspects of *Fusarium* mycotoxicosis in livestock

Gary D. Osweiler, B. S. Hook and G. M. Buening

INTRODUCTION

Fusarium spp. are common throughout the world as pathogens of plants and contaminants in feeds and grains. They may produce many different chemical toxins leading to wide concern, particularly since World War II, with the recognition that alimentary toxic aleukia in man and similar diseases in animals may be caused by *Fusarium* mycotoxins. The class of mycotoxins known as trichothecenes comprise more than 30 derivatives, mainly produced by *Fusarium* spp., characterized by the presence of a tetracyclic 12,13-epoxytrichothec-9-ene nucleus. Major concern within this group has centred around three particular toxins: T-2 toxin, diacetoxyscirpenol (DAS) and deoxynivalenol (DON). Several excellent reviews of the sources, formation, chemistry, and detection of these toxins are available (CAST, 1979; Mirocha, 1979; NAS, 1979; Pier, 1981; Purchase, 1974; Rodericks *et al.*, 1977; WHO, 1979; Wyllie and Morehouse, 1977).

SIGNIFICANCE OF TRICHOTHECENE MYCOTOXINS IN FOOD ANIMALS

Clinical effects produced by trichothecene mycotoxins can be grouped into five categories: (1) feed refusal, (2) dermal necrosis, (3) gastroenteric effects, (4) coagulopathy, and (5) immunological dysfunction.

Clinical signs of trichothecene intoxication of livestock may depend on the route of exposure (Pier, 1981; Hoerr *et al.*, 1981, 1982a,b). Acute toxicologic effects of trichothecenes are similar in all animal species tested but species vary in the amount of each specific trichothecene needed to induce acute or chronic toxicosis. Acute or lethal poisoning in cattle or swine may not be possible by allowing free choice of mycotoxin-contaminated feeds since both species refuse feed containing any of the three major trichothecene toxins (Forsyth *et al.*, 1977; Osweiler *et al.*, 1981; Pollman and Koch, 1983; Trenholm *et al.*, 1982; Vesonder *et al.*, 1979, 1981; Weaver, 1981). Swine are most sensitive to trichothecenes, refusing feed containing as little as 2 mg/kg DON (Pollman and Koch, 1983). Refusal concentration for DAS and T-2 in swine appear to be approximately 10 and 16 mg/kg respectively (Weaver, 1981; Weaver *et al.*, 1980). Occasionally, field reports have associated enteritis, death and haemorrhaging with low levels (2 mg/kg or less) of T-2 toxin in cattle and swine (Hibbs *et al.*, 1976; Hsu *et al.*, 1972), but experimental evidence has not confirmed that these concentrations would be acutely toxic. Other circumstances of exposure to toxicants may have altered toxicity in such instances, or sampling of feeds may not have been entirely representative. Obtaining an appropriate sample indicative of exposure to mycotoxins is extremely difficult under field conditions.

Experimental toxicosis in food animals has generally been via direct dosing, either orally or parenterally. Clinical signs usually include moderate to severe depression, occasional ataxia or weakness, enteritis and/or diarrhoea, leukopenia and lymphoid depletion. Coagulopathy has been demonstrated by laboratory means in poultry and cattle but clinical haemorrhaging has not occurred (Doerr *et al.*, 1974, 1981; Gentry and Cooper, 1983; Hook *et al.*, 1983; Weaver *et al.*, 1980) although DAS has been experimentally associated with haemorrhagic bowel syndrome in swine (Weaver *et al.*, 1981).

Lesions in affected animals involve the skin, mouth, intestine or colon, bone marrow and lymphoid tissue (Hoerr *et al.*, 1982a,b; Hayes *et al.*, 1980; Kosuri *et al.*, 1971; Lutsky *et al.*, 1978; Pier, 1981; Osweiler *et al.*, 1981). Swine have also developed skin irritation and necrosis from standing in feed pans contaminated with T-2 toxin (Pier, 1981). Epithelial necrosis occurs primarily at the commissures of the beak in poultry consuming T-2-contaminated feed. Recently the pattern of oral necrosis in poultry associated with trichothecene consumption has been described in detail (Chi and Mirocha, 1978; Hoerr *et al.*, 1982b). Some investigators have reported erosion and loss of ruminal papillae after consumption of experimentally contaminated rations (Pier, 1981). Lymphoid depletion and necrosis, involving the spleen, thymus, mesenteric lymph nodes and Peyer's patches after T-2 administration appears to be a consistent lesion in all animal species studied.

Because of its known effect on protein synthesis and lymphoid depletion, T-2 has been suggested as an immunosuppressive agent (Hsieh, 1979). A summary of reported effects of T-2 toxin on immune function are presented in Tables 1–3.

NATURAL OCCURRENCE OF TRICHOTHECENES

An informal review of veterinary diagnostic laboratories in the midwestern United States indicates that T-2 toxin is only occasionally found in grain or feed samples. Published surveys indicate that of the *Fusarium* toxins DON occurs with greatest frequency followed by zearalenone. These may reach

Table 1 Immune related lesions for trichothecene mycotoxins

Toxin	Animal	Dose	Response to toxin	Reference
T-2 Toxin	Chicken	16 mg/kg in feed	Smaller size of spleen and bursa	Boonchuvit *et al.*, 1975
T-2 Toxin	Cat	0.06–0.10 mg/kg body weight (1.2–2.0 mg/kg in feed)	Thymus reduced in size; bone marrow hypoplasia	Lutsky *et al.*, 1978
T-2 Toxin	Turkey and chicken	2–10 mg/kg in feed	Thymic weight reduced in turkeys at 10 mg/kg	Richard *et al.*, 1978
T-2 Toxin	Chicken	5 mg/kg body weight (72 mg/kg in feed)	Bursa of Fabricius had reduced lymphoid follicles and altered deck epithelium	Terao *et al.*, 1978
T-2 Toxin	Chicken	1.25 mg/kg body weight (18 mg/kg in feed)	Thymus and bursa had atrophy of cortex	Hoerr *et al.*, 1979
Fusarium extract with T-2	Mouse	¼–½ LD_{50}	Reduced thymic weight at 1–7 days; lymphoid depletion in spleen and thymus	Lafarge-Frayssinet *et al.*, 1979
T-2 Toxin	Mouse	2–3 mg/kg body weight (25 mg/kg in feed)	Thymic weight decreased	Rosenstein *et al.*, 1979
T-2 Toxin	Mouse	20 mg/kg in feed	Thymic size reduced; thymic, lymphoid, splenic and bone marrow tissues hypoplastic	Hayes *et al.*, 1980
T-2 Toxin	Cattle	0.3–0.6 mg/kg body weight (10–20 mg/kg in feed)	Dose-related decrease in thymic weight; cortical atrophy	Osweiler *et al.*, 1981
Diacetoxyscirpenol	Chicken	2.5–3.5 mg/kg	Lymphoid and bone marrow necrosis; lymphatic tissue depletion	Hoerr *et al.*, 1982

Table 2 Response of humoral immune systems to toxin treatment

Species	Dose and duration	Agent	Response	Reference
Chicken	16 mg/kg in feed, 21 days	*Salmonella sp.*	Agglutinin titres normal in face of lowered resistance to *Salmonella*	Boonchuvit *et al.*, 1975
Mouse	3 mg/kg body weight, 14 days	Sheep RBC	Slight increase in HA titre to SRBC	Masuko *et al.*, 1977
Chicken and turkey	2–10 mg/kg in feed, 28 days	*Pasteurella multocida*	No change in tube agglutination titre to *P. multocida*	Richard *et al.*, 1978
Mouse		*Listeria*, tetanus toxoid	Reduced clearance of Listeria; rise in antibody (tetanus anti-toxin)	Wiskott, 1978
Mouse	1/12–1/2 LD_{50}, 2–14 days	Sheep RBC	*Fusarium* extract containing T-2 toxin lowered HA titres to SRBC; effect was transient	Lafarge-Frayssinet *et al.*, 1979
Mouse	0.5–1 mg/kg body weight (7–28 mg/kg in feed), 14–90 days	Sheep RBC	Suppressed antibody response to SRBC; response recovered to 6 days after T-2 stopped	Rosenstein *et al.*, 1979
Cattle	0.3–0.6 mg/kg body weight (10–20 mg/kg in feed), 42 days	*Anaplasma*	Reduced globulin IgM and IgA C_3 depressed; no change in IgG or response to anaplasma antigen	Osweiler *et al.*, 1981; Buening *et al.*, 1982; Mann *et al.*, 1982

Table 3 Effect on T-2 toxin on cell-mediated immunity

Species	Dose and duration	Agent	Response	Reference
Mouse	3 mg/kg body weight (43 mg/kg in feed) 14 days	Sheep RBC	Enhanced delayed hypersensitivity if T-2 given after SRBC sensitizing dose	Masuko *et al.*, 1977
Mouse	3 mg/kg body weight (43 mg/kg in feed) 7 days	Sheep RBC	T-2 reduced tolerance of induction of delayed hypersensitivity if given after sensitization to SRBC	Otokawa *et al.*, 1979
Mouse	$\frac{1}{12}$–$\frac{1}{2}$ LD_{50} 2–14 days	Mitogens PHA and LPS	Reduced lymphoblastogenesis for thymus but not spleen cells using PHA and LPS	Lafarge-Frayssinet *et al.*, 1979
Mouse	0.5–2 mg/kg body weight (7–28 mg/kg in feed) 14–90 days	Skin graft rejection	Prolonged period for skin graft rejection at 0.75 mg/kg	Rosenstein *et al.*, 1979
Cattle	0.3–0.6 mg/kg body weight (10–20 mg/kg in feed) 42 days	Anaplasma antigen. Mitogens (PHA, Con-A, PWM)	Lower CMI response to anaplasma; depressed mitogen response in peripheral lymphocytes	Buening *et al.*, 1982

clinically significant levels in years when weather and other conditions favour fungal growth (Bainton and Jones, 1977; Burmeister *et al.*, 1972; Mirocha *et al.*, 1976; Mirocha, 1979). It appears that clinical toxicoses in food animals from T-2 or DAS are very rarely documented. Recent evidence (Tables 1–3) indicates that subclinical levels could be a factor in immune dysfunction. However, data on the frequency of occurrence of T-2 toxin and on levels found in various limited surveys and our own experience suggest that T-2 or other major trichothecenes in this review are not more than an occasional threat to livestock health. Although these toxins are potent inhibitors of protein synthesis and could be classified as highly toxic, there is little evidence of their presence at significant concentrations in livestock feeds within the United States.

SUBCLINICAL EFFECTS OF T-2 TOXIN IN CATTLE

T-2 toxin affects certain parameters of immunity in animals and poultry (Pier *et al.*, 1979). Detrimental effects have included increased lethality in rodents and poultry when challenged with pathogenic organisms (Boonchuvit *et al.*, 1975; Wiskott, 1978). Lymphoid tissues, spleen, bone marrow and bursa of Fabricius have been depleted in poultry given T-2 toxin, (Hoerr *et al.*, 1981, 1982a; Terao *et al.*, 1978) and thymic involution in chickens and turkeys has been demonstrated after exposure to T-2 (Richard *et al.*, 1978). Although immunological effects have been demonstrated in laboratory animals and poultry, the effect of T-2 toxin on the immunological competence of economically important food animals is still an important question. The consideration of whether or not immunodeficiency develops in the absence of clinical toxicosis has not been fully explored.

Recent work at the University of Missouri has concentrated on the potential subtle effects of *Fusarium* mycotoxins in cattle and swine. Based on evidence that T-2 toxin is an immunosuppressor in laboratory animals, we chose to investigate this potential in cattle exposed to amounts of toxin not causing clinical effects. Our particular interest was feed refusal, growth and immune function. Objectives of our work summarized here were as follows:

1. Establish a dose of T-2 toxin which would not cause overt clinical effects by casual observation.
2. Evaluate effects of mild clinical toxicosis and subclinical T-2 administration on parameters of haematology and immunology.

MATERIALS AND METHODS

T-2 toxin for initial studies was obtained from a commercial source. Identity and purity of the toxin was confirmed by melting-point determination, gas–liquid chromatographic analysis, and mass spectroscopy. In subsequent studies, T-2 toxin was produced by inoculating either *Fusarium tricinctum* NRRL 3299 or *F. sporotrichiella* MC 72083 on white corn grits and incubating at 15 °C for 3 weeks (Burmeister, 1971). Toxin was then extracted with chloroform : acetone, purified through acetone and crystallized in hexane–acetone–toluene. Toxin produced was verified by gas–liquid chromatography and mass spectroscopy (H. M. Stahr, Iowa State University, Ames, Iowa).

Doses of 0.1, 0.3 and 0.6 mg/kg body weight were established for experimental dosing (Osweiler *et al.*, 1981). The toxin dissolved in acetone was added dropwise to a small amount of corn meal in gelatin capsules. After evaporation of the acetone, these capsules were used to administer appropriate toxin doses daily per os.

In Trial 1, nine mixed-breed beef calves weighing 140–200 kg were treated as shown in Table 4. Trial 2, mixed-breed beef calves weighing 85–120 kg were exposed to T-2 as shown in Table 5.

Table 4 Trial 1

Group	No. of animals	T-2 dosage
Control	3	0
Low dose	3	0.1 mg/kg
Intermediate dose	3	0.3 mg/kg

Table 5 Trial 2

Group	No. of animals	T-2 dosage
Full-fed controls	2	0
Pair-fed	3	0
High dose T-2	3	0.6 mg/kg

Pair-fed calves were fed similarly to calves receiving T-2 toxin. The experimental period for each trial was 6 weeks. Temperature, pulse and respiration were evaluated individually on each calf during both trials. Calves were weighed prior to the trial and weekly thereafter. In Trial 1 calves were offered hay *ad lib* and fed a 16% protein calf starter at the rate of 1 kg/100 kg body weight. In Trial 2 concentrate was fed to supply 80% of calculated dietary needs, with the balance supplied by limited feeding of alfalfa hay. All calves in both trials had functional rumens.

Calves were bled by jugular venipuncture prior to the experiment and weekly thereafter. Haematological and clinical chemistry tests were performed by a commercial laboratory (Consolidated Biomedical Laboratories, Wichita, Kansas) as shown in Table 6.

At the end of the experiment complete gross necropsy was performed and organ weights were recorded for liver, kidney, spleen, thymus, adrenals, and heart. Tissues collected for microscopic examination included brain, liver, kidney, peripheral nerve, skin, intestine, abomasum, rumen, adrenals, thyroid, heart, lungs, thymus, spleen, Peyer's patches, and popliteal and prescapular lymph nodes. Samples were fixed in 10% buffered formalin, sectioned at 6 μm, and stained with haematoxylin–eosin.

Humoral immune function was evaluated by response to anaplasmosis vaccine and differential determination of serum protein fractions. Samples were taken 1 week prior to toxin administration and weekly thereafter. Serum protein fractions were determined by electrophoresis on cellulose acetate plates (Buening, 1974). Immunoglobulin levels (IgG, IgA, IgM) were evalu-

Table 6 Haematological and clinical chemistry tests performed

Haematology	Clinical chemistry
Erythrocyte Count	Serum Total Protein
Haemoglobin	Calcium
Packed Cell Volume	Phosphorus
Leukocyte Count	Glucose
Differential Leukocyte Count	Blood urea nitrogen (BUN)
	Creatinine
	Cholesterol
	Serum triglyceride
Coagulation studies	Serum Albumin
Prothrombin Time	Serum Globulin
Partial Thromboblastin Time	Albumin/Globulin (AG) Ratio
Thrombocyte Count	Serum Glutamic Oxalacetic Transaminase
Fibrinogen	Serum Glutamic Pyruvic Transaminase
	Gamma Glutamyl Transferase
	Lactate Dehydrogenase
	Alkaline Phosphatase
	Total Bilirubin
	Sodium
	Potassium
	Chloride

ated with commercial reagent kits (Miles Laboratories, Elkhart, Indiana) using radial immunodiffusion (Mancini *et al.*, 1965). Serum titres of anaplasma specific antibodies were determined by complement fixation assay (Anon, 1973).

Cell-mediated immunity was evaluated in peripheral blood lymphocytes. Blood was collected in preservative-free heparin and lymphocytes separated by the procedures of Buening (1974). Methods for evaluating the responses of lymphocytes from controls and principles were adaptations of the procedures described by Muscoplat *et al.* (1974) and Soper *et al.* (1978) using optimum concentrations of phytohaemagglutinin (PHA), concanavalin-A (Con-A), and pokeweed mitogen (PWM). Procedures utilizing specific antigens were similar to the mitogen studies, except specific antigen was cultured with the lymphocytes.

Statistical analysis of data was by ANOVA and LSD procedures.

RESULTS

Clinical effects were apparent only in the high-dose group (Trial 2). Beginning on the second day of the trial, calves refused to eat the concentrate portion of their diet. Diarrhoea occurred on an intermittent basis, ranging from soft tan stools to dark semi-fluid faeces. Table 7 summarizes the effect of T-2 toxin on feed refusal in calves. The high-dose group refused feed consistently, while the intermediate-dose group (0.3 mg/kg from Trial 1) showed partial feed refusal. Feed refusal appeared dose-related. Calves given 0.3 mg/kg gained weight as rapidly as controls, while high-dose calves lost body weight during the experimental period. Pair-fed calves in Trial 2 gained less than controls, but more than high-dose calves. Weekly body weights are summarized in Tables 8 and 9.

Humoral immunity was altered in the high-dose T-2 toxin group. Total serum protein was decreased 7 days after administration of T-2 toxin.

Table 7 Feed refusal in calves (total days)[a]

Treatment (mg/kg body weight)	*N*	Week of trial 1	2	3	4	5	6
Control	3	5	0	0	0	1	—
0.1	3	4	4	5	13	10	—
0.3	3	17	18	21	20	19	—
0.6	3	20	21	21	21	18	19

[a] Total possible refusal days was 21.

Table 8 Body weight changes (kg) in calves given subclinical doses of T-2 toxin

Treatment (mg/kg body weight)	*N*	Week of trial 1	2	3	4	5	6
Control	3	297.0 ± 48.9[a]	297.0 ± 47.7	306.7 ± 50.8	312.7 ± 53.4	298.7 ± 49.9	326.3 ± 54.9
0.1	3	305.5 ± 31.4	309.0 ± 29.2	322.7 ± 33.7	321.3 ± 31.9	311.7 ± 30.2	336.0 ± 32.6
0.3	3	297.3 ± 13.3	295.7 ± 13.2	318.0 ± 17.1	311.7 ± 10.3	299.3 ± 14.8	311.0 ± 11.1

[a] Standard error of mean.

Table 9 Body weight changes (kg) in calves given T-2 toxin

Treatment	*N*	Week of trial 1	2	3	4	5	6	7
Control	2	227.5 ± 19.5[a]	240.5 ± 20.5	243.0 ± 27.0	250 ± 22.0	258 ± 25.0	263 ± 19.0	274 ± 18.7
Pair-fed	3	227.0 ± 16.2	227.0 ± 16.9	234.0 ± 16.4	234 ± 16.5	231 ± 19.2	230 ± 18.1	239 ± 16.7
0.6 mg/kg body weight	3	223.0 ± 23.6	220.0 ± 25.3	215.3 ± 24.3	209 ± 24.0	208 ± 22.6	206 ± 20.9	211 ± 20.6

[a] Mean ± standard error.

Cumulative group means of the high-dose T-2 toxin group were significantly lower ($p < 0.05$) than the control group for gamma, alpha, $beta_1$ and $beta_2$ serum protein fractions. Albumin/globulin (A/G) ratio was increased for the toxin-treated group, and the cumulative group mean was significantly elevated for high-dose T-2 toxin compared to both full-fed and pair-fed controls. Serum IgM and IgA levels in high-dose T-2 calves were depressed, compared to controls. C_3 levels in high-dose T-2 calves were lower than the control group. Serum titres for anaplasma specific antibodies were not changed. Immunoglobulin and C_3 results are summarized in Table 10.

Response to mitogens was altered at 0.6 mg/kg. PHA response was significantly decreased at 1, 8, and 29 days after toxin administration began. Response to Con-A and PWM was significantly decreased at day 29 only.

T-2 toxin calves had loss of body weight and dose-related reduced thymic weight. Significant ($p < 0.05$) organ weight changes occurred only in thymus and adrenals at 0.6 mg/kg. High-dose calves had thymic weights significantly less than pair-fed controls. Relative organ weights expressed as percentage of body weight are given in Tables 11 and 12.

Microscopic changes occurred in thymus of calves of the intermediate- and high-dose T-2 groups. There was thinning of the thymic cortex with loss of

Table 10 Immunoglobin and C_3 cumulative group means

Treatment	IgG (mg/dl)	Igm (mg/dl)	IgA (mg/dl)	C_3
Control	3016[a]	316.6[a]	48.0[a]	1.396[a]
Pair-fed	2733[a]	302.2[a]	42.8[a]	1.089[a]
T-2 toxin	2611[a]	215.7[b]	24.2[b]	0.753[b]

Column means with same letter are not significantly different ($p < 0.05$).

Table 11 Organ weights (kg) of calves given subclinical doses of T-2 toxin

Treatment (mg/kg body weight)	Thymus	Spleen	Liver	Kidney
Control	0.069 ± 0.007	0.140 ± 0.010	1.149 ± 0.054	0.209 ± 0.016
0.1	0.060 ± 0.016	0.146 ± 0.008	1.114 ± 0.026	0.224 ± 0.012
0.3	0.026 ± 0.013	0.137 ± 0.016	1.148 ± 0.049	0.247 ± 0.018

[a]Given as percentage of average pair-fed value.

Table 12 Organ weights (kg) of calves given a clinically effective dose of T-2 toxin

Treatment	N	Thymus	Spleen	Liver	Kidney	Heart	Adrenal
Control	2	0.1944 ± 0.082^a	0.1739 ± 0.0159^a	1.330 ± 0.0455^a	0.3035 ± 0.0015^a	0.4000 ± 0.001^a	0.0047 ± 0.0001^a
Pair-fed	3	0.0645 ± 0.003^a	0.1589 ± 0.0122^a	1.092 ± 0.0678^a	0.2847 ± 0.0319^a	0.3947 ± 0.0083^a	0.0056 ± 0.0005^a
0.6 mg/kg body weight	3	0.0189 ± 0.004^b	0.1237 ± 0.0090^a	1.291 ± 0.060^a	0.2623 ± 0.0187^a	0.4120 ± 0.0056^a	0.0072 ± 0.0003^b

Means with same letters are not significantly different ($p > 0.05$).

mature lymphocytes and increased prominence of thymic corpuscles. Individual thymic lobules were separated by wide bands of fibrous connective tissue and localized areas of oedema. Changes in high-dose calves included widespread fibrosis and loss of normal thymic architecture. Pair-fed controls had some thinning of thymic cortex and slight decrease in density of mature lymphocytes within the cortex; however, normal thymic architecture was maintained.

DISCUSSION

Subclinical T-2 effects in calves appeared to occur between 0.3 and 0.6 mg/kg body weight. Subclinical in this study was an animal which consumed food voluntarily but at a reduced rate compared to controls. Furthermore, animals had no outward signs of clinical disease as might be readily noticed by a livestock producer during casual observation. By these criteria, 0.3 mg/kg was a subclinical dosage, while 0.6 mg/kg was a clinically effective dose. There was marked difference in body weight gain of the high-dose toxin group when compared to other animals in the experiment. With feed consumption at 3% of body weight, a dose of 0.3 mg/kg is equivalent to approximately 10 mg/kg T-2 toxin in the diet. The high-dose level, equivalent to 20 mg/kg, is considered a refusal concentration for swine. There appears to be no good evidence that cattle would voluntarily consume T-2-contaminated feed in the 10–20 mg/kg range for sufficient time to induce clinical effects. Such information would be useful although actual levels of sensitivity might be altered by synergistic effects of other factors present in a natural mycotoxin contamination.

Weight gain is considered a sensitive, although general, indicator of toxic effects. It is apparent that even 0.6 mg/kg T-2 in calves can effect weight gain adversely.

Only the highest dosage of T-2 toxin was associated with changes in haematology or clinical chemistry. Many of these changes, while differing from control values, were not altered sufficiently to be considered an indicator of clinical disease. It is interesting that leukopenia occurred in these calves, since certain other species are reported to experience bone marrow depression from T-2. There was no coagulopathy at these dosages of T-2. Recently other investigators have shown coagulopathy at high doses of T-2, but clinical haemorrhage has not occurred (Gentry and Cooper, 1983; Hook *et al.*, 1983; Weaver *et al.*, 1980).

Some aspects of humoral immunity were altered by T-2 toxin. The drop in total serum protein was due primarily to decreased globulin fraction. Reductions observed in immunoglobulins IgA and IgM, as well as C_3, are considered potential serious insults to the immune status of an animal (Mann *et al.*, 1982).

Consistent decrease in mitogen responsiveness was observed only for PHA in the high-dose toxin group (Buening *et al.*, 1982). Several studies in mice have indicated reduced lymphoblastogenesis and reduced skin graft rejection in response to T-2 administration (Masuko *et al.*, 1977; LaFarge-Frayssinet *et al.*, 1979; Rosenstein *et al.*, 1979; Otokawa *et al.*, 1979).

Thymic atrophy was marked in both the intermediate- and high-dose groups given T-2 toxin. This finding is consistent with other reports where T-2 toxin reduced thymic size in poultry and mice (Hayes *et al.*, 1980; Hoerr *et al.*, 1982a,b; Richard *et al.*, 1978). This should suggest further studies to determine the relationship of age, period of toxin administration, and species of animal to the importance of the thymus in immune competence.

REFERENCES

Anonymous (1973). *A Microtiter Technique for the Complement Fixation Test for Anaplasmosis*, pp. 41–47. US Department of Agriculture, Animal and Plant Health Inspection Service, Beltsville, MD.

Bainton, J., and Jones, B. D. (1977). Mycotoxins in foods and feeds – their occurrence and significance, *Ann. Nutr. Alim.*, **31**, 415–424.

Boonchuvit, B., Hamilton, P. B., and Burmeister, H. R. (1975). Interaction of T-2 toxin with *Salmonella* infections of chickens, *Poult. Sci.*, **54**, 1693–1696.

Buening, G. M. (1974). Hypopolidemia and hypergammaglobulinemia associated with experimentally induced anaplasmosis in calves, *Amer. J. Vet. Res.*, **35**, 371–374.

Buening, G. M., Mann, D. D., Hook, B., and Osweiler, G. D. (1982). The effect of T-2 toxin on the bovine immune system: cellular factors, *Vet. Immunol. Immunopathol.*, **3**, 411–417.

Burmeister, H. R. (1971). T-2 Toxin production by *Fusarium tricinctum* on solid substrate, *Appl. Microbiol.*, **21**, 739–742.

Burmeister, H. R., Ellis, J. J., and Hesseltine, C. W. (1972). Survey for *Fusaria* that elaborate T-2 toxin, *Appl. Microbiol.*, **23**, 1165–1166.

CAST (1979). *Aflatoxin and Other Mycotoxins: An Agricultural Perspective*. CAST Report no. 80. Memorial Union, Ames, Iowa.

Chi, M. S., and Mirocha, C. J. (1978). Necrotic oral lesions in chickens fed diacetoxyscirpenol, T-2 toxin, and crotocin, *Poult. Sci.*, **57**, 807–808.

Doerr, J. A., Huff, W. E., Tung, H. T., Wyatt, R. D., and Hamilton, P. B. (1974). A survey of T-2 toxin, ochratoxin, and aflatoxin for their effects on the coagulation of blood in young broiler chickens, *Poult. Sci.*, **53**, 1728–1734.

Doerr, J. A., Hamilton, P. B., and Burmeister, H. R. (1981). T-2 toxicosis and blood coagulation in young chickens, *Tox. Appl. Pharmacol.*, **60**, 157–162.

Forsyth, D. M., Yoshizawa, T., Morooka, N., and Tuite, J. (1977). Emetic and refusal activity of deoxynivalenol to swine, *Appl. Environ. Microbiol.*, **34**, 546–552.

Gentry, P. A., and Cooper, M. L. (1983). Effect of intravenous administration of T-2 toxin on blood coagulation in calves, *Am. J. Vet. Res.*, **44**, 741–746.

Hayes, J. A., Bellamy, J. E. C., and Scheifer, H. B. (1980). Subacute toxicity of dietary T-2 toxin in mice: morphology and haematological effects, *Can. J. Comp. Med.*, **44**, 203–218.

Hibbs, C. M., Osweiler, B. D., Buck, W. B., and MacAfee, G. P. (1976). Bovine Hemorrhagic Syndrome Related to T-2 Mycotoxin. Proceedings 17th Annual Meeting, American Association of Veterinary Laboratory Diagnosticians, Roanoke, VA, pp. 305–310.

Hoerr, F. J., Carlton, W. W., and Yagen, B. (1981). Mycotoxicosis caused by a single dose of T-2 toxin or diacetoxyscirpenol in broiler chickens, *Vet. Pathol.*, **18**, 652–664.

Hoerr, F. J., Carlton, W. W., Yagen, B., and Joffe, A. Z. (1982a). Mycotoxicosis produced in broiler chickens by multiple doses of either T-2 toxin or diacetoxyscirpenol, *Avian Pathol.*, **11**, 369–383.

Hoerr, F. J., Carlton, W. W., Tuite, J., Vesonder, R. F., Rohwedder, W. K., and Sziget, G. (1982b). Experimental trichothecene mycotoxicosis produced in broiler chicks by *Fusarium sporotrichiella* var. *sporotrichioides*, *Avian Pathol.*, **11**, 385–405.

Hook, B. S., Johnson, G. A., and Osweiler, G. D. (1983). Unpublished data. University of Missouri, Columbia, Missouri.

Hsieh, D. P. H. (1979). Basic metabolic effects of mycotoxins. In *Interactions of Mycotoxins in Animal Production*, pp. 43–55. National Academy of Science, Washington, DC.

Hsu, I.-C., Smalley, E. B., Strong, F. M., and Ribelin, W. E. (1972). Identification of T-2 toxin in moldy corn associated with a lethal toxicosis in dairy cattle, *Appl. Microbiol.*, **24**, 684–690.

Kosuri, N. R., Smalley, E. B., and Nichols, R. E. (1971). Toxicologic studies of *Fusarium tricinctum* (Cd.) Snyder et Hansen from moldy corn, *Amer. J. Vet. Res.*, **32**, 1843–1850.

Lafarge-Frayssinet, C., Lespinats, G., LaFong, P., Loisillier, F., Mousset, S., Rosenstein, Y., and Frayssinet, C. (1979). Immunosuppressive effects of *Fusarium* extracts and trichothecenes, *Proc. Soc. Exp. Biol. Med.*, **160**, 302–311.

Lutsky, I., Mor, N., Yagen, B., and Joffe, A. Z. (1978). The role of T-2 toxin in experimental alimentary toxic aleukia: a toxicity study in cats, *Tox. Appl. Pharmacol.*, **43**, 111–124.

Mancini, G., Barbonara, A.D., and Heremand, J. F. (1965). Immunochemical quantitation of antigens by single radial immunodiffusion, *Immunochemistry*, **2**, 235–254.

Mann, D. D., Buening, G. M., Hook, B., and Osweiler, G. D. (1982). The effect of T-2 toxin on the bovine immune system: humoral factors, *Infect. Immun.*, **36**, 1249–1252.

Masuko, H., Ueno, Y., Otokawa, M., and Yaginuma, K. (1977). The enhancing effect of T-2 toxin on delayed hypersensitivity in mice, *Jpn. J. Med. Sci. Biol.*, **30**, 159–163.

Mirocha, C. J. (1979). Trichothecene Toxins Produced by *Fusarium*. Proceedings, Conference on Mycotoxins in Animal Feeds and Grains Related to Animal Health (PB-300 300). 8 June 1979. FDA, Washington, DC, pp. 289–373.

Mirocha, C. J., Pathre, S. V., Schauerhamer, B., and Christensen, C. M. C. (1976). Natural occurrence of *Fusarium* toxins in feedstuffs, *Appl. Env. Microbiol.*, **32**, 553–556.

Muscoplat, C. C., Alhagi, I., Johnson, D. W., Pomeroy, K. A., Olson, J. M., Larson, V. L., Stevens, J. B., and Sorensen, J. K. (1974). Characteristics of lymphocyte responses to phytomitogens and comparison of responses of lymphocytes from normal and lymphocytic cows, *Amer. J. Vet. Res.*, **35**, 1053–1055.

NAS (National Academy of Sciences) (1979). *Interaction of Mycotoxins in Animal Production*. NAS, Washington, DC.

Osweiler, G. D., Hook, B. S., Mann, D. D., Buening, G. M., and Rottinghaus, G. E. (1981). Effects of T-2 toxin in cattle, *Proc. U.S.A.H.A.*, **85**, 214–231.

Otokawa, J., Shibahara, Y., and Egashira, Y. (1979). The inhibitory effect on T-2 toxin on tolerance induction of delayed-type hypersensitivity in mice, *Jpn. J. Med. Sci. Biol.*, **32**, 37–45.

Pier, A. C. (1981). Mycotoxins and animal health, *Adv. Vet. Sci. Comp. Med.*, **25**, 185–243.

Pier, A. C., Richard, J. L., and Thurston, J. R. (1979). The influence of mycotoxins on resistance and immunity. In *Interactions of Mycotoxins in Animal Production*, pp. 56–66. National Academy of Sciences, Washington, DC.

Pollman, D. S., and Koch, B. A. (1983). Effect of scabby wheat in starter pig diets. Joint USDA/FDA Workshop on Vomitoxin and Other Fusarium Toxins. Washington, DC, Jan. 12–14, 1983.

Purchase, I. H. F. (ed.) (1974). *Mycotoxins*. Elsevier, Amsterdam.

Richard, J. L., Cysewski, S. J., Pier, A. C., and Booth, G. D. (1978). Comparison of effects of dietary T-2 toxin on growth, immunogenic organs, antibody formation and pathologic changes in turkeys and chickens, *Amer. J. Vet. Res.*, **39**, 1674–1679.

Rodericks, J. V., Hesseltine, C. and Mehlman (eds) (1977). *Mycotoxins in Human and Animal Health*. Pathotox, Park Forest South, Illinois.

Rosenstein, Y., Lafarge–Frayssinet, C., Lespinats, G., Loisillier, F., LaFont, P. and Frayssinet, C. (1979). Immunosuppressive activity of fusarium toxins, *Immunology*, **36**, 111–117.

Soper, J., Muscoplat, C. C., and Johnson, D. W. (1978). *In vitro* stimulation of bovine peripheral blood lymphocytes: analysis of variation of lymphocyte blastogenic response in normal dairy cattle. *Amer. J. Vet. Res.*, **39**, 1039–1042.

Terao, K., Kera, K., and Yazima, T. (1978). The effects of trichothecene toxins on the bursa of Fabricius in day-old chicks. *Virchows Arch. B. Cell Pathol.*, **27**, 359–370.

Trenholm, H. L., Friend, D. W., Hamilton, R. M. G., and Thompson, B. K. (1982). *Vomitoxin and Zearalenone in Animal Feeds*. Publication 1745E, Agriculture Canada, Ottawa.

Vesonder, R. F., Ciegler, A., Burmeister, H. R., and Jensen, A. H. (1979). Acceptance by swine and rats of corn amended with trichothecenes, *Appl. Environ. Microbiol.*, **38**, 344–346.

Vesonder, R. F., Ellis, J. J., and Rohwedder, W. K. (1981). Swine refusal factors elaborated by *Fusarium* strains and identified as trichothecenes, *Appl. Environ. Microbiol.*, **41**, 323–324.

Weaver, G. A. (1981). Experimental mycotoxicoses in food-producing animals. Doctoral dissertation, University of Minnesota, St Paul, Minnesota.

Weaver, G. A., Kurtz, H. J., Mirocha, C. J., Bates, F. Y., Behrens, J. C., Robinson, T. S., and Swanson, S. P. (1980). The failure of purified T-2 mycotoxin to produce hemorrhaging in dairy cattle, *Can. Vet. J.*, **21**, 210–213.

Weaver, G. A., Kurtz, H. J., and Bates, F. Y. (1981). Diacetoxyscirpenol toxicity in pigs, *Res. Vet. Sci.*, **31**, 131–135.

Wiskott, U. (1978). Immunosuppressive effects of *Fusarium* toxins. Doctoral dissertation, Ludwig-Maximilian University, Munich.

WHO (World Health Organization) (1979). *Environmental Health Criteria*, **11**, 1–127.

Wyllie, T. D., and Morehouse, L. G. (eds) (1977). *Mycotoxic Fungi, Mycotoxins, Mycotoxicoses: An Encyclopedic Handbook.*, Marcel-Dekker, New York.

Trichothecenes and Other Mycotoxins
Edited by J. Lacey

CHAPTER 42

The possible role of *Pithomyces chartarum* in the aetiology and pathogenesis of ovine hepatogenous photosensitivity diseases in South Africa

J. A. W. Coetzer, T. S. Kellerman, G. C. A. Van der Westhuizen, Cecilia Roux, W. F. O. Marasas, J. A. Minné, G. F. Bath and P. A. Basson

INTRODUCTION

Ovine hepatogenous photosensitivity diseases are of major economic importance in South Africa. Several hepatotoxic plants, an alga, *Microcystis aeruginosa* (Steyn, 1949; Van Tonder *et al.*, 1972; Kellerman *et al.*, 1983; Coetzer and Bergh, 1983) and two fungi, namely *Phomopsis leptostromiformis* (Van Warmelo *et al.*, 1970; Marasas, 1978) (the cause of lupinosis), and *Pithomyces chartarum* (Marasas *et al.*, 1972; Kellerman *et al.*, 1980) have been implicated in their aetiologies. Although the aetiologies of the majority of these diseases have been partially or completely elucidated, those of a few remain unresolved. This chapter will discuss the possible role of *P. chartarum* in the aetiology and pathogenesis of the three ovine hepatogenous photosensitivity diseases in South Africa; namely, facial eczema, geeldikkop and *Panicum* grass photosensitivity.

FACIAL ECZEMA

Facial eczema or pithomycotoxicosis caused by the fungus *P. chartarum*, which produces the toxin sporidesmin, was first established to be a mycotoxicosis in New Zealand (Mortimer *et al.*, 1978), and has since been reported in only a few countries including Australia and South Africa. In South Africa it was first reported in 1972 in sheep in the Humansdorp district on ryegrass/

clover pastures (Marasas *et al.*, 1972), and subsequently diagnosed several times in sheep and cattle.

P. chartarum is widely distributed geographically in South Africa. This ubiquitous, saprophytic fungus occurs in the cool, relatively moist climate of the coastal belt, in the hot, semi-arid climate of the Karoo, and in the warm temperate grasslands of the Transvaal Highveld (Kellerman *et al.*, 1980). It is therefore adapted to grow under a wide range of conditions of temperature, moisture and substrate. However, fewer than a quarter of the isolates assayed were capable of producing sporidesmin in measurable quantities in culture (Kellerman *et al.*, 1980). The disease has been thoroughly studied in New Zealand and this chapter will only emphasize the main features of the disease as they relate to geeldikkop.

GEELDIKKOP

Geeldikkop or yellow thick head is a seasonal photosensitivity disease of sheep and goats grazing on predominantly *Tribulus terrestris* (dubbeltjies) pastures in the Karoo, a semi-arid, almost desert-like area covering about one-third of the country. The plant is a nutritious annual herb that sporadically becomes toxic under certain conditions, for example when wilted during hot, dry spells following summer rains. In severe outbreaks, hundreds of thousands of sheep living in the area can be affected. No other syndrome of ovine photosensitization, apart possibly from facial eczema in New Zealand, can match it in the magnitude of the outbreaks (Kellerman *et al.*, 1980).

Previously, geeldikkop was experimentally reproduced by feeding *T. terrestris* plants to sheep (Theiler, 1918; Quin, 1928, 1929; Van Tonder *et al.*, 1972), but other similar trials were often unsuccessful and positive results were obtained only in endemic areas during outbreaks of the disease. The fact that *T. terrestris* only sporadically becomes toxic led to the speculation that the wilted plant could sometimes produce a labile hepatotoxin (Theiler, 1918; Quin, 1928, 1929; Van Tonder *et al.*, 1972; Bath *et al.*, 1978). Nevertheless no toxin capable of causing geeldikkop has ever been isolated from *T. terrestris*. Recently geeldikkop was induced with lyophilized alcoholic extracts of toxic *T. terrestris*, but the toxic principle has not yet been identified (G. F. Bath, unpublished data, 1979).

There has long been speculation that fungi might be involved in the aetiology of geeldikkop (Theiler, 1918). Mycotoxicological research into the aetiology of ovine hepatogenous photosensitization in South Africa was given fresh impetus by reports from New Zealand that facial eczema was a mycotoxicosis. *P. chartarum* was the only fungus known to be capable of causing outbreaks of photosensitivity on a scale similar to that of geeldikkop. Consequently, despite the fact that *P. chartarum* was not known to occur in the Karoo, and that pithomycotoxicosis is usually associated with rye grass/clover pastures, and notwithstanding differences in pathology between the two diseases, the possibility that *P. chartarum* was responsible for geeldikkop was investigated. Additional incentive was provided by the diagnosis of facial eczema at Humansdorp and near Johannesburg.

Interaction of *T. terrestris* and *P. chartarum*

T. terrestris was collected during an outbreak of geeldikkop at Aberdeen in the Karoo. An isolate of *P. chartarum* was obtained from this *T. terrestris* which was cultured on semi-synthetic broth under near-ultraviolet light at 20 °C for 10–20 days. After sporulation, toxin production was demonstrated by thin-layer chromatography and the whole culture was dosed to a sheep which subsequently became photosensitive. Thus, for the first time a possible mechanism was demonstrated by which *T. terrestris* could cause photosensitivity (Kellerman *et al.*, 1980).

At the Veterinary Research Institute, Onderstepoort, this isolate induced only facial eczema with the characteristic hepatic lesions. These included necrosis and marked thickening of the intra- and extrahepatic bile ducts as well as a cholecystitis and cystitis in sheep grazing on lucerne. By contrast, approximately similar doses in Karoo sheep grazing *T. terrestris* pastures caused lesions with features of both facial eczema and geeldikkop. When the level of sporidesmin was decreased to the estimated amounts ingested during natural outbreaks of facial eczema in New Zealand (*c.* 0.4–0.7 mg/kg) (Mortimer *et al.*, 1978), the sheep grazing *T. terrestris* in the Karoo developed geeldikkop while similarly dosed sheep grazing on veld with almost no *T. terrestris* developed only facial eczema (Kellerman *et al.*, 1980).

Another interesting observation was that sheep grazing *T. terrestris* contracted geeldikkop at supposedly subclinical doses (*c.* 0.25 mg/kg) of sporidesmin. This, together with the observation that more sheep became photosensitive on *T. terrestris* than at comparable doses on the veld, suggests that *T. terrestris* plants could enhance the ability of sporidesmin to cause photosensitivity or, possibly, the converse (Kellerman *et al.*, 1980).

While ryegrass in facial eczema acts only as a vehicle for the ingestion of spores, *T. terrestris* seems to play a more complex role in the aetiology of geeldikkop. *T. terrestris* thus interacts with sporidesmin in some way to induce lesions of geeldikkop, while sporidesmin alone causes facial eczema.

Ovine hepatogenous or secondary photosensitization is caused by the action of certain plant poisons or mycotoxins on either the parenchyma and/or the biliary system of the liver (Coetzer *et al.*, 1983). In both instances the damage results in phylloerythrin, a photodynamic agent normally rapidly excreted via the liver, being retained.

Pathological changes

In both facial eczema and geeldikkop, the biliary system of the liver is primarily affected (Coetzer *et al.*, 1983). It is known that sporidesmin causes necrosis of the bile duct epithelium. The lesions progress to necrosis of the entire wall, followed by severe periductal concentric lamellar fibrosis and granulation tissue proliferation, culminating in the occlusion of some bile ducts with cellular debris and/or scar tissue (Mortimer *et al.*, 1978). The occluding mechanism in geeldikkop, on the other hand, appears to be the blockage of especially the larger intra- and extrahepatic bile ducts with crystalloid material (Kellerman *et al.*, 1980; Coetzer *et al.*, 1983). The material aggregates, often in plates, to form dense casts or microliths, which can ultimately impede the excretion of phylloerythrin.

Crystalloid material in bile ducts, Kupffer cells and hepatocytes, however, is not unique to geeldikkop; it occurs also in *Panicum* grass photosensitivity or dikoor, a syndrome indistinguishable from geeldikkop (Kellerman *et al*, 1980; Coetzer *et al.*, 1983). Crystalloid material has also been reported in hepatogenous photosensitization in Texas, induced by the plants *Agave lechuguilla* and *Nolina texana* (Mathews, 1938, 1940) but is usually rare in other hepatic conditions in ruminants.

In facial eczema the latent period (the time between ingestion of sporidesmin and the manifestations of photosensitivity) correlates with the time required for the obliterative cholangitis to develop (Mortimer *et al.*, 1978). The latent period of roughly 10–30 days is often longer than that recorded in geeldikkop, which can be as short as 2 days (Van Tonder *et al.*, 1972; Kellerman *et al.*, 1980; Coetzer *et al.*, 1983). We believe that the sudden blockage of the bile ducts with microliths early on in the development of geeldikkop may explain the short latent period usually seen in the disease.

Apart from the different occluding mechanisms operating in facial eczema and geeldikkop, it has been shown that sheep can be made photosensitive within 2–3 days by ligation of the common bile duct (T. S. Kellerman and J. A. W. Coetzer, unpublished data, 1983).

Obstruction of especially the larger bile ducts—whether by necrosis and scarring as in facial eczema, by microliths as in geeldikkop and dikoor or by ligation of the common bile duct—results in very similar lesions in the liver (Coetzer *et al.*, 1983). The most important change is concentric, lamellar fibrosis around the bile ducts, which is an indication of obstructive cholangitis. Other changes include bile duct proliferation, oedema and fibroplasia in the portal triads, foci of periportal hepatocellular necrosis (bile infarcts) and diffuse parenchymal degeneration.

It has been shown that sporidesmin can interact with *T. terrestris* to induce lesions of geeldikkop while sporidesmin alone results in facial eczema (Kellerman *et al.*, 1980). The specific requirements for crystalloid formation in the liver are not clear, but it would seem that they do not form in response to all types of liver damage in sheep grazing on *T. terrestris* in the Karoo. In a limited experiment, sheep running on *T. terrestris* and dosed with *Senecio retrorsus*, or made photosensitive by the administration of either *Phomopsis leptostromiformis* cultures or dried *Microcystis aeruginosa*, showed typical lesions of the respective intoxications without crystalloid material (J. A. W. Coetzer and T. S. Kellerman, unpublished data, 1979).

The chemical nature of the crystalloid material is unknown, but it is apparently not composed of common bile salts such as cholesterol, cholic acid, sodium glycocholate or sodium taurocholate (L. A. P. Anderson, VRI, Onderstepoort, unpublished data, 1978). All the evidence suggests that it is a component of *T. terrestris*. These components may well vary in the plant according to locality, growth stage and physiological condition of the plant. The occurrence of geeldikkop, therefore, might be linked with the status of the crystalloid-promoting factor(s) in *T. terrestris*. Energy dispersive analysis of X-rays showed that the crystalloid material contains high concentrations of sulphur and lower levels of sodium and potassium (J. A. W. Coetzer, unpublished data, 1981).

Just as cholesterol gallstones form in laboratory animals and man with high cholesterol intakes, it appears that *T. terrestris* constitutes a lithogenic diet, responsible for saturation and supersaturation of the bile with a crystalloid factor(s) or its products (Coetzer *et al.*, 1983). These crystalloid factors can conceivably precipitate spontaneously. However, experimental results have indicated that the formation of microliths can be triggered by low levels of sporidesmin, thereby enhancing the ability of *T. terrestris* as well as of sporidesmin to cause photosensitivity (Kellerman *et al.*, 1980).

PANICUM GRASS SENSITIVITY

Panicum grass photosensitivity (dikoor) occurs sporadically among sheep, especially on cultivated lands in the Orange Free State and on the Transvaal Highveld in South Africa (Kellerman *et al.*, 1980). Like *T. terrestris*, *Panicum* grass is also a nutritious fodder which only sporadically becomes toxic when wilted during dry hot spells following summer rains. The epidemiology is thus identical to that of geeldikkop. Outbreaks often occur on reaped wheat lands with clinical signs, macroscopical and microscopical pathology indistinguishable from geeldikkop (Kellerman *et al.*, 1980; Coetzer *et al.*, 1983). *P. chartarum* has been isolated from lands where outbreaks occurred. The only difference is that geeldikkop occurs on *T. terrestris* and dikoor on *Panicum* grass. *P. chartarum* may thus play a similar role to that in geeldikkop.

CONCLUSION

Finally it must be emphasized that geeldikkop is not facial eczema but a more complex disease that can be caused by the simultaneous ingestion of sporidesmin and *T. terrestris* plants. Research should now be directed in identifying the crystalloid factor(s) in the plant and at establishing the conditions affecting their formation and levels in the plant.

REFERENCES

Bath, G. F., Van Tonder, E. M., and Basson, P. A. (1978). Geeldikkop: preservation of toxic material, *J. South African Vet. Assoc.*, **49**, 23–25.

Coetzer, J. A. W., and Bergh, T. (1983). Photosensitivity in South Africa. IV. Pathological changes in the liver in ovine photosensitivity caused by the plant *Asaemia axillaris* (Thunb.) Harv. Ex Jackson, *Onderstepoort J. Vet. Res.*, **50**, 53–55.

Coetzer, J. A. W., Kellerman, T. S., Sadler, W., and Bath, G. F. (1983). The excretion of phylloerythrin and bilirubin by calves and sheep, *Res. Vet. Sci.*, **21**, 12–18.

Kellerman, T. S., Coetzer, J. A. W., Schneider, D. J., and Welman, W. G. (1983). Photosensitivity in South Africa. III. Ovine hepatogenous photosensitivity caused by the plant *Athantasia trifurcata* L. (Asteraceae), *Onderstepoort J. Vet. Res.*, **50**, 47–52.

Kellerman, T. S., Van der Westhuizen, G. C. A., Coetzer, J. A. W., Roux, C., Marasas, W. F. O., Minné, J. A., Bath, G. F., and Basson, P. A. (1980). Photosensitivity in South Africa. II. The experimental production of the ovine hepatogenous photosensitivity disease geeldikkop (Tribulosis ovis) by the simultaneous ingestion of *Tribulus terrestris* plants and cultures of *Pithomyces chartarum* containing the mycotoxin sporidesmin, *Onderstepoort J. Vet. Res.*, **47**, 231–261.

Marasas, W. F.O. (1978). Lupinosis in cattle. In *Mycotoxic Fungi, Mycotoxins and Mycotoxicoses. An Encyclopedic Handbook* (eds T. D. Wyllie and G. L. Morehouse), Vol. II, pp. 73–84. Marcel Dekker Inc., New York.

Marasas, W. F. O., Adelaar, T. F., Kellerman, T. S., Minné, J. A., Van Rensburg, I. B. J., and Burroughs, G. W. (1972). First report of facial eczema in sheep in South Africa, *Onderstepoort J. Vet. Res.*, **39**, 107–112.

Mathews, F. P. (1938). Lechuguilla (*Agave lechuguilla*) poisoning in sheep and goats, *J. Am. Vet. Med. Assoc.*, **93**, 168–175.

Mathews, F. P. (1940). Poisoning in sheep and goats by sacahuiste (*Nolina texana*) buds and blooms. Texas Agricultural Experimental Station, College Station, Brazos Country, Texas. Bulletin No. 585.

Mortimer,P. H., White, E. P., and Di Menna, M. (1978). Pithomycotoxicosis 'facial eczema' in sheep. In *Mycotoxic Fungi, Mycotoxins, Mycotoxicoses. An Encyclopedic Handbook* (eds T. D. Wyllie and G. L. Morehouse), Vol. II, pp. 195–203. Marcel Dekker Inc., New York.

Quin, J. I. (1928). Recent investigations into geeldikkop affecting sheep and goats in the Cape Province, *J. South African Vet. Med. Assoc.*, **1**, 43–45.

Quin, J. I. (1929). Further investigations into geeldikkop (*Tribulosis ovis*), *Rep. Vet. Res., Union of South Africa*, **15**, 765–767.

Steyn, D. G. (1949). *Vergiftiging van mens en dier*. Van Schaik, Pretoria.

Theiler, A. (1918). Geeldikkop in sheep (*Tribulosis ovis*), *Rep. Vet. Res., Union of South Africa*, **7**, and **8**, 1055.

Van Tonder, E. M., Basson, P. A., and Van Rensburg, I. B. J. (1972). 'Geeldikkop' experimentally induced by feeding *Tribulus terrestris* L. (Zygophyllaceae), *J. South African Vet. Med. Assoc.*, **43**, 363–375.

Van Warmelo, K. T., Marasas, W. F. O., Adelaar, T. F., Kellerman, T. S., Van Rensburg, I. B. J., and Minné, J. A. (1970). Experimental evidence that lupinosis of sheep is a mycotoxicosis caused by the fungus *Phomopsis leptostromiformis* (Kühn) Bufak, *J. South African Vet. Med. Assoc.*, **41**, 235–247.

Trichothecenes and Other Mycotoxins
Edited by J. Lacey

CHAPTER 43

Responses of Syrian hamsters to diets of copper acetate, mixed aflatoxins alone or both in combination

G. C. Llewellyn, J. D. Morgan and W. V. Dashek

INTRODUCTION

Aflatoxins are metabolites of *Aspergillus flavus* and *A. parasiticus*, fungi which often grow upon economically important crop plants (Dashek and Llewellyn, 1982). Aflatoxins are hepatocarcinogens, mutagens, teratogens and toxins (Carnaghan *et al.*, 1963; Butler, 1965; Wogan and Busby, 1980). Copper, a trace element, is a co-factor for many enzymes involved in oxidations, including tyrosinase, ceruloplasmin, amine odixase and cytochrome oxidase (Schwartz, 1975). Copper can also inhibit neoplastic growth, protecting rat livers from damage by 2-acetamidofluorene, α-naphthylisothiocyanate, DL-ethionine, thioacetamide, 3-methoxy-4-aminoazobenzene or its *N*-methyl derivatives (Kamamoto *et al.*, 1973; Carlton and Price, 1973; Fare and Woodhouse, 1963). It has been suggested that the protective effect may be due to competitive binding of copper for available protein sites in the liver (Fare and Howell, 1964). On the other hand, Goodall (1964) was able to demonstrate that the inhibition of az-dye-hepatocarcinogenesis by excess copper is not a general phenomenon.

Aflatoxins may affect trace metal levels in animal tissues and organs (Doyle *et al.*, 1977) while trace metal supplements may affect some aspects of developing aflatoxicosis (Barber *et al.*, 1969; Neatherly *et al.*, 1981; Llewellyn, 1982). The present study reports an attempt to determine whether copper acetate in the diet can provide protection and alter recovery from induced aflatoxicosis in Syrian hamsters, *Mesocricetus auratus*.

MATERIALS AND METHODS

Animal maintenance

Weanling, male Syrian hamsters (*Mesocricetus auratus*; Engle Laboratory Animals Inc., Farmsburg, IL) were allotted randomly to one of four treatment groups consisting of six animals each. Their mean weight was 37.6 g. The hamsters were housed individually in transparent, polystyrene, Model E-0210 Econo-Cages (Maryland Plastics, Inc., Federalsburg, MD). The cage lids of clear polystyrene sheets (Novi American Inc., Novi, MI) were perforated with numerous airholes. Distilled water was supplied from conventional water bottles and an identical cage, without an animal, was used to correct for loss. Initially, animals were acclimatized by feeding a control diet and water *ad libitum*. Feed consumption was determined daily, and water consumption and whole body weights weekly. Feed consumption was adjusted when spilled feed was recovered during weekly bedding changes using an aluminium wire sifter.

The hamsters were maintained on a 12–12 h., light–dark photoperiod from 9.00 a.m. to 9.00 p.m. at 24 ± 5 °C and a relative humidity of 68 ± 3%. General health and activity were noted each day visually.

Diets

Animals were fed one of four diets all based on Purina Laboratory Meal (copper content 18 mg/kg) for rats, hamsters and mice (Ralston Purina, St Louis, MO). These were: the control group (c), unsupplemented meal; the copper group (CuAc), meal supplemented with 0.75% (w/w) copper acetate; the toxin group (AFT), meal supplemented with 30.3 mg/kg mixed AFB_1, AFB_2, AFG_1, AFG_2; double-insult group (AFT + CuAc), meal supplemented with both 0.75% copper acetate and mixed aflatoxins. The hamsters were kept on these diets for 45 days following the acclimatization period and then all groups were returned to a control diet for a further 123 days.

Aflatoxins were produced from cultures of *A. parasiticus* NRRL 2999 on rice as described by Llewellyn *et al.*, (1973). After 12 days incubation, cultures were killed by adding 100 ml $CHCl_3$. Mycelium was removed 24 h later by filtration through Whatman No. 1 filter paper in a Büchner funnel. The filtrate was refrigerated and samples were analysed by combined thin-layer chromatography with a visual dilution technique sensitive to 2 mg/kg (Horowitz *et al.*, 1975a).

Stock diet was prepared in a chemical hood by moistening 200 g of Purina Laboratory Meal with 360 ml of extract using a 3 ml disposable syringe with a 25 gauge needle. Work areas were decontaminated with sodium hypochlorite.

The slurry was dried on aluminium trays for 72 h in a ventilated chamber and subsequently stored at room temperature in a light-proof container. The contaminated diet was prepared by thoroughly mixing 14 g of stock diet with 196 g of uncontaminated Purina Meal in a 'Skil-Craft Corp.' tumbler (Skil-Craft Corp., Chicago, IL). The stock diet contained 457.5 mg/kg total aflatoxins and the contaminated diet used for feeding contained 30.3 mg/kg total aflatoxins (AFB_1, 13.8 mg/kg; AFB_2, 0.4 mg/kg; AFG_1, 15.6 mg/kg; AFG_2, 0.5 mg/kg). The copper diet was prepared by mixing reagent grade copper acetate with fresh or contaminated Purina Laboratory Meal to give a 0.75% copper acetate–Purina mixture with or without 30.3 mg/kg total aflatoxin.

Organs and histopathology

At sacrifice, animals were anaesthetized with Metofane (2,2-dichloro-1,1-difluoro-ethyl-methyl ether) in an enclosed metal chamber for 4 min, and dissected. Blood was collected by heart puncture of the left ventricle into heparinized vacutainers (Clay Adams Inc., Parsipanny, NJ) and stored under refrigeration until copper levels were quantified by atomic absorption spectrophotometry (Horowitz *et al.*, 1975b). The whole liver, the left lung, the heart, both kidneys, the spleen and both testes were excised, rinsed in fresh Ringer's solution, weighed and fixed in 10% formalin neutralized with Na_2CO_3 (5 g/l). The gall bladder was removed from the liver prior to weighing of the latter. The entire stomach and a segment of the large intestine were removed, washed and fixed. A portion of the liver was quick-frozen for copper assay. The adrenal glands were excised, washed, weighed and fixed in 10%, pH 7.0, 0.02 M phosphate-buffered formalin for histology. To ensure penetration by the preservative, the organs were either cut or punctured. Haematoxylin and eosin-stained liver and other organ sections were prepared and read using light microscopy. Another portion of the liver was utilized for lipid analyses.

Other analyses

Bone measurement

The left femur was removed, dipped into hot water for 30 s, and the muscle tissue was dissected away to expose the bone. Bone length was measured under a dissecting microscope using a millimetre rule. Bone diameter was measured with an industrial micrometer at a constant location near the centre of the bone as dictated by a flat bone configuration on both sides of the micrometer arms.

Haematocrit percentages

Following sacrifice, blood was drawn into a capillary tube which was sealed at the distal end with 'Sel-a-Ease' putty. The tube was centrifuged in a micro-haematocrit at 2000 rpm for 5 min. The height of the packed cell volume was read from a 'Spirocrit' device (Lourdes Instrument Co., Brooklyn, NY) and the haematocrit percentage recorded.

Haemoglobin

Twenty millilitres of whole blood was drawn into a haemoglobin pipette and expelled into an absorption tube containing 5 ml of 1% HCl. The tube was allowed to 'stand' for 60 min. A 'blank' tube consisting of 5 ml 1% HCl was used to 'zero' a Turner Model 350 Spectrophotometer (Turner Assoc., Palo Alto, CA) at 525 nm. A calibration chart was used to give g haemoglobin/100 ml blood (Kimbrough and Llewellyn, 1973).

Lipid and water analyses

Immediately after death, a portion of the excised liver was weighed and then homogenized with an equal weight of distilled water in a tissue homogenizer. The dry weight of 0.5–2.0 ml of the homogenate was determined, and dry and wet weights were compared to test for oedema (Hamilton and Garlich, 1972; Hamilton *et al.*, 1972).

The 'crude' lipid content of the liver was determined gravimetrically by extracting 2 ml of the homogenate with 10 ml of chloroform : methanol (2 : 1) using a tissue homogenizer and drying the extract at 26 ± 2 °C in a tared weighing dish.

Statistical analyses

Data are means and standard deviations. Significant differences between treatments were sought with an analysis of variance (Bruning and Kintz, 1968; Downie and Heath, 1970).

RESULTS

General health and lethality

No deaths occurred during the 23-day acclimation and growth period or during the 123-day recovery period, but there were three deaths during the 45-day diet treatment period, one each from the control, copper, and toxin groups. The general health and physical activity of control and copper-fed animals appeared normal throughout the experiment, although copper-fed animals tended to be more passive when handled.

The double-insult and toxin groups both developed inappetance during the first few days upon their diets, perhaps a consequence of its taste. Animals in the toxin group appeared very active during the first half of the diet period, frequently gnawing upon the cage walls, but subsequently they exhibited mild ataxia and became progressively lethargic and recumbent. On return to the control diet the condition of the toxin group animals improved noticeably and by the end of the experiment they appeared normal. All animals gained weight steadily during the acclimation and growth periods, as expected during this period of maturation. During the diet period, although both the control and copper groups continued to gain weight (but at a slower rate), both the double-insult and toxin groups showed decreased mean body weight (Figure 1). On day 1 of the diet period mean body weights for the control, copper, double-insult and toxin groups were 85.0 ± 5.4, 83.6 ± 7.1, 82.7 ± 8.9 and 83.4 ± 6.8 g, respectively. By contrast, the mean body weights on day 45 for

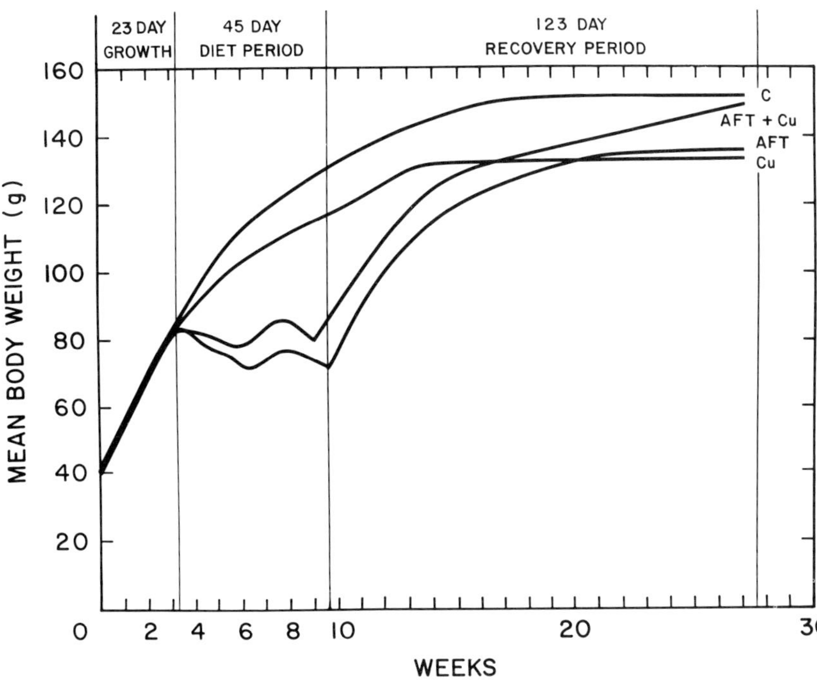

Figure 1 Mean body weight for control (c), copper-supplemented (Cu), aflatoxin + copper (AFT + Cu), and aflatoxin (AFT) groups for each treatment period.

the same sequence of animal groups were 130.0 ± 14.3, 116.6 ± 15.0, 83.1 ± 10.3, and 72.5 ± 9.6 g, respectively. The differences in mean body weights from days 1 to 45 for each group were: +52.94% (control), +39.47% (copper), +0.48% (double-insult), and −13.07% (aflatoxin). Body weights did not differ significantly up to the 23rd day of the experiment, but at day 45 significant differences ($p = 0.001$) were confirmed between the control and double-insult groups, control and toxin groups, copper and double-insult groups and copper and toxin groups.

Feed consumption

Changes in weekly feed consumption for each group with time are presented in Figure 2. Mean food consumption for the toxin group declined 36.1% from 5.72 g (day 1) to 36.5 g (day 16) and rose to 61.6 g (day 30) during the 45-day diet period. At this time, mean feed consumption dropped sharply again. The changes in weekly feed consumption for the double-insult group showed similar, although smaller, variations, to those for the toxin group. The decrease in feed consumption for this group from days 1 to 18 was 61.1 g to 44.7 g (26.6%). From days 18 to 32 consumption increased to 44.7 g (26.8%) followed by an increase to 60.3 g by day 32.

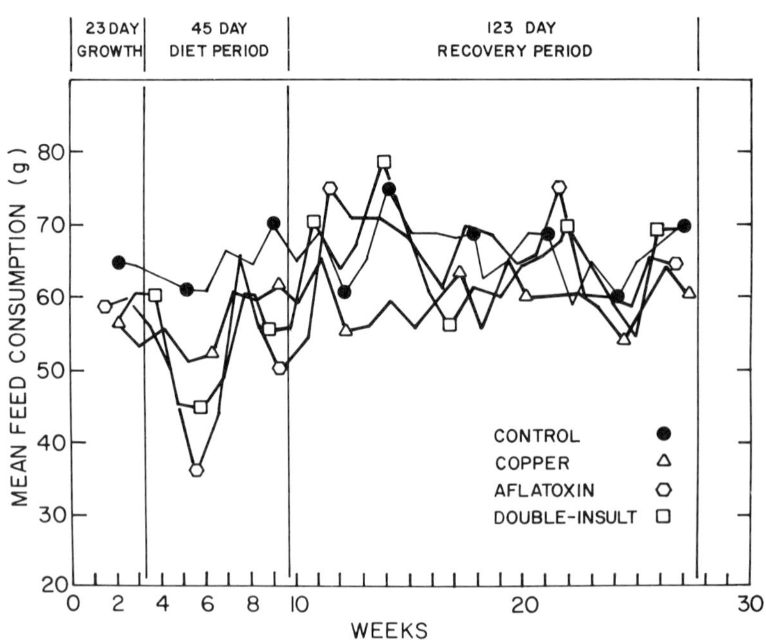

Figure 2 Mean feed consumption for all four treatment groups for each treatment period.

Table 1 summarizes the consumption of AFB_1, total aflatoxins and copper acetate within the diet by each of the groups. The mean AFB_1 consumption for the double-insult group was 4.84 mg and for the toxin group 4.40 mg. Mean copper acetate consumptions were 2.63 and 2.69 mg for the double-insult and copper groups, respectively.

At the end of the diet period the control and toxin groups displayed the maximum difference (94.19 g) in diet consumption. In contrast, the greatest differences (141.96 g) at the termination of the experiment were between the control and copper groups.

Table 1 Mean total feed, water and related diet supplement contents consumed by each of the four treatment groups[a]

Treatment groups	Diet period	Recovery period	Total
Control			
Diet consumption (g)	413.16	1155.74	1568.90
Aflatoxin B_1 (mg)	0.00	0.00	0.00
Total aflatoxins (mg)	0.00	0.00	0.00
Copper acetate (g)	0.00	0.00	0.00
Water consumption (g)	482.86	1292.26	1775.12
Copper			
Diet consumption (g)	358.28	1068.66	1426.94
Aflatoxin B_1 (mg)	0.00	0.00	0.00
Total aflatoxins (mg)	0.00	0.00	0.00
Copper acetate (g)	2.69	0.00	2.69
Water consumption (g)	421.52	1149.08	1570.60
Double-insult			
Diet consumption (g)	350.90	1161.57	1512.47
Aflatoxin B_1 (mg)	4.84	0.00	4.84
Total aflatoxins (mg)	10.63	0.00	10.63
Copper acetate (g)	2.63	0.00	2.63
Water consumption (g)	1136.78	1455.92	2592.70
Aflatoxin			
Diet consumption (g)	318.97	1160.43	1479.40
Aflatoxin B_1 (mg)	4.40	0.00	4.40
Total aflatoxins (mg)	9.66	0.00	9.66
Copper acetate (g)	0.00	0.00	0.00
Water consumption (g)	739.98	1416.40	2156.38

[a] All diets had 18 mg/kg copper and no detectable levels of aflatoxins prior to copper and/or aflatoxin addition.

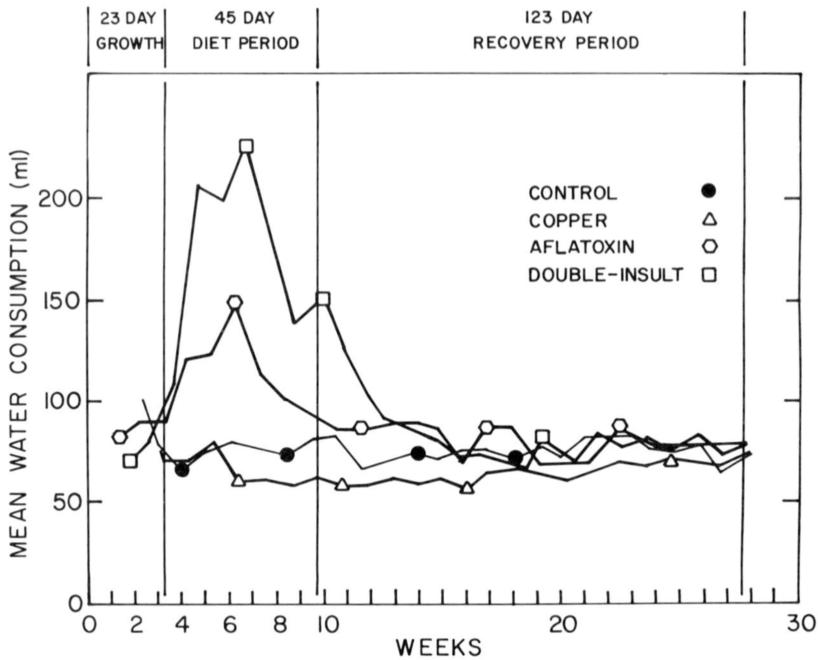

Figure 3 Mean water consumption for all four treatment groups for each treatment period.

Water consumption

Changes in weekly mean water consumption with time are illustrated in Figure 3. Consumption by the toxin and double-insult groups was maximal on days 23 and 25 of the diet period at 147.8 ml and 227.8 ml, respectively. On day 21 of the diet period mean water consumption for the copper group dropped to 50.0 ml/week and then remained close to this level until week 6 of the recovery period.

Total water consumption differed between groups more than total feed consumption (Table 1). Total mean water consumption during the diet period decreased relative to the controls by 12.70% in the copper groups but increased by 135.42 and 53.25% in the double-insult and toxin groups, respectively. During the recovery period copper-treated animals continued to consume less water (11.07%) than control animals while the total mean water consumptions for the double-insult and toxin group animals were higher than the control group by only 12.64 and 9.60%, respectively. When the mean water consumption was expressed as an additive value, the visual ranking of groups with respect to cumulative consumption appears to be double-insult > control > aflatoxin > copper groups (Figure 4).

Feed and water demand

The copper group exhibited a decreasing demand for water relative to feed as the total quantity of diet ingested during the diet period increased. By the third week, demand for water was less than that for food, but when returned to the control diet the mean water : food ratio rose slightly above 1 : 1. The toxin and double-insult groups both showed noticeably enhanced water demand, greater in the latter group. Although the double-insult group consumed approximately the same amount of copper acetate and total aflatoxins as either the copper or toxin groups, its water demand surpassed both of these groups. The mean water : food ratio increased to day 13 of the diet period, when it was respectively 270.8, 202.7 and 56.1% greater than that for control, copper-fed and toxin-fed groups.

Diminution in the relative demand for water by the double-insult and toxin groups paralleled each other during the diet period and also during the first 4 weeks of the recovery period.

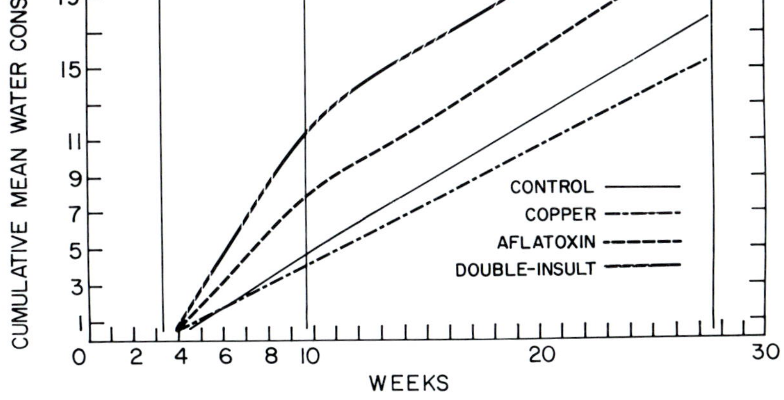

Figure 4 Mean accumulative water consumption for all four treatment groups for the duration of the study.

The copper group animals experienced a steady decrease in water demand throughout the diet period, which continued for the first 4 days of the recovery period.

Organ weights

The weights of the kidneys, liver, spleen, testes, heart, left lung and adrenals as percentages of gross body weight are shown in Table 2. Relative to the controls, the copper groups showed decreased liver and lung weights and increases in the weights of all other excised organs. The toxin-fed groups showed increased liver, kidney, spleen, heart and adrenal weights but decreases in weights of both the testes and left lung. The liver weight of the double-insult group was also markedly increased (by 9.2%). Although kidney weights in both the copper and toxin-fed groups were increased, a slight decline was found when copper and toxin were fed together (− 0.48%). Spleen enlargement was found in copper and toxin-fed groups and, to a lesser extent, the double-insult group. Copper and aflatoxin appeared to act synergistically both in increasing adrenal weight and in decreasing left lung weight.

Histopathology

Control animals—liver

Hepatocytes and nuclei within the cells of all animals were normal in size. Livers in two of the five animals showed slight lymphocytic infiltration, which in one had progressed to cholangitis.

Table 2 Responses of organ weights for the four treatment groups based on mean ± SD organ weight in g/100 g of body weight

Organs	Control	Copper	Toxin	Double-insult
Liver[a]	4.20 ± 0.272	3.84 ± 0.404[b]	4.64 ± 0.398[b]	4.58 ± 0.471[b]
Kidneys	0.92 ± 0.052	1.02 ± 0.126	0.95 ± 0.095	0.92 ± 0.026
Spleen[a]	0.08 ± 0.010	0.10 ± 0.013	0.11 ± 0.019	0.09 ± 0.005
Testes[a]	2.43 ± 0.441	3.01 ± 0.125	2.54 ± 1.000	2.58 ± 0.307
Heart	0.38 ± 0.044	0.40 ± 0.028	0.44 ± 0.059	0.42 ± 0.038
Left lung	0.20 ± 0.035	0.19 ± 0.023	0.19 ± 0.016	0.17 ± 0.017
Adrenals	0.03 ± 0.004	0.03 ± 0.009	0.03 ± 0.007	0.03 ± 0.009

[a] Organ with significant *F*-ratio
[b] Group with significant *F*-ratio (Scheffe test) relative to the controls.

Copper acetate-fed animals—liver

Cells mostly appeared normal but two of the animals displayed signs of leukocytic infiltration within their periportal regions.

Aflatoxin-fed animals—liver

Histopathological changes in the animals examined ranged from a weak rank one response (two animals) to close to rank two (two animals). Some meglahepatocytes, bile ductule hyperplasia, and nuclear inclusion bodies were observed. In one animal, changes extended into midzone area. Three of the six livers showed a toxic–inflammatory reaction in addition to hyperplasia, including a light to heavy cellular infiltration of either neutrophils, lymphocytes or leukocytes into the periportal region. The overall rank was just above category one with some characteristics of category two apparent.

Aflatoxin and copper acetate-fed animals (double-insult—liver)

Table 3 gives a progressive ranking for liver histopathology of male hamsters following aflatoxin treatment. All livers from animals fed both aflatoxin and copper acetate were ranked one. Two animals were definitely in this category but two others were only marginally so with only slight changes. Bile ductule proliferation and nuclei with inclusion bodies were apparent in all animals while some leukocytic infiltration of the periportal areas occurred in two.

Control animals—kidneys

The kidney cells of most animals were without morphological abnormalities, although one had a few tubules with pyknotic nuclei and a second showed some tubular necrosis within the cortex.

Copper acetate-fed animals—kidney

The kidney cells were mostly normal but one animal showed scattered tubular necrosis within the cortex. Kidneys of another animal had regions with degenerating tubular epithelium within the cortex and a third displayed numerous small necrotic areas.

Aflatoxin-fed animals—kidney

Mild nephritis was seen in two of the six animals with scattered coagulative necrosis of the tubules, most damage being associated with the convoluted tubules. Two animals had tubules within the cortex with pyknotic nuclei, and

Table 3 A progressive numerical ranking describing liver histopathology in male hamsters receiving aflatoxins

Numerical value/rank	Descriptive histopathology
0	Controls—normal cells. An occasional hepatocyte was found that showed vacuolization. Some hepatocytes were binucleated but few were observed during cell division.
1	Hepatic cells in the peripheral (periportal) area of the lobules were swollen and could be considered meglahepatocytes. They were usually about twice normal size. Occasionally these cells had cytoplasmic inclusions within the nucleus. There was a slight amount of bile duct proliferation (hyperplasia) in the periportal areas.
2	Many of the lobules were outlined by proliferating bile duct cells. There was some fibrous tissue in the periportal zone. Not all areas of the liver section showed proliferation. There was an increased number of large hepatocytes having intranuclear inclusion bodies. Occasionally some of these cells were seen in the midzone and the central vein areas. Many of the hepatic cells had greatly enlarged nuclei.
3	There was a greater number of lobules affected in the peripheral areas. Zones of bile duct proliferation were twice that of number 2 above. Bile duct proliferation has extended into the midzone areas with the lobules having a shrunken appearance.
4 and greater	Lobules have been nearly replaced by bile duct proliferation. Lobular size was reduced by 25%. The hepatic cells in the central vein zone were undergoing fatty change.

two others resembled the controls in having slightly congested and coagulated tubules. Overall, kidney damage was greater in these animals than in those comprising the three other treatment groups.

Aflatoxin and copper acetate-fed animals (double-insult)—kidney

Three of the four animals contained some necrotic tubules within the cortex, but cells of the fourth animal appeared normal.

Other analyses

Haematocrit and haemoglobin

Haematocrit percentages from the toxin group were close to those of the control group (Table 4) but those of the copper and double-insult groups

Table 4 Means and standard deviation for haematocrit percentage and haemoglobin for the four treatment groups

Treatment groups	Haematocrit (%)	Haemoglobin (g/100 ml blood)
Control	43.20 ± 6.90	5.98 ± 1.18
Copper	48.50 ± 5.55	7.20 ± 0.99
Double-insult	55.20 ± 6.76	10.60 ± 5.89
Aflatoxin	44.60 ± 11.05	7.06 ± 1.66

were, respectively, 1.1 and 1.3 times greater. However, these values did not differ significantly from the controls. Haemoglobin levels for the copper, double-insult and toxin groups were increased, respectively, 1.2, 1.8 and 1.2-fold times over those of the control group. Statistically significant differences between the treatment and control groups were not observed.

Crude lipid content

Crude lipid content was quantified to determine whether elevations in lipid could result in liver enlargement. The percentage of lipids rose from a control value of 6.11 to 6.46% in the double-insult, 6.13% in the copper and 7.35% in the toxin groups, or an increase of 5.8% in the double-insult and 20.0% in the toxin groups.

Bone measurements

No significant differences between the groups in either diameter or length occurred.

Liver oedema

The occurrence of oedema within the liver was evaluated by calculating the dry to wet weight ratios of the excised liver fractions from the different treatment groups. The mean values were 0.110, 0.123, 0.094 and 0.105, respectively, for the control, copper, double-insult and toxin groups. The proximity of these values suggests that oedema was not a major factor contributing to enhanced liver size.

DISCUSSION

Aflatoxin diets were toxic to the animals but insufficiently so to kill more than one in the toxin group and none in the double-insult group. It was evident from body weight, organ weight and histopathology that animals in all toxin-treated groups developed aflatoxicosis. The decreased body weights in the aflatoxin group suggest that if treatment had continued for much more than 45 days, more animals would have succumbed. Even though hamsters are reportedly not nearly as sensitive as Fischer rats to aflatoxins (Butler, 1964), the dietary level of 30.3 mg/kg caused them severe problems.

By the end of the diet period the double-insult group had consumed about the same quantity of copper acetate as the copper group and about as much aflatoxin as the toxin group (Table 1). Thus neither copper acetate nor aflatoxin in the presence of the other markedly influenced feed consumption, although the presence of either chemical in the diet decreased consumption in comparison with the control group by up to 23%.

Consumption patterns during the experimental stages had little effect on feed consumption during the recovery period. Only the copper group, which established a mean feed intake of approximately 10 g below the norm early in the experiment (week 2 of 23-day growth period) continued this throughout the experiment, resulting in a slightly lagging total feed consumption.

Neither the considerable differences in mean body weights at the end of the diet period nor the rapid recovery of mean body weight in both the toxin and double-insult groups can be attributed to differences in total diet consumption.

The differences in mean body weights between the toxin and non-toxin treatment groups after 45 days of the diet period were statistically significant. The ratio of mean body weight to weight of feed consumed (feed efficiency) gave values commensurate with body weight changes characterizing each group. Thus, although the total quantity of feed ingested is of minor importance, the utilization of the feed is, and this is affected by the 'contaminants'. Water consumption is similarly affected. As with feed consumption, the copper group consumed less water than other groups throughout the experiment. Both the toxin and double-insult groups immediately increased their water intake when fed a contaminated diet.

Decreased water gains and losses are not unusual phenomena in aflatoxin-treated animals. Hastings and Llewellyn (1973) observed that feed consumption had little influence on weight changes in Mongolian gerbils. Llewellyn *et al.* (1974) found weight gains were decreased in Syrian hamsters fed either 1 or 10 mg/kg AFB_1. Growth rate inhibition has also been reported in pigs fed diets containing aflatoxin (Barber *et al.*, 1968; Armbrecht, 1973).

Single classification analysis of variance was applied to mean body weights on days 1, 45 and 123. Mean body weights of some groups differed significantly on day 45 at the end of the diet period but not at either the beginning or the end of the experiment. These significant differences, between the control and double-insult, control and toxin, copper and double-insult and copper and toxin groups, were confirmed using the Scheffe test.

Copper acetate provided some protection from the effects of aflatoxicosis in the double-insult group so that they could maintain themselves slightly better than the toxin group and regained weight to approximate the controls by day 123.

The rapid recovery of body weight in both the double-insult and toxin groups suggests that only limited and perhaps repairable organ damage was inflicted. Although diarrhoeic stools were not seen at any time, differences in relative organ size were seen. For example, the liver, the main target organ, was larger in both groups receiving aflatoxin and smaller in those receiving copper alone, perhaps due to copper toxicity. Liver analyses confirmed that oedema did not contribute largely to enhanced liver size but lipid accumulation may have. Crude lipid content rose 5.8% in the double-insult group and 20.0% in the toxin group. Aflatoxicosis also causes lipid accumulation in ducklings (Shanks and Wogan, 1966), rats (Singh and Venkitasubramanian, 1975) and chickens (Hamilton and Garlich, 1972). This accumulation is probably due to impairment of normal lipid metabolism or inhibition of lipid transport from its site of synthesis, thereby resulting in progressive deposition. This condition is thought necessary for the development of neoplasms, (Halver, 1969).

Splenomegaly probably results from the spleen's role in the lymphatic system and its phagocytic function. Its failure to respond to dietary fat during aflatoxicosis means simply that its enlargement is associated with alterations in protein and nucleic acid syntheses (Hamilton *et al.*, 1972; Smith, 1963).

Histopathological evaluation of both the primary and secondary target organs indicated that copper gave limited protection to these organs against aflatoxicosis, perhaps by necessarily altering the normal sequence of damage. Bile ductule proliferation occurred in both aflatoxin-consuming groups but was substantially less severe in the double-insult group.

Slight increases in haemaglobin levels and haematocrit percentages were noted, although not significant. These were perhaps a response of the bone marrow to low oxygen tension in arterial blood.

Bone lengths and diameters were not significantly affected by any of the treatments.

A major reason for investigating a disease is to find a preventive or ameliorative procedure. A possible means of preventing aflatoxicosis was suggested by the minimal but consistent difference in susceptibility to aflatoxins of groups receiving copper acetate and those not. Our data suggest that hamsters can make a recovery after being fed 30.3 mg/kg aflatoxins for 45 days, especially when 0.75% copper acetate was added. Similar protection by 0.5% copper acetate was found in hamsters maintained continuously on 23 mg/kg aflatoxin for 6 and 11 weeks (Llewellyn *et al.*, 1981).

Since consistent rather than significant differences between toxin and double-insult groups were observed, it is premature either to speculate upon the economic importance of our results or to theorize over mechanisms by which copper might prevent aflatoxicosis. Instead, further studies in which a variety of copper acetate concentrations is employed seem warranted.

ACKNOWLEDGEMENTS

The assistance of F. W. Rea, DVM, of the Virginia Veterinary Diagnostic Laboratory, Virginia Department of Agriculture and Commerce, was appreciated. He was instrumental in evaluating the histopathological aspects of the study. Aflatoxin analyses were kindly confirmed by T. Eadie of the Mycotoxin Laboratory, Virginia Division of Consolidated Laboratory Services. Technical assistance was provided by J. Duncan. We thank Ms. Mary Wesley for clerical assistance.

REFERENCES

Armbrecht, B. H. (1973). Assessment of growth and other responses in growing pigs fed aflatoxin, *Environ. Physiol.*, **1**, 198–208.

Barber, R. S., Braude, R., Mitchell, K G., Harding, J. D. J., Lewis, G., and Loosmore, R. M. (1968). The effects of feeding toxic groundnut meal to growing pigs and its interaction with high copper diets, *Brit. J. Nutr.*, **22**, 535–554.

Bruning, J. L., and Kintz, B. L. (1968). *Computational Handbook of Statistics*. Scott, Foresman, and Co., Glenview, IL.

Butler, W. H. (1964). Liver injury and aflatoxin. In *Mycotoxins in Foodstuffs* (ed. G. N. Wogan), pp. 175–186. MIT Press, Cambridge, MA.

Carlton, W. W., and Price, P. S. (1973). Dietary copper and the induction of neoplasms in the rat by acetylaminofluorene and dimethylnitrosamine, *Food Cosmet. Toxicol.*, **2**., 827–840.

Carnaghan, R. B. A., Hartley, R. D., and O'Kelly, J. (1963). Toxicity and fluorescence properties of the aflatoxins, *Nature*, **200**, 1101–1102.

Dashek, W. V., and Llewellyn, G. C. (1982). Aflatoxin and plants, *Postepy Microbiologii*, **2**, 65–84.

Downie, N. M., and Heath, R. W. (1970). *Basic Statistical Methods*, 3rd edn. Harper and Row, New York, NY.

Doyle, J. J., Stearman, W. C., III, Norman, J. O., and Petersen, H. D. C. (1977). Effects of aflatoxin B_1 on distribution of Fe, Cu, Zn, and Mn in rat tissues, *Bull. Environ. Contam. Toxicol.*, **17**, 33–39.

Fare, G., and Howell, J. S. (1964). The effect of dietary copper on rat carcinogenesis by 3-methoxy dyes. I. Tumors induced at various sites by feeding 3-methoxy-4-aminoazobenzene and its N-methyl derivatives, *Cancer Res.*, **24**, 1279–1283.

Fare. G., and Woodhouse, D. L. (1963). The distribution of copper in the tissue of rat: the effects of age and feeding p-dimethylaminoazobenzene with and without copper acetate, *Brit. J. Cancer*, **17**, 775–786.

Goodall, G. M. (1964). Failure of copper to inhibit carcinogenesis by 2-aminofluorene, *Brit. J. Cancer*, **18**, 777–781.

Halver, J. E. (1969). Aflatoxicosis and trout hepatoma. In *Aflatoxin, Food, Science and Technology* (ed. L. A. Goldblatt), pp. 265–306. Academic Press, New York, NY.

Hamilton, P. B., and Garlich, J. D. (1972). Failure of vitamin supplementation to alter the fatty liver syndrome caused by aflatoxin, *Poult. Sci.*, **51**, 688–692.

Hamilton, P. B., Tung, Hsi-Tang, Harris, J. R., Gainer, J. H., and Donaldson, W. E. (1972). The effect of dietary fat on aflatoxicosis in the turkey, *Poult. Sci.*, **51**, 165–170.

Hastings, W. S., and Llewellyn, G. C. (1973). The effect of aflatoxin B_1 on growth rate and iron metabolism in juvenile Mongolian gerbils, *Environ. Physiol. Biochem.*, **3**, 213–220.

Horowitz, W., Senzel, A., and Reynolds, H. (1975a). Natural Poisons. In *Official Methods of Analysis*, pp. 465–473. Assoc. Offic. Anal. Chemists, Washington, DC.

Horowitz, W., Senzel, A., and Reynolds, H. (1975b). Metal analysis. In *Official Methods of Analysis*, pp. 18–26. Assoc. Offic. Anal. Chemists, Washington, DC.

Kamamoto, Y., Makiura, S., Hiasa, Y., Arai, M., and Nobuyuki, I. (1973). The inhibitory effect of copper on DL-ethionine carcinogenesis, *Cancer Res.*, **23**, 1129–1135.

Kimbrough, T. D., and Llewellyn, G. C. (1973). *An Analysis of Human Function. A Laboratory Guide and Handbook for Human Physiology*. Kendall/Hunt Pub. Co. Dubuque, IO.

Llewellyn, G. C. (1982). Aflatoxin–metal interactions and responses in laboratory and agricultural animals: a review, *Dev. Ind. Microbiol.*, **19**, 237–246.

Llewellyn, G. C., Carlton, W. W., Robbers, J. E. and Hansen, W. G. (1973). A rapid and simplified method for the preparation of pure aflatoxin B_1, *Dev. Ind. Microbiol.* **14**, 325–330.

Llewellyn, G. C., Carlton, W. W., Robbers, J. E., and Hansen, W. G. (1974). The response of the Syrian hamster to chronic administration of aflatoxin B_1 and 17-B estradiol, *Dev. Ind. Microbiol.*, **15**, 358–367.

Llewellyn, G. C., Thomen, L. E., and Katzen, J. S. (1981). Effects of dietary copper on developing aflatoxicosis in Syrian hamsters, *J. Environ. Sci. Health*, **B16** (2), 211–225.

Neatherly, M. W., Moos, W. H., Wyatt, R. D., Miller, W. J., Gentry, R. P., and George, L. W. (1981). Effects of dietary aflatoxin on performance and zinc metabolism in dairy calves, *J. Dairy Sci.*, **63**, 789–799.

Schwartz, M. K. (1975). Role of trace elements in cancer, *Cancer Res.*, **35**, 3481–3487.

Shanks, R. C., and Wogan, G. N. (1966). Acute effects of aflatoxin B_1 on liver consumption and metabolism in the rat and duckling, *Toxicol. Appl. Pharmacol.*, **9**, 468.

Singh, N., and Venkitasubramanian, T. A. (1975). Effect of aflatoxin B_1 on lipids of rat tissues, *Environ. Physiol. Biochem.*, **5**, 147–157.

Smith, R. L. (1963). The influence of toxins of *Aspergillus flavus* on the incorporation of C-14 leucine into protein, *Biochem. J.*, **88**, 50P–51P.

Wogan, G. N., and Busby, W. F. (1980). Naturally occurring carcinogens. In *Toxic Constituents of Plant Foodstuffs* (ed I. E. Liener), pp. 329–369. Academic Press, New York, NY.

Trichothecenes and Other Mycotoxins
Edited by J. Lacey

CHAPTER 44

Equine aflatoxicosis: a brief review

R. L. Asquith

INTRODUCTION

There is little information on the incidence of equine aflatoxicosis. This does not mean that the problem does not exist, but rather a lack of awareness and diagnostic capabilities. Our present knowledge of the effects of this toxin on horses is woefully lacking. Results of research trials at our laboratory and elsewhere, together with a few reported clinical cases of naturally occurring equine aflatoxicosis, provide the meagre information available on this subject.

CASE REPORTS

Greene and Oehme (1976) first described a possible case of equine aflatoxicosis in which a 15-year-old Arabian stallion died. The horse exhibited icterus, anorexia, and rapid weight loss for 4 days prior to death. Upon necropsy the liver was described as being black, enlarged, and having a firm consistency. Histopathological changes were bile duct hyperplasia, deposits of haemosiderin in epithelial cells of the renal tubules and congestion of renal vessels. The feed was assayed for aflatoxins using the method of Pons *et al.* (1968) and an aflatoxin B_1 equivalent concentration of 58.4 μg/kg was found. Two other fungal metabolites were detected, but could not be identified. Upon culture, an estimated 30% of the colonies was *Aspergillus flavus*.

Later, McGavin and Knake (1977) cited the death of a horse that had been fed mouldy hay. The horse was in poor physical condition and demonstrated bilateral congestion of the conjunctiva. The horse died 2 days after the onset of colic, which appeared to have been induced by eating the mouldy hay. Bile duct hyperplasia, midzonal hepatic necrosis, degeneration of the renal tubules and fatty degeneration of the heart were reported. Assay results ot the suspect hay were not presented.

We studied several horses in a training stable in which a 7-year-old thoroughbred stallion died from aflatoxicosis. Clinical signs presented were: icterus, depression, shifting lameness, and subcutaneous haemorrhage. The horse died 3 days after the initial examination. These animals were being fed a commercial horse ration, hay and 2.7 kg of maize daily. The maize was examined using ultraviolet black light and fluorescence suggested the presence of aflatoxin. Samples were then assayed and 300 μg/kg aflatoxin were found. Other horses in the stable were also affected, but all made a gradual recovery after the contaminated corn was removed from the diet (Asquith and Edds, 1981).

Angsubhakorn *et al.* (1981) reported the death of 12 yearling colts on a breeding farm outside Bangkok, Thailand. Illness began after a new lot of feed, composed of stored ground maize and stored peanut meal, was added to the diet. Reduced feed intake was the only pre-mortem symptom. Although samples identical to those consumed by the animals were not available, representative feed samples were selected later in the course of the disease. The aflatoxin B_1 assay results were as follows: the ground corn contained 325 μg/kg, while the peanut meal contained 6500 μg/kg.

These authors also reported a case occurring in southern Georgia involving adult horses. Several horses were becoming depressed and two 2-year-old and one 7-year-old mare died. The mixed ration contained 1978 maize from Georgia which was found to contain 840 μg/kg aflatoxin B_1, while one mixed ration had 417 μg/kg and a second 940 μg/kg. Aflatoxin B_1 was detected in liver specimens from two of the horses.

Blood samples from one of the horses in Georgia was examined prior to death. These revealed an increase in number of red blood cells and haematocrit and a marked decrease in white blood cells, notably in lymphocytes. Decreased serum glucose and elevated total serum lipid and cholesterol were found in chemical examinations.

Gross lesions observed at necropsy during these two episodes included: swollen kidneys, a haemorrhagic enteritis, myocardial necrosis, epicardial petechiae, and swollen, fatty livers. In some of the animals the brain was swollen with compression of the sulci.

Microscopic lesions found, were both focal and diffuse degenerative changes of myocardial fibres in the heart, damage of the proximal tubules in the kidneys, and centrolobular necrosis of the liver with pronounced bile duct hyperplasia. Interestingly, involvement of the central nervous system included: cerebral oedema, degeneration of neurons in the cortex, focal malacia, and some degeneration of the cerebellar Purkinje fibres.

Research trials

Asquith and Edds (1981) reported a pilot trial with a 6-month-old pony that had received 5 mg/kg of purified aflatoxin B_1 per os 68 hours before death. Body temperature increased from 100.8 °F up to 4 hours after dosing to a maximum of 106.6 °F at 29 hours. It then fell to 100.2 °F just prior to death. The heart rate remained normal; 44–60 per minute for the first 24 hours and then accelerated to 140 per minute before becoming erratic. Respiration also remained normal, 16–20 per minute for 24 hours, then rapidly increased to 116 per minute and finally became irregular. Anorexia did not occur until temperature peaked. Lethargy and reluctance to move were then observed. Approximately 19 hours after dosing, the pony became ataxic. Prior to death the pony went down into lateral recumbency with tetanic convulsions. Later, the pony became comatose and died 80 minutes after tremors began. Changes in blood values and serum enzyme levels are shown in Table 1.

A controlled experiment utilizing 12 weanling ponies was conducted following the pilot trial. All of the ponies that received a single dose of aflatoxin B_1 at a level of 2 mg/kg died, two within 76 hours of dosing. Clinical signs observed during hourly monitoring throughout this period were: anorexia, elevated temperatures, increased heart rates, ataxia, convulsions, icterus, colic, bloody faeces, abdominal straining, and both rapid and delayed death (Asquith and Edds, 1981).

Of the several enzyme evaluations used for assessment of liver damage, iditol dehydrogenase (ID), also known as sorbitol dehydrogenase (SDH), is

Table 1 Haematological changes and serum enzyme levels after administering a single dose of purified aflatoxin B_1 at 5 mg/kg to a pony

	Hours after dosing			
Blood, enzyme parameters	0	24	30	33
Haematocrit (Pct)	37.1	37.5	33.5	37.7
Haemoglobin (g/100 ml)	13.1	12.9	12.0	13.4
White blood cells ($10^3/\mu\ell$)	10.147	11.039	9.375	8.25
Red blood cells ($10^6/\mu\ell$)	9.32	9.19	8.36	9.44
Blood urea nitrogen (mg/dℓ)	16.0	12.1	11.6	12.4
Prothrombin time (s)	12.0	15.9	22.1	25.2
Alkaline phosphatase (IU/ℓ)	211.0	247.0	213.0	231.0
Gamma glutamyltransferase (IU/ℓ)	14.0	19.0	32.0	32.0
Aspartate transaminase (IU/ℓ)	156.0	172.0	894.0	1,260.0
Iditol dehydrogenase (IU/ℓ)	4.5	452.0	1,800.0	3,677.0

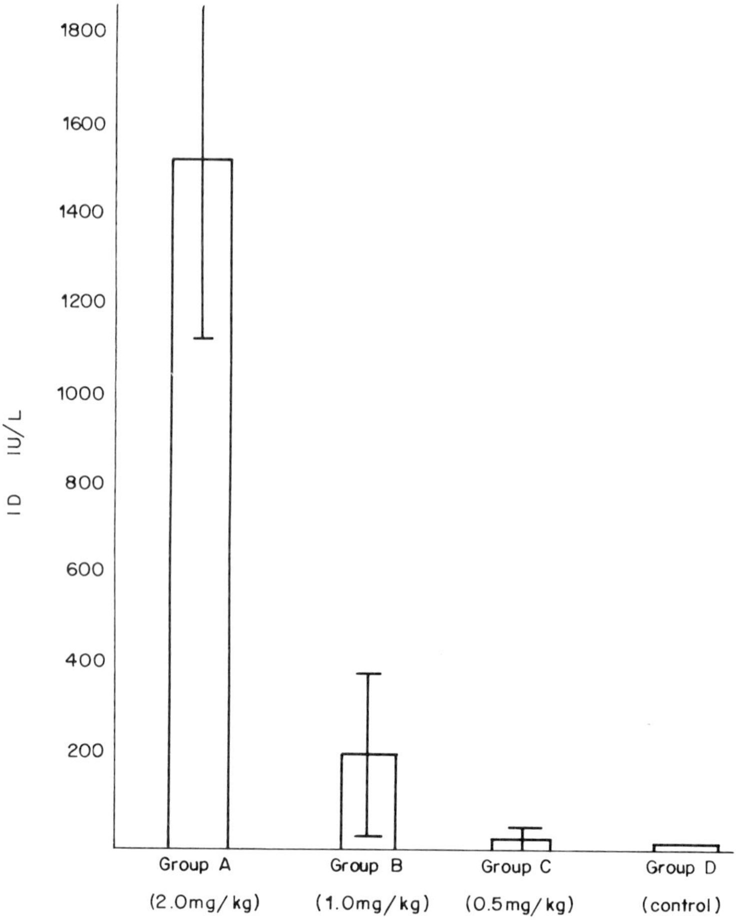

Figure 1 Iditol dehydrogenase responses in young ponies to aflatoxin B_1 (mean peak responses ± SEM).

more specific for hepatocyte damage than others which may also be associated with lesions in skeletal muscle and myocardial damage. The research of Asquith *et al.* (1980) demonstrated extremely high concentrations of ID in ponies treated with varying single doses of aflatoxin B_1 (Figure 1). The post-dosing concentrations in relation to time are shown in Figure 2. These findings suggest that monitoring this enzyme activity may serve as a prognostic aid during the course of the toxicosis.

Bortell (1981) treated weanling ponies with a single weighed dose of aflatoxin B_1 in aqueous suspension given via nasogastric intubation. Ponies were divided into six dosage groups: group 1—0 mg/kg, group 2—0.5 mg/kg, group 3—1 mg/kg, group 4—2.0 mg/kg, group 5—4.0 mg/kg, and group 6—6.0 mg/kg of body weight.

The single ponies in Groups 6 and 5 died after 34 and 46 hours, respectively; two of the four ponies in Group 4 died after 68 and 76 hours. One other pony in Group 4 and one control pony (Group 1) died 32 and 70 days after dosing, respectively, apparently from pneumonia and not aflatoxicosis. Aflatoxins B_1 and M_1 were recovered in large quantities from faeces and the gastrointestinal contents of treated animals. In other tissues, livers contained intermediate amounts of aflatoxin with the kidneys and skeletal muscle containing smaller amounts. The ponies that died from pneumonia had no detectable aflatoxins in their tissues. At necropsy, those young ponies that died from acute aflatoxicosis exhibited visceral petechial haemorrhages, focal lesions in the liver, and various degrees of haemorrhage in skeletal muscle.

Prothrombin time was significantly elevated in those groups receiving 2 mg/kg or greater, and lesser elevations were seen in those receiving 1 mg/kg. The group of ponies given 2 mg/kg showed significant elevations in packed cells volume and haemoglobin. Red blood cell values were significantly

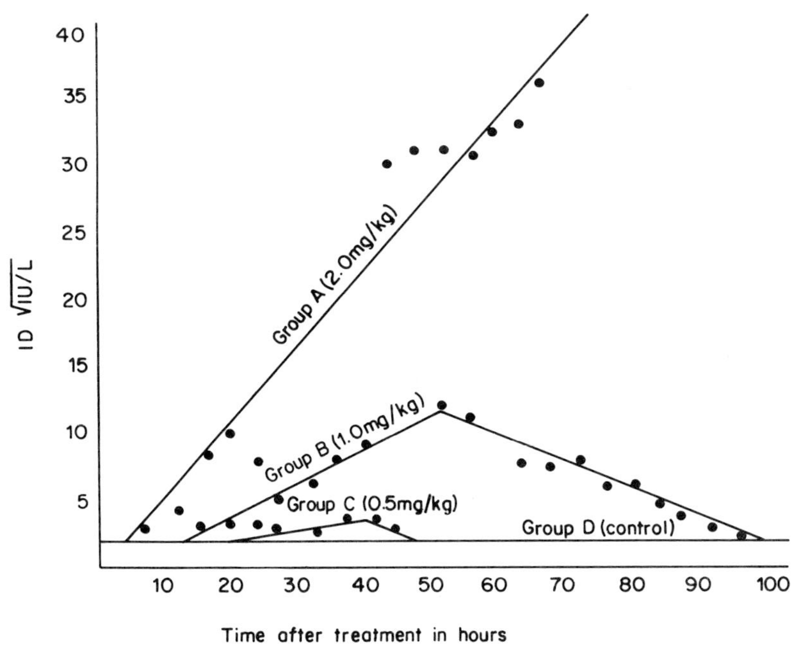

Figure 2 Iditol dehydrogenase (ID) responses in young ponies to aflatoxin B_1 (post-dosing means, square root transformation *vs* time).

depressed. The other treatment groups in this trial showed no significant changes in these haematologic values.

There was significant elevation of aspartate aminotransferase (AST), alanine aminotransferase (ALT), and gamma glutamyl transferase (γGT) when aflatoxin B_1 was administered at 2 mg/kg or higher. Those ponies receiving 0.5 mg/kg and 1 mg/kg of aflatoxin B_1 demonstrated proportionally lower elevations in AST and γGT. No changes in alkaline phosphatase nor ALT were seen in these groups. Serum urea nitrogen was not elevated in any of the treatment groups.

Six adult male Shetland ponies were assigned to three treatmen groups with two animals in each group by Cysewski *et al.* (1982). Ponies were given aflatoxin daily at levels of 0.3, 0.15, or 0.075 mg/kg. Deaths occurred at 17 and 16 days in the group receiving 0.3 mg/kg, 26 and 32 days in the group given 0.15 mg/kg, and in those receiving 0.075 mg/kg at 37 and 39 days. Prothrombin time, sulphobromophthalein clearance time, total plasma bilirubin, and icterus index increased markedly during the last 7–10 days before death. Neither blood urea nitrogen nor the plasma activity of creatine phosphokinase were elevated during the experiment. There were substantial increases in plasma activity of ornithine carbamoyl transferase. Levels of AST activity were less but still significantly increased.

Histopathological findings were: centrolobular fatty change with hepatic cell necrosis, and periportal fibrosis in animals given 0.3 mg/kg. These changes were greater in animals given 0.075 mg/kg and were also accompanied by mild bile duct proliferation. Bile stasis and irregularity in size and shape of hepatocytes and binucleate hepatocytes were also seen.

Gross lesions observed were: generalized icterus, haemorrhages, and brown to tan livers in those ponies that died. Kidneys were consistently dark in colour at necropsy.

An experiment designed to simulate possible field conditions of aflatoxicosis was conducted by Aller *et al.* (1981). Sixteen yearling ponies were randomly assigned to four groups and treated for 21 days at levels of 0.045, 0.030 and 0.015 and 0 mg of aflatoxin B_1/kg or a total dose of 0.945, 0.630, 0.315 and 0 mg/kg. During this treatment period all animals remained clinically normal with no significant variations in haematological or enzymatic parameters. The ponies were allowed to recover for 3 months and then reassigned to four treatment groups such that each group during the second trial was composed of one pony from each group from the first trial. The ponies were treated for 5 days at 0.4, 0.2, 0.1 and 0 mg/kg aflatoxin B_1 per day or a total dose of 2.0, 1.0, 0.5 and 0 mg/kg. During the second trial two of the ponies in the highest (2.0 mg/kg) treatment group demonstrated significantly elevated serum enzyme levels of ID and γGT. These ponies had been in the high (0.945 mg/kg) and median (0.630 mg/kg) treatment groups during the first experiment. All ponies subsequently recovered and returned to normal health.

DISCUSSION

There has been no systematic study employing pathological lesions, haematological and enzymatic parameters, along with clinical observations to establish pathognomonic characteristics of equine aflatoxicosis. This brief review describes case reports and research trials that have recorded similar findings. Field cases are handicapped by insufficient information and are often masked by concurrent disease processes. This is a stark contrast to data gathered from a research laboratory with scientific controlled experiments; and while these case reports should be carefully evaluated before any interpretation is made, their additions serve as an important contribution to our understanding of the natural occurrence of this toxicosis.

The total macroscopic lesions described in all reports varied, but the most common finding was hepatic damage. In addition, bile duct proliferation appeared to be the most frequent histopathological lesion observed in this toxicosis, and is consistent with the findings in other species that have died from aflatoxicosis.

Substantial decreases in the number of white blood cells were seen by Angsubhakorn *et al.* (1981) and Asquith and Edds (1981) in animals receiving high concentrations of aflatoxin B_1. This haematological change was not reported in trials where the aflatoxin concentration was low. Increases in red blood cells were noted by Angsubhakorn *et al.* (1981) and Bortell (1981) in equine aflatoxicosis. Marked increases in prothrombin time were demonstrated in trials conducted by Asquith and Edds (1981), Bortell (1981) and Cysewski *et al.* (1982).

Clinical chemistry determinations have shown significant increases of AST in the trials of Asquith and Edds (1981), Bortell (1981) and Cysewski *et al.* (1982). Aller *et al.* (1981) reported significant elevations of γGT in yearling ponies treated with aflatoxin B_1 over an extended period of time, and Bortell (1981) described this increased enzyme activity in weanling ponies given a single dose of aflatoxin B_1. Information obtained from the research of Asquith *et al.* (1980) suggests that monitoring the enzyme activity of ID may be a useful prognostic aid during the course of this disease. Aller *et al.* (1981) confirmed that the rise of ID concentrations in the blood is only temporary in non-lethal aflatoxicosis.

The differences in aflatoxin concentrations causing toxicosis in the equine appear to be due to modifying influences on intoxication. Younger animals are more susceptible. The nutritional plane, particularly lesser amounts of dietary protein, will decrease the amount of aflatoxin necessary to elicit toxicosis. These effects have been reported in other species. With these considerations it may be difficult to define even a single-dose LD_{50} of aflatoxin B_1 without several qualifying conditions. This points out that further research is necessary to characterize the biological effects of this mycotoxin in the equine species.

REFERENCES

Aller, W., Edds, G. T., and Asquith, R. L. (1981). Effects of aflatoxins in young ponies, *Am. J. Vet. Res.*, **42**, 2162–2164.

Angsubhakorn, S., Poomvises, P., Romruen, K., and Newberne, P. M. (1981). Aflatoxicosis in horses, *J. Am. Vet. Med. Assoc.*, **178**, 274–278.

Asquith, R. L., and Edds, G. T. (1981). Investigations in equine aflatoxicosis, *Proc. Am. Assoc. Equine Pract..*, **26**, 193–200.

Asquith, R. L., Edds, G. T., Aller, W. W., and Bortell, R. (1980). Plasma concentrations of iditol dehydrogenase (sorbitol dehydrogenase) in ponies treated with aflatoxin B_1, *Am. J. Vet. Res.*, **41**, 925–927.

Bortell, R. (1981). Clinical chemistry and hematology of weanling ponies exposed to aflatoxin B_1. *M.S. thesis, University of Florida, Gainesville.*

Cysewski, S. J., Pier, A. C., and Baetz, A. (1982). Experimental equine aflatoxicosis, Toxicol. Appl. Pharmacol., **65**, 354–365.

Greene, H. J., and Oehme, F. W. (1976). A possible case of equine aflatoxicosis, *Clin. Toxicol.*, **9**, 251–254.

McGavin, M. D., and Knake, R. (1977). Hepatic midzonal necrosis in a pig fed aflatoxin and a horse fed moldy hat, *Vet. Pathol.*, **14**, 182–187.

Pons, W. A., Jr., Cucullu, A. F., Franz, A. O., Jr., and Goldblatt, L. A. (1968). Improved objective fluorodensitometric determination of aflatoxins in cottonseed products, *J. Assoc. Off. Anal. Chem.*, **45**, 694–699.

Trichothecenes and Other Mycotoxins
Edited by J. Lacey

CHAPTER 45

In utero transfer of aflatoxin and selected effects in neonatal pigs

A. C. Pier, M. E. McLoughlin, J. L. Richard, A. Baetz and R. R. Dahlgren

INTRODUCTION

Aflatoxins cause both acute and chronic intoxications in swine (Cysewski *et al.*, 1968; Patterson, 1973; Pier, 1981). As with other animal species, young swine appear to be more susceptible than older mature animals. While the liver is the primary target organ in aflatoxicosis, considerable evidence indicates that the aflatoxins are immunosuppressive through their actions on the thymus, the cell-mediated immune system, and non-specific humoral pathways (Pier *et al.*, 1979). Cysewski *et al.* (1978) demonstrated an impaired immune response to *Erysipelothrix rhusiopathiae* vaccines in swine receiving aflatoxins at a daily oral dose of 0.04 mg B_1 activity/kg. Joens *et al.* (1981) demonstrated reduced resistance to experimental *Treponema hyodysenteriae* infections in swine given 0.04 mg B_1 activity/kg/day. As with other species, reduced weight gain, elevated serum enzyme activities indicative of hepatic necrosis and an altered serum protein profile are early biological indicators of aflatoxicosis. Most of the existing data on aflatoxicosis in swine has been obtained with purified aflatoxin B_1 or mixtures of aflatoxin B_1 with small amounts of G_1, B_2 and G_2. No information on the effects of known doses of M_1 toxin in swine are available.

Aflatoxin has been shown to be teratogenic when given by parenteral injection into pregnant female laboratory rodents (DiPaolo *et al.*, 1967); similar effects were shown by Arora *et al.* (1981) with orally administered aflatoxin B_1. These studies demonstrated the ability of biologically effective quantities of toxin to cross the relatively simple rodent placenta.

The purpose of the present study was to determine whether aflatoxin given orally to pregnant sows in terminal pregnancy could be detected in tissues of pigs at birth, and whether such pigs had altered immunological or other biological responsiveness. A further objective was to determine effects of aflatoxin M_1, the metabolite commonly found in milk and urine of intoxicated animals, in neonatal pigs born of normal sows.

METHODS

To determine the effect of aflatoxin in paranatal swine, two series of experiments were conducted. In the first phase, pregnant sows were fed measured doses of aflatoxin during their final 2 weeks of pregnancy. At parturition a third of the piglets in each litter was sacrificed, before they could suckle their dam, and tissues were collected for histopathological examination and chemical testing for aflatoxin residues. Selected tissues included liver, kidney and skeletal muscle for aflatoxin analysis, and thymus, liver and kidney for histopathology. Peripheral blood was collected for lymphocyte harvest and lymphoblastogenesis analysis. The remaining two-thirds of the litter was injected intravenously with sheep red blood cells (SRBC) and subcutaneously with killed *Mycobacterium paratuberculosis* for later assessment of immunological responsiveness. These pigs were divided equally before suckling; half of them were placed on their own dam and the other half were placed on a normal dam who had just farrowed and had not been given aflatoxin. The litters of the normal sows were divided and treated in identical fashion to that of the aflatoxin-treated sows. Details of the experiment are presented below.

Animals

Pairs of Yorkshire sows from the National Animal Disease Center (NADC) herd were selected with farrowing dates estimated for the same 24-hour period. One sow of each pair was given aflatoxin starting 14 days before anticipated farrowing. The sows were placed in raised farrowing crates that prevented access of the piglets to the sow after parturition. When farrowing time was imminent the sows were monitored on closed-circuit television so the time of farrowing was known and the piglets could be tested shortly after birth. This experiment was replicated twice using two pairs of sows and a total of 47 piglets.

Aflatoxin

Partially purified aflatoxin was prepared and administered as previously described (Cysewski *et al.*, 1978). The toxin preparation was estimated to have activity equivalent to 65.3% aflatoxin B_1 (Cysewski *et al.*, 1968). Daily doses of 0.06 mg B_1 equivalence/kg body weight were placed on top of part of the daily ration; the remainder of the ration was given after the toxin-treated portion was consumed. The basal ration was NADC Farrowing ration No. 541 (14% protein, 3% fat, 68% total digestible nutrients). This ration was tested and found to be free of detectable aflatoxin. On the day of farrowing, aflatoxin treatment was discontinued.

Immunological methods

Selected immunological reactions were tested on the day of birth and after 2 weeks. Peripheral blood lymphocytes were harvested from approximately one-third of each litter at birth and from the remainder of the pigs after 2 weeks. In the first experiment lymphocytes were purified by Ficol Hypaque separation, cultured and tested for lymphoblastogenic responsiveness to phytohaemagglutinin (PHA-P 1 : 80) as described by Pier *et al.* (1977). However, it was necessary to decrease the specific gravity of the Ficol Hypaque to 1.070 to separate lymphocytes from the erythrocytes of day-old piglets. In the second experiment lymphoblastogenesis was tested in whole blood after the method of Hintz (1981). Because the pigs in the second experiment were sensitized with *Mycobacterium paratuberculosis* cells, lymphoblastogenesis was determined following stimulation with PHA-P and purified protein derivative of *M. paratuberculosis*.

On the day of birth, two-thirds of each litter was injected intravenously with 1.0 ml washed SRBC (packed cells diluted with equal volume of phosphate buffered saline (PBS)). In the second experiment these pigs were also injected subcutaneously with 0.2 ml killed *M. paratuberculosis* cells suspended in oil (Larsen *et al.*, 1978). After 2 weeks the anti-SRBC titre was determined (SRBC lysis in the presence of 2 units guinea pig complement) and delayed cutaneous hypersensitivity was determined by skin testing in the flank with 0.1 ml johnin.

Clinical chemical tests

Sera taken on the day of birth were tested for total protein, aspartate aminotransferase (AST), sorbitol dehydrogenase (SDH), glycocholic acid (GC) and glycodeoxycholic acid (GDC). Aspartate aminotransferase was determined using automated equipment (Technicon Auto Analyzer, Technicon Instruments Corp., Ardsley NY) by the manufacturer's recommended procedure (Auto Analyzer Technical Methods: Glutamic Oxalacetic Transaminase by Fluorometer, N-53, Technicon Instruments Corp., Ardsley, NY, 1964). Sorbitol dehydrogenase was determined on the automated analyser using the method of Baetz and Pier (1976). Glycocholic acid and glycodeoxycholic acid were determined using diagnostic radioimmunoassay kits following the instructions provided by the manufacturer (Nordiclab, Uusikatu 92,SF90120, Oulu 12, Finland).

On the day of birth (day 1) all the pigs were weighed. Blood samples were obtained from one-third of the litter and these pigs were then immediately sacrificed before they had suckled their dam and the gluteal muscle mass, liver, thymus, and kidneys were harvested. Liver, muscle, and kidney tissues were analysed for aflatoxin B_1 and M_1 residues following Stubblefield and Shotwell (1981). The remaining two-thirds of the litter were immunized as described above and divided between the sows as described below. Histopathological examination was conducted on specimens of thymus, liver and kidney which were collected from all pigs and formalized on the day of necropsy.

Pigs born to the normal dam and killed on day 1 were designated N; half of the remainder were left on the normal sow throughout the experiment and were designated NN. The others were transferred to the aflatoxin sow at day 1 and were designated NA. Similarly, pigs born to the aflatoxin sow and killed on day 1 were designated A; those left on the aflatoxin sow were designated AA and those transferred to the normal sow at day 1 were designated in AN. Data from the first experiment are represented as I and that for second as II (Tables 1 and 2).

Preparation of M_1 toxin

The second phase of these studies was to examine selected effects of aflatoxin M_1 in neonatal pigs. Partially purified aflatoxin, totalling 52 mg B_1 equivalence, was given orally by capsule daily for 3 to 5 consecutive days to each of three calves surgically prepared with indwelling urinary catheters. Urine was collected continuously from 24 to 168 hours after the first dose of aflatoxin.

A portion of each urine collection was assayed for M_1 by adding 45 ml methanol to 30 ml of urine and partitioning with two 55 ml volumes of chloroform as described by Richard *et al.* (1979). The chloroform layers were collected, evaporated to dryness, redissolved in a minimal volume of chloroform, transferred quantitatively to a small vial and evaporated to dryness in a steam bath under flowing nitrogen. The residue was then dissolved in 100 μl of 90 : 10 (v/v) benzene : acetonitrile for TLC (thin-layer chromatography) analysis. Three 20 μl spots of the urine extract were placed on a silica gel 60 plate (Brinkman Instruments, Des Plaines, IL 60016). Additionally, three spots of M_1 toxin standard (5 ng/spot) were placed on the plate. The plate was developed in 97 : 2 : 1 (v/v/v) chloroform : methanol : formic acid and then M_1 toxin was quantified by fluorescent densitometry (Vis-UV Chromatogram Analyzer, Farrand Optical Co., Valhall, NY 10595).

The general procedure for purification of M $_t$oxin from the urine was to mix two volumes of urine with three volumes of absolute methanol. This mixture was extracted twice with four volumes of chloroform for each extraction using 1 minute of agitation. The chloroform layers were combined, filtered through anhydrous sodium sulphate in Whatman No. 1 filter paper and evaporated to dryness. The residue was redissolved in a minimal amount of chloroform and the extract was streaked on silica gel 60 plates. The plates were developed in 97 : 2 : 1 (v/v/v) chloroform : methanol : formic acid and the M_1 band was scraped from each plate. The silica gel containing M_1 toxin was ground lightly to break clumps of silica gel in a mortar and pestle using sufficient $CHCl_3$ to cover the silica gel to reduce aerosol hazard during grinding. This mixture was transferred quantitatively to a thimble in a Soxhlet apparatus and extracted with chloroform for 60 hours. The resultant extract was evaporated to dryness, redissolved in a minimal but known volume of chloroform. A 100 μl aliquot of this volume was removed for TLC analysis. For TLC analysis, appropriate dilutions of the aliquot were made in chloroform and three spots of 10 μl each were placed on a silica gel 60 plate. The plate was developed in 97 : 2 : 1 (v/v/v) chloroform : methanol : formic acid. The M_1 on the developed plate was then quantified by fluorescence densitometry.

The amounts of M_1 toxin recovered at different times after the start of dosing with aflatoxin were recorded for each calf. The major portion of M_1 toxin appeared in the urine between 48 and 144 hours after beginning aflatoxin dosing; peak excretion was at 72 hours. A total of 10.3 mg M_1 toxin was recovered from the urine of the three calves.

Treatment of pigs with aflatoxin M_1

Three littermate pigs were weighed on their first day of life and treated orally with weighed doses of M_1 toxin in capsules for 5 consecutive days. Two additional littermate pigs were observed without treatment; one as a placebo control, the other as a normal control. Three daily dose schedules were employed including 0.05, 0.1 and 0.3 mg/kg. The pigs were weighed daily and sacrificed on day 6 for histopathological examination. Tissues collected included liver, kidney and thymus.

RESULTS

Effects of *in utero* transfer of aflatoxin

Data from the two replicates (I and II) of the *in utero* transfer of aflatoxin study are presented in Tables 1 and 2. There was generally poor survival in the AN group in both experiments. There were no apparent real differences between groups of equal age in body weight, AST, and SDH activity levels,

Table 1 Serum and tissue chemical observations in paranatal swine born of normal and aflatoxin-treated sows

Group	N	Age (days)	AST (IU)	SDH (IU)	GC (μM/ℓ)	GDC (μM/ℓ)	Σ PROT (g/dℓ)	AFT (μg/kg)
N I	4	1	31	4	—	—	5.1	0
II	3	1	87	6	4.2	0	—	0
A I	5	1	72	15	—	—	2.7	tr B_1
II	4	1	75	11	1.5	0.3	—	0.1 B_1 tr M_1
NN I	3	12	30	9	—	—	8.9	—
II	4	15	34	4	1.6	4.0	—	—
NA I	3	12	54	37	—	—	3.5	—
II	3	15	46	2	—	5.8	—	—
AN I	0	—	—	—	—	—	—	—
II	2	15	30	3	1.9	4.9	—	—
AA I	2	12	44	4	—	—	4.6	—
II	4	15	88	8	2.7	7.5	—	—

AST = Aspartate amino transferase; SDH = sorbitol dehydrogenase; GC = glycocholic acid; GDC = glycodeoxycholic acid; Σ PROT = total serum protein; AFT = aflatoxin residue: liver.

or serum glycocholic acid (GC) levels. Serum glycocholic acid was low in both groups of 1-day-old pigs as compared to their 2-week-old littermates. Total serum proteins appeared to be low in all groups exposed to aflatoxin before birth or through nursing. No differences were seen between groups in anti-SRBC titre or lymphoblastogenic response. However, delayed cutaneous hypersensitivity tests (DCHT) indicated a diminished response in all groups exposed to aflatoxin as compared to the normal control group NN. This difference was statistically significant ($p < 0.01$) when the combined aflatoxin group data (AA, AN and NA) was compared with that of the normal control group (NN).

Chemical analysis of tissues taken on day 1 before nursing showed detectable residues of aflatoxin B_1 (0.1 μg/kg) and M_1 (trace) in livers of pigs born of aflatoxin-treated sows but not in livers from pigs born of the normal sows. Histopathological changes were confined to the livers of the A group where mild to moderate fatty changes and perioportal lymphoids infiltration were observed.

Effects of aflatoxin M_1in neonatal pigs

Extraction of urine of calves fed aflatoxin B_1 proved to be an effective method for deriving M_1 toxin for experimental animal use. Approximateliy 25 parts of

Table 2 Body weight and immunological reactivity of paranatal swine born of normal and aflatoxin-treated sows

Group		N	Age (days)	Body weight kg	Body weight %NN	Anti-SRBC	DCHT (mm)
N	I	4	1	1.0	—	0	—
A	I	5	1	1.1	—	0	—
NN	I	3	12	2.2	100	45	—
	II	4	15	4.9	100	52	2.8[a]
NA	I	3	12	2.0	91	27	—
	II	3	15	3.9	80	19	.7
AN	I	0	—	—	—	—	
	II	2	15	5.1	104	5	1.0
AA	I	2	12	1.5	68	40	—
	II	4	15	4.9	100	64	1.5

[a] Significantly different from combined data of NA, AN and AA ($p < 0.01$).
%NN = Body weight as percentage of NN (normal) littermates; Anti SRBC = sheep and blood cell antibody titre; DCHT = delayed cutaneous hypersensitivity test responses.

B_1 was needed to produce 1 part of M_1. The majority of the M_1 toxin was excreted between 48 and 144 hours after toxin administration with a peak excretion at 72 hours. Body weight data are presented in Figure 1. The pig receiving 0.3 mg M_1/kg/day had a markedly reduced weight gain as compared to control pigs or those receiving 0.1 and 0.05 mg M_1/kg/day.

Histopathological changes were seen in the thymus and liver of the pig receiving 0.3 mg M_1/kg but not in those receiving smaller doses. These are depicted in Figures 2 and 3. The thymic changes indicated marked lymphoid aplasia. The cortex was coposed of reticular cells and macrophages; there was a complete loss of lymphocytes in the cortex. The medulla contained thymic corpuscles which were increased in size and filled with necrotic debris. Many corpuscles were surrounded by large epithelial and reticulum cells. Inflammatory cells, primarily neutrophils and eosinophils, were present within both the cortex and medulla.

In the liver the lobular architecture was intact. The sinusoids were filled with reticulo-endothelial cells and macrophages which contained bile pigments. The hepatocytes surrounding the central vein had early necrotic changes which included cytoplasmic and nuclear swelling, eosinophilic granulation of cytoplasm and nuclear pyknosis. Neutrophilic infiltration was present in areas of hepatic necrosis. In many lobules the necrotic changes

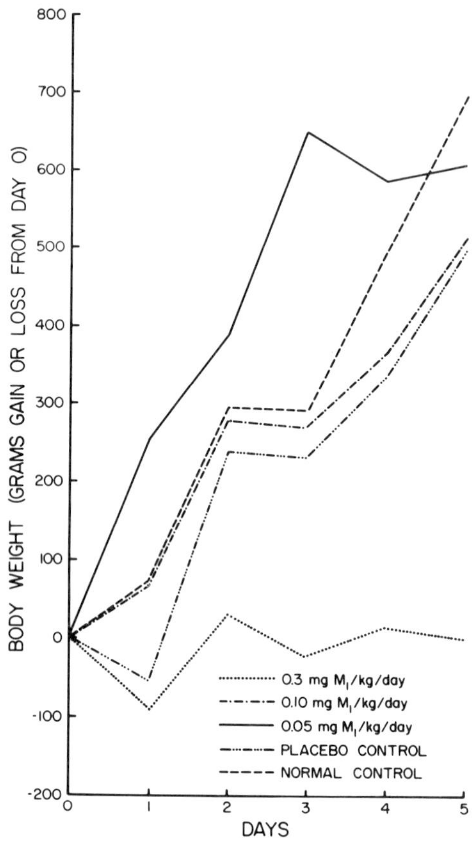

Figure 1 Daily changes in body weight in pigs given varying oral doses of aflatoxin M_1.

extended into the midzonal area. Occasional bile ducts showed mild hyperplasia while the other structures of the portal tract remained intact.

DISCUSSION

It was evident from the replicated experiments that aflatoxin was transferred across the porcine placenta. Detectable aflatoxin residues were present in the livers of pigs in both replicates at the time of birth before they had nursed the sow. These residues were transferred in sufficient quantity to exert apparent biological effects on the neonatal pigs. Pigs born to the aflatoxin-treated sow had low serum protein levels and poor vitality as evidenced by a low survival

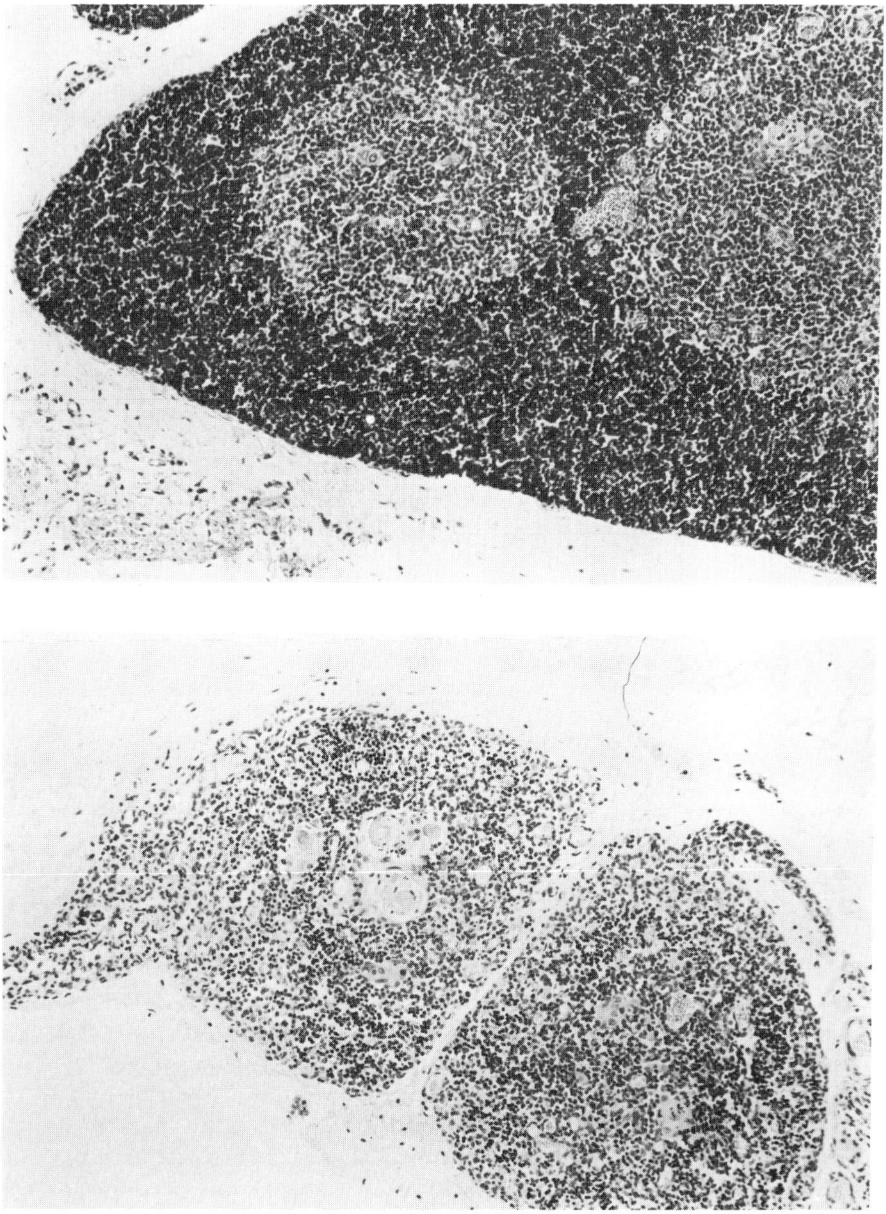

Figure 2 Thymus of normal littermate pig above, and that of pig receiving 0.3 mg M_1/kg/day, below. Note severe thymic aplasia. (H and E stain ×100.)

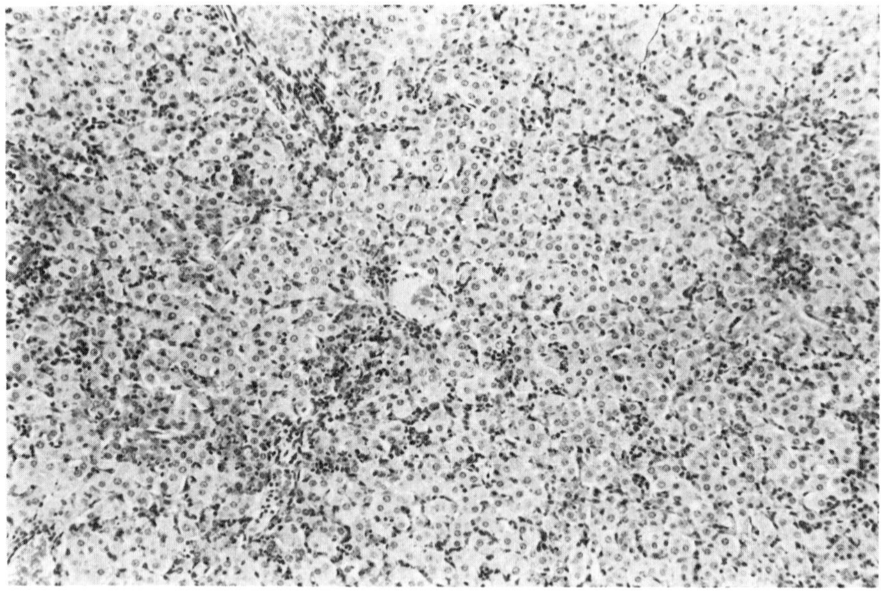

Figure 3 Liver of pig receiving 0.3 mg M_1/kg/day. Sinusoids were occluded with RE cells and macrophages containing bile pigments. Hepatic cell necrosis was seen about the central vein (centre) and extending into the midzonal area. (H and E stain ×100.)

rate when transferred to the normal lactating sows. Furthermore pigs born to or transferred to the aflatoxin-treated sows had reduced delayed cutaneous hypersensitivity responses as compared to those of the normal (NN) pigs. This is a finding of considerable interest in terms of the health of neonatal pigs. Its importance would be all the greater if these results can be extrapolated to neonatal animals of other species. It has been demonstrated previously that aflatoxin causes teratogenic effects in paranatal rodents (DiPaolo *et al.*, 1967; Arora *et al.*, 1981). However the rodent placenta is a relatively rudimentary structure (haemoendothelial) with a single layer of tissue separating the foetal and maternal blood pools. Adamesteanu *et al.* (1974) found liver changes in calves born to cows that had consumed aflatoxin-contaminated feed. This implied that aflatoxin might cross the relatively complex (epitheliochorial) bovine placenta but attempts at experimental reproduction of the condition have been unsuccessful. The porcine placenta is among the most complex (epithelio-chorial) and has six tissue layers separating the maternal and foetal blood pools. Thus if aflatoxin can transfer across the porcine placenta as we have shown, one might expect it to transfer *in utero* in any placentate species and be capable of exerting biological effects on the foetus. Because the very young of most species are considered more susceptible to aflatoxin than more mature animals, the foetus would be highly susceptible and could be expected to show the biological effects once removed from the dam's placental support. Besides liver disease *per se*, suppression of native defence mechanisms and immunogenesis are important considerations in neonatal survival.

The studies on the effects of M_1 toxin in neonatal pigs showed that short-term exposure to 0.3 mg M_1/kg/day caused marked biological effects as demonstrated by cessation of weight gain, hepatic necrosis and severe thymic aplasia. This was a very large dose by comparison to B_1 toxin; it approximates one-half of an LD_{50} dose of B_1 toxin in older swine (Patterson, 1973) and was approximately 5 times the dose of B_1 found to cause immunosuppression and reduced resistance in weanling swine (Cysewski *et al.*, 1978; Joens *et al.*, 1981). More extended treatment might have shown biological effectiveness by a lesser dose. M_1 toxin is important to neonatal animals because it is the usual form of aflatoxin found in milk. Of practical importance was the relative ease with which usable quantities of M_1 toxin could be obtained from the urine of calves treated with aflatoxin B_1. This method offers considerable promise in enabling future studies of M_1 toxin on neonatal animals.

ACKNOWLEDGEMENTS

The authors gratefully acknowledge the assistance of R. E. Fichtner, R. Lyon, J. Hendricks and I. McFadden for animal studies, and L. Maki for statistical evaluation of data.

REFERENCES

Adamesteanu, I., Adamesteanu, C., Baba, I., Danielescu, N., Maldovanu, N. A., and Rotura, O. (1974). Leberzirrhose bei neugebornen Kalbern durch plazentare Übertragung von Aflatoxin, *Deutsche Tierarz. Woch.*, **81**, 141–144.

Arora, R. G., Frolen, H., and Nilsson, A. (1981). Interference of mycotoxins with prenatal development of the mouse. 1. Influence of aflatoxin B_1, ochratoxin A and zearalenone, *Acta Vet. Scand.*, **22**, 524–534.

Baetz, A. L., and Pier, A. C. (1976). An automated fluoremetric method for the determination of sorbitol dehydrogenase and its application to the assessment of liver damage in bovine mycotoxicosis, *Clin. Chem.*, 28th Nat'l. Meet, p. 337.

Cysewski, S. J., Pier, A. C., Engstrom, G. W., Richard, J. L., Doughtery, R. W., and Thurston, J. R. (1968). Clinical pathologic features of acute aflatoxicosis of swine, *Am. J. Vet. Res.*, **29**, 1577–1590.

Cysewski, S. J., Wood, R. L., Pier, A. C., and Baetz, A. L. (1978). Effects of aflatoxin on the development of acquired immunity to swine erysipelas, *Am. J. Vet. Res.*, **39**, 445–448.

Di Paolo, J. A., Elis, J., and Erwin, H. (1967). Teratogenic response by hamsters, rats and mice to aflatoxin B_1. *Nature (London)*, **215**, 638–639.

Hintz, A. M. (1981). Comparative lymphocyte stimulation studies on whole blood from vaccinated and nonvaccinated cattle with paratuberculosis, *Am. J. Vet. Res.*, **42**, 507–510.

Joens, L. A., Pier, A. C., and Cutlip, R. C. (1981). Effects of aflatoxin consumption on the clinical course of swine dysentery. *Am. J. Vet. Res.*, **42**, 1170–1172.

Larsen, A. B., Moyle, A. I., and Himes, E. M. (1978). Experimental Vaccination of cattle against paratuberculosis (Johnes disease) with killed bacterial vaccines. A controlled field study. *Am. J. Vet. Res.*, **39**, 65–69.

Patterson, D. S. P. (1973). Metabolism as a factor in determining the action of the aflatoxins in different animal species. *Food Cosmet. Tox.*, **11**, 287–294.

Pier, A. C. (1981). Mycotoxins and animal health. *Adv. Vet. Sci. and Comp. Med.*, **25**, 185–243.

Pier, A. C., Fichtner, R. E., and Cysewski, S. J. (1977). Effects of aflatoxin on the cellular immune system. *Ann. Nutr. Ailment.*, **31**, 781–788.

Pier, A. C., Richard, J. L., and Thurston, J. R. (1979). Effects of aflatoxin on the mechanisms of immunity and native resistance. *Med. Mycol. Zent. Blat. Bakt.*, Suppl. 8, pp. 301–309.

Richard, J. L., Thurston, J. R., and Pier, A. C. (1979). Laboratory diagnosis of mycotoxicoses of veterinary importance in the United States. *Proc. 83rd Ann. Meet. U.S. Anim. Health Assoc.*, pp. 205–225.

Stubblefield, R. D., and Shotwell, O. L. (1981). Determination of aflatoxins in animal tissues. *J. Assoc. Off. Anal. Chem.*, **64**, 964–968.

Trichothecenes and Other Mycotoxins
Edited by J. Lacey

CHAPTER 46

Mycotoxic suppression of immunity

A. C. Pier and M. E. McLoughlin

INTRODUCTION

Inadvertent modification of the mechanisms of native resistance and immunity may be caused by a number of substances found in the environment of animals and man (FDA, 1980). Among these substances are heavy metals and industrial or agricultural chemical pollutants that find their way into the food supply or environment through inattentive use or disposal. Another group of substances is formed by naturally occurring products of microbiological activity. The mycotoxins are such a group of compounds. They are produced as secondary metabolites by a wide variety of fungi growing on grains, oil seeds, nuts, spoiled fruits, standing forages and stored foods and feeds (Hesseltine, 1981; Pier, 1981; Wyllie and Morehouse, 1977). Their natural occurrence, seasonal variation, regional prevalence and substrate preference create several very real and scientifically important problems in human and animal health. The effects exerted by mycotoxins on mechanisms of immunity and resistance have been the subject of several reviews (Pier *et al.*, 1979, 1980a,b; Richard *et al.*, 1975, 1978). It is the purpose of this chapter to update the accumulated information on these important health-related effects on animals and man.

Three groups of mycotoxins are prominently associated with immunosuppressive activity; namely the aflatoxins, ochratoxin A and certain of the trichothecene toxins. Much of their toxicity is referable to their inhibition of protein synthesis, but each acts at a different site of protein formation. Aflatoxin binds to DNA, suppresses DNA-dependent RNA production and thus interferes with transcription (Clifford and Rees, 1966; Hsieh *et al.*, 1977). T-2 toxin, representative of several of the trichothecene toxins, affects initiation of translation in protein production (Hsieh *et al.*, 1977; McLaughlin *et al.*, 1977). Ochratoxin A inhibits protein synthesis through inhibition of phenylalanyl-tRNA synthetase (McLaughlin *et al.*, 1977). Current information indicates that these mycotoxins affect immunogenesis in different ways.

AFLATOXINS

Immunosuppressive action by aflatoxin has been more widely studied than that of T-2 toxin or ochratoxin A (Dreismann and Roschenthaler, 1982). Aflatoxin consumption during and after experimental immunization was shown to impair actively acquired immunity in turkeys; however the effect(s) were reversible when toxin consumption was discontinued before immunization (Pier and Heddleton, 1970). Numerous subsequent investigations have demonstrated five key points:

1. Aflatoxin impaired immunogenesis without suppressing antibody formation in most studies.
2. Aflatoxin suppressed formation of non-specific humoral substances related to resistance and immunity, namely complement (C_4) and interferon.
3. Aflatoxin suppressed phagocytosis by macrophages.
4. Aflatoxin caused thymic aplasia (effects on the bursa of Fabricius were variable but generally minimal).
5. Aflatoxin suppressed cell-mediated immunity, notably delayed cutaneous hypersensitivity; lymphoblastogenesis and leukocyte migration were also variably suppressed.

These findings are summarized in Tables 1–3 with references to individual investigations.

The exact mechanism(s) of aflatoxin-induced immunosuppression remain obscure. The effects on complement, interferon and serum proteins were presumably the result of liver injury and inhibition of protein synthesis. However the effects on cell-mediated immune (CMI) mechanisms were unexplained. Presumably they arose from differences in peripheral blood lymphocyte populations, lymphocyte function, lymphokine production or macrophage function.

Recent investigations have shed additional light and some confusion on the issue of aflatoxin suppression of cell-mediated immune responses. McLoughlin (1982) has shown that delayed cutaneous hypersensitivity in guinea pigs was dose-dependent. Aflatoxin doses of 0.04 mg B_1/kg/day significantly decreased skin test response but decreased weight gain, and increased serum glycocholic acid levels occurred at only 0.02 and 0.01 mg B_1/kg/day respectively and thus were more sensitive measures of aflatoxin effect. McLoughlin (1982) also studied adoptive transfer of cutaneous hypersensitivity and

Table 1 Effects of aflatoxin on experimentally acquired immunity and resistance

Effect	Model	Reference
Immunosuppression (reversible)	Turkey (*P. multocida*)	Pier and Heddleston 1970
Normal immunity	Turkey (Newcastle disease)	Pier *et al.*, 1971
Immunosuppression	Pigs (*E. rhusiopathiae*)	Cysewski *et al.*, 1978
Susceptibility increased	Chicken (*C. albicans*)	Hamilton and Harris, 1971
Susceptibility normal	Chicken (*S. gallinarum*)	Smith *et al.*, 1969
Susceptibility increased	Chicken (three species *Salmonella*)	Boonchuvit and Hamilton, 1975
Susceptibility normal	Turkey (*A. fumigatus*)	Richard *et al.*, 1973
Susceptibility normal	Hamster (*M. paratuberculosis*)	Larsen *et al.*, 1975
Susceptibility increased	Chicken (*E. tenella*)	Edds *et al.*, 1973
Susceptibility increased	Chicken (*E. tenella*)	Wyatt *et al.*, 1975
Inflammation increased	Bovine (*Streptococcus* and *Staphylococcus* mastitis)	Brown *et al.*, 1981
Susceptibility increased	Pig (*T. hyodysenteriae*)	Joens *et al.*, 1981
Susceptibility increased	Chicken (Infectious bursal disease)	Bryden, 1983

Table 2 Effects of aflatoxin on cell-mediated immune reactivity

Effect	Animal model	Reference
Thymic aplasia (bursal normal)	Turkeys	Pier *et al.*, 1972
Thymic and bursal hypoplasia	Chicken	Thaxton *et al.*, 1974
Delayed cutaneous hypersensitivity (DCH) depressed	Guinea pig	Pier *et al.*, 1977
DCH depressed (dose-dependent); peripheral blood B and T cell populations normal (bacterial binding reduced)	Guinea pig	McLoughlin, 1982
DCH depressed; PHA skin test normal	Turkey	Schultz, 1983
Adoptive transfer DCH reduced	Guinea pig	McLoughlin, 1982
DCH and graft versus host depressed	Chicken	Giambrone *et al.*, 1978
DCH depressed *in utero*	Pig	Pier *et al.*, chapter 45 in this volume
Lymphoblastogenesis depressed (*in vitro*)	Human	Savel *et al.*, 1970
Lymphoblastogenesis depressed (*in vitro*)	Bovine	Paul *et al.*, 1977
Lymphoblastogenesis depressed (*in vivo*)	Guinea pig	Pier *et al.*, 1977
Phagocytosis impaired	Chicken	Chang and Hamilton, 1979
Phagocytosis impaired	Rabbit	Richard and Thurston, 1975
Leukocyte migration reduced	Guinea pig	Pier *et al.*, 1977

circulating lymphocyte populations in normal and aflatoxin-treated guinea pigs. Adoptive transfer of delayed hypersensitivity by peritoneal exudate cells (PEC) was reduced substantially but not significantly in aflatoxin-treated PEC recipients and when PEC were derived from aflatoxin-treated and sensitized animals. Studies of circulating lymphocytes, using fluorescent antibody (anti-IgG and anti-T cell) methods and bacterial binding, failed to show differences in absolute B cell and T cell populations in peripheral blood of aflatoxin-treated and normal guinea pigs. Thus it was demonstrated that aflatoxin treatment caused impaired delayed cutaneous hypersensitivity responsiveness and adoptive transfer of that hypersensitivity without causing detectable changes in the T and B cell populations of the peripheral blood.

These findings would appear to imply that the CMI suppression might be caused by (1) suppressor cell activity that did not significantly alter the peripheral blood T and B lymphocyte numbers; (2) changes in macrophage function relating to antigen transfer; or (3) differences in lymphokine activity.

The CMI suppressive effects of aflatoxin have been shown to transfer *in utero* into neonatal pigs born to aflatoxin-treated sows (Pier *et al.*, chapter 45 in this volume). Such pigs had suppressed delayed cutaneous hypersensitivity responses. Thus the suppressive effects of aflatoxin on specific CMI responses have been demonstrated *in vitro* in lymphocytes treated directly with aflatoxin (Savel *et al.*, 1970; Paul *et al.*, 1977); *in vivo* by direct action on guinea pigs and turkeys (Pier and Heddleton, 1970; Pier *et al.*, 1977; Schultz *et al.*, 1982)

Table 3 Effects of aflatoxin on humoral elements of immunity

Effect	Model	Reference
Antibody suppression	Mouse (typhoid)	Galikeev *et al.*, 1968
Antibody normal	Turkey (*P. multocida*)	Pier and Heddleton, 1970; Pier *et al.*, 1972
Antibody normal	Turkey (Newcastle DV)	Klinkert *et al.*, 1981
Antibody normal	Chicken (*S. pullorum*)	Adinaryanaiah *et al.*, 1973
Antibody lag	Chicken (Sheep rbc)	Michael *et al.*, 1973
Antibody suppressed	Pig (*T. hyodysenteriae*)	Joens *et al.*, 1981
Antibody normal	Guinea pig (*B. abortus*)	Thurston *et al.*, 1974; Richard *et al.*, 1974
IgG suppression (IgM normal)	Chicken	Tung *et al.*, 1975
IgA and IgG suppression (IgM normal)	Chicken	Giambrone *et al.*, 1978
α and β globulin suppression (α globulin normal)	Pig	Cysewski *et al.*, 1978
Interferon lag	Turkey	Pier *et al.*, 1971
Complement reduced	Guinea pig	Thurston *et al.*, 1972, 1974; Richard *et al.*, 1974
Complement (C_4) reduced	Guinea pig	Thurston and Richard, 1979

and *in vivo* by transplacental migration with an *in utero* effect on neonatal swine.

Many of the above experiments utilized partially purified aflatoxin mixtures that contained assayed quantities of aflatoxins B_1, B_2, G_1 and G_2. Some recent investigations using chemically purified aflatoxin B_1 have produced results that appear to be in partial conflict with these findings (Schultz *et al.*, 1982; Schultz, 1983). Using a broad range of immunological parameters immunosuppression could not be demonstrated when purified aflatoxin B_1 was mixed into feeds of swine and turkey poults at concentrations up to 4000 μg/kg. There appeared to be considerable binding of this aflatoxin in the feed which decreased chemical recovery of toxin, and several biological parameters (e.g. weight gain) were unaffected. When the study was repeated, giving the purified B_1 in capsules at doses of approximately 200 μg/kg dietary aflatoxin, significantly depressed delayed cutaneous hypersensitivity to mycobacterial antigens was demonstrated but cutaneous response to injection of the mitogen PHA was unaltered, as were other immunological parameters.

These later results are of interest from three standpoints: (1) Is aflatoxin B_1 the immunosuppressive factor or does it require one or more of the other toxins to be present? (2) Can purified aflatoxin B_1 be bound in the intestine by feeds as it was in the mixed feed, thus reducing its biological activity? (3) The fact that intradermal injections of PHA produced no differences in cutaneous response while injections of mycobacterial sensitins showed significant suppression of delayed cutaneous hypersensitivity in the aflatoxin recipients suggests problems in antigen processing rather than in lymphocyte populations, lymphokine production or other lymphocyte function. This agrees with previous reports (McLoughlin, 1982; Pier *et al.*, 1977).

Significant direct *in vitro* suppression of lymphoblastogenesis has been reported for both human (Savel *et al.*, 1970) and bovine (Paul *et al.*, 1977) lymphocytes treated with aflatoxin B_1. However a third report (Lafarge-Frayssinet *et al.*, 1980) indicates that aflatoxin B_1 had no appreciable effect on DNA and protein synthesis by lymphocytes of Balb C mice and leukaemic rats but did suppress these functions in hepatic cells. The source and purity of aflatoxin B_1 in the latter report were not stated; but those used on human and bovine lymphocytes (Savel *et al.*, 1970; Paul *et al.*, 1977) were both different from the purified B_1 that was found to be suppressive only in delayed cutaneous hypersensitivity (Schultz, 1983) or the partially purified aflatoxin that was found immunosuppressive in turkeys, guinea pigs and swine (Pier and Heddleton, 1970; Pier *et al.*, 1977; Cysewski *et al.*, 1978). It is possible that different preparations may vary in their biological activity and small impurities may contribute to the apparent CMI-based immunosuppressive activity ascribed to aflatoxin B_1. In a very recent study (Yang, 1983) different purified aflatoxin moieties were used in direct *in vitro* tests on porcine lymphocytes. Aflatoxins B_1, B_{2a}, M_1, M_2 and Q_1 caused dose-dependent suppression of lymphoblastogenetic responses to PHA. Aflatoxin Q_1 was suppressive in 1 μg/ml, B_{2a} at 5 μg/ml while the others required 10–25 μg/ml for suppression. Aflatoxins B_2, G_1, G_{2a} and P_1 had no effect at 25 μg/ml.

TRICHOTHECENES

Immunosuppressive activity of several toxic trichothecenes has been reported and recently reviewed (Pier *et al.*, 1980b). The major immunosuppressive agents in this complex group of mycotoxins appear to be T-2 toxin, diacetoxyscirpenol (DAS) and stachybotryotoxin (SBT). These agents have been associated with alterations in serum proteins and immunoglobulin profiles, reduced antibody formation, thymic aplasia, generally reduced CMI responsiveness but enhanced delayed cutaneous hypersensitivity and impairment of bacterial clearance and acquired immunity. These responses and specific references are included in Table 4. Some interesting comparisons are possible between the immunosuppressive activities of aflatoxins and T-2 toxin.

The immunosuppressive action of both T-2 toxin and the aflatoxins appears to be associated with inhibition of protein synthesis. However, obviously different mechanisms are at play. Aflatoxins inhibit protein synthesis by impeding transcription while T-2 toxin affects initiation of translation. Thus one sees different lesions in the immune mechanisms.

Aflatoxins inhibit complement formation, primarily C_4; diminish serum proteins, primarily albumin, alpha and beta globulins, and only at very high levels do they affect immune globulins (IgA and IgG but not IgM). Aflatoxins cause thymic aplasia but have not been associated with any demonstrable changes in circulating lymphocyte populations. However, the most constantly demonstrated defect in CMI response has been the reduction in delayed cutaneous hypersensitivity reaction. The aflatoxins also inhibit phagocytosis. T-2 toxin has many similar effects; it too reduces complement formation; but primarily C_3. It diminishes serum immune globulin levels (IgA and IgM but not IgG), and diminishes antibody production. Like aflatoxin, T-2 and DAS cause thymic aplasia but a diminution in T suppressor cell activity has been identified as the basis for an increased delayed cutaneous hypersensitivity reaction in T-2-treated mice. Aflatoxins have been associated with diminished lymphoblastogenic responsiveness and graft-versus-host reactivity while T-2 toxin diminishes graft rejection, reduces lymphoblastogenic responsiveness and inhibits neutrophil migration. It seems likely that these two mycotoxins would have additive or synergistic effects if combined within a susceptible animal model. Of particular significance is the fact that both toxins are capable of transfer across the placenta and affecting the foetus *in utero* (Pier *et al.*, chapter 45 in this volume; Lafarge-Frayssinet *et al.*, 1980).

Table 4 Effects of trichothecene mycotoxins on mechanisms of immunity and resistance

Toxin	Effect	Reference
T-2 toxin	Turkey: thymic aplasia (*P. multocida* antibody normal)	Richard *et al.*, 1978
	Mouse: thymic aplasia (anti SRBC and graft versus host reduced)	Rosenstein *et al.*, 1979
	Mouse: reduced clearance (*Listeria*); *rise* in tetanus antitoxin	Wiskott, 1979
	Mouse: enhanced delayed cutaneous hypersensitivity; T suppressor cell reduction	Masuko *et al.*, 1977; Ueno, 1980
	Mouse: transplacental thymic aplasia; reduced lymphoblastogenesis	Lafarge-Frayssinet *et al.*, 1980; Ueno, 1980
	Cattle: antibody (*A. marginale*) normal; IgA, IgM reduced; IgG normal; C_3 reduced; lymphoblastogenesis and leucocytic migration reduced	Osweiler *et al.*, 1981; Buening *et al.*, 1982; Mann *et al.*, 1982
	Mouse: antibody (SRBC) reversibly reduced; thymic aplasia; reduced graft versus host	Rosenstein *et al.*, 1979
	Mouse: enhanced susceptibility to HSV-1	Friend *et al.*, Chapter 40, this volume
DAS	Mouse: antibody (tetanus antitoxin) reduced; clearance (*Listeria*) reduced; immune response (*Pasteurella*) reduced	Wiskott, 1979
	Mouse: thymic aplasia; antibody (SRBC) reduced; reduced graft versus host	Rosenstein *et al.*, 1979
	Mouse, rat: *in vitro* lymphocyte suppression	Robbana Barnat *et al.*, 1982
Stachybotryotoxin	Mouse: antibody (SRBC) reduced	Danko and Krasznia, 1976
	Guinea pig: delayed cutaneous hypersensitivity; (BCG) reduced	Danko and Krasznia, 1976

OCHRATOXIN

There is growing evidence of the immunosuppressive activity of ochratoxin A. Ochratoxin affects protein formation through the inhibition of phenylalanyl tRNA synthetase. Suspicion of the potent immunosuppressive potential of this mycotoxin was raised when a generalized necrosis of lymph nodes was described in dogs and swine (Szczech *et al.*, 1973a,b). This effect was largely invested in gut-associated lymph tissues. Lymphocytopenia (Chang *et al.*, 1979; Gupta *et al.*, 1979), inhibition of macrophage migration (Klinkert *et al.*, 1981) and reductions in immune globulins (IgM and IgG) (Creppy *et al.*, 1982) and antibody production (Haubeck *et al.*, 1981; Prior and Sisodia, 1982) were expected sequelae of that lesion. Ochratoxin also is capable of crossing the placenta and exerting biological effects on the unborn foetus (Munro *et al.*, 1973; Szczech and Hood, 1981). The effects of ochratoxin A on the immune system and immunogenesis are summarized in Table 5. It is interesting that ochratoxin appears to have a spectrum of immunosuppressive actions much different from that of aflatoxin, and to some extent T-2 toxin. The major effects appear to be related to immune globulin synthesis and antibody production. Complement production and cell-mediated immune mechanisms appear to be unaltered.

Table 5 Effects of ochratoxin A on mechanisms of immunity and resistance

Animal	Effect	Reference
Dog, pig	Lymphoid necrosis	Szczech *et al.*, 1983a,b
Mouse	Antibody (*B. abortus*) reduced; (dose- and route-dependent)	Prior and Sisodia, 1979, 1982
Mouse	Lymphocytosis, neutrophilic and monocytic leukopenia	Gupta *et al.*, 1979
Mouse	IgM and IgG suppression (SRBC)	Creppy *et al.*, 1982; Haubeck *et al.*, 1981
Mouse	Macrophage migration inhibition	Klinkert *et al.*, 1981
Rat	*In vitro* lymphocyte suppression	Robhana Barnat *et al.*, 1982
Chicken	Leukopenia (lymphocytopenia and monocytopenia); heterophil phagocytosis suppression	Chang *et al.*, 1979; Chang and Hamilton, 1980
Turkey	Thymic hypoplasia; lymphocytopenia	Chang *et al.*, 1981

DISCUSSION

The evidence for, and the mechanisms of, mycotoxic immunosuppression grow clearer with each added piece of information. It is well established that the aflatoxins, ochratoxin A, and at least three of the toxic trichothecenes (T-2, DAS and SBT) are capable and sometimes potent modifiers of immune reactions. Because these are natural products of microbial metabolism that occur in the food supply they present an especially insidious threat to the health of both animals and man. Recent information shows that all three of these major mycotoxins (aflatoxin B_1, ochratoxin A and T-2 toxin) can pass placental barriers and affect the unborn foetus. Once this foetus has been removed from the life support system of the dam it can be expected to have altered immune competence and this in fact has been demonstrated with aflatoxin in neonatal swine. Young animals are generally more susceptible to mycotoxic actions than are the mature, and the demands on their immune systems are greater since they meet nearly all challenges *de novo*. Further they may be affected by vertically transferred mycotoxins that were insufficient to cause noticeable clinical illness in their dams. This is a particularly insidious aspect of the immunosuppressive actions of mycotoxins. Because the effects may be exerted by levels of toxin below those causing a clearly defined clinical toxicosis, the clinically identifiable disease is usually the infectious process to which the mycotoxin predisposed the host. Thus the true underlying cause may go undetected.

While we know much about the mechanisms of some actions of these mycotoxins, we have considerably more to learn about their exact sites of action and effects on the very complex immune systems. Some of their actions appear to be reversible; others appear to remain over extended periods of time. In the case of ochratoxin A, many of the effects can be prevented by phenylalanine supplementation (Kleinkert *et al.*, 1982; Haubeck *et al.*, 1981). Until we learn additional means of recognition, prevention and control, the immunosuppressive effects of these mycotoxins promise to exert continuing economically and medically important effects on the health of both animals and man.

REFERENCES

Adinaryanaiah, G. L., Nanjiah, R. D., and Keshavamurthy, B. S. (1973). *Indian Vet. J.*, **50**, 297–303.

Boonchuvit, B., and Hamilton, P. B. (1975). *Poult. Sci.*, **54**, 1567–1573.

Brown, R. W., Pier, A. C., Richard, J. L., and Krogstad, R. E. (1981). *Am. J. Vet. Res.*, **42**, 927–933.

Bryden, W.(1983). *Abstr. International Mycotoxin Symposium*, Sydney, Australia.

Buening, G. M., Mann, D. D., Hook, B., and Osweiler, G. D. (1982). *Vet. Immunol. Immunopathol.*, **3**, 411–417.

Chang, C. F., and Hamilton, P. B. (1980). *Appl. Environ. Micro.*, **39**, 572–575.
Chang, D. V., and Hamilton, P. B. (1979). *Poult. Sci.*, **58**, 562–566.
Chang, C. F., Doerr, J. A., and Hamilton, P. .B. (1981). *Poult. Sci.*, **60**, 114–119.
Chang, C. F., Huff, W. E., and Hamilton, P. B. (1979). *Poult. Sci.*, **58**, 555–558.
Clifford, J. .L., and Rees, K. R. (1966). *Nature*, **209**, 312–313.
Creppy, E. E., Lorkowski, G., Roschenthaler, R., and Dirheimer, G. (1982). *Proc. V IUPAC Symposium on Mycotoxins and Phycotoxins*, Vienna, pp. 289–292.
Cysewski, S. J., Wood, R. L., Pier, A. C., and Baetz, A. L. (1978). *Am. J. Vet. Res.*, **39**, 445–448.
Danko, G., and Krasznia, G. (1976). *Magyar Allatorvosok Lapja*, **31**, 233–244.
Dreismann, H., and Roschenthaler, R. (1982). *Proc. V IUPAC Symposium on Mycotoxins and Phycotoxins*, Vienna, pp. 293–296.
Edds, G. T., Nair, K. P. C., and Simpson, C. F. (1973). *Am. J. Vet. Res.*, **34**, 819–826.
FDA (Food and Drug Administration) (1980). Proc. IV Science Symposium: *Inadvertent Modifications of the Immune Response* (ed. J. Asher). FDA/OHA.
Galikeev, K. H. L., Raipov, O. R., and Manyasheva, R. A. (1968). *Byul. Eksper. Biol. Med. (USSR)*, **65**, 88–90.
Giambrone, J. J., Ewert, D. L., Wyatt, R. D., and Edison, C. S. (1978). *Am. J. Vet. Res.*, **39**, 305–308.
Gupta, M., Bandopadhyay, S., Paul, B., and Majunder, S. K. (1979). *Toxicology*, **14**, 95–98.
Hamilton, P. B., and Harris, J. R. (1971). *Poult. Sci.*, **50**, 906–912.
Haubeck, H. D., Lorkowski, G., Kolsh, E., and Roschenthaler, R. (1981). *Appl. Environ. Micro.*, **41**, 1040–1042.
Hesseltine, C. W. (1974). *Mycopathologia*, **53**, 141–153.
Hsieh, D. P. H., Wong, Z. A., Wong, J. J., Michas, C., and Ruebner, B. H. (1977). In *Mycotoxins in Human and Animal Health* (eds J. Rodericks, C. Hesseltine and M. Mehlman), pp. 37–50. Pathotox Publ., Park Forest South, IL.
Joens, L. A., Pier, A. C., and Cutlip, R. (1981). *Am. J. Vet. Res.*, **42**, 1170–1172.
Klinkert, W., Lorkowski, G., Creppy, E. E., Dirheimer, G., and Roschenthaler, R. (1981). *Toxicol. Eur. Res.*, **3**, 185–189.
Lafarge-Frayssinet, C., Chakor, K., Lafont, P., and Frayssinet, C. (1980). *Proc. 4th International IUPAC*. Chem. Rund. 35: Abstr. 704.
Lafarge-Frayssinet, C., Lespinats, G., Lafont, P., Loisillier, F., Mousset, S., Rosenstein, Y. and Frayssinet, C. (1979). *Proc. Soc. Exp. Biol. Med.*, **160**, 302–311.
Larsen, A. B., Cysewski, S. J., and Miller, J. M. (1975). *Am. J. Vet. Res.*, **36**, 1545–1547.
Mann, D. D., Buening, G. M., Hook, B. S., and Osweiler, G. D. (1982). Infect. Immun., **36**, 1249–1252.
Masuko, H., Ueno, Y., Otokawa, M., and Yaginuma, K. (1977). *Jap. J. Med. Sci.*, **30**, 159–163.
McLoughlin, M. E. (1982). Ph.D. thesis, Iowa State University, Ames, IA.
McLaughlin, C. S., Vaughan, M. H., Campbell, I. M., Wei, C. M., Stafford, M. E., and Hansen, B. S. (1977). In *Mycotoxins in Human and Animal Health*. (eds J. Rodericks, C. Hesseltine and M. Mehlman), pp. 263–264. Pathotox Publ., Park Forest South, IL.
Michael, G. Y., Thaxton, P., and Hamilton, P. B. (1973). *Poult. Sci.*, **52**, 1206–1207.
Munro, I. C., Scott, P. M., Moodie, C. A., and Wilkes, R. F. (1973). *J. Am. Vet. Med. Assoc.*, **163**, 1269–1273.
Osweiler, G. D., Hook, B. S., Mann, D. D., Buening, G. M., and Rottinghaus,

G. D. (1981). *Proc. 85th Ann. Meet. U.S. An. Health Assoc.*, **36**, pp. 214–231.
Paul, P. S., Johnson, D. W., Mirocha, C. J., Soper, F. F., Thoen, C. O., Mucscoplat, C. C., and Weber, A. F. (1977). *Am. J. Vet. Res.*, **38**, 2033–2035.
Pier, A. C. (1981). *Adv. Vet. Sci. Comp. Med.*, **25**, 185–243.
Pier, A. C., and Heddleston, K. L. (1970). *Avian Dis.*, **14**, 797–809.
Pier, A. C., Fichtner, R. E., and Cysewski, S. J. (1977). *Ann. Nutr. Aliment.*, **31**, 781–787.
Pier, A. C., Heddleston, K. L., Boney, W. A., and Lukert, P. D. (1971). *Proc. XIX Cong. Mundial de Med. Vet. Zootech.*, **1**, 216–219.
Pier, A. C., Heddleston, K. L., Cysewski, S. J., and Patterson, J. M. (1972). *Avian Dis.*, **16**, 381–387.
Pier, A. C., Richard, J. L., and Thurston, J. R. (1979). In *Interactions of Mycotoxins in Animal Production*, pp. 56–66. Nat. Acad. Sci., Washington, DC.
Pier, A. C., Richard, J. L., and Thurston, J. R. (1980a). In *Medical Mycology. Zbl. Bakt.* Suppl. 8, G. Fischer Verlag, Stuttgart, 301–309.
Pier, A. C., Richard, J. L., and Thurston, H. R. (1980b). In *Natural Toxins*. (eds D. Eaker and T. Wadstrom), pp. 691–699. Pergamon, Oxford.
Prior, M. G., and Sisodia, C. S. (1979). 6th Int. Symp. on Animal, Plant and Microbiol Toxins, *Toxicon*, **17**, 146.
Prior, M. G., and Sisodia, C. S. (1982). *Can. J. Comp. Med.*, **46**, 91–96.
Richard, J. L., and Thurston, J. R. (1975). *Appl. Microbiol.*, **30**, 44–47.
Richard, J. L., Cysewski, S. J., Pier, A. C., and Booth, G. D. (1978). *Am. J. Vet. Res.*, **39**, 1674–1679.
Richard, J. L., Pier, A. C., Cysewski, S. J., and Graham, C. K. (1973). *Avian Dis.*, **17**, 111–121.
Richard, J. L., Thurston, J. R., Deyoe, B. L., and Booth, G. D. (1974). *Appl. Microbiol.*, **29**, 27–29.
Richard, J. L., Thurston, J. R., and Pier, A. C. (1975). In *Microbiology–1975*, pp. 388–396.
Richard, J. L., Thurston, J. R., and Pier, A. C. (1978). In *Toxins: Animal, Plant and Microbial.*, pp. 801–817. Pergamon, Oxford.
Robbana Barnat, S., Lafarge-Frayssinet, C., and Frayssinet, C. (1982). *Proc. V IUPAC Symposium on Mycotoxins and Phycotoxins*, Vienna, pp. 269–272.
Rosenstein, Y., Lafarge-Frayssinet, C., Lespinats, G., Loisillier, R., Lafont, P., and Frayssinet, C. (1979). *Immunology*, **36**, 111–117.
Savel, H., Forsyth, B., Schaeffer, W., and Cardella, T. (1970). *Proc. Soc. Exp. Biol. Med.*, **134**, 1112–1115.
Schultz, R. D. (1983). Personal communication. Madison, WI.
Schultz, R D., Yang, W. C., Pangangala, V. S., and Diener, U.L. (1982). Abstr. 158, 63 Ann. Meet. Conf. Res. Workers An. Dis., Chicago, p. 29.
Smith, J. W., Prince, W. R., and Hamilton, P. B. (1969). *Appl. Microbiol.*, **18**, 946–947.
Szczech, G. M., and Hood, R. D. (1981). *Tox. Appl. Pharm.*, **57**, 127–137.
Szczech, G. M., Carlton, W. W., and Tuite, J. (1973a). *Vet. Path.*, **10**, 219–231.
Szczech, G. M., Carlton, W. W., and Tuite, J. (1973b). *Vet. Path.*, **10**, 347–364.
Thaxton, J. P., Tung, H. T., and Hamilton, P. B. (1974). *Poult. Sci.*, **53**, 721–725.
Thurston, J. R., and Richard, J. L. (1979). *Am. J. Vet. Res.*, **40**, 1206–1208.
Thurston, J. R., Deyoe, B. L., Baetz, A. L., Richard, J. L., and Booth, G. D. (1974). *Am. J. Vet. Res.*, **35**, 1097–1110.
Thurston, J. R., Richard, J. L., Cysewski, S. J., Pier, A. C., and Graham, C. K. (1972). *Proc. Soc. Exp. Biol. Med.*, **138**, 300–303.

Tung, H. T., Wyatt, R. D., Thaxton, P., and Hamilton, P. B. (1975). *Tox. Appl. Pharmacol.*, **34**, 320–326.

Ueno, Y. (1980). *Adv. Nutr. Sci.*, **3**, 301–353.

Wiskott, U. (1979). Doctoral dissertation, Ludwig-Maximilian, University of Munich.

Wyatt, R. D., Huff, M. D., and Page, R. K. (1975). *Avian Dis.*, **19**, 730–740.

Wyllie, T. D., and Morehouse, L. D. (1977). *Mycotoxic Fungi, Mycotoxins and Mycotoxicoses: An Encyclopedic Handbook*, Vols 1–3. Dekker, New York.

Yang, W. C. (1983). Ph.D thesis, Auburn University, Auburn, Alabama.

Trichothecenes and Other Mycotoxins
Edited by J. Lacey

CHAPTER 47

An assessment of liver cancer risk posed by aflatoxin M_1 in the Western world

Dennis P.H. Hsieh

INTRODUCTION

A shocking episode of food contamination occurred recently in the U.S.A. where commercial milk was labelled as 'poison' by panicky consumers because samples were found to contain aflatoxin M_1 (AFM) at levels greater than 0.5 μg/kg, the action level set forth by the U.S. Food and Drug Administration (FDA, 1977). In certain regions hundreds of gallons of the contaminated milk were dumped (Kobbe *et al.*, 1979). Valuable food was thus destroyed and agricultural fields were polluted by a flood of organic matter.

AFM is a hydroxylated metabolite of aflatoxin B_1 (AFB) which is the most potent hepatocarcinogen known in the rat and rainbow trout (Wogan, 1973). When feed contaminated with AFB is ingested by dairy cattle, up to 1% will appear in the milk as AFM (Rodricks and Stoloff, 1977). Concern arose regarding the carcinogenicity of AFM because its acute toxicity is similar to that of AFB in several assay systems (Holzapfel *et al.*, 1966; Purchase, 1967; Pong and Wogan, 1971), and limited studies indicated that AFM is genotoxic and carcinogenic though less potent than AFB (Wong and Hsieh, 1976; Green *et al.*, 1982; Sinnhuber *et al.*, 1970; Lutz, 1979; Lutz *et al.*, 1980). Until recently, however, the carcinogenicity of AFM has not been well established due to the very limited supply of pure AFM for animal testing. It is not feasible to obtain AFM from contaminated milk because the concentration is very low and the purification procedure tedious. Therefore, despite all concerns, the cancer risk due to ingestion of AFM in milk at 0.5 μg/kg has remained unclear.

CARCINOGENICITY OF AFM IN THE FISCHER RAT

We have overcome the supply problem by producing sufficient quantities of AFM by a fermentation method which uses a rice culture of *Aspergillus flavus*

(Hsieh *et al.*, 1983), a fungal isolate known to produce relatively large amounts of AFM (Stubblefield *et al.*, 1970). The AFM extracted from the fungal cultures was separated from AFB, which was present in an amount about 30 times greater than that of AFM, with normal-phase column chromatography. The final purification was achieved by reverse-phase column chromatography. In this system AFM is eluted ahead of AFB to ensure a product completely free of AFB contamination.

The AFM produced from fungal cultures was confirmed by its co-chromatography with a reference compound produced by hydroxylation of AFB using a rat liver mixed function oxygenase system (Rice and Hsieh, 1982) and by its ultraviolet, mass, and proton nuclear magnetic resonance spectra. The laevorotatory form of the compound as it is found in milk was verified by its optical rotation. Using the Ames *Salmonella*/microsome mutagenicity test system (Ames *et al.*, 1975), the AFM so produced was 1/300 as mutagenic as AFB.

A long-term feeding experiment was conducted to determine the carcinogenicity of AFM in the male Fischer rat, a sensitive animal model for the study of the carcinogenicity of AFB (Wogan, 1973). AFM was mixed into a semi-synthetic agar-based diet (Wogan and Newberne, 1967) at 0, 0.5, 5, and 50 μg/kg and fed to groups containing 62 animals for 21 months. During this period, AFB at 50 and 100 μg/kg was fed to groups containing 42 animals as positive controls, and 23 animals were maintained on regular laboratory rodent chow as a negative control group. Throughout the experiment the body weight, feed consumption, and general behaviour of the animals was continuously monitored. At appropriate time intervals, animals from each group were sacrificed to examine pathological changes related to the development of hepatocellular carcinomas. At the end of 17 months, multiple hepatic tumours were visually evident in 20 out of 20 animals (100% incidence) taken from the AFB 50 μg/kg group, whereas no grossly visible tumours were found in any of the animals taken from the AFM 50 μg/kg group (0% incidence) or other AFM and control groups. However, preneoplastic signs indicated by histochemically altered foci were found in all the AFB animals and some of the AFM 50 μg/kg animals. Based on the difference in the incidence of gross tumours and the degree of disease development, it is concluded that AFM is two orders of magnitude less potent than AFB as a hepatocarcinogen.

LEVELS OF EXPOSURE TO AFM

If AFM is two orders of magnitude less potent than AFB, then in order for the cancer risk posed by AFM to be as significant as the risk posed by AFB, a population must ingest AFM at a level which is hundreds of times greater than the exposure level of AFB. This is not the case, as illustrated by the

Table 1 Level of exposure to AFB_1 in the U.S.A.

Major source	Average concentration (μg/kg)	Rate of consumption (g/capita/day)	Daily intake of AFB_1 (ng/kg body weight/day)
Peanut products	0.19	7.6	0.03
Corn	9.5	135	25.7

Data from USDA (1981) and Stoloff (1977).
Assuming per capita body weight = 50 kg.

Table 2 Level of exposure to AFM_1 in the U.S.A.

Major source	Milk and dairy products
Consumption rate	242 g/capita/day (milk)
Frequency of contamination	63%
AFM levels	0.1–0.7 μg/kg (Mean = 0.27 μg/kg)
Daily intake	Average = 0.82 ng/kg BW/day Worst case = 3.4 ng/kg BW/day

Data from USDA (1981) and Stoloff (1980).
Assuming per capita body weight = 50 kg.

comparison of the levels of exposure to AFB and AFM in the U.S., summarized in Tables 1 and 2.

Food surveys indicated that the major sources of AFB in food in the U.S.A. are peanut products and corn (Stoloff, 1977). The average concentrations of AFB were 0.19 μg/kg in peanut products and 9.5 μg/kg in corn. The average concentrations were calculated by multiplying the average concentration of the contaminated samples by the percentage of samples that were contaminated. The average per capita consumption rates are 7.6 g/capita/day for peanut products and 135 g/capita/day for maize, based on the USDA statistics for 1981 (USDA, 1981a). Assuming that per capita body weight is 50 kg, the daily intake of AFB was calculated to be 0.03 ng/kg body weight (BW)/day from peanut products and 25.7 ng/kg BW/day from maize. It should be noted that though these are only rough estimates, nonetheless they provide a base for comparison.

Similarly, the major food sources of AFM ingested by humans are milk and dairy products. The average per capita consumption rate of milk in the

U.S.A. is 242 g/capita/day (USDA, 1981b). The frequency of contaminated milk samples is 63% (Stoloff, 1980). The AFM concentrations in the samples ranged from 0.1 to 0.7 μg/kg. Again, assuming that per capita body weight is 50 kg, the worst case of daily intake of AFM, in which all (100%) the milk samples were assumed to contain 0.7μg/kg of AFM, was calculated to be 3.4 ng/kg BW/day ($242 \times 0.7 \div 50$).

CANCER RISK POSED BY AFM AND AFB

Using the average daily intake of AFB and the worst-case intake of AFM for comparison, the level of exposure to AFB is calculated to be at least 5.9 times greater than the level of exposure to AFM. Considering that AFB is two orders of magnitude more potent than AFM and is also being consumed at a rate which is at least 5.9 times greater, it is obvious that the cancer risk posed by AFM is only a very small fraction of the risk posed by AFB.

The cancer risk posed by AFB can be assessed by using the epidemiological data obtained from 10 regions in South-east Asia and Africa for the correlation between liver cancer and exposure to aflatoxins. The data on the male populations in these regions were compiled and analysed by Carlborg (1979). A linear dose–response relationship was proposed. The dose was expressed as daily intake of aflatoxin in ng/kg BW/day. Here 'aflatoxin' can be represented by AFB because it is the principal member of the aflatoxin family and is also the most potent. The response was expressed as lifetime risk; that is the number of deaths per 100,000 due to liver cancer in the life span of a population. Based on the linear dose–response relationship, the following risk values were determined:

Background risk of liver cancer	121 deaths/100,000
Increment in risk due to exposure to 1 ng/kg BW/day	10 deaths/100,000

These values, however, were not consistent with the one calculated for the male population in the U.S.A., suggesting a difference in the susceptibility to the carcinogenic effect of aflatoxin between populations. As shown in Table 3, the 1977 statistics (McKay, 1982) indicated that there were a total of 3418 deaths due to bile passage and liver cancer in U.S. males for that year. The total U.S. male population was 105.6 million (U.S. Bureau of Census, 1982), giving an annual death rate due to liver cancer of 3.24/100,000. This rate is translated to a lifetime risk of 227/100,000 due to liver cancer, assuming the life span of U.S. males to be 70 years. Based on the risk values shown above, this level of lifetime risk (227/100,000) corresponds to an exposure level of 10.6 ng/kg BW/day. This expected exposure level is less than one-half that of the estimated level of exposure to AFB in the U.S.A. (25.7 ng/kg BW/day)

Table 3 Bile passage and liver cancer in U.S. males (1977)

Total deaths due to disease	3418
Total male population	105.6 million
Death rate due to disease	3.24/100,000
Life-time risk	227/100,000
Corresponding exposure level from epidemiological data	10.6 ng/kg BW/day
Estimated exposure level from food survey	25.7 ng/kg BW/day

Data from McKay, 1982 and U.S. Bureau of Census, 1982.

elaborated in the preceding section. The difference in the expected and the estimated exposure levels suggests that U.S. males are twice as resistant to the hepatocarcinogenicity of AFB than the males in those regions where the epidemiological studies were conducted, assuming a cause–effect relationship.

FACTORS AFFECTING SUSCEPTIBILITY TO LIVER CANCER

The difference in the response to AFB exposure between the U.S. males and the males in South-east Asia and Africa may be attributable to the following possibilities: (a) inaccuracy in the estimation of exposure levels; (b) inaccuracy in the epidemiological data or in the dose–response relationship derived from the data; (c) differences in lifestyle, which includes environmental and nutritional conditions; and (d) factors other than aflatoxin being involved in the aetiology of human primary liver cancer. It has been well documented that infection by hepatitis B virus is highly correlated with the incidence of liver cancer (Lutwick, 1979).

Despite all these arguments, the risk of liver cancer in the U.S.A. is rather small, compared to the risk of other forms of cancer (McKay, 1982). The risk posed by AFM in milk should be even smaller in the U.S.A., suggesting that there was no health-related reason that warranted destruction of the large quantities of raw cow milk mentioned in the beginning of this chapter.

EXPOSURE TO AFM IN OTHER WESTERN COUNTRIES

The situation with AFM in other Western countries, where milk is a significant diet constituent, is probably very similar to that of the U.S.A. This is due to the similarity in the social, economic and nutritional conditions of these countries and the comparable levels of exposure to AFM. The consumption rates of milk and the levels of AFM found in milk in some

Table 4 Consumption of milk in 1980

Country	Consumption g/capita/day	%
New Zealand	373	118
Canada	323	102
U.S.A.	316	100
Denmark	301	95
France	282	89
Great Britain	266	84

From statistics year books (1981).

Western countries are shown in Tables 4 and 5. The consumption rates of milk range from New Zealand being 18% greater, to Great Britain being 16% less than that of the U.S.A., all values being in the same order of magnitude. The geometric means of the AFM content in the contaminated milk samples in these countries are also in the same order of magnitude, with the highest frequency of contamination found in the U.S.A. Therefore, it is reasonable to conclude that the cancer risk posed by AFM in these Western countries is probably as small as the risk in the U.S.A.

It should be noted that the risk assessment presented above is based on data obtained from either adult animals or adult humans. How large a fraction of the observed risk is attributable to ingestion of AFM by infant human populations needs to be evaluated separately, because milk represents the major constituent of their diet and infant animals have been found to be considerably more susceptible to AFB carcinogenicity than adult animals

Table 5 Levels of AFM in market milk

Country	Year	No. of samples examined	Percentage positive	AFM_1 level (μg/kg) Range	Mean[a]
South Africa	1968	21	24	0.1–0.2	0.14
U.S.A.	1976	302	63	0.1–0.7	0.27
Belgium/Netherlands	1975	68	62	0.1–0.5	0.22
West Germany	1976	419	19	0.1–0.5	0.22

[a] Geometric mean. Data taken from Stoloff (1980).

(Vesselinovitch *et al.*, 1972). Also, the contribution of AFB and AFM to the risk of cancer in organs other than the liver should receive more attention in view of the potent animal carcinogenicity of AFB and the surprisingly low risk of liver cancer in the Western world.

ACKNOWLEDGEMENT

The author thanks Linda Beltran for assistance in the preparation of this chapter. The author's research in this area is supported in part by USPHS Grant CA 27426, the U.S. Western Regional Research Project W122, and the Dairy Council of California.

REFERENCES

Ames, B. N., McCann, J., and Yamasaki, E. (1975). Methods for detecting carcinogens and mutagens with the Salmonella/mammalian-microsome mutagenicity test, *Mutat. Res.*, **31**, 347–364.

Carlborg, F. W. (1979). Cancer, mathematical models and aflatoxin, *Food Cosmet. Toxicol.*, **17**, 159–166.

FDA (Food and Drug Administration) (1977). Criteria and procedures for evaluating assays for carcinogenic residues in edible products of animals. *Fed. Registrar*, **42**, 61630; Guideline 7406.06.

Green, C. E., Rice, D. W., Hsieh, D. P. H., and Byard, J. L. (1982). The comparative metabolism and toxic potency of aflatoxin B_1 and aflatoxin M_1 in primary cultures of adult rat hepatocytes, *Food Chem. Toxicol.*, **20**, 53–60.

Holzapfel, C. W., Steyn, P. S., and Purchase, I. F. H. (1966). Isolation and structure of aflatoxins M_1 and M_2, *Tetrahedron Lett.*, **25**, 2799–2803.

Hsieh, D. P. H., Beltran, L. M., and Rice, D. W. (1983). The production of aflatoxin M_1 for toxicologic studies, *Toxicologist*, **3**, 152.

Kobbe, B., Hsieh, D. P. H., and Dunkley, W. L. (1979). Aflatoxin in milk—What is it, how does it get there? *The Dairyman*, July, A6–A8.

Lutz, W. K. (1979). *In vivo* covalent binding of organic chemicals to DNA as a quantitative indicator in the process of chemical carcinogenesis, *Mutat. Res.*, **65**, 289–356.

Lutz, W. K., Jaggi, W., Leuthy, J., Sagelsdorff,P., and Schlatter, C. (1980). *In vivo* covalent binding of aflatoxin B_1 and aflatoxin M_1 to liver DNA of rat, mouse and pig, *Chem-Biol. Interact.*, **32**, 249–256. (1980).

Lutwick, L. (1979). Relation between aflatoxin, hepatitis-B virus and hepatocellular carcinoma, *Lancet*, **1**, 755–757.

McKay, F. W. (1982). Biliary passages and liver. In *Cancer Mortality in the United States, 1950–1977* (eds M. R. Hanson and R. W. Miller) National Cancer Institute Monograph No 59, p. 116.

Pong, R. S., and Wogan, G. N. (1971). Toxicity and biochemical and fine structural effects of synthetic aflatoxins M_1 and B_1 in rat liver, *J. Natl. Cancer Inst.*, **47**, 585–592.

Purchase, I. F. H. (1967). Acute toxicity of aflatoxins M_1 and M_2 in one-day-old ducklings, *Food Cosmet. Toxicol.*, **5**, 339–342.

Rice, D. W., and Hsieh, D. P. H. (1982). Aflatoxin M_1: *in vitro* preparation and comparative *in vitro* metabolism versus aflatoxin B_1 in the rat and mouse, *Res. Commun. Chem. Path. Pharmacol.*, **35**, 467–490.

Rodricks, J. V., and Stoloff, L. (1977). Aflatoxin residues from contaminated feed in edible tissues of food producing animals. In *Mycotoxins: In Human and Animal Health*, p. 67 (eds J. V. Rodrick, C. W. Hesseltine and M. A. Mehlman) Pathotox, Illinois.

Sinnhuber, R. D., Lee, D. J., Wales, J. H., Landers, M. K., and Keyl, A. C. (1970). Aflatoxin M_1: a potent liver carcinogen in rainbow trout, *Fed. Proc.*, **29**, A568.

Stoloff, L. (1977). Aflatoxins—an overview. In *Mycotoxins: In Human and Animal Health* (eds J. V. Rodrick, C. W. Hesseltine and M. A. Mehlman), p. 7. Pathotox, Illinois.

Stoloff, L. (1980). Aflatoxin M_1 in perspective, *J. Food Prot.*, **43**, 226–230.

Stubblefield, R. D., Shannon, G. M., and Shotwell, O. L. (1970). Aflatoxins M_1 and M_2: Preparation and purification, *J. Am. Oil Chem. Soc.*, **47**, 389–390.

U.S. Bureau of the Census (1982). *Statistical Abstract of the United States: 1982–83* (103rd edition), Sec. 1, Population, No. 30, Total population, by age and sex, 1960 to 1982, p. 31. Washington, DC.

USDA (1981a). Food consumption, prices and expenditures, 1960–1980. In *Economic Research Service*, Statistical Bulletin No. 672, pp. 62 and 75.

USDA (1981b). Food consumption, prices and expenditures, 1960–1980. In *Economic Research Service*, Statistical Bulletin No. 672, p. 9.

Vesselinovitch, S. D., Mihailovich, N., Wogan, C. N., Lombard, L. S., and Rao, K. V. N. (1972). Aflatoxin B_1, a hepatocarcinogen in the infant mouse, *Cancer Res.*, **32**, 2289–2291.

Wogan, G. N., (1973). Aflatoxin carcinogenesis. In *Methods in Cancer Research* (ed. H. Busch), Vol. 7, p. 309. Academic Press, New York.

Wogan, G. N., and Newberne, P. M. (1967). Dose–response characteristics of aflatoxin B_1 carcinogenesis in the rat, *Cancer Res.*, **27**, 2370–2376.

Wong, J. J., and Hsieh, D. P. H.(1976). Mutagenicity of aflatoxins related to their metabolism and carcinogenic potential, *Proc. Natl. Acad. Sci.*, **73**, 2241–2244.

Trichothecenes and Other Mycotoxins
Edited by J. Lacey

CHAPTER 48

The toxicity of cyclopiazonic acid

Joe W. Dorner, Richard J. Cole and Larry G. Lomax

INTRODUCTION

Cyclopiazonic acid (CPA) is a toxic indole-tetramic acid that was first isolated from cultures of *Penicillium cyclopium* Westling (Holzapfel, 1968), later renamed *P. griseofulvum* Dierckx (de Jesus *et al.*, 1981). Subsequently, CPA has been reported to be produced by *P. patulum*, *P. viridicatum*, *P. puberulum*, *P. crustosum*, *P. camembertii*, *Aspergillus flavus*, *A. versicolor*, *A. oryzae*, and *A. tamarii* (Dorner, 1983; Le Bars, 1979a; Leistner and Pitt, 1977; Luke *et al.*, 1977; Ohmomo *et al.*, 1973; Orth, 1977; Still *et al.*, 1978). Many of the *Penicillium* species that produce CPA were isolated from meats such as ham, sausage and frankfurters (Leistner and Pitt, 1977). Other foods and feeds from which CPA-producing fungi have been isolated include cheese, maize, guinea pig ration, turkey ration, mixed feed, walnuts, and peanuts (Gallagher *et al.*, 1978; Le Bars, 1979a,b; Still *et al.*, 1978). Natural occurrence of the toxin has been reported in cheese contaminated with *P. camembertii* (Le Bars, 1979b) and in maize (Gallagher *et al.*, 1978) and peanuts (Lansden and Davidson, 1983) contaminated with *A. flavus*. Gallagher *et al.* (1978) found that 14 of 54 *A. flavus* isolates (26%) produced both aflatoxins and CPA, four (7%) produced aflatoxins only, and 14 (26%) produced CPA only. It is interesting taxonomically that a survey of 47 isolates of the closely related fungus, *A. parasiticus*, showed that none produced the toxin (Dorner *et al.*, 1984). Nevertheless, the ubiquitous nature of fungi that do produce CPA certainly warrants thorough study of the toxicity of this metabolite.

Toxicological studies with CPA have been conducted in several animals including rats, chickens and pigs. This chapter reviews and summarizes the findings of these studies.

RATS

The first toxicological study of CPA was reported in 1971 and involved oral (p.o.) and intraperitoneal (i.p.) administration to rats (Purchase, 1971). CPA was dissolved in 1 N sodium bicarbonate and administered p.o. to male and female Wistar-derived rats and i.p. to male rats. The LD_{50} value for rats dosed i.p. was estimated as 2.3 mg/kg. Rats receiving more than 8 mg/kg died within 2 h. They were disinclined to move, became ataxic and resented being handled. They were immobile in 10–30 min and within the next 90 min died in extensor spasm with cyanotic mucous membranes. The groups receiving lower doses recovered and survived in 10 days. A different syndrome was observed in rats receiving oral administration of CPA. Mortalities occurred up to 6 days after dosing and animals were prostrate for several hours before death. No muscle spasm was observed. The LD_{50} values were 36 mg/kg in males and 63 mg/kg in females.

The difference in toxicity and clinical signs produced by CPA when administered i.p. or p.o. was attributed to the acidic nature of the toxin. Since CPA behaves as a monobasic acid and is insoluble in aqueous solvents with pH below 7.0, oral administration into the acidic stomach rendered the toxin insoluble, with absorption occurring only in the more alkaline small intestine. Thus, oral administration results in a delayed and less dramatic effect when compared to i.p. administratioin which yields rapid absorption producing nervousness and earlier death.

Histopathological examination of organs was conducted on rats receiving oral doses of CPA. Rats that died within 36 h of dosing were all females, whereas rats dying 4–6 days after dosing were males. High doses produced single-cell necrosis in many organs, and at lower doses the major effect was on cells lining ducts. In females there was focal or single-cell necrosis of hepatocytes. Cells lining the bile ducts were enlarged with large nuclei containing a thin marginal band of chromatin. At higher doses the bile duct cells were enlarged to the point of occlusion of the lumen. Degenerative changes in the renal medulla resulted in pyknosis and enlargement of the nucleus with margination of the chromatin. The exocrine and endocrine pancreas underwent changes with particular necrosis of cells in the periphery of the islets of Langerhans. Necrosis of acinar cells and enlargement of nuclei and cytoplasm of duct cells were observed in the exocrine pancreas.

Lesions observed in males were similar to those observed in females. In addition the spleen was atrophied with an irregularly folded capsule. The red and white pulp was less densely packed than controls, and necrosis of lymphocytes was observed. There was also extensive hyaline degeneration of the myocardium associated with necrosis of myocardial fibres with haemorrhage into small necrotic areas. The rats that survived were killed at 10 days, and examination revealed similar, but less severe, changes in all organs.

Animals receiving 50 mg/kg were killed at intervals to observe the progress of lesions. Liver lesions appeared at 8 h and were most severe at 24 h. The renal tubular lesions were observed between 8 and 192 h. Pancreatic changes were first observed at 16 h and were still present at 192 h. Lesions in other organs were recovering by 96 and 192 h.

A teratological study of CPA in rats was conducted using Fischer 344 rats (Morrissey *et al.*, 1984). Rats were dosed daily by gastric intubation with CPA in 1.0 N sodium bicarbonate at concentrations of 0, 1, 5 or 10 mg CPA/kg body weight on days 8–11 or 12–15 of pregnancy. One high-dose rat in each group died prior to term and other dams were killed on day 21. Results of the study indicated that the teratogenic potential of CPA is low. The only consistent finding in term foetuses was a retardation in skeletal development with no significant differences in pup weights, percentage of resorptions, or foetal deaths compared to controls. Signs of maternal toxicity were observed in the high-dose groups, and those with the most severe signs had abnormal spleens and livers with microscopic lesions including focal areas of coagulative necrosis and single-cell necrosis. There were minor lesions in the kidney, salivary glands and duodenum, but no stomach lesions were observed. These results led to the conclusion that CPA is not teratogenic in the Fischer 344 rat.

CHICKENS

A metal chelate of CPA, named flavutoxin, was toxic when administered p.o. in corn oil suspension to 1-day-old chickens (Kirksey and Cole, 1973). The toxin was isolated from *A. flavus* and had an LD_{50} of 19 mg/kg as compared to 3.5 mg/kg for aflatoxin B_1 in the same bioassay.

More recently, feeding trials of 7 weeks duration were conducted in which chickens were given feed contaminated with purified CPA at concentrations of 10, 50 and 100 μg CPA/g feed (Dorner, 1982; Dorner *et al.*, 1983). Clinical signs occurred in chickens given feed with 100 μg/g of CPA and pathological changes occurred in chickens in both the 100 μg/g and 50 μg/g group. Significantly higher mortality occurred in groups given feed with 100 μg/g of CPA and the weight gain of those chickens was greatly reduced compared to controls.

Post-mortem examination of the chickens in the high-dose group (100 μg/g) revealed areas of erosion, ulceration, and hyperaemia of the proventricular mucosa. There was also excess mucus overlying the mucosal surface and dilatation of the proventricular lumen. Chickens in this group also had moderately pale livers and irregular, variable-sized, yellow foci in their livers and spleens. The yellow foci extended into both the hepatic and splenic parenchyma and were associated with the granulomas observed in the microscopic examination of those tissues. The bursa of Fabricius was small in all birds in this group (Dorner *et al.*, 1983). Birds given 50 μg/g of CPA had dilatation of proventricular lumens and apparent thick proventricular mucosal surfaces. In addition, the kidneys were observed to be slightly enlarged in birds from the 100 μg/g and 50 μg/g groups (Dorner, 1982). Gross lesions were not observed in the birds given 10 μg/g of CPA and in the control birds.

A review of the histological lesions in chickens indicated a dose-related response of chickens to CPA-contaminated feed (Dorner, 1982; Dorner *et al.*, 1983). Birds given 100 μg/g of CPA had mucosal epithelial necrosis and inflammation of the proventriculus, crop, and gizzard. The inflammatory changes in the proventriculus also occurred in the birds given 50 μg/g of CPA; however, these changes were not usually associated with mucosal necrosis. Proventricular mucosal hyperplasia occurred more frequently in birds given 50 μg/g of CPA than in birds given 100 μg/g CPA. The livers from birds in the 100 μg/g groups had focal to diffuse areas of hepatocellular vacuolization in peripheral portions of the hepatic lobules. There were also multifocal areas of non-patterned hepatocellular coagulative necrosis, multifocal areas of chronic hepatic portal inflammation, and occasional focal hepatic granulomas. The birds in the 50 μg/g groups had less severe liver lesions. Liver lesions in the birds given 10 μg/g consisted primarily of mild chronic hepatitis. Splenic lesions, which occurred in the 100 and 50 μg/g groups, were characterized by focal to diffuse areas of coagulative necrosis of the splenic parenchyma and were of greater severity in the high-dose group (Dorner *et al.*, 1983). Minimal to mild tubular degeneration of the kidney characterized by vacuolization of the tubular epithelium was observed in both the 50 μg/g and 100 μg/g groups (Dorner, 1982). In addition, there was subacute to chronic inflammation of the epicardium and chronic myocardial inflammation in birds that received feed with 100 and 50 μg/g of CPA (Dorner *et al.*, 1983).

PIGS

The acute toxicity of CPA to pigs was determined in two trials in which weaned pigs were given gelatin capsules containing CPA in quantities sufficient to provide daily doses of 0.01, 0.1, 1.0 and 10.0 mg CPA/kg body weight for 14 days (Lomax *et al.*, 1984). Pigs in the high-dose groups lost weight; had rough hair coats; and became weak, depressed and inappetent. Among the other treatment groups, only the pigs given 1.0 mg/kg CPA had clinical signs. During the second week of theexperiment these pigs developed rough hair coats and were mildly depressed.

Gross post-mortem lesions were observed in pigs in the 10 and the 1.0 mg/kg treatment groups. Characteristic lesions of pigs in the high-dose group were inflammation of the large and small intestine, mucosal ulceration and hyperaemia of the glandular and non-glandular regions of the stomach, and moderately pale livers. The only gross lesion observed in pigs given 1.0 mg/kg of CPA was ulceration of the non-glandular region in the stomach of one pig.

The microscopic lesions indicated that the major target organ was the gastrointestinal tract where lesions were dose-related and more severe at higher dosage levels. This ranged from severe and diffuse necrotizing gastritis in the glandular and non-glandular regions of the stomachs of pigs given 10 mg/kg of CPA to mild focal gastritis in the non-glandular stomachs of pigs given 0.1 mg/kg. Attenuation of villi and diffuse mucosal necrosis and inflammation were characteristic in the duodenum, jejunum and ileum of pigs in the high-dose group, whereas pigs given 1.0 mg/kg of CPA had only villous attenuation in the jejunum and ileum.

Liver and kidney lesions occurred in pigs given 10 mg/kg of CPA. Liver lesions varied from mild hepatocellular vacuolization (lipid) to severe and diffuse hepatocellular necrosis. Renal lesions were characterized by mild, multifocal tubular nephrosis and focal suppurative tubulointerstitial nephritis.

CONCLUSIONS

The toxicity of CPA has been studied in rats, chickens and pigs (Dorner, 1982; Dorner *et al.*, 1983; Lomax *et al.*, 1984; Morrissey *et al.*, 1984; Purchase, 1971). Histological findings in rats dosed with CPA dissolved in 1 N sodium bicarbonate were degeneration and necrosis in the liver, spleen, pancreas, kidney, salivary glands, myocardium and skeletal muscle (Purchase, 1971). In a teratological study of CPA in rats (Morrissey *et al.*, 1984), dams given the high dose of CPA had lesions similar to those described in the study by Purchase (1971), particularly in the liver and spleen. However, it was concluded that CPA was not teratogenic (Morrissey *et al.*, 1984). Chickens given feed contaminated with purified CPA had necrosis and inflammation in the proventriculus, crop, gizzard, liver, kidney and spleen (Dorner, 1982; Dorner *et al*, 1983). In pigs dosed daily for 2 weeks with capsules containing CPA the major target organs were the gastrointestinal tract, kidney and liver (Lomax *et al.*, 1984).

From these studies it can be seen that CPA produces tissue degeneration in a wide range of organs. Although the liver and kidney were the only affected organs common to all three animals discussed here, they usually were not the most severely affected. In the chickens and pig the most devastating effects were observed in the gastrointestinal tract; however this pattern was not seen in the rat. In rats, chicken and pigs hepatocellular necrosis resulted from ingestion of CPA. In rats only, there was enlargement of the cells lining the bile ducts, whereas in chickens and pigs hepatocellular vacuolization was consistently seen in animals receiving high doses of CPA. Kidney lesions, while common to all three animal species, were usually not as destructive when compared to the gastrointestinal lesions in chickens and pigs and hepatic lesions in rats. However, necrosis of tubular epithelial cells is a consistent histological lesion produced by CPA in the three animal species tested thus far.

Therefore, it may generally be said that manifestations of CPA toxicity in different animals can be found in a number of organs, and it appears to act primarily as an entero-nephrotoxin in chickens and pigs, and primarily as a hepatotoxin in rats. It should also be noted that these studies were conducted with purified toxin, and any naturally occurring toxicosis involving CPA would quite possibly be exacerbated by the presence of other toxins, depending on the fungi present in the contaminated feed. Because of the ubiquitous nature of the fungi that produce CPA, it certainly seems plausible that the involvement of CPA in mycotoxicoses has occurred in the past and will continue to be a problem. The knowledge of its toxic effects coupled with improving analytical techniques should improve recognition and control of CPA toxicity as it occurs in the future.

REFERENCES

de Jesus, A. E., Steyn, P. S., Vleggaar, R., Kirby, G. W., Varley, M. J., and Ferreira, N. P. (1981). Biosynthesis of α-cyclopiazonic acid. Steric course of proton removal during the cyclisation of β-cyclopiazonic acid in *Penicillium griseofulvum*, *J. Chem. Soc., Perkin Trans. I*, 3292–3294.

Dorner, J. W. (1982). Production of cyclopiazonic acid by *Aspergillus flavus* and its effects on chickens. M.S. thesis, Auburn University.

Dorner, J. W. (1983). Production of cyclopiazonic acid by *Aspergillus tamarii* Kita, *Appl. Environ. Microbiol.*, **46**, 1435–1437.

Dorner, J. W., Cole, R. J., and Diener, U. L. (1984). The relationship of *Aspergillus flavus* and *Aspergillus parasiticus* with reference to production of aflatoxins and cyclopiazonic acid, *Mycopathologia* **87**, 13–15.

Dorner, J. W., Cole, R. J., Lomax, L. G., Gosser, H. S., and Diener, U. L. (1983). Cyclopiazonic acid production by *Aspergillus flavus* and its effects on broiler chickens, *Appl. Environ. Microbiol.*, **46**, 698–703.

Gallagher, R. T., Richard, J. L., Stahr, H. M., and Cole, R. J. (1978). Cyclopiazonic acid production by aflatoxigenic and non-aflatoxigenic strains of *Aspergillus flavus*, *Mycopathologia*, **66**, 31–36.

Holzapfel, C. W. (1968). The isolation and structure of cyclopiazonic acid, a toxic metabolite of *Penicillium cyclopium* Westling, *Tetrahedron*, **24**, 2101–2119.

Kirksey, J. W., and Cole, R. J. (1973). New toxin from *Aspergillus flavus*, *Appl. Microbiol.*, **26**, 827–828.

Lansden, J. A., and Davidson, J. I. (1983). Occurrence of cyclopiazonic acid in peanuts, *Appl. Environ. Microbiol.*, **45**, 766–769.

LeBars, J. (1979a). Cyclopiazonic acid bioproduction by *Penicillium camemberti* Thom: effect of temperature on individual strains, *Ann. Rech. Vet.* **10**, 601–602.

LeBars, J. (1979b). Cyclopiazonic acid production by *Penicillium camemberti* Thom and natural occurrence of this mycotoxin in cheese, *Appl. Environ. Microbiol.*, **38**, 1052–1055.

Leistner, L., and Pitt, J. I. (1977). Miscellaneous *Penicillium* toxins. In *Mycotoxins in Human and Animal Health* (eds J. V. Rodricks, C. W. Hesseltine and M. A. Mehlman). Pathotox Publishers Inc., Park Forest South, Ill.

Lomax, L. G., Cole, R. J., and Dorner, J. W. (1984). The toxicity of cyclopiazonic acid in weaned pigs, *Vet. Path.* **21**, 418–424.

Luk, K. C., Kobbe, B., and Townsend, J. M. (1977). Production of cyclopiazonic acid by *Aspergillus flavus* Link, *Appl. Environ. Microbiol.*, **33**, 211–212.

Morrissey, R. E., Cole, R. J., and Dorner, J. W. (1984). A teratological study of cyclopiazonic acid in rats. *Toxicologist*, **4**, 84.

Ohmomo, S., Sugita, M., and Abe, M. (1973). Isolation of cyclopiazonic acid, cyclopiazonic acid imine and bissecodehydrocyclopiazonic acid from cultures of *Aspergillus versicolor* (Vuill.) Tiraboschi, *J. Agric. Chem. Soc. (Jpn.)*, **47**, 57–63.

Orth, R. (1977). Mycotoxins of *Aspergillus oryzae* strains for use in the food industry as starters and enzyme producing molds, *Ann. Nutr. Aliment.*, **31**, 617–624.

Purchase, I. F. H. (1971). The acute toxicity of the mycotoxin cyclopiazonic acid to rats. *Toxicol. Appl. Pharmacol.*, **18**, 114–123.

Still, P., Eckardt, C., and Leistner, L. (1978). Bildung von cyclopiazonsäure durch *Penicillium camemberti*—isolate von Käse, *Die Fleischwirtschaft*, **58**, 876.

Trichothecenes and Other Mycotoxins
Edited by J. Lacey

CHAPTER 49

Experimental mycotoxicosis: some observations on the teratopathological effects of a single dose of aflatoxin B_1 and ochratoxin A in CBA/CA mice

Ram Gopal Arora

INTRODUCTION

Mycotoxins have long been recognized as the major causes of feed toxicosis in animals, and may also constitute a serious hazard to human health (Wilson and Hayes, 1973). Aflatoxins and ochratoxins are the mycotoxins isolated most frequently from food and feedstuffs and have been extensively studied in recent years. Because of their diverse chemical, structural and biological properties their effects vary. For instance aflatoxin B_1 (AF B_1) is a potent hepatotoxin and hepatocarcinogen (Newberne and Butler, 1969; Wogan and Shank, 1971 and others) whereas ochratoxin A (OCT A) is a nephrotoxin (Krogh *et al.*, 1973; Elling *et al.*, 1975; Berndt and Hayes, 1979). Several mould metabolites, in addition to causing toxic damage in adults, may also seriously interfere with the growth and development of the foetus (Hood, 1979; Hayes, 1981). Experiments have been conducted in our laboratory to gain additional insight into mycotoxin teratogenesis. This chapter presents some observations on the teratopathological effects of single doses of AF B_1 and OCT A administered orally to mice.

MATERIAL AND METHODS

Female mice of the inbred CBA/CA strain were given a single p.o. dose of either AF B_1 (4 mg/kg) or OCT A (4 mg/kg) on gestation day 8 or 9 (the day following the night of conception as judged from the presence of vaginal plug, was called day 1 of gestation). Doses were prepared in corn oil and administered in a volume equivalent to 1 ml/100 g body weight. Controls were dosed with corn oil only. Foetuses, delivered at maternal necropsy on day 19 of pregnancy, were weighed individually prior to fixation in Bouin's fluid or in 10% formalin. Numbers of live and dead foetuses were counted separately and total implantations and resorption sites were recorded. Dead foetuses were neither weighed nor included in subsequent examination for malformations. External malformations were evaluated with the aid of stereomicroscope using material fixed both in Bouin's fluid and formalin, while visceral lesions were characterized from histological examination of serial sections of material fixed in Bouin's fluid and embedded in paraffin. The sections were routinely stained in haematoxylin and eosin. Skeletal defects were assessed by stereomicroscopy following clearing and staining of formalin fixed material according to the method of Peters (1977).

RESULTS

Data on prenatal growth and development as judged by litter size, pup weights, prenatal deaths and visible malformations are presented in Table 1.

The nature of the response varied according to the type of the mycotoxin and on the stage of embryonic development at the time of maternal exposure. AF B_1 had little effect on prenatal survival and growth but it did produce visible abnormalities such as exencephaly, open eyes and failure of the abdominal wall to close resulting in protrusion of intestines. The incidence of malformed offspring was about 11% after exposure on day 8 of gestation but was zero after exposure on day 9. On the other hand OCT A was foetotoxic and teratogenic for litters treated on either of these days of gestation. However, the effects of dosing on day 9 of gestation were more drastic than those of dosing on day 8 of gestation.

Besides high prenatal mortality and growth inhibition, administration of OCT A on day 9 produced malformations in nearly all the offspring (93 of 101). By contrast similar treatment on day 8 caused malformations only in 11 of 67 (17%) of the offspring. No abnormalities were detected in controls treated with corn oil alone.

Since OCT A was found to be most teratogenic of the two mycotoxins, further experiments were conducted to examine the teratogenicity of this mycotoxin administered at other stages of embryogenesis, i.e. before day 8 or after day 9 of gestation. These revealed no appreciable effect except for severe growth retardation of foetuses whose dams were given OCT A 2 days prior to conception and a high prenatal death rate in litters exposed on day 7 of embryogenesis.

Table 1 Prenatal effects of aflatoxin B_1 (AF B_1) and ochratoxin A (OCT A) on mouse development[a]

	Treatment (mg/kg) and gestation day					
	Controls		AF B_1 (4)		OCT A (4)	
Parameters	8	9	8	9	8	9
No. of litters	19	17	8	7	10	17
No. of implants	153	143	72	57	91	136
No. of dead or resorbed	11	12	11	6	24	35
Mean litter size	7.5 ± 1.23	7.7 ± 1.24	7.6 ± 1.31	7.3 ± 1.28	6.7 ± 1.32	5.9 ± 2.31
Mean pup weight (g)	1.02 ± 0.13	1.05 ± 0.12	1.06 ± 0.12	0.92 ± 0.15	0.93 ± 0.22	0.64 ± 0.14
No. malformed/total examined	0/142	0/131	7/61	0/51	11/67	93/101
Exencephaly			4		7	90
Anophthalmia			—		—	39
Microphthalmia			—		4	26
Open eyelid			3		4	16
Agenesis of nares			—		—	19
Cleft lip			—		—	5
Cleft face			—		—	9
Malformed jaws/short maxilla with protruding tongue			—		1	33
Intestinal protrusion/ gastrochisis			2		—	4
Fused ribs			—		—	11[b]
Fused vertebrae			—		2[b]	7[b]

[a] Test mycotoxins were dissolved in corn oil and administered orally in a volume of 1 ml/100 g body weight on day 8 or 9 of gestation. Controls received equivalent amount of corn oil alone.
[b] About one-half of the total foetuses was cleared and stained for examination of skeletal anomalies.

Survey of malformations and their morphological characteristics

Gross anatomical alterations were, in general, restricted to the cranio-facial region. The most consistently observable defect was exencephaly characterized by partial or complete exposure of the brain. The exposed part(s) of the brain did not show any sign of skin or of the membranous neurocranium. Frequently the eyes in exencephalics were either missing (anophthalmic) or smaller than normal (microphthalmic). Less severe ocular anomalies were represented by poorly developed eyelids which failed to fuse (open eyes). Other abnormal facial features were elevation of the snout and maxilla and protruding tongue. Mid-facial defects ranging from hypoplasia of the jaws and agenesis of the external nares to a partial or total clefting of intermaxillary tissue segments were noticed less often. Examination of cleared and stained skeleton revealed a clear shortening of the skull as a whole, and a lack of cranial cavity. The basal bones of the skull were all present but abnormal in shape, size and structure. The viscerocranium, (particularly the maxillary and premaxillary bones) was consistently affected in cases of midfacial clefts (Figure 1). Other skeletal abnormalities noted sometimes were: a fan-like configuration of the ribs owing to the fusion of two or more adjacent ribs; and a short, curved spine due to anomalous and fused vertebral bodies. These

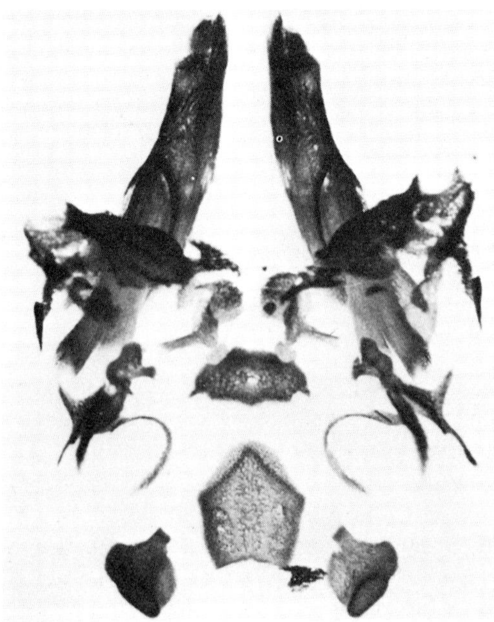

Figure 1 Dorsal view of the skull of an exencephalic foetus with total mid-facial cleft. The two maxillae are widely separated, the cranial vault is missing and the basal bones are severely abnormal.

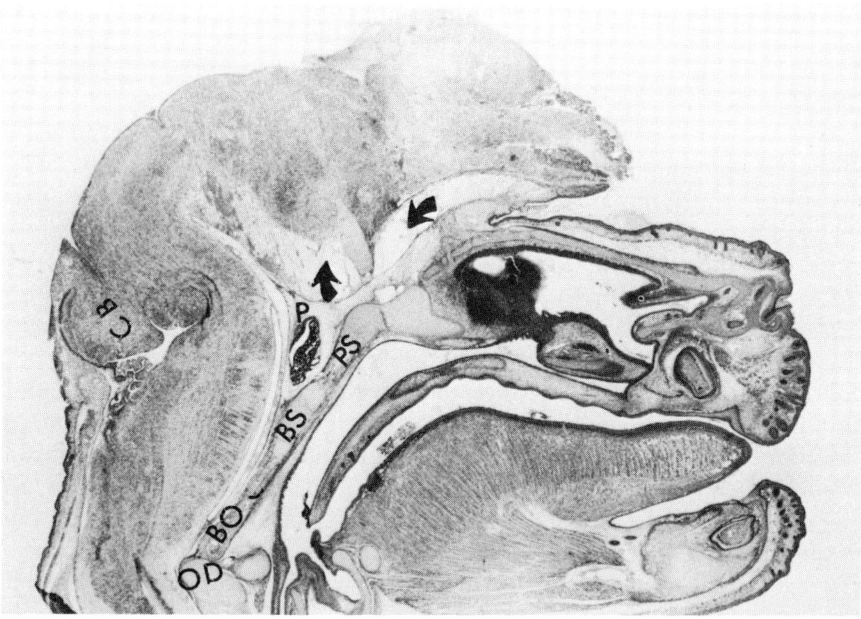

Figure 2 Mid-sagittal section of a foetal head showing degeneration, necrosis, haemorrhage and oedema of the brain tissue (arrows). Note the absence of neurocranium and skin over the exposed parts of the brain; the cerebellum (CB) is hypoplastic and the pituitary (P) is lifted from its normal position. The basicranium is short and abnormal in position relative to the vertebral axis (OD: odontoid process of second vertebra, BO: basioccipital, BS: basisphenoid, PS: presphenoid). (H and E stain, × approx. 15.)

macroanatomical anomalies of the axial skeleton have been described in a previous paper (Arora *et al.*, 1981).

The histological appearance of the lesions in different parts of the brain, including pituitary, is shown in Figure 2. The brain showed extensive degeneration and necrosis not only in the exposed parts but also ventrally (diencephalon, mid-brain and pons) facing the basicranium. Haemorrhages and fluid accumulation in these areas were common. Occasionally the cells from the mantle zone proliferated into the ventricles where they were identified as aggregates admixed with necrotic debris. Microphthalmic eyes sometimes showed a retinal dysplasia and a failure of nerve fibres to exit from the retina, while in anophthalmic eyes only rudimentary remnants of disorganized ocular tissue were detectable. Respiratory tract lesions were characterized by dismorphogenesis of ethmoidal components, alveolar hypoplasia and atelectasis. Cardiac malformations consisted of poorly formed muscle fibres and vacuolated or swollen appearance of the cells. Cellular alterations characteristic of poor cell differentiation were seen in the urinary tubules, adrenal medulla, thymic lymphocytes, liver cells and gut epithelium.

DISCUSSION

Of the two mycotoxins tested on days 8 and 9 of mouse pregnancy OCT A was a much more potent teratogen than was AF B_1. Developmental anomalies were produced by AF B_1 in a small percentage of foetal mice. Although these have not previously been reported to occur in mice (DiPaolo *et al.*, 1967) or rats (LeBreton *et al.*, 1964; Butler and Wigglesworth, 1966), they were similar to those produced in golden hamster (Elis and DiPaolo, 1967). These authors (Elis and DiPaolo, 1967) suggested that the absence of malformations, growth retardation and liver damage in foetal mice could be due to the differences in the metabolism of AF B_1 in mouse and hamster. In view of the conflicting results in different studies it seems difficult to draw any definite conclusion about the teratogenicity of this, which is otherwise one of the most potent toxic and carcinogenic mycotoxins. The findings that OCT A caused increased prenatal mortality, reduced foetal growth and a wide variety of structural abnormalities involving soft as well as hard tissues of head and face, are in close agreement with those reported previously (Arora and Frölén, 1981; Arora *et al.*, 1981; Hayes *et al.*, 1974; Hood *et al.*, 1976). The sensitivity of the embryo to the mycotoxin was greatly influenced by the stage of development at the time of exposure. Structural abnormalities could only be produced by OCT A if exposure occurred on days 8 and 9 of pregnancy with the greatest effect following administration on day 9. Before day 7 and after the end of day 9 of embryogenesis there was little effect of toxin administration. These observations disagree with those of Hayes *et al.* (1974), who found malformations and foetal deaths after i.p. injection of the compound on any day between days 7 and 12 of gestation, with a peak effect on day 8. Whether the discrepancy between the results of these studies is related to the different administration routes, the genetic constitution of animals or to other contributory factors such as external or internal maternal environment, is difficult to ascertain. None the less, it seems reasonable to assume that OCT A could be more effective in causing developmental errors when given to pregnant CBA/CA mice by oral route. However, the reverse was reported for the teratogenicity of another mycotoxin, rubratoxin B (Hayes, 1978). The morphological characteristics of malformations predominantly affecting the cranio-facial complex are suggestive of primary injury by the toxin to the developing nervous tissue leading to the failure of closure of the brain (Arora, 1982).

The observations of tissue damage, atypical and excessive cell proliferation as well as a severe disturbance of the normal differentiation of neural cells under certain circumstances may also imply a potentially increased risk of future tumour development. The histological changes found in the visceral organs suggest that subtle effects, characterized by impairment of maturation and morphological differentiation of thymus, pulmonary alveoli, cardiac muscle, liver, gut, kidney and adrenal, also occur. These observations contrast with the findings of Hood *et al.* (1978), who reported that all tissues other than brain and eyes were within limits accepted as normal. Although the mechanism of mycotoxin teratogenesis is not clearly understood and the studies relevant to the pathomorphogenetic mechanism are still lacking, the demonstration of radioactivity from [^{14}C]-labelled OCT A in embryonic tissue exposed during the susceptible period of teratogenesis, i.e. on day 8–9 but not later (Appelgren and Arora, 1983), support the contention that the prenatal damage is most likely to be due to a direct effect of the toxin on the developing embryo. The possibility of an indirect action due to altered maternal metabolism or placental functions cannot be ruled out, however.

ACKNOWLEDGEMENTS

This work was supported by the Swedish Council for Forestry and Agricultural Research (SJFR).

REFERENCES

Appelgren, L. E., and Arora, R. G. (1983). Distribution of ^{14}C-labelled ochratoxin A in pregnant mice, *Food Chem. Toxic.*, **21**, 563–568.

Arora, R. G. (1982). Abnormal morphogenesis of cephalic neural folds due to ochratoxin A in the mouse. In *Proceedings of V Internat. IUPAC Symposium on Mycotoxins and Phycotoxins*, Vienna, 1–3 September, pp. 232–234.

Arora, R. G., and Frölén, H. (1981). Interference of mycotoxins with prenatal development of the mouse. II. Ochratoxin A induced teratogenic effects in relation to the dose and stage of gestation, *Acta Vet. Scand.*, **22**, 535–552.

Arora, R. G., Frölén, H., and Nilsson, A. (1981). Interference of mycotoxins with prenatal development of the mouse. I. Influence of aflatoxin B_1, ochratoxin A and zearalenone, *Acta Vet. Scand.*, **22**, 524–534.

Berndt, W. O., and Hayes, A. W. (1979). *In vivo* and *in vitro* changes in renal function caused by ochratoxin A in the rat, *Toxicology*, **12**, 5–17.

Butler, W. H., and Wigglesworth, H. S. (1966). The effects of aflatoxin B_1 on the pregnant rat, *Brit. J. Exp. Path.*, **47**, 242–247.

DiPaolo, J. A., Elis, J., and Erwin, H. (1967). Teratogenic response by hamsters, rats and mice to aflatoxin B_1, *Nature*, **215**, 638–639.

Elis, J., and DiPaolo, J. A. (1967). Aflatoxin B_1: Induction of malformations, *Arch. Path.*, **83**, 53–57.

Elling, F., Hald, B., Jacobsen, Chr., and Krogh, P. (1975). Spontaneous toxic nephropathy in poultry associated with ochratoxin A, *Acta Path. Microbiol. Scand.*, Sect. A; **83**, 739–741.

Hayes, A. W. (1978). Mycotoxin teratogenicity. In *Toxins: Animal, Plant and Microbial* (ed. P. Rosenberg), pp. 739–758. Pergamon Press, Elmsford, NY.

Hayes, A. W. (1981). *Mycotoxin Teratogenicity and Mutagenicity*, CRC Press, Boca Raton, Florida.

Hayes, A. W., Hood, R. D., and Lee, H. L. (1974). Teratogenic effects of ochratoxin A in mice, *Teratology*, **9**, 93–98.

Hood, R. D. (1979). Effects of mycotoxins on development. In *Advances in the Study of Birth Defects* (ed. T. V. N. Persaud), Vol. II, pp. 191–210. MTP Press, Lancaster.

Hood, R. D., Kuczuk, M. H., and Szczech, G. M. (1978). Effects in mice of simultaneous prenatal exposure to ochratoxin A and T-2 toxin, *Teratology*, **17**, 25–30.

Hood, R. D., Naughton, M. J., and Hayes, A. W. (1976). Prenatal effects of ochratoxin A in hamsters, *Teratology*, **13**, 11–14.

Krogh, P., Hald, B., and Pedersen, E. J. (1973). Occurrence of ochratoxin A and citrinin in cereals associatd with mycotoxic porcine nephropathy, *Acta Path. Microbiol. Scand.*, Sect. B; **81**, 689–695.

LeBreton, E., Frayssinet, C., Lafarge, C., and DeRecondo, A. M. (1964). Aflatoxin mécanisme de l'action (Mechanism of aflatoxin action), *Food Cosmet. Toxicol.*, **2**, 675–676.

Newberne, P. M., and Butler, W. H. (1969). Acute and chronic effects of aflatoxin on the liver of domestic and laboratory animals: a review, *Cancer Res.*, **29**, 236–250.

Peters, P. W. J. (1977). Double staining of foetal skeletons for cartilage and bone. In *Methods in Prenatal Toxicology: evaluation of embryotoxic effects in experimental animals* (eds D. Neubert, H. J. Merker and T. E. Kwasigroch), pp. 153–154. Thieme, Stuttgart.

Wilson, B. J., and Hayes, A. W. (1973). Microbial toxins. In *Toxicants Occurring Naturally in Foods* (ed. F. M. Strong), 2nd Edition, pp. 372–423. National Academy of Science, Washington, DC.

Wogan, G. N., and Shank, R. N. (1971). Toxicity and carcinogenicity of aflatoxins. In *Advances in Environmental Science and Technology* (eds J. N. Pitts and R. L. Metcalf), pp. 321–350. Wiley-Interscience, New York and London.

Trichothecenes and Other Mycotoxins
Edited by J. Lacey

CHAPTER 50

Cases of nephrotoxic mycotoxicoses in cattle and swine in the United States

W. E. Lloyd, G. N. Daniels and H. M. Stahr

INTRODUCTION

Ochratoxin has been demonstrated naturally and experimentally to cause nephropathies in swine, characterized by the presence of necrosis of the proximal convoluted tubules and fibrosis (Elling and Moller, 1973; Krogh *et al.*, 1974, 1976). Ochratoxin and citrinin have also been shown to cause nephrotoxicosis in poultry and laboratory animals (Huff and Hamilton, 1974; Zimmermann and Carlton, 1976; Kanisawa *et al.*, 1977; Thacker and Carlton, 1979; Roberts and Mora, 1978). Ochratoxin was incriminated as the cause of human nephropathies and cirrhosis in Africa and Yugoslavia (Coady, 1975).

Relatively little information has been published about the toxic effects of ochratoxin on ruminants. Swedish authors, using *in vitro* techniques, reported that ochratoxin A was degraded in forestomach ingesta buffered at pH 7.5, but not in abomasal contents (Hult *et al.*, 1976). They did not report degradation studies in forestomach ingesta at physiological pH values of 4.5–7.0.

The first case of probable citrinin toxicosis in a bovine herd was reported in 1978 (Lloyd *et al.*, 1978). During 1979, three more cases of bovine nephrotoxicosis due to ochratoxin and citrinin were reported (Lloyd and Stahr, 1979). Although porcine ochratoxicosis has been reported to be common in Denmark (FDA, 1979), it has not previously been reported in the U.S.A. This chapter is a compilation of relatively rare bovine and porcine intoxications by citrinin and ochratoxin A in the U.S.A.

CASE HISTORIES—BOVINE

Herd A

Approximately 100 steers weighing approximately 500 kg were sick in a herd of 9600 fattening animals. Approximately 50 animals died during the episode. Clinical signs included anorexia, fever, diarrhoea and uraemia. Gross post-mortem lesions were pneumonia and ulcers in the oesophagus and small intestines, resembling lesions of bovine virus diarrhoea (BVD). Histopathological examination of tissues revealed a nephritis and nephrosis characterized by the presence of hyaline casts and birefringent crystals. BVD was not diagnosed. Tests on blood and serum revealed the following significant changes in uraemic animals: increases in serum urea nitrogen (SUN), white blood cell count (WBC), plasma protein, fibrinogen and serum P; decreases in serum K and Ca.

Herdsmen had been feeding some mouldy high-moisture maize equal to 74% dry matter (DM) and maize silage (35% DM). Chemical analyses of the feeds using thin-layer chromatography (TLC) (Stahr, 1980) revealed 2 and 3 mg citrinin/kg in fresh maize and silage samples, respectively. Within 2 weeks of discontinuing feeding of mouldy feedstuffs, the affected animals had recovered.

Herds B, C and D

Approximately 63 cows and ruminating calves were affected in the three herds of 1190 animals.

Frequently observed clinical signs were depression, hypothermia, dehydration and pneumonia. Post-mortem lesions were inconsistent except for some cases of perirenal oedema and aseptic pneumonia and one case of thromboembolic meningoencephalitis. Microscopic lesions included nephrosis with hyaline casts and birefringent crystals, dilated tubules and fibrosis in the kidneys; fatty changes in the liver were frequently observed. Affected animals died of uraemia with elevated SUN values. TLC analyses (Stahr, 1980) confirmed the presence of citrinin and/or ochratoxin in feedstuffs in all three cases.

Approximate concentrations of citrinin and ochratoxin in mg/kg (ppm), on a DM basis, were respectively: Case A high moisture maize: 3.0; corn silage: 9.0. Case B corn silage: 3.3–90. Case C hay: 4.1; sunflower hulls: 2.1. Case D oatlage: 0.20. In three of the four cases visibly mouldy feedstuffs were being fed to affected animals. Nephrosis and uraemia disappeared in all herds once feeding of the toxic feedstuffs was discontinued.

CASE HISTORIES—PORCINE

Herd A

Approximately 20 animals died from a herd of 90 7-month-old pigs weighing 50–100 kg each. Clinical signs were polyuria, polydipsia, diarrhoea, coughing and slow growth. Gross post-mortem lesions were ulceration of the glandular stomach and oesophagus, anaemia and large, white kidneys with subcapsular oedema. Packed cell volumes ranged from 10 to 14% and SUN concentrations ranged up to 90 mg/dl. Bacteria could not be isolated from tissues. Renal histology revealed extensive chronic interstitial fibrosis, with few functional tubules remaining; glomeruli had degenerated. Histologically, the livers appeared to be normal.

Analysis of kidney tissue revealed that it contained approximately 200 μg/kg of ochratoxin A. The complete ration contained 500 μg/kg of ochratoxin A but no other mycotoxins could be detected. Since 200 μg/kg of ochratoxin A in grain rations can cause nephrotoxicosis in pigs (Krogh *et al.*, 1976), toxin was present in kidney tissue and histological findings were consistent, a diagnosis of ochratoxicosis was made. When a new, non-contaminated ration was fed the general health of the herd improved.

Herd B

Approximately 12 animals died from a herd of approximately 70 fattening swine weighing 30–45 kg each. The clinical signs were cachexia, diarrhoea and weakness. Post-mortem observations of two necropsied pigs revealed a fibrinous pericarditis, mesenteric lymphadenitis and pneumonia. A *Salmonella* sp. C was isolated from the lymph nodes, lung, liver and spleen of one animal, and *Pasteurella multocida* from the lung of another pig but other microbiological tests were negative. Histological examination of tissues revealed the presence of bacterial pneumonia, enteritis and nephrosis with interstitial fibrosis, and tubules which were calcified or contained polarizable crystals.

Ochratoxin A at a concentration of 1–2 mg/kg was detected by chemical analysis of the complete ration being fed to the pigs. A diagnosis of ochratoxicosis, salmonellosis and pasteurellosis was made. Antibacterial administration to the herd and replacement of the ration which contained ochratoxin resulted in improved health of the herd.

CONCLUSION

The incidence of nephrotoxic mycotoxicoses in U.S. cattle and swine may be greater than recognized, especially in animals being fed visibly mouldy feedstuffs. Diagnostic features include consistent renal uraemia and chronic nephrosis, which may exist with several other secondary disease conditions. The diagnosis, nephrotoxic mycotoxicosis, can be substantiated by analysing rations of feedstuffs and kidney tissue for the presence of mycotoxins combined with histological examination of the kidneys.

REFERENCES

Coady, A. (1975). Human mycohepatotoxin problems of developing countries: cirrhosis, cancer, and viral interactions (a view from Africa), *Int. J. Environ. Stud.*, **8**, 213–216.

Elling, F. and Moller, T. (1973). Mycotoxic nephropathy in pigs, *Bull. Wld. Hlth. Org.*, **49**, 411–418.

Huff, W. E., and Hamilton, P. B. (1974). Nephrotoxicity of ochratoxin A in broiler chicks, *Poultry Sci.*, **53**, 1937–1938.

FDA (1979). Conference on mycotoxins in animal feeds and grains related to animal health. BVM, U.S. Government Printing Office, Rockville, Maryland.

Hult, K., Teiling, A., and Gatenbeck, S. (1976). Degradation of ochratoxin A by a ruminant, *Appl. Environ. Microbiol.*, **32**, 443–444.

Kanisawa, M., Suzuku, S., Kozuka, Y., and Yamazaki, M. (1977). Histopathological studies on the toxicity of ochratoxin A in rats. 1. Acute oral toxicity, *Toxicol. Appl. Pharmacol.*, **42**, 55–64.

Krogh, P., Axelsen, N. H., Elling, F., Gyrd-Hansen, N., Hald, B., Hyldgaard-Jensen, J., Larsen, A. E., Madsen, A., Mortensen, H. P., Moller, T., Petersen, O. K., Ravnskov, U., Rostgaard, M., and Aalund, O. (1974). Experimental porcine nephropathy, *Acta Pathol. Microbiol. Scand., Section A, Supplement No. 246.*

Krogh, P., Elling, F., Gyrd-Hansen, N., Hald, B., Larsen, A. E., Lillehoj, E. B., Madsen, A., Mortensen, H. P., and Ravnskov, U. (1976). Experimental porcine nephropathy: changes of renal function and structure perorally induced by crystalline ochratoxin A, *Acta Pathol. Microbiol. Scand., Section A*, **84**, 429–434.

Lloyd, W. E. and Stahr, H. M. (1974). Ochratoxin toxicosis in cattle. In *Proceedings*, 22nd Annual Meeting, American Association of Veterinary Laboratory Diagnosticians, pp. 223–238.

Lloyd, W. E., Holter, J. A., Hill, H. T., Wohlgemuth, K., and Classick, L. (1978). A case of probable citrinin toxicosis in a herd of Iowa feeder cattle. In *Proceedings*, 21st Annual Meeting, American Association of Veterinary Laboratory Diagnosticians, pp. 147–160.

Roberts, W. T., and Mora, E. C. (1978). Toxicity of *Penicillium citrinum* AUA-532 contaminated corn and citrinin in broiler chicks, *Poultry Sci.*, **57**, 1221–1226.

Stahr, H. M. (ed.) (1980). *Supplement to Analytical Toxicology Methods Manual*, 1st edition, pp. 234–238, 256–259 and 262–264. Iowa State University Press, Ames, Iowa.

Thacker, H. L. and Carlton, W. W. (1977). Ochratoxin A mycotoxicosis in the guinea-pig, *Food Cosmet. Toxicol.* 563–574.

Zimmermann, J. L. and Carlton, W. W. (1976). Mycotoxicosis produced in mice by cultural products of an isolate of *Aspergillus ochraceus*, *Food Cosmet. Toxicol.*, 571–575.

Trichothecenes and Other Mycotoxins
Edited by J. Lacey

CHAPTER 51

Toxicology of secalonic acid D

Chada S. Reddy

INTRODUCTION

Penicillia are an important component of the microflora of maize and in midwestern U.S.A. 44% of seeds may be contaminated with these fungi before harvest (Carlton *et al.*, 1968). Some species are known to be toxic. For instance, Scott (1965) found that 22 species of *Penicillium* caused death of ducks within 14 days after the animals were fed a maize meal infected with pure cultures of the organism. Seven were very toxic, six were less toxic and the remaining nine were apparently non-toxic to mice. Carlton *et al.* (1968) tested several species of *Penicillium* for toxicity in mice and found that *P. oxalicum* and *P. viridicatum* were the most toxic. Although *P. viridicatum* extracts produced a variety of liver lesions including focal areas of necrosis, periductular fibrosis, cholangitis and periductular hyperplasia, no histopathological changes were reported for *P. oxalicum*-exposed animals. However, extensive liver necrosis was the major lesion reported by Steyn (1970) when experimental animals were fed maize meal on which pure cultures of *P. oxalicum* had grown. A toxic pigment isolated from these cultures was identified as secalonic acid D (dimethyl-9,9′-dioxo-2,2′-bixanthen-10a,10a′-dicarboxylic acid), a stereo-isomer of the known secalonic acids A, B and C. Secalonic acids E and F have also been isolated recently (Howard and Johnstone, 1973b; Anderson *et al.*, 1977). Besides *P. oxalicum*, *Aspergillus aculeatus* (Anderson *et al.*, 1977), *A. ochraceus* (Yamazaki *et al.*, 1971), *Claviceps purpurea* (Franck *et al.*, 1966) and *Phoma terrestris* (Howard and Johnstone, 1973a) are known to produce one or more of the secalonic acids in cereal grains with *A. aculeatus* and *C. purpurea* producing secalonic acid D (Anderson *et al.*, 1977; Howard and Johnstone, 1973a).

Secalonic acid D has antibacterial activity against *Bacillus megaterium* (Demain *et al.*, 1976) and *B. subtilis* (Yamamto, 1975). While screening compounds for antitumour activity, Yamamoto (1975) demonstrated that SAD carried some immunopotentiating capacity in addition to normalization of increased sugar transport through the membranes of chicken embryo cells infected with Rous sarcoma virus (increased sugar transport is associated with cell transformation). Although tetrahydropyranyl derivatization of the alcoholic OH groups of SAD reduced toxicity 6–10 times, the immunopotentiating capacity remained intact. Administration of pyranyl-SAD along with 10^6 ascites sarcoma cells into mice resulted in regression of the initial tumour mass as well as protection from future challenges by the same tumour cells. Against Ehrlich's ascites tumour, i.p. injection of 8 mg/kg of pyranyl-SAD on days 1 and 3 gave very similar survival curves to that of mitomycin C (1 mg/kg) on days 1, 2, 3, 4 and 5 after tumour cell injection.

There had been few toxicity studies on the compounds comprising the secalonic acid family before our investigation. There was one report of an LD_{50} of secalonic acid D (42 mg/kg, i.p.) given as a suspension in aqueous carboxymethyl cellulose to white mice (Steyn, 1970) and one on the local inflammatory activity of secalonic acid A in rats (Harada *et al.*, 1974). The following results describe the nature of the toxicologic effects of SAD in mice and allow for speculation of possible importance of contamination of grain by SAD in animal and human health.

MATERIALS AND METHODS

Toxin

Secalonic acid D was produced from cultures of *P. oxalicum* grown in cracked maize as described by Reddy *et al.* (1979).

Animals

Species, strain and the source of the animals used in these investigations are listed under each investigation. All animals were housed (unless specified) in plastic cages in temperature-controlled (72 ± 2 °F) and artificially illuminated (12 h/day) rooms free from known sources of toxic contaminants. Commercial feed (Ralston Purina, St Louis, MO) and fresh water were provided *ad libitum*.

Acute and multiple dose toxicity of secalonic acid D in mice and rats

Adult male and female CD-1 mice (Charles River Breeding Laboratories, Wilmington, MA), adult male Texas (ICR) random-bred mice (Texas Inbred Mice Co., Houston, TX) and adult male SCH ARS (CF-1) outbred non-Swiss albino mice (Sprague-Dawley, Madison, WI) were used (Reddy *et al.*, 1979).

Adult rats (Sprague-Dawley, Charles River Breeding Laboratories) were mated to produce neonates and weanlings. One female was housed with two males overnight and then the female was removed to a separate cage. In the case of neonates used to establish LD_{50} values, pups were given a single oral dose of SAD on the first day postpartum and then returned to the mother. For the weanling studies, rat pups reached 45–55 g before use. All animals were observed daily for morbidity and mortality during the 21-day experiment. Tissues from all animals that died during the experiment and those that were sacrificed at 21 days were examined for histopathology by routine methods.

Effects of phenobarbital, 3-methylcholanthrene and SKF-525A on acute SAD toxicity

Male Charles River CD-1 mice weighing 26–28 g were pretreated i.p. with phenobarbital (75 mg/kg/day for 3 days), 3-methylcholanthrene (25 mg/kg/day for 3 days) or SKA-525A (50 mg/kg once). At intervals of 24, 24 and 0.75 h after these pretreatments, respectively, groups of 10 mice were given 0, 10, 20, 40 and 80 mg/kg of SAD i.p. (in 1 : 9 DMSO and 5% sodium bicarbonate). Mice given 20 mg/kg of SAD were weighed on days 0, 1, 3, 7 and 11 after treatment and survivors were sacrificed on day 14 for histopathological examination. Body weights of each treatment group were compared with those of appropriate controls by the Student's *t*-test.

Groups of at least four male CD1 mice were treated with normal saline, phenobarbital or SKF-525A followed by 15 mg/kg of ^{14}C-secalonic acid D, i.v. The mice were anaesthetized and bile duct cannulated as described by Reddy *et al.* (1981c). Bile was collected for a period ranging between 30 and 60 min and the volume measured. Bile samples were analysed for parent compound and metabolites by a combination of HPLC (Reddy *et al.*, 1981b) and radioassay following radioassay of a 50 μl aliquot of each sample for total radioactivity (Reddy *et al.*, 1981c).

Teratogenicity of secalonic acid D in mice

Methodology used in this investigation is described by Reddy *et al.* (1981a). The day a vaginal plug was detected was considered day 1 of pregnancy. Pregnant CD1 mice were given i.p. doses of 5–15 mg/kg of SAD in either 5% $NaHCO_3$ alone or a mixture of DMSO and $NaHCO_3$ (DS) on days 7–15 of pregnancy. Mothers that went full-term were sacrificed on day 19 with an overdose of ether. The uterus was opened and the number of implants, live foetuses and resorptions were recorded. Stillborn foetuses and both early and late resorptions were included in the category of resorptions. Live foetuses were blotted dry, weighed and divided into groups of 2 to 1 and processed for visceral and skeletal examination (Wilson, 1965).

Dose level of 25 mg/kg of SAD in $NaHCO_3$ was tested on days 11, 12, 13 or 14 of pregnancy to determine the critical period of teratogenicity. Also, a dose–response between 15 and 30 mg/kg of SAD was studied on day 12 of pregnancy.

Mutagenicity of secalonic acid D *in vitro* and *in vivo*

Three tests—a bacterial (Ames *Salmonella* reverse) mutation system and two rodent tests, one each for somatic (micronucleus test) and germ cell (dominant lethal) mutations, were selected. The methodology used is as described in McCann and Ames (1976), Schmid (1975) and Generoso (1973), respectively. Modifications of these methods adapted in our investigations were reported by Reddy *et al.* (1980).

Disposition and metabolism of ^{14}C-secalonic acid D in mice

The methodology used to produce ^{14}C-SAD is as described by Reddy *et al.* (1981c).

Groups of three male CD-1 mice were housed in each stainless steel metabolism cage designed to allow separate collection of faeces and urine. Following i.v. administration of 15 mg/kg of ^{14}C-SAD three groups of three mice each were sacrificed at 0.083, 0.25, 0.5, 1, 3, 5, 12, 24, 48, 72 and 168 h.

One millilitre of blood was collected from each mouse at the time of sacrifice by cardiac puncture. The blood was centrifuged and plasma separated. Urine and faeces were collected at designated intervals. Tissues were excised, weighed, homogenized in 3–5 volumes of cold distilled water and the volume of homogenate measured.

Duplicate aliquots of body fluids and tissue homogenates, dry faeces and urine were oxidized in a Packard 306 Tri-Carb oxidizer and radioassay and quench correction performed as described by Reddy *et al.* (1981c).

Characterization of the radioactive species was achieved by cannulation of common bile duct, radioassay of 50 μl aliquot of bile, direct injection of bile samples into a reverse phase HPLC system (Reddy *et al.*, 1981b) and radioassay of 10 s fractions and reconstruction of radioactivity elution profiles, and calculation of fractions of radioactivity eluting as parent compound and metabolites. The fraction of metabolites accounted for by glucuronides and sulphates was determined by digestion of bile samples with glucuronidase and sulphatase (Unger *et al.*, 1979) and determination of the increase in the recovery of parent compound following enzyme treatment. Details of these methods are published (Reddy *et al.*, 1981c).

RESULTS

Toxicity of secalonic acid D (SAD) was examined using lethality, growth retardation and histopathology as indices (Reddy *et al.*, 1979). The i.p. single-dose LD_{50} values of 37, 31 and 27 mg/kg were obtained for Charles River CD-1, Texas (ICR) and Sprague-Dawley (CF-1) strains of mice, respectively. The i.p. LD_{50} was 52 mg/kg in female CD-1 mice. The i.v. LD_{50} was 25 mg/kg in CD-1 male mice. Oral LD_{50} values of 400 mg/kg in male CD-1 mice, and 25 and > 400 mg/kg in Sprague-Dawley day-old and weanling (21 days) rats of both sexes, respectively, were obtained. All i.p. doses of SAD caused pulmonary atelectases and focal peritonitis in male CD-1 mice. The latter involved surfaces of abdominal viscera and produced limited subcapsular necrosis of hepatic parenchyma. Exposure to a single lethal dose i.v. (25 mg/kg or more) of SAD caused limited hepatic portal necrosis but no peritonitis or other associated local effects observed in CD-1 male mice after i.p. exposure. Death resulting from cardiac and/or pulmonary insufficiency was suggested by atelectasis, pulmonary haemorrhages and oedema, and massive atrial dilation in mice that died after lethal i.p. or i.v. doses of SAD (Table 1). Five daily sublethal doses in CD-1 male mice resulted in dose-dependent mortality (LD_{50}, 11.5 mg/kg) indicating cumulative effects (Reddy *et al.*, 1979).

Experiments designed to evaluate the drug-metabolizing enzyme modulating effect of SAD, by studying alterations in pentobarbital-induced narcosis in mice, suggested lack of such an effect (Reddy *et al.*, 1983). Pretreatment of mice with SKF 525-A increased toxicity, whereas phenobarbital pretreatment tended to protect against SAD toxicity (Table 2). Pretreatment with SKF 525-A resulted in reduced rate of biliary excretion of ^{14}C-SAD compared to that in non-pretreated mice, whereas phenobarbital pretreatment appeared to have no effect. Neither pretreatment altered the ratio of parent compound to metabolites in the biliary radioactivity. Although inconclusive, these data suggested that interactions other than alterations in SAD-metabolizing enzyme activity may be involved (Reddy *et al.*, 1983).

The effect of i.p. SAD treatment on days 7–15 of pregnancy on the survival and growth and development of post-implantation embryos were studied in mice (Reddy *et al.*, 1981a). A dose-dependent reduction in weight gain of mothers receiving all doses of SAD and an increase in resorptions of implanted embryos of dams treated with more than 5 mg/kg of SAD, i.p., occurred. A corresponding decrease in the percentage of live foetuses and a decrease in the average foetal body weight on day 19 of pregnancy also occurred. In $NaHCO_3$ alone, the minimum teratogenic dose of SAD was only 6 mg/kg. Major malformations included cleft palate, cleft lip, open eyelids, missing phalangeal ossification centres and shortened mandibles. Data indicated that, although day 11 or day 12 appear to be more sensitive than day 13

Table 1 Lesions of CD-1 mice receiving single intraperitoneal doses of secalonic acid D[a]

Dose (mg/kg)	Number necropsied		Percentage of mice examined showing lesions				
				Focal necrosis of	Pulmonary		Severe atrial
	Died[b]	Killed[c]	Peritonitis	liver surface	Atelectasis	Haemorrhage	dilation
10 or 20		20	100	10	40	0	0
30	3		100	33	66	100	100
	(5–8 days)	7	100	28	57	0	0
40	4		100	25	100	100	100
	(5–7 days)	3	100	33	100	0	0
50	7		100	100	100	100	100
	(4–7 days)	1	100	100	100	0	0
75	5		100	100	100	100	100
	(2–7 days)						
100	3		100	100	100	100	100
	(2 days)						

[a] Dimethyl sulphoxide was used as the solvent vehicle.
[b] Time of death (days) after administration is given in parentheses.
[c] All survivors killed 21 days after administration.

Table 2 Effect of phenobarbital, 3-methylcholanthrene and SKF 525-A on the acute toxicity of secalonic acid D in mice

Treatment	Dose (mg/kg)	LD_{50} of secalonic acid D	95% confidence limits	Percentage change
None		34.8	28.18–43.02	
Phenobarbital	75 × 3	45.9	35.24–59.92	+32.8
3-Methylcholanthrene	25 × 3	35.7	27.69–45.14	+2.6
SKF 525-A	50 × 1	21.4[a]	17.09–26.88	−38.5

[a] Significantly different from the value for non-pretreated mice, $p < 0.05$.

or day 14 of pregnancy, foetal weight reduction tended to be more severe in mice treated on day 11 than in mice treated on day 12 (Table 3).

Dose–response was studied by giving pregnant mice 0, 15, 20, 25 and 30 mg/kg of SAD on day 12. It was found that these dose levels caused cleft palate in 0.8, 14, 40, 51 and 37%, respectively, of liveborn foetuses. There was no increase in foetal resorptions and no decrease in foetal weights in mice treated with 20 mg/kg or less of SAD; however, higher doses tended to increase foetal resorptions as well as to decrease foetal weights (Table 4).

Data obtained in mutagenicity studies (Reddy *et al.*, 1980) indicated that the positive control compound triethylene melamine (TEM) induced dominant lethal mutations in germinal cells in male mice and also resulted in a greatly increased population of micronucleated polychromatic (immature) erythrocytes in bone marrow of treated male mice (not shown). SAD, however, failed to induce dominant lethal mutations but did produce a slight but statistically significant increase in the number of micronucleated polychromatic erythrocytes in mouse bone marrow at maximally tolerated doses. SAD was negative in the Ames' test. It is, therefore, concluded that SAD may be weakly mutagenic in a mouse somatic cell mutation system.

Elimination of ^{14}C-SAD-derived radioactivity from plasma was biphasic (Reddy *et al.*, 1981c). The rapid phase with a half-life of 24.3 min and the slower phase with a half-life of 20.5 h, were identified (Figure 1). The volume of distribution was equivalent to the total body water. Liver, kidney, gastrointestinal tract and the muscle contained major quantities of radioactivity. Concentration of radioactivity in liver and kidney exceeded that in plasma at all times beginning 1 h after administration. Urinary excretion was essentially complete by 48 h after administration, accounting for 5% of the total radioactivity. Excretion by faecal route appears to be predominant with 73% of the radioactivity administered being recovered in faeces by 7 days. High-pressure liquid chromatography of bile samples indicated that only 13%

Table 3 Embryonic survival and foetal growth in CD-1 mice following exposure to 25 mg/kg of SAD on days 11, 12, 13 or 14 of pregnancy

	Day(s) of treatment ($N = 5$)				
Parameter	Day 11	Day 12	Day 13	Day 14	Control (11–14)
No. of implants	9.2[b] ± 4.32	14.0 ± 2.74	13.0 ± 1.87	14.8 ± 1.30	13.6 ± 1.67
No. of resorptions	1.2 ± 0.84	3.0 ± 2.55	1.0 ± 1.00	0.6 ± 0.89	0.9 ± 0.92
No. of live foetuses	8.0[b] ± 3.74	11.0 ± 2.00	12.0 ± 1.87	14.2 ± 1.79	12.7 ± 1.66
Foetal body weight (g)[a]	1.198 ± 0.2658	1.274 ± 0.1036	1.324 ± 0.1240	1.186 ± 0.1126	1.389 ± 0.1236
No. of foetuses with cleft palate	6.2 ± 4.44	5.6 ± 4.72	4.8 ± 2.68	5.6 ± 5.32	0.1 ± 0.32
(Percentage)	(77.5)	(50.9)	(40.0)	(39.4)	(0.8)

[a] Each value is a mean of all individual foetuses in each litter within each group, ± standard deviation.
[b] Significantly different ($p < 0.01$) from the values for the same parameter on different days of gestation.

Table 4 Embryonic survival and foetal growth in CD-1 mice following maternal exposure to 0 to 30 mg/kg of SAD on day 12 of pregnancy

	Dose (mg/kg) ($N = 5$)				
Parameter	0	15	20	25	30
No. of implants	13.6 ± 1.67	12.0 ± 1.41	12.8 ± 2.49	14.0 ± 2.74	13.6 ± 1.99
No. of resorptions	0.9 ± 0.92	0.6 ± 0.89	0.8 ± 0.84	3.0 ± 2.55	1.7 ± 1.11
No. of live foetuses	12.7 ± 1.66	11.4 ± 1.14	12.0 ± 2.92	11.0 ± 2.00	11.9 ± 2.12
Foetal body weight (g)[a]	1.389 ± 0.1236	1.294 ± 0.1632	1.308 ± 0.1250	1.274 ± 0.1036	1.243 ± 0.1009
No. of foetuses with cleft palate	0.1 ± 0.32	1.6 ± 3.05	4.8 ± 5.17	5.6 ± 4.72	4.4 ± 5.03
(Percentage)	(0.8)	(14.0)	(40.0)	(50.9)	(37.0)

[a] Each value is a mean of all individual foetuses in each litter within each group, ± standard deviation.

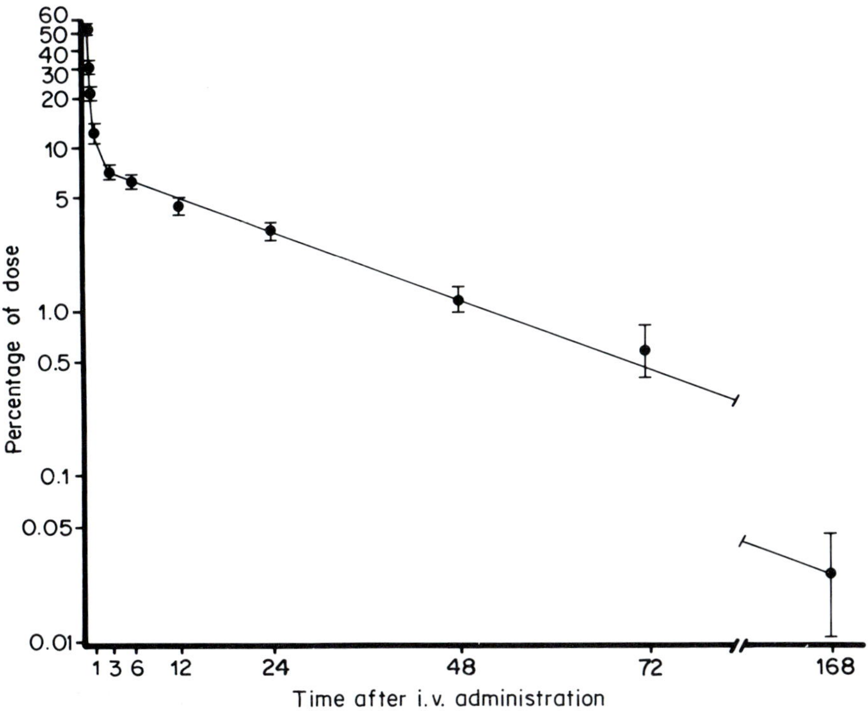

Figure 1 Disappearance (percentage of administered dose *vs* time) of radioactivity from the plasma of mice receiving 15 mg/kg of ^{14}C-secalonic acid D by intravenous route. Each point represents data from nice mice.

of radioactivity excreted in bile was accounted for by the parent compound, 10% by sulphate conjugates and the rest by at least three unidentified organosoluble metabolites. No glucuronides were detected in bile (Reddy *et al.*, 1981c).

Pretreatment with phenobarbital did not alter either the rate of biliary excretion of ^{14}C-SAD-derived radioactivity or the ratio of parent compound to metabolites in the bile (Reddy *et al.*, 1983). Although pretreatment with SKF 525-A reduced the rate of bile flow as well as biliary excretion of SAD-derived radioactivity, no alterations in the ratio of parent compound to metabolites were found (Table 5).

DISCUSSION

Besides confirming earlier reports of acute toxicity of SAD, our studies demonstrated teratogenic effects of SAD in mice. Although the kinetic data

Table 5 Effect of pretreatment with phenobarbital and SKF 525-A on the bile flow and biliary excretion of secalonic acid and its metabolites

		Biliary excretion of radioactivity as		
Pretreatment	Bile flow (ml/h)	Total (μg equivalents/h)[a]	Parent compound (%)	Metabolites (%)
Normal saline	0.30 ± 0.028	31.4 ± 6.06	16.8 ± 4.46	83.2 ± 4.46
Phenobarbital	0.35 ± 0.031	24.3 ± 2.86	16.9 ± 1.28	83.1 ± 1.28
SKF 525-A	0.18 ± 0.015[b]	9.3 ± 1.92[b]	12.7 ± 2.24	87.3 ± 2.24

[a] μg equivalent is the amount of radioactivity representing 1 μg of ^{14}C-secalonic acid D (420 cpm).
[b] Significantly different from the normal saline pretreated value in the same column by analysis of variance ($p < 0.05$).

indicate liver and kidney as possible target organs for SAD effects, the histopathological observations suggest that mortality may be associated with atrial dilation and pulmonary haemorrhages and oedema indicative of cardiac failure.

The critical period of SAD teratogenicity appears to be day 11 and 12 of pregnancy. Although the incidence of cleft palate is higher on day 11, cleft palate in pups born to mothers treated with up to 20 mg/kg SAD on day 12 was associated with no reduction in foetal body weights and no increase in foetal resorptions. The incidence of cleft palate in 40% pups born at this dose × time treatment suggests the value of this compound as a model to study pathogenesis of cleft palate. The results of Mayura *et al.* (1982) also clearly demonstrate the teratogenic effect of SAD in rats. The spectrum of malformations, however, is different from that in mice. Major malformations seen were tracheo-oesophageal fistula, hydronephrosis, renal agenesis, anophthalmia and exencephaly. Differences in species as well as the route of exposure (s.c.) used in these investigations appear to be involved in the observed differences.

The development of HPLC method of analysis (Reddy *et al.*, 1981b) has allowed quantitation of SAD as a natural contaminant in maize dust from grain storage elevators in concentrations of 0.3–4.5 mg/kg dust (Ehrlich *et al.*, 1982). Also, Sorenson *et al.* (1982) have shown that a single intratracheal instillation of 2 mg/kg dose of SAD in rats causes, progressively, the inflammatory, granulomatous and polypoid lesions in the respiratory part of the pulmonary tree typical of bronchiolitis obliterans in humans. This suggests that SAD contamination of grain may present a hazard to human health in grain-processing and feed-manufacturing industries.

Besides the possible respiratory and teratogenic effects, the growth-depressant effect of SAD may be important in domestic animal production, by retarding growth especially in predominantly maize-consuming livestock. SAD, alone or in combination with other mycotoxins occurring simultaneously in contaminated corn, may also cause acute toxicoses of corn-consuming livestock and humans.

ACKNOWLEDGEMENT

These studies are partially supported by grant No. 1R23DE06196-01 from the National Institute of Dental Research, and some aspects of this work were performed at the Department of Veterinary Preventive Medicine, the Ohio State University, Columbus, Ohio.

REFERENCES

Anderson, R., Buchi, G., Kobbe, B., and Demain, A. L. (1977). Secalonic acids D and F are toxic metabolites of *Aspergillus aculeatus*, *J. Org. Chem.*, **42**, 352–353.

Carlton, W. W., Tuite, J., and Mislivec, P. (1968). Investigations of the toxic effects in mice of certain species of *Penicillium*, *Toxicol. Appl. Pharmacol.*, **13**, 372–387.

Demain, A. L., Hunt, N. A., Malik, V., Kobbe, B., Hawkins, H., Matsuo, K., and Wogan, G. N. (1976). Improved procedure for production of cytochlasin E and tremorgenic mycotoxins by *Aspergillus clavatus*, *Appl. Environ. Microbiol.*, **31**, 138–140.

Ehrlich, K. C., Lee, L. S., Ciegler, A., and Palmgren, M. S. (1982). Secalonic acid D: natural contaminant of corn dust, *Appl. Environ. Microbiol.*, **44**, 1007–1008.

Franck, B., Goottschalok, E. M., Ohnsorge, V., and Huper, F. (1966). Trennung, Struktur und absolute Konfiguration der diasteriomeren secalonsäuren A, B and C, *Chem. Ber.*, **99**, 3842–3862.

Generoso, W. M. (1973). Evaluation of chromosome oberration effects of chemicals on mouse germ cells, *Environ. Health Perspect.*, **6**, 13–22.

Harada, M., Yano, S., Watanabe, H., Yamazaki, M., and Miyaki, K. (1974). Phlogistic activity of secalonic acid A. *Chem. Pharm. Bull.*, **22**, 1600–1606.

Howard, C. C., and Johnstone, R. A. H. (1973a). Fungal metabolites. Part III. Isolation of secalonic acids from *Phoma terrestris*, *J. Chem. Soc.*, *Perkin Trans.*, **1**, 2440–2444.

Howard, C. C., and Johnstone, R. A. H. (1973b). Isolation of a new secalonic acid, *J.C.S. Chem. Commun.*, **1**, 464–467.

Mayura, K., Hayes, A. W., and Berndt, W. O. (1982). Teratogenicity of secalonic acid D in rats, *Toxicology*, **25**, 311–322.

McCann, J., and Ames, B. N. (1976). A simple method for detecting environmental carcinogens as mutagens. *Ann. N.Y. Acad. Sci.*, **271**, 5–13.

Reddy, C. S., Hayes, A. W., Williams, W. L., and Ciegler, A. (1979). Toxicity of secalonic acid D, *J. Toxicol. Environ. Health*, **5**, 1159–1169.

Reddy, C. W., Reddy, R. V., Chan, P. K., and Hayes, A. W. (1980). Mutagenicity of secalonic acid D in mice, *J. Environ. Pathol. Toxicol.*, **4**, 31–37.

Reddy, C. S., Reddy, R. V., and Hayes, A. W. (1981a). Teratogenicity of secalonic acid D in mice, *J. Toxicol. Environ. Health*, **7**, 445–455.

Reddy, C. S., Reddy, R. V., and Hayes, A. W. (1981b). High performance liquid chromatography of secalonic acid D and its application to biological fluids, *J. Chromatogr.*, **208**, 17–26.

Reddy, C. S., Reddy, R. V., Hayes, W. A., and Berndt, W. O. (1981c). Disposition of (^{14}C) secalonic acid D in male mice, *Toxicol. Appl. Pharmacol.*, **61**, 399–406.

Reddy, C. S., Reddy, R. V., and Hayes, A. W. (1983). Interaction of secalonic acid D with phenobarbital, 3-methyl cholanthrene and SKF 525-A in mice, *J. Toxicol. Environ. Health*, **12**, 687–694.

Schmid, W. (1975). The micronucleus test, *Mutation Res.*, **31**, 9–15.

Scott, D. B. (1965). Toxigenic fungi isolated from cereal and legume products, *Mycopathol. Mycol. Appl.*, **25**, 213–222.

Sorenson, W. G., Green, F. H. Y., Vallyathan, V., and Ciegler, A. (1982). Secalonic acid D toxicity in rat lung, *J. Toxicol. Environ. Health*, **9**, 515–525.

Steyn, P. S. (1970). The isolation, structure and absolute configuration of secalonic acid D, the toxic metabolite of *Penicillium oxalicum*, *Tetrahedron*, **26**, 51–57.

Unger, P. D., Hayes, A. W., and Mehendale, H. M. (1979). Hepatic uptake, disposition and metabolism of rubratoxin B in isolated perfused rat liver, *Toxicol. Appl. Pharmacol.*, **47**, 529–536.

Wilson, J. G. (1965). Methods for administering agents and detecting malformations in experimental animals. In *Teratology: Principles and Techniques* (eds J. G. Wilson and J. Warkany), pp. 309–310. University of Chicago Press, Chicago.

Yamamoto, T. (1975). Antitumor antibiotics carrying immunopotentiating capacity. In *Host Defense Against Cancer and Its Potentiation* (eds D. Mizuno *et al.*), pp. 379–388. University Park Press, Baltimore.

Yamazaki, M., Maebayashi, Y., and Miyaki, K. (1971). The isolation of secalonic acid A from *Aspergillus ochraeus* cultured on rice, *Chem. Pharm. Bull.*, **19**, 199–201.

Index